BASICS OF BIOSTATISTICS

BASICS OF BIOSTATISTICS
A Manual for the Medical Practitioners

Jatinder Bali
MBBS MS (Ophth) CDM PGDHHM STT (Pediatric Ophth)
WHO Fellowship (VR&ROP) MBA (Operations Research and
Management Information Systems)
Member-Secretary
Institutional Ethics Committee
Hindu Rao Hospital and North Delhi Municipal Corporation (NDMC)
Medical College, New Delhi
Member-Secretary
Scientific Research Committee
Hindu Rao Hospital, New Delhi
Formerly, Nodal Officer (Information Technology) and
Assistant DNB Coordinator
Hindu Rao Hospital, New Delhi, India

Anil Kant
BSc (Statistics), Kurukshetra University
MBA (International Business), Banaras Hindu University
Head (Sales and Marketing), Max Speciality Films Ltd, New Delhi, India
Formerly, Director, Wifag-Polytype India, Noida, Uttar Pradesh, India
AVP, Sakata Inx India Ltd, Gurgaon, Haryana, India

Foreword
Barun K Nayak

The Health Sciences Publisher
New Delhi | London | Philadelphia | Panama

Jaypee Brothers Medical Publishers (P) Ltd.

Headquarters
Jaypee Brothers Medical Publishers (P) Ltd.
4838/24, Ansari Road, Daryaganj
New Delhi 110 002, India
Phone: +91-11-43574357
Fax: +91-11-43574314
E-mail: jaypee@jaypeebrothers.com

Overseas Offices

J.P. Medical Ltd.
83, Victoria Street, London, SW1H 0HW (UK)
Phone: +44-20 3170 8910
Fax: +44(0) 20 3008 6180
e-mail: info@jpmedpub.com

Jaypee-Highlights Medical Publishers Inc.
City of Knowledge, Building 235, 2nd Floor
Clayton, Panama City, Panama
Phone: +1 507-301-0496
Fax: +1 507-301-0499
E-mail: cservice@jphmedical.com

Jaypee Medical Inc.
325, Chestnut Street
Suite 412, Philadelphia, PA 19106, USA
Phone: +1 267-519-9789
E-mail: support@jpmedus.com

Jaypee Brothers Medical Publishers (P) Ltd.
17/1-B, Babar Road, Block-B
Shaymali, Mohammadpur
Dhaka-1207, Bangladesh
Mobile: +08801912003485
E-mail: jaypeedhaka@gmail.com

Jaypee Brothers Medical Publishers (P) Ltd.
Bhotahity, Kathmandu, Nepal
Phone: +977-9741283608
E-mail: kathmandu@jaypeebrothers.com

Website: www.jaypeebrothers.com
Website: www.jaypeedigital.com

© 2017, Jaypee Brothers Medical Publishers

The views and opinions expressed in this book are solely those of the original contributor(s)/author(s) and do not necessarily represent those of editor(s) of the book.

All rights reserved. No part of this publication may be reproduced, stored or transmitted in any form or by any means, electronic, mechanical, photocopying, recording or otherwise, without the prior permission in writing of the publishers.

All brand names and product names used in this book are trade names, service marks, trademarks or registered trademarks of their respective owners. The publisher is not associated with any product or vendor mentioned in this book.

Medical knowledge and practice change constantly. This book is designed to provide accurate, authoritative information about the subject matter in question. However, readers are advised to check the most current information available on procedures included and check information from the manufacturer of each product to be administered, to verify the recommended dose, formula, method and duration of administration, adverse effects and contraindications. It is the responsibility of the practitioner to take all appropriate safety precautions. Neither the publisher nor the author(s)/editor(s) assume any liability for any injury and/or damage to persons or property arising from or related to use of material in this book.

This book is sold on the understanding that the publisher is not engaged in providing professional medical services. If such advice or services are required, the services of a competent medical professional should be sought.

Every effort has been made where necessary to contact holders of copyright to obtain permission to reproduce copyright material. If any have been inadvertently overlooked, the publisher will be pleased to make the necessary arrangements at the first opportunity.

Inquiries for bulk sales may be solicited at: jaypee@jaypeebrothers.com

Basics of Biostatistics: A Manual for the Medical Practitioners
First Edition: **2017**
ISBN: 978-93-86150-71-4
Printed at Rajkamal Electric Press, Plot No. 2, Phase-IV, Kundli, Haryana.

Dedicated to

*Our teachers, parents and families
Padma Shri Prof (Dr) Surendra Singh Yadav
and
Dr Barun K Nayak
for his missionary efforts to guide a whole generation of researchers*

Foreword

Research is very important in today's era of evidence-based medicine. I am of the opinion that if there is no evidence, then research should be carried out and the evidence should be created. Similarly, if there is evidence, then it should be found out from the available literature and applied wherever it seems fit. However, to do this, understanding the procedure is very necessary—both "conducting" and "applying". *Basics of Biostatistics: A Manual for the Medical Practitioners* covers the different aspects of conducting an in-depth research.

This well-documented book starts by covering the two important aspects—bioethics and evidence-based medicine—in this evidence-based era. Most people in the medical profession are not well conversed with various aspects of research, which include planning, understanding of statistics, conducting a research and finally drawing the conclusion.

This book, authored by Jatinder Bali and Anil Kant, has been written in response to a 'timely perceived requirement' and deals with the various aspects of research in a lucid manner, wherein there will be something—right from the beginners to an accomplished researchers—to grasp.

I have seen that medical practitioners and clinicians feel vulnerable when confronted with statistics. This is a mindset and can be easily overcome after going through this book. I would particularly like to make a mention of the chapters which are dedicated to statistical concepts for medical administration, where indexing, statistical quality control, and decision theory have been elucidated.

There have been numerous instances when researches have never been taken up or left partially done because of lack of funds or sponsorship. At such times, it is heartbreaking not only for the person conducting the research but also for all of us, because we are keeping humankind away from another revelation, which may go a long way to help us in one way or the other. Taking this into account, the last section of this book also deals with report writing and grant applications, which is the backbone of research.

Needless to say, the authors have done a wonderful job. I am truly impressed with the topics and quality of contents and how they have, in a studious manner, gone about laying the flow of the topics making them easy-to-understand and relate to the practical use. I am of the opinion that the book should not be missed and should be kept handy for ready reference, as all the doctors need to contribute to the progress of medical science.

I am thankful to the authors for giving me the honor of writing the Foreword for this book because it is a subject which is very close to my heart and is a need of the hour.

Barun K Nayak
Honorary General Secretary
All India Ophthalmology Society
Head, Department of Ophthalmology
PD Hinduja National Hospital and Medical Research Center
Mumbai, Maharashtra, India
Editor, Journal of Clinical Ophthalmology and Research
Past Editor, Indian Journal of Ophthalmology
President, Maharashtra Ophthalmological Society
Past President, Bombay Ophthalmologists' Association

Preface

This effort started off as replies to students' queries when their research projects were being evaluated. Biomedical research driven by the large databases is already a reality. In fact, some of the biggest data centers in the world today host the complex information stores in their databases. However, research is not limited to these large datasets alone. It is conducted by a large proportion of practitioners and students of medicine also. Medical professionals often do not train in mathematics and statistics. Medical experts who conduct a large quantum of research often feel at ease using these datasets, when it comes to handling the data. Due to limited exposure to the various concepts and tools of statistics coupled with the fear of getting the procedure wrong, they often run the risk of improper application and incorrect inferences. These tools can be used in daily life as well as for researches.

Influenced by the lucid style of the workshops of Dr Barun K Nayak—one of the inspirations behind this effort—we made an effort to provide a reference manual in nonintimidating language in the field of basic statistical concepts related to medicine. It is written with the understanding that medical professionals are not mathematical experts. The prime purpose of the treatise is to remove the fear of statistics as an extra-terrestrial object and to foster the scientific way of thinking—being cynical and not agreeing until clearly proved. Pure mathematical approach has been limited to minimum possible with focus on the concepts. A complete overview of the statistical methods is provided in a concise and precise manner to build the strong foundation for understanding advanced tools and techniques. Concepts have been compared wherever possible so that the fine differences are captured and conveyed. The chapter *Back to School* to review elementary concepts such as fractions, logs, set theory, permutation and combination, etc. is added to recapitulate the concepts that were left behind in high school by many of us.

Basics of Biostatistics: A Manual for the Medical Practitioners is one of the few books on statistics written to provide the practical help to the medical professionals. The chapter *Measurement and Error Analysis* was incorporated because of the medical mindset of the authors whose focus is often research papers and clinical trials. Additional topics such as life table, Kaplan-Meier curves, interpolation, time series, etc. also have been included. Subjects such as statistical quality control, decision theory, index number, which find use in medical administration, also get a fair treatment. The authors felt that without these applied aspects, the treatise would be incomplete. Everyone will use some of the topics at some stages in his career and all topics will rarely be used by any set of practitioners. The appeal of this book lies in the fact that everyone—from the students to the hardened primary investigators—will find this book very useful. And the useful chapters will keep on changing with the level of expertise developed. We decided to go from the known to the unknown. So every time the readers come back to the book, they will find something new. It was an onerous task which took over a year in preparation. We hope it enriches the lives of our readers.

Special mention must be made of the efforts of Mr Hitesh Mendiratta, who stuck it out with me—decoding the conspicuously bad handwriting. We gave up a large number of assignments for this labor of love. It was an onerous task but the one which was enjoyed. Last but not least, our thanks to all those who are reading this book. Hope you find it educative.

Jatinder Bali
Anil Kant

Acknowledgments

I am thankful to Dr V Devgan, who is a great person and was an important source of inspiration for *Basics of Biostatistics: A Manual for the Medical Practitioners* by suggesting that I take it up in real earnest. Speaking of encouragement, special mention must be made of Dr Barun K Nayak, who mentored me in my learning years. He taught me the value of hard work and diligence. He is a fantastic teacher and I can only be grateful to have met him. Another special thank to Ms Shabry Bakshi, the author and the journalist, who was there with lots of words of help and each of those words turned out to be helpful. So finally, for me, one part of my quest is over.

Jatinder Bali

The place of a *Guru* is of highest order in our culture. I could very well realize this while undertaking this journey. In this respect, I thank Dr (Mrs) Anita Goyal, Kurukshetra University, Haryana, India, who taught and guided me through the maze of statistics. Life beyond classrooms teaches a lot and one can consider himself privileged if one gets professional *gurus,* who always motivate one to challenge and strive for the best. Thanks to Mr BS Kampani, Mr VK Seth and Mr K Alparslan for believing in me over the years. The *padyatra* i.e. the journey continues…

Anil Kant

First and foremost, we would like to thank our families Dr (Mrs) Renu, Ojasvini, Nandan, Mrs Meeta and Mahir as well as our parents, who put up with us through the writing of *Basics of Biostatistics: A Manual for the Medical Practitioners*. They have been anchors, who allowed us to weather many a storms in this odyssey. They knew this book meant a lot to us. They sacrificed many hours of happiness at the altar of the keyboard. We look forward to future family gatherings, where we will be able to discuss this book, which we hope, they all will read. Not really, but we hope, they find it useful if they ever chance upon this path in future.

Special thanks is also due to Mr KP Singh and the entire team of M/s Jaypee Brothers Medical Publishers (P) Ltd, New Delhi, India. We would like to thank Shri Jitendar P Vij (Group Chairman), Mr Ankit Vij (Group President), Mr Tarun Duneja (Director–Publishing), Ms Samina Khan (Executive Assistant to Director–Publishing), Mr KK Raman (Production Manager), Mr Mohit Bhargava (Production Coordinator) and the other staff. They were prompt and professional in their responses. They have elucidated the value of time. They almost had the job finished even before the impossible deadline. They sure know how to hire good talent. They have been exceptional in the quality of artwork and back-end support for complicated projects. The future looks bright for Indian skill transfer publishing. We hope they keep it up like this only.

While on the subject of sticking it out with high-quality work, we must bring on board Mr Hitesh Mendiratta, a wizard with scientific data, if ever, there was one. He did medical terminology and complex mathematical equations with the same consummate ease. At times, he had picked out the mistakes in the printouts even before we read the drafts. He is one man we will have on our team for any prolonged writing campaign. He never tires. And, finally, thanks to all the students and principal investigators, who inspired us to undertake the project in the first place.

Contents

Zero Point: Approach Road to Learning 1
Statistics in a Nutshell 8

Section 1: Bioethics, Evidence-based Medicine

1. **Bioethics in Clinical Research** 13
   *Major Divisions of Ethics 13; Bioethics 14; Main Branches of Bioethics 14
   Etymology, History and the Famous Nonethical Experiments 14
   Dachau and Auschwitz Concentration Camp Hypothermia Experiments 15
   Animal Ethics 34; Environmental Ethics 35; Anthropocentrism 36*

2. **Evidence-based Medicine and Levels of Evidence** 38
   *Evidence-based Medicine 38; Fundamental Principles 38
   Components of Evidence-based Practices 39*

3. **Introduction to Statistics** 49
   *Etymology 49; History 49; Notable Contributors 50
   India in Modern Context 51; Applications 51; Summing Up 52; Biostatistics 52*

Section 2: Research Methodology and 'Back to School'

4. **Research Methodology** 57
   *Objectives of Research 57; Types of Research 58
   Approaches to Research 59; Research Methods 60; Research Methodology 60*

5. **Inside Research Methodology: The Nuts and Bolts** 63
   What Constitutes a Research? 63; Research Plan 72

6. **Back to School** 74
   *Greek Alphabets 74; Common Mathematical Operators/Symbols 75
   Modulus 76; Basics of Set Theory and Venn Diagrams 76; Indices 79
   Progressions: Arithmetic and Geometric Progressions 79
   Logarithms and Antilogarithms 80; The Exponential Function 'e' 81
   Ratio 81; Proportion 82; Factorial; 82; Permutations and Combinations 83*

Section 3: Describing the Data

7. **Frequency Distribution** 87
   *Classification of Data 87; Variables: Discrete and Continuous 89
   Frequency Distributions: Discrete and Continuous 89
   Relative Frequency Distribution 93
   Bivariate or Two Way Frequency Distribution 93; Marginal Distribution 94*

8. **Measures of Central Tendency** 95
   *Central Value Concept 95; Measures of Central Tendency 95
   Types of Measures of Central Tendency 96; Median (M_d) 100
   Quartiles, Deciles and Percentiles 103; Mode (M_0) 105
   Harmonic Mean 109; Which Measure to Use? 111*

9. **Measures of Dispersion (or Variation)** 114
   *Understanding Dispersion 114; Types of Measures of Dispersion 115
   Quartile Deviation 116; Percentile Deviation 116; Mean Deviation 116
   Standard Deviation 117; Lorenz Curve 119
   Coefficient of Dispersion and Its Measures (Also referred as a Relative Measure of
   Variation) 120; Which Measure of Dispersion to Use? 121*

10. **Skewness, Moments and Kurtosis** — 122
   Skewness 122; Kurtosis 125; Moments 126; Significance 127

Section 4: Probability, Probability Distribution and Sampling

11. **Theory of Probability** — 131
   Terminology in Probability Theory 132; Approaches to Probability 133
   Conditional Probability 137; Bayes Theorem 139
12. **Theoretical Discrete Probability Distributions** — 142
   Underlying Concepts and Terminology 142
   Types of Theoretical Discrete Distributions 147
13. **Theoretical Continuous Probability Distributions** — 163
   Normal Distribution 163; Rectangular Distribution 174
   Lognormal Distribution 176; Exponential Distribution 176
14. **Sampling Theory** — 178
   What is a Sample and why it is Required? 179; Types of Sampling Methods 180
   Advantages and Disadvantages of Sampling 182
   What should be the Sample Size? 183; Sampling Errors 185
   Standard Error 186; Underlying Basis of Sampling Theory 186

Section 5: Correlation, Regression, Partial and Multiple Correlation and Regression, Theory of Attributes

15. **Correlation** — 191
   What is Correlation? 191; Correlation Measures 192
16. **Regression Analysis** — 205
   Components of Regression Analysis 205; Types of Regression 206
   Linear Regression Analysis and Line of Best Fit 207; Logistic Regression 212
17. **Partial and Multiple Correlation and Regression** — 216
   Partial Correlation; 216; Multiple Correlation 217
   Partial and Multiple Regression 218; Multicollinearity 221
18. **Theory of Attributes** — 225
   Terminology of Attribute Theory 225; Association of Attributes 228

Section 6: Statistical Testing

19. **Hypothesis Testing** — 235
   Basic Concepts 236;
   Practical Considerations in Framing and Testing Hypothesis 246
20. **Tests of Significance (Z and Student's 't'-Test)** — 248
   Large Sample Tests 249; Small Sample Tests (Student's t-tests) 256
   What is P-value? 267
21. **F-Test and Analysis of Variance and Covariance** — 269
   Constants and Properties of F-Distribution 269; Applications 270
22. **Chi-square Test** — 286
   Degrees of Freedom 286; Salient Points of Chi-square 287
   Chi-square Distribution 288; Applications of χ^2 Distribution 289
   Chi-square Test of Homogeneity of Correlation Co-efficient 298
   Remarks on Chi-Square 300
23. **Nonparametric Tests** — 301
   Advantages 301; Disadvantages 302
   Run Test: Test for Randomness in a Series of Observations 302
   The Signed-rank Test for Testing Specified Mean or Median of a Population 303
   Wilcoxon Signed Rank Test for Paired data 304

Kolmogorov-Smirnov Test for One Sample 307
Kolmogorov-Smirnov Test (For Comparing Two Populations) 308
Mann-Whitney 'U' Test for Equality of Two Means 310
Wilcoxon-Wilcox Test for Comparison of Multiple Treatment on a Series 311
Kruskal-Wallis Rank Sum Test (H-Test) 313
Friedman Test (Two-way ANOVA) 314
Test of Significance of Spearman's and Kendall Rank Correlation 317
McNemar's Test for Paired Samples 317; Cochran Q Test 318

24. **Which Statistical Test to Use?** 321
Summary of Statistical Tests 322

25. **Special Considerations about Inferential Issues in Biostatistics** 327
Cause-Effect Relationship: Is it Really True? 328; Bias 329
Confounding 329; Spurious Relationship 331; Effect Modification 331
Bradford Hill Causality Criteria 332; Exposure-Disease Association 333
Odds Ratio 334; Attributable Risk 335

26. **Measurement and Error Analysis** 337
Measurement 337; Levels of Measurement 338
Evaluating Measurement 340; Reliability v/s Validity 343
Assessment of Validity: Specificity and Sensitivity 344
Errors in Measurement 350

Section 7: Other Statistical Concepts Frequently Used in Research

27. **Vital (Demographic) Statistics and Life Tables** 357
Basics of Vital Statistics 357; Life Tables 367
Survival Analysis and Kaplan-Meier Curves 374; Log-Rank Test 378

28. **Time Series Analysis** 381
Time Series Components 382; Measuring the Components of Time Series 384

29. **Interpolation, Extrapolation and Forecasting Methods** 398
Terminology 399; Methods of Interpolation 400
Extrapolation and Forecasting 406

Section 8: Statistical Concepts for Medical Administration

30. **Index Numbers** 411
Key Decisions in Creating Index Number 412; Weighted Index Numbers 417
Quantity and Value Index Numbers 423; Chain Index Numbers 423
Base Shifting and Splicing 426; Index Number Tests 427

31. **Statistical Quality Control** 429
Control Charts 430; Types of Control Charts 431
Charts based on Actual Measurements 432
'C-Control Chart' for Number of Defects Per Unit 435
P-Control Chart (Fraction Defectives or Proportion of Defects) 436
Acceptance Sampling 437

32. **Decision Theory** 440
Notation/Terminology 441; Decisions Methods 443
Principles of Decision Making 444; Decision Tree Analysis or Tree Diagram 449

Section 9: Report Writing and Applying for Grants

33. **Research Report Writing: How to do?** 453
Organization of the Research Report 453
Manuscript Preparation for Journals 460

34. Applying for Grants 467
Preparing for Application Process 468
Ground Reality: The Background Check 469
Main Application: Contents and Pitfalls 470
Background and Significance or Introduction 471
Preliminary Studies/Progress Report 471;   Research Design and Methods 471
Results 472;   Monitoring and Evaluation 472
Protection of Human Subjects and Institutional Review Boards 473
Bibliography 474;   Contracts 474;   Abstracts 474;   Title 474
Biographical Sketches 474;   Budget 475
Formatting and Instructions to Applicants 475
Cover Letter 476;   Decision Process Explained 476;   Pay Lines 477
Outcomes 477;   Common Problems Encountered 478;
Revision of Application 479;   Voila! Manna from Heaven: The Funding 480
Records and Accounting 480

Appendix 481
Oxford Centre for Evidence-Based Medicine 2011 Levels of Evidence 482
Random Number Table 484;   Binomial Coefficients 485
Binomial Probabilities 486
$e^{-\lambda}$ Values for Calculating Poisson Probabilities for ($0 < \lambda < 1$) 490
Area Under the Standard Normal Curve 491
Area Under the Standard Normal Curve 493
Area Under the Standard Normal Curve 495
Critical Values of 't' 497;   Critical Values of Chi Square (χ^2) 498
Critical Values of F (0.05 or 5%) 500;   Critical Values of F (0.01 or 1%) 503
Factors for Construction of Control Charts (Statistical Quality Control) 506
Studentized Range: Critical Values 508
Studentized Range: Critical Values 510
Pearson's Correlation Coefficient (r): Critical Values 512
Spearman Rank Correlation Coefficient (ρ): Critical Values 513
Run Test: Acceptance Region for Values of 'R' 514
Sign Test: Critical Values of 'T' 515
Kolmogorov-Smirnov Test: Critical Values of 'D' 517
Wilcoxon-Wilcox Two-Sided Test: Critical Values 518
Mann-Whitney Test (U Test): Critical Values of 'U' 521
Critical Values of τ: Kendall Rank Correlation Significance Test 522
Wilcoxon Signed Rank Test 'W': Lower and Upper Critical Values 523
Common Logarithms ($\log_{10} x$) 524;   Antilogarithms 530

Index 537

Zero Point:
Approach Road to Learning

Research is essence of human life. It has brought mankind to the stage it occupies today. The pursuit to seek knowledge, move beyond frontiers and charter into the unknown has helped development of science and society. Advancements in any field— scientific or artistic, largely depend on the experimental approach. Whether it is a renowned musician experimenting with a new note or a medical researcher experimenting with a new treatment, the objective is to gain access into the unexplored—creative and inventive realm.

Scientific research is an advanced step of knowledge generation, as it propounds use of scientific methods to conduct research. And use of 'Statistics' is an integral part of scientific research, irrespective of the amount of data involved.

Statistics is a specialized subject in itself. The question whether it is a 'science' or 'an art' has been around since ages, with no clear cut answers to it even today. Instead of debating this, it is more appropriate to appreciate what statistics can do to both artistic creativity and scientific analysis and interpretation. True, statistics neither operates on very exact laws and principles similar to those in basic sciences, nor has access to a painter's white canvas to create a 'master piece'. But it supports all disciplines of knowledge and learning, and allows them to grow further.

The field of medical science relies a lot on statistics. Like statistics, medical field also has to deal with complex situations, but of possibly higher order as it has multitude of factors and outcomes involved. Our body is one of the most complex systems, which is yet to be deciphered fully. It is true that in other sciences too, lot of things are still to be deciphered, like the sub-atomic particle etc, but one needs to accept that a failed experiment in physics and medical science are not comparable. The price of failure in practicing medical science results in loss of human life. A failure in finding a treatment in time can result in epidemic and so on. In other disciplines of study, a failed experiment would be a 'cost,' but in medical science it is a 'price' and a steep one at times.

As the subject matter of medical research is 'life,' which is precious, it needs a proper scientific support for its research endeavors. Statistics as a specialized subject provides medical field with most of the necessary tools and techniques for conducting research.

Apart from its own development, other subjects have too been instrumental in the development of the science of statistics. For example, analysis of variance, as a method was developed for agricultural science, probability originated through solving gambling problems, etc. These methods once developed found immense application in other fields of science, which led to the development of statistics. This resulted in the subject of statistics getting evolved into a vast repertoire of methods and techniques. With these methods, same data can be evaluated in different ways, to arrive at completely different results. For example, Mode and Mean are both measure of central tendency. Statistics does not preach which one to use and leaves it to the user to decide. Such leeway may be of interest to those who want misuse statistics, but for medical and scientific researchers, it can cause confusion. The abstractness and diversity in statistical tools and methods also makes understanding of the subject a bit different and at times difficult, especially for professionals from other fields.

Reports suggest that human brain is able to better handle and grasp orderly information as compared to random information. It is also suggested that brain tries to form patterns of order or some relational parameter, in order to see through the flux. This thought, coupled with acknowledgment of statistical concepts being complex and diverse, prompted us to introduce the "Zero Point."

In science, everything should have a reason. When this chapter was conceived it was felt that it should be called zero—not because it came before the first chapter or that it was India's contribution to the number theory. The reason for naming it 'zero point' was more esoteric. Both authors grew up in a hill station and hill stations generally have a point at a height with an outward protrusion, generally called 'zero point' which allows a wide angle view of the mountain peaks/landscape in front. The same is the essence of this chapter—to enable understanding of the contours better. The fact that zero was India's contribution to the world and scientific advancement would not have been possible if this was not available also played a bit of a supportive role in naming this as Chapter Zero—nothing by itself, but advancement is not possible without it. Of course, shunya means more than zero but then that is not within the scope of the present treatise.

Therefore 'Zero Point' aims to provide in a concise and precise manner, an overall view of the subject, chapters contained in this book and along with necessary remarks on them. This layout will help readers to understand the broad contours of statistics as a subject too.

Zero Point: Approach Road to Learning

SECTION 1: Bioethics, Evidence-based Medicine		
Chapter	*Title*	*Description*
1	Bioethics in Clinical Research	The book opens with the initial chapters on 'Bioethics' and 'Evidence Based Medicine.' These discuss origins of bioethics and the importance of evidence based medicine and the levels of evidence. Collectively, they impress upon the need for the application of scientific research methodology on the data collected in an ethical manner in consonance with the principles of biology. Scientific research method encompasses proper data collection, its analysis conducting appropriate test, and drawing proper inferences. Scientific research relies upon statistical techniques. The third chapter briefly and formally introduces the subject of Statistics.
2	Evidence-based Medicine and Levels of Evidence	
3	Introduction to Statistics	
SECTION 2: Research Methodology and 'Back to School'		
Chapter	*Title*	*Description*
4	Research Methodology	Before venturing into the core statistical concepts, it is imperative to understand, what is research; its process, methods, design etc. This also broadly discusses what constitutes a good research, what is research design and plan; and practical considerations in defining research.
5	Inside Research Methodology: The Nuts and Bolts	
6	Back to School	The objective is to refresh the relevant basic mathematical concepts studied in high school. Without knowing Permutation and Combination, it is difficult to understand Probability and Theoretical Distributions like Binomial, Normal, Poisson etc. Without understanding Modulus, drawing inferences in statistical tests may pose a challenge. Many Greek alphabets are used as standard notations in Statistics; therefore the list is also included in this chapter. Though this chapter could have been in appendix, but we strongly feel that reading through this chapter is important and should be done before statistical concepts are taken up for study.
SECTION 3: Describing the Data		
Chapter	*Title*	*Importance*
7	Frequency Distribution	Voluminous raw data needs to be condensed so that statistical tools can be applied. Frequency distribution details how data is classified, what is a variable, whether it is discrete or continuous. It converts raw data into data on which statistical techniques can be applied.
8	Measures of Central Tendency	Data occupies large space and is scattered. Measures of Central Tendency are the various methods through which we can calculate a 'Central Value' or 'Single Value' which is considered as representative of the entire data. This chapter details the various types, their merits and demerits, etc.
9	Measures of Dispersion (or Variation)	Central Value provides a single value, which is important but does not tell us anything more. Dispersion is the spread or scatter of the values around their central value, and measures of dispersion are the various methods to study this quantum of scatter (dispersed state). It gives a more holistic picture of the data under review.

(Contd...)

(Contd...)

Chapter	Title	Importance
10	Skewness, Moments and Kurtosis	Measures of Dispersion tell us the quantum of spread, but they do not convey anything about the direction of the spread, i.e. on which side (> or <) of the central value is this spread more. Concept of 'Skewness' answers this question. Further, two sets of data can have same central value, dispersion and similar directional spread, but may still differ in terms of the 'peakedness' of their distribution curves. Roughly, it translates into the magnitude of the central value and other values in its vicinity. This measure of 'peakedness' characteristics of curve are the subject matter of the concept of 'Kurtosis.'

SECTION 4: Probability, Probability Distribution and Sampling Theory

Chapter	Title	Importance
11	Theory of Probability	Statistical Testing uses Probability to draw inferences. The Theory of Probability forms the basis of the various Probability Distributions including Normal Distribution—a distribution which is the underlying assumption in most of the statistical tests. Understanding the concept and theory of Probability is, therefore, critical.
12	Theoretical Discrete Probability Distributions	The various theoretical probability distributions which a discrete random variable can follow are the subject matter here. These include the Binomial and Poisson Distribution. Binomial deals with probabilities when there are only two possible (mutually exclusive) outcomes; whereas Poisson concerns probabilities of rare events.
13	Theoretical Continuous Probability Distributions	These concern probability distributions of a continuous variable, and include the most widely used 'Normal Distribution.' Unlike Discrete, where the variable can only take discrete values, in continuous distribution the variable can take any intermediate fractional value too. Normal Distribution is the basis of the Tests of Significance and hence an important base of Inferential Statistics.
14	Sampling Theory	When it is not possible to study every individual object which is part of the research, it is better to draw a sample and study it. On the basis of the sample, an estimate can be made about all the objects. This process is called Sampling and its theory includes the various methods by which samples can be drawn, the errors involved etc. If sampling process is invalid, the results and conclusions of the research become invalid.

SECTION 5:
a. Correlation, Regression, Partial and Multiple Correlation and Regression
b. Theory of Attributes

Chapter	Title	Importance
15	Correlation	Correlation concerns the measure of relationship two sets of variables exhibit. Are the two variables directly or inversely related? Correlation value between two variables signifies the 'degree and direction' of relationship between them.

(Contd...)

(Contd…)

Chapter	Title	Importance
16	Regression Analysis	Two variables may be correlated, but correlation looks only at linear relationships and is exquisitely sensitive to the terminal points. Any other type of association is not captured by correlation and this necessitates use of other methods of association. A measure of association which explains how much of the dependent variable is explained by the independent variable is the subject matter of regression. Two variables will have one value of correlation, but can have two regression equations depending on which variable is considered dependent and independent. Regression equation answers the query—how much change is seen in a variable, for a unit change in the other variable. It ignores the extreme sensitivity correlation shows with reference to the extreme ends of the data. The remainder variability which is not explained by the independent variable/s is grouped together under error term. The basic objective is to reduce the quantum of the error term.
17	Partial and Multiple Correlation and Regression	In Correlation and Regression stated above, the number of variables involved was two only. In case, the number of variables is more than two, Partial and Multiple—Correlation and Regression techniques are used to determine the relationships' strength, direction and cause-effect among the variables.
18	Theory of Attributes	Attributes are characteristics which are qualitative, cannot be quantified, but their presence or absence is identifiable. Attribute Theory relates to handling of such data for determining correlation and associations among the attributes being studied.

SECTION 6: Statistical Testing
a. **Hypothesis and Statistical Testing Process**
b. **Statistical Tests- Z, t, F, Chi-Square and various Non-Parametric Tests**
c. **Which Statistical Test to use?**
d. **Inferential Issues, Measurement and Error Analysis**

Chapter	Title	Importance
19	Hypothesis Testing	Every research aims to cross new barriers to knowledge. Hypothesis is the objective of scientific research, which the researcher wants to prove or disprove. The data collected as a part of the research may or may not support the hypothesis. Setting up hypothesis and the procedure of testing hypothesis is an important component of 'Statistical Significance' or carrying out the 'Tests of Significance.' This topic covers the process and the underlying fundamentals in Hypothesis testing, and stops short of discussing actual tests, which are covered in subsequent chapters.
20	Tests of Significance (Z and Student's 't'-Test)	Subject to fulfillment of underlying assumptions, it is expected that a random variable, in all its randomness would be scattered around its central value, but within a theoretical limit. If the actual data observations indicate it to be beyond these limits, then the deviations cannot be attributed to its randomness alone, but are said to have been caused by factors beyond randomness. Another perspective is that in research, samples are studied to determine the population characteristics. Tests of significance help in determining the significant differences observed between two samples or between the sample and population parameters etc.

(Contd…)

(Contd...)

Chapter	Title	Importance
		As these limits and the process of testing are based on statistical theory, the differences are either termed as "Statistically Significant" or "Statistically Not Significant." Z-test and Student's t-test are classical parametric tests concerning testing the significance of large and small samples, and are covered in this chapter.
21	F-Test and Analysis of Variance and Covariance	F-test is employed to test whether the two independent samples have been drawn from normal populations with same variance or not; or whether the two independent estimates of population variance are homogenous or not. ANOVA (Analysis of Variance)/ANCOVA, a very widely used test is used to determine if two, or more than two, samples have equal means or no, allows understanding the sources of variation in samples.
22	Chi-square Test	It is one of the most widely used Non-Parametric Tests as it does not involve any population parameters. Its applications include—to test goodness-of-fit, to test independence of attributes, and to test for homogeneity.
23	Non-Parametric Tests	Also known as 'Distribution Free Tests' as these tests do not assume anything about the form and distribution characteristics of the population from which the samples have been drawn. Some of these tests can be done to test for randomness, ranks, association or independence etc.
24	Which Statistical Test to Use?	Given the large number of statistical tests available, it becomes difficult to decide on the appropriate test. This chapter summarizes all the tests in this text along with brief description, giving a bird's overview. This will allow the user to evaluate, compare and decide on suitable test. Also provided are most commonly used statistical tests in medical field.
25	Special Considerations about Inferential Issues in Biostatistics	In medical field, it is not easy to understand the cause-effect relationships and associations, due to multiplicity of factors involved. This chapter discusses the broad contours to assess Bias, Confounding, Effect Modification and the Exposure-Disease Association.
26	Measurement and Error Analysis	This chapter discusses the theory of 'Measurement' and what can be referred to as good measurement. To complete the circle, it also takes up the Errors in Measurement and what leads to the errors.
SECTION 7: Other Statistical Concepts Frequently used in Research a. Vital Statistics, Life Tables, Kaplan Meier Curves and Log Rank Test b. Time Series Analysis c. Interpolation and Extrapolation		
Chapter	Title	Importance
27	Vital (Demographic) Statistics and Life Tables	Demographic study is interlinked with medicine. Vital statistics deals with the demographic statistics. Life Tables, by nomenclature, refer to human life mortality rates and life expectancy, but are widely used in all studies when the subject event can be equated as a 'failure' (Mortality = Failure of Life). Kaplan-Meier curves are used widely in medical research, along with the Log-Rank Tests which tests for difference between two or more life curves.

(Contd...)

(Contd...)

Chapter	Title	Importance
28	Time Series Analysis	Time series analysis refers to study where the subject of study can be considered as dependent on time. Time series techniques are used to understand the long-term trend, the seasonal components, cyclical impacts and random components of a data which can be plotted against time. It can also be used as a forecasting tool.
29	Interpolation, Extrapolation and Forecasting Methods	Interpolation refers to systematic estimating of a missing or an intermediate value in the data series. If the value to be estimated lies outside the range for which the values are available, then it becomes a case of extrapolation.
SECTION 8: Statistical Concepts for Medical Administration		
Chapter	Title	Importance
30	Index Numbers	Index numbers as a tool helps in understanding the percentage change between the various values of the data series. In Index numbers, the values of a 'data series' are expressed as a percentage of 'another value', after equating this 'another value' as 100. The data series can consist of one variable value or may include data on many variables which are considered collectively. Consumer Price Index, Wholesale Price Index, Sensex are very visible Index Numbers. Purchase Index, Index on number of patients treated over the years or in different types of institutions, etc are possible indices which can be used in medical administration. The concept can be applied to create index specific to ones needs and is of immense help to reflect the relative changes over time. This is of use to investors and healthcare administrators. For clinicians, global risk indices and app based risk stratification, though attractive, is still in its infancy.
31	Statistical Quality Control	Six Sigma is widely used in the goal to achieve consistent quality. Medical Science today not only offers treatment, but its definition has widened and now it offers 'Health Services' where treatment is a part of this offering. Therefore, Statistical Quality Control as a measure can be applied to improve the services like diagnostic tests, food services to admitted patients, pharmacy services, OPD patients' experience while availing services etc. It also includes 'Acceptance Sampling' for consumables and other materials which form a big proportion of cost of delivering healthcare.
32	Decision Theory	Decision Theory deals with the basic principles of decision making, especially in risk situations, through the use of concepts of probability concepts and mathematical expectation. Decision Tree Analysis helps in representing the various scenarios and the probabilistic pay-offs for them.
SECTION 9: Report Writing and Applying for Grants		
Chapter	Title	Importance
33	Research Report Writing: How to do?	Reporting is as important as the actual recording and analysis of observations. The communication process should be comprehensive and concise. It is critical that accurate and proper reports are submitted. Though an integral part of research, it also helps in three ways—the research gets properly documented, which can be used by other researchers. It enhances the researcher's reliability and most importantly, it sustains the sponsor's interest to continue providing grants to the research community.

(Contd...)

(Contd...)

Chapter	Title	Importance
34	Applying for Grants	Grants are required to carry out research and, accordingly, the organization extending the grant requires proper documentation to be submitted. Getting grants to fund a research is not easy. Many a times, a good research idea does not get taken up due to rejection of the funding application. Therefore, it is necessary to ensure that the application for grants is proper and well written. This chapter discusses some of the 'do's and don'ts' while applying for grants. Though it may not guarantee funding, but it will ensure that the application does not get rejected for correctable reasons.

STATISTICS IN A NUTSHELL

The Pictorial Representation on the next page provides an overview of all major statistical concepts. A proper study of this will help in getting an overview of the subject. Many a times, readers are confused due to lack of guidance in contextual relationships among the various statistical concepts for application to biomedical or biological research. The Matrix shown is a simplified way to obviate this confusion and also help the reader in the understanding the flow of the concepts.

Zero Point: Approach Road to Learning

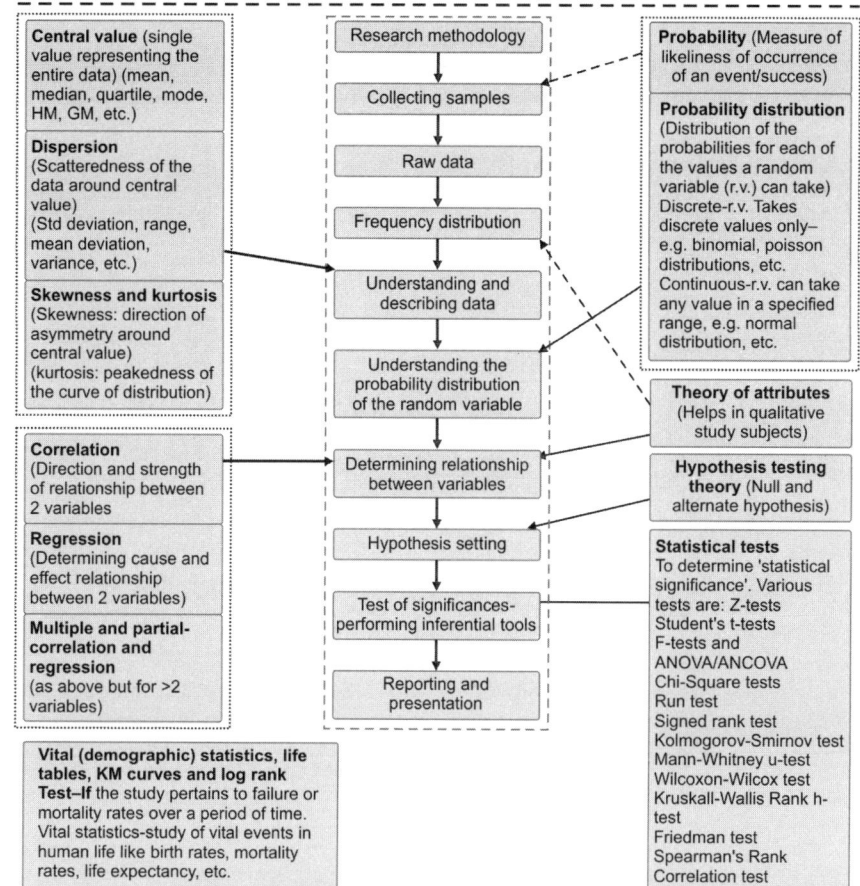

Section 1

Bioethics, Evidence-based Medicine

- Bioethics in Clinical Research
- Evidence-based Medicine and Levels of Evidence
- Introduction to Statistics

CHAPTER 1

Bioethics in Clinical Research

Ethics refers to *the body of moral principles or values governing or distinctive of a particular culture or group*. Ethics traces its root to the Greek word 'ethikos' or 'ethos' which roughly means habit or custom. Ethical behavior is that behavior which is *good*. The field of ethics or moral philosophy involves the development, defense and recommendation of concepts pertaining to right and wrong behavior.

MAJOR DIVISIONS OF ETHICS

There are three major divisions of study of ethics:
1. Normative ethics
2. Metaethics
3. Applied ethics

Normative ethics studies the practical methods of finding a moral course of action.

Metaethics studies the theoretical meaning and reference of moral propositions. It deals with determination of truth values of moral propositions.

Applied ethics deals with the courses of action a person is obligated or permitted to follow in a particular situation or domain of action.

Rapid advances in science like cloning, stem cell research and organ transplant have raised concerns about community's safety and the socially appropriate use of emerging medical processes and procedures. The legal positions on abortion, euthanasia, the use of animals in research and use of gene modified foods are also not clear in most societies. All this has lead to a debate to address the concerns of the communities regarding their positions on the correct course of action regarding research and clinical practices to be adopted for their community. Traditional ethical theories are inadequate in dealing with these emerging complex problems, and hence, the need for evolving bioethics as a dynamic philosophy and practice. All human interactions are governed by ethics but medicine needs to be driven more by these because of the emerging nature of knowledge in this field. This evolution of knowledge itself has been accelerated by the information revolution, and it further needs to be guided by principles proven to be correct or thought to be correct from an ethical perspective.

BIOETHICS

Bioethics is an emerging interdisciplinary science providing the disciplinary framework for moral questions and issues surrounding the life sciences. These issues encompass human beings, animals and nature. It is a way of ethical reasoning and decision making which integrates empirical data from relevant natural sciences (e.g. medicine in case of medical ethics) and considers other disciplines of applied ethics such as research ethics, information ethics, social ethics, economic ethics, political ethics, and legal ethics for resolving the dilemmas. It examines discussions and debates from different perspectives to elaborate on important arguments involved in ethical guidance in that particular area of human endeavor.

Bioethics is the *field of study concerned with the ethics and philosophical implications of certain biological and medical procedures, technologies, and treatments, as also organ transplants, genetic engineering, and care of the terminally ill.*

MAIN BRANCHES OF BIOETHICS

Bioethics is broadly divided into three main branches:
1. Medical ethics
2. Animal ethics
3. Environmental ethics.

Medical Ethics

The origin of the notion of bioethics can be traced to the Hippocratic Oath (500 BC) for medical ethics. Prior to that the Code of Hammurabi (1750 BC) also contained some written provisions related to medical practice. Fritz Jahr, a German theologian, used the German term "Bio-Ethik" (which translates as "Bio-Ethics") for the first time in 1927. He advocated the establishment of a new academic discipline and use of a new, more civilized approach to issues concerning human beings and the environment. Other authors have traced these origins to VR Potter's *Bioethics: The Science of Survival* [Potter VR. Bioethics: The Science of Survival, Perspectives in Biology and Medicine. 1970;14(2):127-53]; D Callahan's *Bioethics as a Discipline* [Callahan D. Bioethics as a Discipline. Hastings Center Studies. 1973;1(1):66-73] and discussions between Shriver and Hellegers on the establishment of what would be the 'Joseph and Rose Kennedy Center for the Study of Human Reproduction and Bioethics' (1971). Potter focused on the global movement to foster concern for the environment and ethics using the term. Callahan called for establishment of a new discipline. Shriver and Hellegers discussions centered on an institution where researchers could examine and analyze medical dilemmas in relation to moral philosophy. Thus different aspects of the branch of study were emphasized by different authors.

ETYMOLOGY, HISTORY AND THE FAMOUS NONETHICAL EXPERIMENTS

In the 1800s, bioethics was synonymous with 'medical ethics', a term coined by Thomas Percival in 1803. Animal and Environmental ethics emerged much later. The Nazis conducted many research experiments on human subjects which lead to a furore which had to be put out by the Nuremberg Code in 1947. Some of these gruesome experiments on nonwilling subjects in Nazi control included:

Josef Mengele's Experiments on Twins

Josef Mengele performed experiments on nearly 1,500 sets of imprisoned twins at Auschwitz in 1943–44. Survival rate in these experimental subjects was less than 7% and the experiments ranged from injection of different dyes into the eyes to see the effect on eye color to sewing twins together to create conjoined twins. Twin children in concentration camps were used to study the similarities and differences in genotype and phenotypic modifications on the genotype in twins.

Ravensbrück Experiments

Experiments at Ravensbrück concentration camp in 1942–43 were conducted to study bone, muscle and nerve regeneration as well as allogenic bone transplant, i.e. from one person to another. Sections of bones, muscles, and nerves were removed from the subjects without any anesthesia. The victims suffered intense agony, mutilation and even permanent disability. When the stories of the survivors reached the world the people were shocked at the scale of cruelty and demands for regulation of scientific activities gained momentum.

Baranowicze Head Injury Experiment

A preteenage boy was strapped into a chair so that he could not move in Baranowicze in Poland in 1942. A mechanized hammer came down and hit his head every few seconds. The boy was driven insane from the torture.

DACHAU AND AUSCHWITZ CONCENTRATION CAMP HYPOTHERMIA EXPERIMENTS

The Luftwaffe conducted experiments to discover methods for preventing and treating hypothermia in 1941. Ernst Holzlöhner and Dr Sigmund Rascher immersed subjects in cold water at Dachau concentration camp. Naked victims were kept in open air for several hours with temperatures as low as –6 °C (21 °F). Different methods of re-warming survivors were assessed and some victims were even thrown in boiling water for re-warming. These freezing/ hypothermia experiments were used to understand the effects of circumstances on the cold Eastern Front. Germans were not well-prepared for the cold weather on the Eastern front. They wondered whether genetic differences were responsible for the superior resistance to cold that they observed in the Russians. Dr Sigmund Rascher presented these experiments in 1942 at a medical conference with the name "Medical Problems Arising from Sea and Winter". Approximately 100 people died due to these experiments. In Dachau, Luftwaffe and German Experimental Institution for Aviation Physicians also conducted high-altitude experiments using low-pressure chamber for finding maximum altitude from which damaged aircraft crews could parachute to safety. Prisoners were subjected to a low-pressure chamber simulating conditions at altitudes of up to 20,000 m (66,000 ft). Out of 200 subjects 80 died; the others were executed and vivisections of the brains was carried out.

Dachau Potable Sea Water Experiment

In 1944 Dr Hans Eppinger deprived prisoners, especially Roma gypsies, of food and gave them nothing but sea water to drink to study various methods of making sea water drinkable at Dachau concentration camp. The prisoners were so dehydrated that they licked freshly mopped floors to get drinkable water. These experiments moved the people who learnt of them after the war.

Malaria Experiments

Professor Claus Schilling tested synthetic malaria drugs and malaria immunization in Dachau injecting healthy prisoners with high and even lethal doses in 1939–45, after being infected by mosquitoes or by injections of extracts of the mucous glands of female mosquitoes. Drugs tested included quinine, pyrifer, neosalvarsan, antipyrin, pyramidon and a drug called 2516 Behring among others. Professor Schilling had been working on an anti-malaria vaccine in Italian mental homes prior to this. Professor Gerhard Rose tested malaria drugs in mentally unwell Russian prisoners of war in Thuringia. Malaria experiments were also conducted in Buchenwald. This was a most horrible torture of unwilling subjects.

Immunization Experiments

At the concentration camps of Sachsenhausen, Dachau, Natzweiler, Buchenwald and Neuengamme scientists tested immunization compounds and sera were tested for malaria, typhus, tuberculosis, typhoid fever, yellow fever and infectious hepatitis. Many non-consenting prisoners died in these experiments.

Mustard Gas Experiments

In period between 1939 and 1945, Sachsenhausen, Natzweiler and other camps saw prisoners being exposed to mustard gas and phosgene. These gases inflicted severe chemical burns on the subjects. The objective of the investigators was to find the best therapy for them. People inhaled these agents or were exposed or injected with them. It resulted in agony and death of a large proportion of the nonconsenting subjects enrolled.

Ravensbrück Sulfonamide Experiments

In 1942–43, wounds were inflicted on subjects and infected with bacteria such as *Streptococcus*, *Clostridium perfringens* and *Clostridium tetani* to be treated with sulfonamides to test efficacy. At Ravensbrück, blood circulation to the wound was interrupted by tying off blood vessels at both ends and filling of wood shavings with or without ground glass into the wounds to simulate a battlefield wound. This was then treated with sulfonamide and other drugs. The morbidity and mortality in that era was with the existing conditions was heartrending.

Sterilization Experiments

In 1933, Germany passed the Law for the Prevention of Genetically Defective Progeny. It called for sterilization of persons with hereditary diseases. These diseases included weak-mindedness, schizophrenia, alcohol abuse, insanity, blindness, deafness and physical deformities. 300,000 patients were sterilized by 1937. Dr Carl Clauberg studied methods of sterilization suitable for sterilizing millions of people with minimum of time and effort at Auschwitz, Ravensbrück and Dachau besides other places. X-ray, surgery and various drugs were tested. Thousands of victims were sterilized. The Nazis also sterilized around 400,000 people under compulsory sterilization program. Intravenous injections of solutions containing iodine and silver nitrate were reported to be successful, but came with side effects such as vaginal bleeding, severe abdominal pain and cervical cancer. Radiation was thus the favored method of sterilization. Specific amounts of radiation destroyed a victim's ability to produce ova or sperm. The prisoners were asked to complete forms in a room, which took two to three minutes, and the radiation was administered without their knowledge, rendering them completely sterile. Severe radiation burns were also reported in the process.

Buchenwald Experiments with Poison and Incendiary Bombs

In 1943–44, experiments using poisoned food, poison on bullets and phosphorus containing incendiary bombs were carried out at Buchenwald to investigate the effect of various poisons and the best method of treatment of phosphorus burns. Poisoned food was administered secretly. Prisoners either died of poisoning or were killed immediately to carry out autopsies. Same procedure was followed with poisonous bullets. The subjects were forced into experiments where the outcomes were preordained-death.

Unit 731

The imperial Japanese Army performed secret biological and chemical warfare research experiments in Harbin in 1939–45. It was called Unit 731. Russian and Chinese men, women, children, infants, pregnant women and even old persons were subjected to atrocious crimes such as removal of organs from a live body, amputation for studying blood loss, germ warfare and even live testing of weapons. "Some prisoners even had their stomachs surgically removed and their esophagus reattached to the intestines." The accounts were moving and left little to imagination. These received less attention than the Nazi atrocities but were equally inhuman.

Robert Allen (Laud) Humphreys' Tearoom Sex Study

Laud Humphreys, a US sociologist observed that the public and the law-enforcement authorities held highly simplistic stereotyped belief about men who committed impersonal sexual acts with one another in public restrooms. This oral sex was called "Tearoom Sex". It was felt that these people were mostly homosexuals.

For his Washington University dissertation Humphreys acted as a "watchqueen". A watchqueen kept watch and raised an alarm by coughing when a cop or stranger came near. Humphreys observed hundreds of acts of fellatio. After gaining their confidence, he interviewed many of the participants who were willing to talk about their lives to him. Those who were willing to talk openly tended to be among the better-educated members of "Tearoom Sex Escapades". To avoid bias, Humphreys secretly followed some of the other men randomly. He recorded the license numbers of their cars. A year later, a disguised Humphreys went to their homes claiming to be a health-service interviewer. He interviewed them about their marital status, race, job and other variables. He found that 54% of his subjects were married; 38% were Catholics; 24% were clearly bisexual, happily-married, well-educated and economically successful; only 14% fit the stereotype of homosexual and fringe elements. This research broke the stereotypes held by the public and law-enforcement agencies. In the middle 1960s, there were no institutional review boards in existence. Only Humphreys' PhD committee reviewed the dissertation proposal. After completion of the research, the other members of the Sociology Department learnt of it. Some of these members of the department objected that Humphreys' research had unethically invaded the privacy and threatened the social standing of the subjects. They held that this research was unethical. He had observed acts of homosexuality by posing as a voyeur. He had not taken the subjects' consent. He had tracked down names and addresses through license plate numbers clandestinely and interviewed the men in their homes in disguise and under false pretenses. These faculty members petitioned the President of Washington University and advocated rescinding Laud Humphreys' PhD degree. The matter came to a head and even a fist fight broke out among faculty members following this development. This culminated in the exodus of about half of the department members to positions at other universities. Robert Allen (Laud) Humphreys' retained his PhD degree in the face of much opposition. Public at large was also outraged when the affair became public. Journalist Nicholas von Hoffman got to know of the case through angered members of the Sociology Department. He condemned the intrusion saying, "We overlook the social scientists behind the hunting blinds who are also peeping into what we thought were our most private and secret lives. But there they are, studying us, taking notes, getting to know us, as indifferent as everybody else to the feeling that to be a complete human involves having an aspect of ourselves that's unknown." This was one of the key landmark events that pushed the debate on bioethics to the fore in addition to the Belmont Report which we will now discuss.

Tuskegee Syphilis Study

"Tuskegee Study of Untreated Syphilis in the Negro Male" started in 1932. The Public Health Service and the Tuskegee Institute began this study to record the natural history of syphilis. It initially enrolled 600 black men—399 with syphilis and 201 without it. Researchers told the men, they were being treated for "bad blood." This was a term used in local parlance to describe several ailments including syphilis, anemia and fatigue. In exchange for participating in the study, they received free medical examinations, free meals and burial insurance. But they were not given the proper treatment required to cure their illness. The participants did not know this.

It was originally projected to last 6 months. However, the study actually went on till 1972. In 1947, penicillin became the drug of choice for syphilis. But the researchers did not offer it to the enrolled participants. The reviewing panel found nothing in the records to show that subjects were ever given the choice of quitting the study even when this new, highly effective treatment became widely used. It was found that the subjects had agreed freely to be examined and treated. However, there was no evidence that researchers had informed them of the study or its real purpose. In fact, the men had been misled. All the facts required to provide informed consent was not made available to them. The study was terminated in 1972. In 1973 a class-action lawsuit was filed on behalf of the study participants and their families and a $10 million out-of-court settlement was reached the next year. The US Government undertook to give lifetime medical benefits and burial services to all living participants and the Tuskegee Health Benefit Program (THBP) was started to provide these services. The wives, widows and offsprings were extended these benefits in 1975. The last study participant died in January 2004. The last widow receiving benefits died in January 2009. The Belmont Report (1978) emerged from the findings of this investigation and today forms the bedrock of modern bioethics by giving three of the four principles guiding bioethical perspective- respect for persons (i.e. autonomy), beneficence and justice. In 1983, Beauchamp and Childress' textbook on the principles of biomedical ethics adopted the three principles of the Belmont Report and added a fourth principle, that of non-maleficence. Subsequently, the UNESCO Universal Declaration on Bioethics and Human Rights in 2005 gave the framework adopted by over 190 member states. This evolution can be traced from the ancient ethical paradigms to the Prussian informed consent processes through the documents developed through the 20th century.

Nuremberg Trials

After the Second World War, the United States authorities held trials in occupation zone in Nuremberg, Germany. The 'United States of America versus Karl Brandt et al' also known as Doctors' trial was the first of 12 trials for war crimes of German doctors. It was held before US military courts and not before the International Military Tribunal in the same rooms at the Palace of Justice. The 12 trials were collectively called the "Trials of War Criminals before the Nuremberg Military Tribunals" or the "Subsequent Nuremberg Trials". The following were charged:

Karl Brandt, Rudolf Brandt, Hermann Becker-Freyseng, Wilhelm Beiglböck, Helmut Poppendick, Kurt Blome, Gerhard Rose, Viktor Brack, Fritz Fischer, Karl Gebhardt, Waldemar Hoven, Karl Genzken, Siegfried Handloser, Joachim Mrugowsky, Herta Oberheuser, Adolf Pokorny, Hans Wolfgang Romberg, Paul Rostock, Siegfried Ruff, Konrad Schäfer, Oskar Schröder, Wolfram Sievers and Georg August Weltz.

Josef Mengele escaped this trial. Mengele was transferred from Auschwitz to Gross-Rosen concentration camp on 17 January, 1945. The Soviet Army captured Auschwitz on 27 January, 1945. Mengele fled Gross-Rosen on 18 February, 1945. The Soviet army reached a week later. He hurried west with his unit and was taken prisoner of war by the Americans in June, 1945. Mengele was initially registered under his own name. He did not have the SS blood group tattoo. He was not identified as a major war criminal and was let off in July, 1945. He procured false identification documents with the name "Fritz Ullman" or "Fritz Hollmann". He continued to live in Germany till 1949 working as a farm laborer near Rosenheim. In 1949, he escaped to Argentina traveling through Genova. His wife, whom he subsequently divorced in 1954, refused to accompany him. He lived in Argentina till 1959 where a young maiden's death due to a botched up abortion resulted in his cover getting compromised and then he eloped to Paraguay and later to Brazil. In 1979, Mengele drowned while swimming. He was buried under a false name. The remains were disinterred and identified by forensic examination after his death.

These doctors were involved in several human experiments and over 3,500,000 sterilizations of German citizens. The trials began on December 9, 1946 in Nuremberg. They stated in their defence, the fact their experiments differed little from pre-war ones. There was no law to segregate legal from illegal experiments. Dr Leo Alexander, the chief medical advisor to Telford Taylor, the US Chief of Counsel for War Crimes had submitted a draft of six points defining legitimate medical research. The verdict adopted these points adding an extra four. Together these ten points constituted the "Nuremberg Code". General Telford Taylor felt that the US judge, Harold Sebring, was the author of the Code. Leo Alexander and Andrew Ivy, American physicians who helped prosecute the Nazi doctors have each been given credit too. From the proceedings of the trial, it can be concluded that the credit for authorship needs to be shared. The famous 10 principles of the Nuremberg Code developed from the trial itself. The Nuremberg code includes the principles of informed consent, absence of coercion, properly formulated scientific experimentation and beneficence towards participants.

Ten Points of Nuremberg Code

The ten points of the Nuremberg Code enumerated from the original document as follows:

"The voluntary consent of the human subject is absolutely essential.

This means that the person involved should have legal capacity to give consent; should be so situated as to be able to exercise free power of choice, without the intervention of any element of force, fraud, deceit, duress, over-reaching, or other ulterior form of constraint or coercion; and should have sufficient knowledge and comprehension of the elements of the subject matter involved, as to enable him to make an understanding and enlightened decision. This latter element requires that, before the acceptance of an affirmative decision by the experimental subject, there should be made known to him the nature, duration, and purpose of the experiment; the method and means by which it is to be conducted; all inconveniences and hazards reasonably to be expected; and the effects upon his health or person, which may possibly come from his participation in the experiment."

The duty and responsibility for ascertaining the quality of the consent rests upon each individual who initiates, directs or engages in the experiment. It is a personal duty and responsibility, which may not be delegated to another with impunity.

The experiment should be such as to yield fruitful results for the good of society, unprocurable by other methods or means of study, and not random and unnecessary in nature.

The experiment should be so designed and based on the results of animal experimentation and knowledge of the natural history of the disease or other problem under study, that the anticipated results will justify the performance of the experiment.

The experiment should be so conducted as to avoid all unnecessary physical and mental suffering and injury.

No experiment should be conducted, where there is an *a priori* reason to believe that death or disabling injury will occur; except, perhaps, in those experiments where the experimental physicians also serve as subjects.

The degree of risk to be taken should never exceed that determined by the humanitarian importance of the problem to be solved by the experiment.

Proper preparations should be made and adequate facilities provided to protect the experimental subject against even remote possibilities of injury, disability, or death.

The experiment should be conducted only by scientifically qualified persons. The highest degree of skill and care should be required through all stages of the experiment of those who conduct or engage in the experiment.

During the course of the experiment, the human subject should be at liberty to bring the experiment to an end, if he has reached the physical or mental state, where continuation of the experiment seemed to him to be impossible.

During the course of the experiment, the scientist in charge must be prepared to terminate the experiment at any stage, if he has probable cause to believe, in the exercise of the good faith, superior skill and careful judgment required of him, that a continuation of the experiment is likely to result in injury, disability, or death to the experimental subject".

"Trials of War Criminals before the Nuremberg Military Tribunals under Control Council Law No. 10", Vol. 2, pp. 181-2. Washington, DC.: US Government Printing Office, 1949.

What Changed from Hippocratic Oath?

Hippocrates of Cos (460–365 BC) was an ancient Greek physician. He is traditionally regarded the Father of Medicine and revered as an ideal physician. About 60 odd medical writings survive bearing his name. Most of them are believed to have been written by others. He is revered to this day for his contribution to ethical standards in medical practice—mainly for the ubiquitous Hippocratic Oath. It is suspected that even this was written by his pupils though the credit is given to him to this day. This Hippocratic Oath is historically taken by physicians at the time of initiation into the practice. The original oath was written in Ionic Greek and was part of the Hippocratic Corpus. The modern version is credited to Louis Lasagna, the Academic Dean of the School of Medicine at Tufts University.

The Hippocratic ethics centered around the maxim *primum non nocere* or *first do no harm*. This phrase is not a part of the Hippocratic Oath. It was in the first book 'Epidemics, that it was said,'"Practice two things in your dealings with disease: either help or do not harm the patient". In the Hippocratic Oath, the physician swears, "I will, according to my ability and judgment, prescribe a regimen for the health of the sick; but I will utterly reject harm and mischief" which is taken by many as an expression of the same feelings.

The judges at Nuremberg felt that more stringent steps were required to protect human research subjects. These principles, now called the Nuremberg Code, include a new, comprehensive and absolute requirement of informed consent (Principle 1) and a new right of the subject to withdraw from participation in an experiment

(Principle 9). The traditional Hippocratic Oath deals with a doctor–patient relationship in which the patient is silent and dutifully obedient to a beneficent and trusted physician. Once a patient agrees to be treated, he entrusts that the physician with a responsibility to act in his interest. He believes that the doctor's actions will benefit him or at least will do no harm. Research is outside the beneficent context of the physician–patient relationship. Here the physician's primary goal is not treatment but testing a scientific hypothesis by following an algorithm or protocol regardless of the subject's best interest. Hippocratic ethics even with informed consent tend to submerge the subject's right to decide for himself into what the physician-investigator thinks is best for the subject. Hippocratic view of medical research can potentially blind us to the risks to which research subjects are exposed which are many times greater than the risks to which patients who are merely being treated get exposed.

Informed consent, the core of the Nuremberg code, is a protection of subjects' human rights. Nuremberg code merged Hippocratic ethics with the protection of human rights. The Nuremberg code requires that physician-researchers protect the best interests of their subjects (Principles 2–8 and 10). It also bestows upon the subjects the choice to actively protect themselves (Principles 1 and 9). In Hippocrates' code the subject relies on the physician to decide when it is in the subject's best interest to end his or her participation in an experiment. In the Nuremberg code the subject is given an equal authority to end the experiment before its conclusion (Principle 9).

The Universal Declaration of Human Rights was adopted by the General Assembly of the United Nations in 1948. Helsinki Declaration of the World Medical Association in 1964 is the fundamental international document in biomedical research ethics. It forms the basis of most international, regional and national legislation and codes of conduct. The General Assembly adopted the International Covenant on Civil and Political Rights in 1966. Article 7, of the Covenant states "No one shall be subjected to torture or to cruel, inhuman or degrading treatment or punishment. In particular, no one shall be subjected without his free consent to medical or scientific experimentation". This statement expresses the fundamental human value that should govern all research involving human subjects.

Declaration of Helsinki (1964)

These recommendations guiding physicians in biomedical research involving human subjects were adopted by the 18th World Medical Assembly, Helsinki, Finland, June 1964, amended by the 29th World Medical Assembly, Tokyo, Japan, October 1975, and the 35th World Medical Assembly, Venice, Italy, October 1983.

'The Declaration of Geneva of the World Medical Association binds the physician with the words, "The health of my patient will be my first consideration," and the International Code of Medical Ethics declares that, "A physician shall act only in the patient's interest when providing medical care which might have the effect of weakening the physical and mental condition of the patient."

The purpose of biomedical research involving human subjects must be to improve diagnostic, therapeutic and prophylactic procedures and the understanding of the etiology and pathogenesis of disease.

Medical progress is based on research which ultimately must rest in part on experimentation involving human subjects. In the field of biomedical research a fundamental distinction must be recognized between medical research in which the aim is essentially diagnostic or therapeutic for a patient and medical research the essential object of which is purely scientific and without implying direct diagnostic or therapeutic value to the person subjected to the research.

Special caution must be exercised in the conduct of research which may affect the environment, and the welfare of animals used for research must be respected.

Because it is essential that the results of laboratory experiments be applied to human beings to further scientific knowledge and to help suffering humanity, the World Medical Association has prepared the following recommendations as a guide to every physician in biomedical research involving human subjects. They should be kept under review in the future. It must be stressed that the standards as drafted are only a guide to physicians all over the world. Physicians are not relieved from criminal, civil and ethical responsibilities under the law of their own countries.

"I. Basic Principles

1. Biomedical research involving human subjects must conform to generally accepted scientific principles and should be based on adequately performed laboratory and animal experimentation and on a thorough knowledge of the scientific literature.
2. The design and performance of each experimental procedure involving human subjects should be clearly formulated in an experimental protocol which should be transmitted to a especially appointed independent committee for consideration, comment and guidance.
3. Biomedical research involving human subjects should be conducted only by scientifically qualified persons and under the supervision of a clinically competent medical person. The responsibility for the human subject must always rest with a medically qualified person and never rest on the subject of the research, even though the subject has given his or her consent.
4. Biomedical research involving human subjects cannot legitimately be carried out unless the importance of the objective is in proportion to the inherent risk to the subject.
5. Every biomedical research project involving human subjects should be preceded by careful assessment of predictable risks in comparison with foreseeable benefits to the subject or to others. Concern for the interests of the subject must always prevail over the interests of science and society.
6. The right of the research subject to safeguard his or her integrity must always be respected. Every precaution should be taken to respect the privacy of the subject and to minimize the impact of the study on the subject's physical and mental integrity and on the personality of the subject.
7. Physicians should abstain from engaging in research projects involving human subjects unless they are satisfied that the hazards involved are believed to be predictable. Physicians should cease any investigation, if the hazards are found to outweigh the potential benefits.

8. In publication of the results of his or her research, the physician is obliged to preserve the accuracy of the results. Reports of experimentation not in accordance with the principles laid down in this Declaration should not be accepted for publication.
9. In any research on human beings, each potential subject must be adequately informed of the aims, methods, anticipated benefits and potential hazards of the study and the discomfort it may entail. He or she should be informed that he or she is at liberty to abstain from participation in the study and that he or she is free to withdraw his or her consent to participation at any time. The physician should then obtain the subject's freely given informed consent, preferably in writing.
10. When obtaining informed consent for the research project the physician should be particularly cautious, if the subject is in dependent relationship to him or her or may consent under duress. In that case, the informed consent should be obtained by a physician who is not engaged in the investigation and who is completely independent of this official relationship.
11. In case of legal incompetence, informed consent should be obtained from the legal guardian in accordance with national legislation. Where physical or mental incapacity makes it impossible to obtain informed consent, or when the subject is a minor, permission from the responsible relative replaces that of the subject in accordance with national legislation. Whenever the minor child is, in fact, able to give a consent, the minor's consent must be obtained in addition to the consent of the minor's legal guardian.
12. The research protocol should always contain a statement of the ethical considerations involved and should indicate that the principles enunciated in the present declaration are complied with.

II. Medical Research Combined with Professional Care (Clinical Research)

1. In the treatment of the sick person, the physician must be free to use a new diagnostic and therapeutic measure, if in his or her judgment it offers hope of saving life, re-establishing health or alleviating suffering.
2. The potential benefits, hazards and discomfort of a new method should be weighed against the advantages of the best current diagnostic and therapeutic methods.
3. In any medical study, every patient including those of a control group, if any— should be assured of the best proven diagnostic and therapeutic method.
4. The refusal of the patient to participate in a study must never interfere with the physician-patient relationship.
5. If the physician considers, it essential not to obtain informed consent, the specific reasons for this proposal should be stated in the experimental protocol for transmission to the independent committee.
6. The physician can combine medical research with professional care, the objective being the acquisition of new medical knowledge, only to the extent that medical research is justified by its potential diagnostic or therapeutic value for the patient.

III. Nontherapeutic Biomedical Research Involving Human Subjects (Nonclinical Biomedical Research)

1. In the purely scientific application of medical research carried out on a human being, it is the duty of the physician to remain the protector of the life and health of that person on whom biomedical research is being carried out.
2. The subjects should be volunteers—either healthy persons or patients for whom the experimental design is not related to the patient's illness.
3. The investigator or the investigating team should discontinue the research, if in his/her or their judgment it may, if continued, be harmful to the individual.
4. In research on man, the interest of science and society should never take precedence over considerations related to the well-being of the subject".

[World Medical Organization. Declaration of Helsinki. British Medical Journal (7 December) 1996;313(7070):1448-9].

WMA International Code of Medical Ethics

Adopted by the 3rd General Assembly of the World Medical Association, London, England, October 1949 and amended by the 22nd World Medical Assembly, Sydney, Australia, August 1968 and the 35th World Medical Assembly, Venice, Italy, October 1983 and the 57th WMA General Assembly, Pilanesberg, South Africa, October 2006.

Duties of Physicians in General

A physician shall	Always exercise his/her independent professional judgment and maintain the highest standards of professional conduct.
	Respect a competent patient's right to accept or refuse treatment.
	Not allow his/her judgment to be influenced by personal profit or unfair discrimination.
	Be dedicated to providing competent medical service in full professional and moral independence, with compassion and respect for human dignity.
	Deal honestly with patients and colleagues, and report to the appropriate authorities those physicians who practice unethically or incompetently or who engage in fraud or deception.
	Not receive any financial benefits or other incentives solely for referring patients or prescribing specific products.
	Respect the rights and preferences of patients, colleagues, and other health professionals.
	Recognize his/her important role in educating the public but should use due caution in divulging discoveries or new techniques or treatment through non-professional channels.
	Certify only that which he/she has personally verified.
	Strive to use healthcare resources in the best way to benefit patients and their community.
	Seek appropriate care and attention, if he/she suffers from mental or physical illness.
	Respect the local and national codes of ethics.

Duties of Physicians to Patients

A physician shall	Always bear in mind the obligation to respect human life.
	Act in the patient's best interest when providing medical care.
	Owe his/her patients complete loyalty and all the scientific resources available to him/her. Whenever an examination or treatment is beyond the physician's capacity, he/she should consult with or refer to another physician who has the necessary ability.
	Respect a patient's right to confidentiality. It is ethical to disclose confidential information when the patient consents to it or when there is a real and imminent threat of harm to the patient or to others and this threat can be only removed by a breach of confidentiality.
	Give emergency care as a humanitarian duty unless he/she is assured that others are willing and able to give such care.
	In situations, when he/she is acting for a third party, ensure that the patient has full knowledge of that situation.
	Not enter into a sexual relationship with his/her current patient or into any other abusive or exploitative relationship.

Duties of Physicians to Colleagues

A physician shall	Behave towards colleagues as he/she would have them behave towards him/her.
	Not undermine the patient-physician relationship of colleagues in order to attract patients.
	When medically necessary, communicate with colleagues who are involved in the care of the same patient. This communication should respect patient confidentiality and be confined to necessary information.

Universal Declaration on Bioethics and Human Rights

Universal Declaration on Bioethics and Human Rights was adopted by 191 member states in October, 2005 to:

"a. Provide a universal framework of principles and procedures to guide states in the formulation of their legislation, policies or other instruments in the field of bioethics;
b. To guide the actions of individuals, groups, communities, institutions and corporations;
c. To promote respect for human dignity and protect human rights, by ensuring respect for the life of human beings, and fundamental freedoms, consistent with international human rights law;
d. To recognize the importance of freedom of scientific research and the benefits derived from scientific and technological developments, while stressing the need for such research and developments to occur within the framework of ethical principles set out in this Declaration and to respect human dignity, human rights and fundamental freedoms;
e. To foster multidisciplinary and pluralistic dialogue about bioethical issues between all stakeholders and within society as a whole;
f. To promote equitable access to medical, scientific and technological developments as well as the greatest possible flow and the rapid sharing of knowledge concerning those developments and the sharing of benefits, with particular attention to the needs of developing countries;

g. To safeguard and promote the interests of the present and future generations;
h. To underline the importance of biodiversity and its conservations as a common concern of human kind".

It has 15 basic principles, viz:
1. Human dignity and human rights
2. Benefit and harm
3. Autonomy and individual responsibility
4. Consent
5. Persons without the capacity to consent
6. Respect for human vulnerability and personal integrity
7. Privacy and confidentiality
8. Equality, justice and equity
9. Nondiscrimination and nonstigmatization
10. Respect for cultural diversity and pluralism
11. Solidarity and cooperation
12. Social responsibility and health
13. Sharing of benefits
14. Protecting future generations
15. Protection of the environment, the biosphere and biodiversity biosphere and biodiversity.

This is nonbinding on nation states but it is likely that because of the document the ethical principles enunciated here will find their way into legislations of the developing member nations. This is the first document to have a large following and is unique insofar as it incorporates ethics and human rights. It has been argued that without human rights there can be no ethics to guard.

> *"Article 3: Human dignity and human rights*
> 1. Human dignity, human rights and fundamental freedoms are to be fully respected.
> 2. The interests and welfare of the individual should have priority over the sole interest of science or society.
>
> *Article 4: Benefit and harm*
> In applying and advancing scientific knowledge, medical practice and associated technologies, direct and indirect benefits to patients, research participants and other affected individuals should be maximized and any possible harm to such individuals should be minimized.
>
> *Article 5: Autonomy and individual responsibility*
> The autonomy of persons to make decisions, while taking responsibility for those decisions and respecting the autonomy of others, is to be respected. For persons who are not capable of exercising autonomy, special measures are to be taken to protect their rights and interests.
>
> *Article 6: Consent*
> 1. Any preventive, diagnostic and therapeutic medical intervention is only to be carried out with the prior, free and informed consent of the person concerned, based on adequate information. The consent should, where appropriate, be express and may be withdrawn by the person concerned at any time and for any reason without disadvantage or prejudice.
> 2. Scientific research should only be carried out with the prior, free, express and informed consent of the person concerned. The information should be adequate, provided in a comprehensible form and should include modalities for withdrawal of consent. Consent may be withdrawn by the person concerned at any time and for any reason without any disadvantage or prejudice. Exceptions to this principle should be made only in accordance with ethical and legal standards adopted by States, consistent with the principles and provisions set out in this Declaration, in particular in Article 27, and international human rights law.

Contd...

Contd...

3. In appropriate cases of research carried out on a group of persons or a community, additional agreement of the legal representatives of the group or community concerned may be sought. In no case, should a collective community agreement or the consent of a community leader or other authority substitute for an individual's informed consent.

Article 7: Persons without the capacity to consent

In accordance with domestic law, special protection is to be given to persons who do not have the capacity to consent:

(a) Authorization for research and medical practice should be obtained in accordance with the best interest of the person concerned and in accordance with domestic law. However, the person concerned should be involved to the greatest extent possible in the decision-making process of consent, as well as that of withdrawing consent;

(b) Research should only be carried out for his or her direct health benefit, subject to the authorization and the protective conditions prescribed by law, and if there is no research alternative of comparable effectiveness with research participants able to consent. Research which does not have potential direct health benefit should only be undertaken by way of exception, with the utmost restraint, exposing the person only to a minimal risk and minimal burden and, if the research is expected to contribute to the health benefit of other persons in the same category, subject to the conditions prescribed by law and compatible with the protection of the individual's human rights. Refusal of such persons to take part in research should be respected.

Article 8: Respect for human vulnerability and personal integrity

In applying and advancing scientific knowledge, medical practice and associated technologies, human vulnerability should be taken into account. Individuals and groups of special vulnerability should be protected and the personal integrity of such individuals respected.

Article 9: Privacy and confidentiality

The privacy of the persons concerned and the confidentiality of their personal information should be respected. To the greatest extent possible, such information should not be used or disclosed for purposes other than those for which it was collected or consented to, consistent with international law, in particular international human rights law.

Article 10: Equality, justice and equity

The fundamental equality of all human beings in dignity and rights is to be respected so that they are treated justly and equitably.

Article 11: Nondiscrimination and nonstigmatization

No individual or group should be discriminated against or stigmatized on any grounds, in violation of human dignity, human rights and fundamental freedoms.

Article 12: Respect for cultural diversity and pluralism

The importance of cultural diversity and pluralism should be given due regard. However, such considerations are not to be invoked to infringe upon human dignity, human rights and fundamental freedoms, nor upon the principles set out in this Declaration, nor to limit their scope.

Article 13: Solidarity and cooperation

Solidarity among human beings and international cooperation towards that end are to be encouraged.

Article 14: Social responsibility and health

1. The promotion of health and social development for their people is a central purpose of governments that all sectors of society share.
2. Taking into account that the enjoyment of the highest attainable standard of health is one of the fundamental rights of every human being without distinction of race, religion, political belief, economic or social condition, progress in science and technology should advance:

Contd...

Contd...

(a) Access to quality health care and essential medicines, especially for the health of women and children, because health is essential to life itself and must be considered to be a social and human good;
(b) Access to adequate nutrition and water;
(c) Improvement of living conditions and the environment;
(d) Elimination of the marginalization and the exclusion of persons on the basis of any grounds;
(e) Reduction of poverty and illiteracy.

Article 15: Sharing of benefits
1. Benefits resulting from any scientific research and its applications should be shared with society as a whole and within the international community, in particular with developing countries. In giving effect to this principle, benefits may take any of the following forms:
   (a) Special and sustainable assistance to, and acknowledgement of, the persons and groups that have taken part in the research;
   (b) Access to quality health care;
   (c) Provision of new diagnostic and therapeutic modalities or products stemming from research;
   (d) Support for health services;
   (e) Access to scientific and technological knowledge;
   (f) Capacity-building facilities for research purposes;
   (g) Other forms of benefit consistent with the principles set out in this Declaration.
2. Benefits should not constitute improper inducements to participate in research.

Article 16: Protecting future generations
The impact of life sciences on future generations, including on their genetic constitution, should be given due regard.

Article 17: Protection of the environment, the biosphere and biodiversity
Due regard is to be given to the interconnection between human beings and other forms of life, to the importance of appropriate access and utilization of biological and genetic resources, to respect for traditional knowledge and to the role of human beings in the protection of the environment, the biosphere and biodiversity.

Application of the principles

Article 18: Decision-making and addressing bioethical issues
1. Professionalism, honesty, integrity and transparency in decision-making should be promoted, in particular, declarations of all conflicts of interest and appropriate sharing of knowledge. Every endeavor should be made to use the best available scientific knowledge and methodology in addressing and periodically reviewing bioethical issues.
2. Persons and professionals concerned and society as a whole should be engaged in dialogue on a regular basis.
3. Opportunities for informed pluralistic public debate, seeking the expression of all relevant opinions, should be promoted.

Article 19: Ethics committees
Independent, multidisciplinary and pluralist ethics committees should be established, promoted and supported at the appropriate level in order to:
(a) Assess the relevant ethical, legal, scientific and social issues related to research projects involving human beings;
(b) Provide advice on ethical problems in clinical settings;
(c) Assess scientific and technological developments, formulate recommendations and contribute to the preparation of guidelines on issues within the scope of this Declaration;
(d) Foster debate, education and public awareness of, and engagement in, bioethics.

Article 20: Risk assessment and management
Appropriate assessment and adequate management of risk related to medicine, life sciences and associated technologies should be promoted.

Contd...

Contd...

Article 21: Transnational practices
1. States, public and private institutions, and professionals associated with transnational activities should endeavor to ensure that any activity within the scope of this Declaration, undertaken, funded or otherwise pursued in whole or in part in different States, is consistent with the principles set out in this Declaration.
2. When research is undertaken or otherwise pursued in one or more States (the host State(s)) and funded by a source in another State, such research should be the object of an appropriate level of ethical review in the host State(s) and the State in which the funder is located. This review should be based on ethical and legal standards that are consistent with the principles set out in this Declaration.
3. Transnational health research should be responsive to the needs of host countries, and the importance of research contributing to the alleviation of urgent global health problems should be recognized.
4. When negotiating a research agreement, terms for collaboration and agreement on the benefits of research should be established with equal participation by those party to the negotiation.
5. States should take appropriate measures, both at the national and international levels, to combat bioterrorism and illicit traffic in organs, tissues, samples, genetic resources and genetic-related materials.

Promotion of the Declaration
Article 22: Role of states
1. States should take all appropriate measures, whether of a legislative, administrative or other character, to give effect to the principles set out in this Declaration in accordance with international human rights law. Such measures should be supported by action in the spheres of education, training and public information.
2. States should encourage the establishment of independent, multidisciplinary and pluralist ethics committees, as set out in Article 19.

Article 23: Bioethics education, training and information
1. In order to promote the principles set out in this Declaration and to achieve a better understanding of the ethical implications of scientific and technological developments, in particular for young people, States should endeavor to foster bioethics education and training at all levels as well as to encourage information and knowledge dissemination programmes about bioethics.
2. States should encourage the participation of international and regional intergovernmental organizations and international, regional and national nongovernmental organizations in this endeavor.

Article 24: International cooperation
1. States should foster international dissemination of scientific information and encourage the free flow and sharing of scientific and technological knowledge.
2. Within the framework of international cooperation, States should promote cultural and scientific cooperation and enter into bilateral and multilateral agreements enabling developing countries to build up their capacity to participate in generating and sharing scientific knowledge, the related know-how and the benefits thereof.
3. States should respect and promote solidarity between and among States, as well as individuals, families, groups and communities, with special regard for those rendered vulnerable by disease or disability or other personal, societal or environmental conditions and those with the most limited resources".

Principalism

The four-principle approach developed by Tom Beauchamp and James Childress (1978, latest edition 2009) has become the most preferred and important approach to medical ethics. It is also called Principalism. It consists of four universal mid-level ethical principles, which are merely meant to act as a guide and thus leave a lot of scope for interpretation:

1. Autonomy,
2. Nonmaleficence,
3. Beneficence,
4. Justice.

Respect for autonomy refers to respect for the decision-making capacities of autonomous persons capable of making logical informed choices. This enjoins upon the practitioners an effort to enable capable individuals to make their own choices with information being provided.

Beneficence refers to the act of balancing benefits of treatment against the risks and costs entailed therein. The healthcare professional should act in a way that benefits the patient.

Non-maleficence is the avoidance of causing harm. A medical practitioner should not harm a patient or a subject knowingly by his actions. In other words, the side effects or adverse effects should not be disproportionately large as compared to the benefits accrued from therapy for most patients. In case the margin of error in benefit is slim, it should be so conveyed to the patients and they should be involved in the decision.

Justice refers to fair distribution of benefits, risks and costs. Patients with similar conditions should be treated similarly, and if the benefits are not to be extended to a given community the medicine should not be tried on subjects from there.

This can be made the starting point and constraining framework of ethical reasoning and decision making or "common morality". "The common morality is the set of norms shared by all persons committed to morality. The common morality is not merely a morality, in contrast to other moralities. The common morality is applicable to all persons in all places, and we rightly judge all human conducts by its standards."—Tom Beauchamp and James Childress.

Even though the four-principle approach is the most prevalent, authoritative and widely used bioethical approaches, there are three most important objections to it, which are:

1. There exists no master principle in case of conflict among the principles. Therefore, there is a lack of ethical guidance to break a tie.
2. The problem of bias in cross-cultural contexts.
3. The fact that it may be construed as a mere checklist of considerations and so it is methodologically unsound.

Basic Ethical Principles in Research Ethics

All researches involving human subjects is governed by four basic ethical principles:
1. Respect for persons covered under the head of autonomy
2. Beneficence
3. Nonmaleficence and
4. Justice.

Respect for Persons

Respect for persons has two principal components:
1. Respect for autonomy and protection of persons with impaired or diminished autonomy. Respect for autonomy means that those who are capable of

deliberating about their personal choices should be treated with deference for their capacity for self-determination.
2. Protection of persons with impaired or diminished autonomy means that those who are vulnerable or dependent or incapable of deliberating about their personal choices should be accorded security against harm or abuse.

Beneficence and nonmaleficence

Beneficence is the obligation to maximize benefit and/or minimize harm. This leads to development of guidelines and norms requiring the risks of research to be weighed against the expected benefits, the research design to be sound, the investigators to be competent in conducting research and safeguarding the welfare of their research subjects. Beneficence enjoins doing good to subjects. It also encompasses non-maleficence which prohibits inflicting harm on persons. Since all treatments involve some adverse effects, this principle can be seen as an extension of the former principle. Simply put, it means that either benefit the subject or at least do not harm with your actions. This is sometimes called the principle of non-maleficence (do no harm) also [cf Hippocratic Oath]. In the case of research ethics, the research question should have *clinical equipoise* meaning that the answer is not fully known and each of the outcome possibilities has similar chances of being true. Thus the investigator has no way of knowing that he is not causing any obvious harm. The subject may or may not gain from the study. Thus this limitation of the principle must be borne in mind when it is applied. In simple words, it is a condition of *bounded rationality* where the outcome is not fully known. The researcher should not resort to a course of action or treatment known to be disproportionately harmful compared to the benefits accrued.

Justice

Justice refers to morally right and proper treatment of each person. It requires that each person gets what is due to him. In research ethics, it refers to distributive justice or equitable distribution of burdens and benefits of participation in research. This can be given a go by only in case of vulnerability. "Vulnerability" is a substantial inability in protecting one's own interests owing to impediments like lack of capability to give informed consent, lack of alternative means of obtaining medical care or other expensive necessities, or being a junior or subordinate member of a hierarchical group. Justice requires that the research be responsive to the health conditions or needs of vulnerable subjects. The subjects selected should be the least vulnerable necessary to accomplish the purposes of the research. Accordingly, special provision must be made for the protection of the rights and welfare of vulnerable persons.

Informed Consent

Individual informed consent can be traced to the case of Dr Albert Neisser (1896) who publicly announced his concern about the possible dangers to the experimental subjects whom he vaccinated with an experimental immunizing serum. Prussian and German bureaucratic regulations of 1900–1901 expounded further on the idea. (Zentralblatt der gesamten Unterrichtsverwaltung in Preussen, 1901:188). A review of the deaths of 75 German children in connection with experimental tuberculosis vaccines in 1931 revealed that the mandatory informed consent had not been obtained (Rundschreiben des Reichsministers des Inneren 28.2.1931, in: Sass 1989:362-366). "The informed consent doctrine was thus originally a regulatory innovation created by Prussian bureaucrats." It appears to be a German solution to issues arising out of advancement of German biomedical and allied sciences. These actions were nonbinding and the Germans themselves did not adhere to these practices in the years that followed till the Nuremberg Code was established to correct the wrongs.

In the 1950s, it was increasingly being realized that there was a serious need for thinking about complex moral issues in medical and allied sciences. It facilitated the creation of the new academic discipline of medical ethics also known as 'bioethics, New technological changes in life sciences increased survival chances of the sick patients. These advancements came from innovation and experimentation. They brought with them problems of distribution. For example, access to scarce dialysis and intensive care units caused much debate regarding resource allocation. Bioethics developed as a discipline and got initial institutionalization in 1970s in the US. The Institute of Society, Ethics and the Life Sciences (1969) [later the Hastings Center], and the Joseph and Rose Kennedy Center for the Study of Human Reproduction and Bioethics (1971) were created. They conducted research in medical ethics and published high quality academic journals—the Hastings Center Report and the Kennedy Institute of Ethics Journal. Many bioethics programs and degrees were started at universities in the USA during the 1970s and 1980s. The Rockefeller Foundation, the Russell Sage Foundation and the Ford Foundation were notable contributors in bioethics grant funding in the early years.

Morals and ethics are often believed to be synonymous. However, there is a subtle philosophical and semantic difference between the two. Morals refer to the principles on which one judges right and wrong. Ethics refer to the principles of right conduct. The two are closely related and often used interchangeably. Morals are more abstract, subjective and person or religion-based. Ethics are more practical having been conceived as shared principles to promote fairness in social interactions. A famous example often cited is of a politician's sex scandal which involves a moral lapse (a subjective judgment) compared to another politician accepting money from a company, he is supposed to regulate (ethical issue). 'Ethics are the science of morals, and morals are the practice of ethics.'

Command Responsibility

Command responsibility or Yamashita standard or Medina standard or superior responsibility is the doctrine of hierarchical accountability. It was used for war crimes committed during wartime originally. The term broadly refers to the duty to supervise subordinates and liability in case of failure to do so in government, military, corporations and trusts. This doctrine can be traced back to the Hague Conventions of 1899 and 1907. Its initial application by German Supreme Court at the Leipzig

War Crimes Trials in 1921 resulted in Emil Müller being sentenced to six months imprisonment for failure to maintain decent living conditions at the Flavy de Martel Prisoners of War Camp. This lead to many deaths due to dysentery. It was ruled that he had failed to prevent crimes and punish the perpetrators of physical violence against prisoners. "Yamashita Standard" refers to the verdict in the case of Japanese General Tomoyuki Yamashita by US Supreme Court in 1945. He was prosecuted for atrocities committed by troops under his command in the Philippines during World War II. He was found guilty of "unlawfully disregarding and failing to discharge his duty as a commander to control the acts of members of his command by permitting them to commit war crimes." "Medina standard" refers to 1971 prosecution of US Army Captain Ernest Medina for the 'My Lai Massacre' in the Vietnam War. 'A Commanding Officer or Supervisor will be held criminally liable, if he does not take action upon becoming aware of a human rights violation or a war crime.' This was not fully proved in this case and Medina was acquitted of all charges.

Medical Torture

Medical torture refers to the involvement and sometimes participation of medical personnel in torture of prisoners. They may be in helping in judging what the victims can endure, in applying treatments to enhance torture or as torturers. "Medical torture or medical interrogation involves the use of their expert medical knowledge to facilitate interrogation or corporal punishment, in the conduct of torturous human experimentation or in providing professional medical sanction and approval for the torture of prisoners. Medical torture also covers torturous scientific (or pseudo-scientific) experimentation upon unwilling human subjects." Medical torture is a violation of medical ethics. World Medical Association Declaration of Tokyo (1975) states clearly that, "The doctor shall not countenance, condone or participate in the practice of torture". The 'UN Convention against Torture' also applies to medical personnel (in addition to law enforcement officers, military personnel, politicians and other persons acting in an official capacity). It prohibits the use of torture under any circumstance. Article 2(2) of the Convention states that, "No exceptional circumstances whatsoever, whether a state of war or a threat of war, internal political instability or any other public emergency, may be invoked as a justification of torture." Health professionals are expected to adhere to a higher standard of behavior". The UN Principles of Medical Ethics are not enforceable when governments are complicit in violations of human rights. This higher standard is reflected in principles of beneficence, nonmaleficence (above all do no harm), autonomy, justice, dignity and informed consent. These are not covered comprehensively by the UN Convention against Torture.

ANIMAL ETHICS

In old times, only men of a particular social status were part of the moral community. After a long and hard social struggle of hundreds of years, women achieved equal status with men. The idea of animals being part of the moral community evolved from the 19th century ethics of utilitarianism, lead notably by J Bentham. He argued that it 'does not matter morally whether animals can reason but rather whether they can suffer.' Animal rights groups developed a new awareness of sensitivity towards cruelty against animals. The scientific findings of Darwin's evolutionary theory provided convincing empirical evidence of kinship between animals and human beings as the humans developed through the former. Ethical argument defending animals broadens the scope of the traditional position by its claim that the 'ability to suffer is the key point, and hence sentient beings should be protected as part of the moral community.' Peter Singer in 1975 'argued for a utilitarian animal ethics based on the equal consideration of interests of sentient beings in combination with the criterion of the ability to feel pain.' In 1983, Regan claimed that 'sentient beings who are able to see themselves as "subjects of life" do have an "inherent value" which provides them with strong defensible moral rights that implicate prima facie duties for human beings towards animals.' In the works of the proponents, animals possess characteristics to get a moral authority to be treated as sentient beings like humans immaterial of their rank in the sentient hierarchy. The movement has gained a lot of support from developed societies. The other species have as much a right to be on the planet as the humans.

ENVIRONMENTAL ETHICS

Environmental ethics deals with the moral dimension of the relationship between human beings and nonhuman nature—animals and plants, local populations, natural resources and ecosystems, landscapes, as well as the biosphere and the cosmos. Strictly speaking, human beings are, of course, part of nature and it seems somewhat odd to claim that there is a contrast between human beings and nonhuman nature. At second glance, however, it seems reasonable to make this distinction because human beings are the only beings who are able to reason about the consequences of their actions which may influence the whole of nature or parts of nature in a positive or negative way.

Ideas about the "right" conduct concerning the environment are as old as humankind but the establishment of environmental ethics as an academic discipline dates back to the 1970s when issues of vital importance emerged, such as the global threat to the natural basis of existence, the growing number of extinct species, the destruction of ecosystems and natural resources, as well as the more recognized dangers of technological inventions—for example, nuclear power, including its radioactive waste, and the new biotechnologies like genetic engineering. The exploitation of the environment was first justified by the religious teachings of the Old Testament (such as the stewardship of the environment in the Bible) and, during the secular period of the enlightenment, supported by Francis Bacon's scientific program to (rigorously) disclose all the secrets of nature. René Descartes' famous and influential dualism of rational beings, on the one hand, and soulless matter, on the other hand, led to the debasement of nature, including animals, since the objects of morality were by nature rational beings only. The first serious counter-movement can be traced back to the Romantic philosophies of nature of the 18th and 19th centuries. In the non-Western context, the idea of respect for and valuing nature is more prevalent and at least 2500 years old, referring to the general teachings of Hinduism and Buddhism which influenced the Western view in Europe in the 18th and 19th centuries (for example, Schopenhauer). Of course, contemporary environmentalists, particularly feminist ethicists and supporters of the idea of natural aesthetics, have refined the criticism of the traditional view by claiming that animals and nature are not valueless but deserving of moral protection.

It is possible to make the following broad distinctions regarding environmental ethics. Environmental ethics is commonly divided into two distinct areas:
1. Anthropocentrism and
2. Nonanthropocentrism (or Physiocentrism).

ANTHROPOCENTRISM

Anthropocentric approaches such as virtue ethics and deontology stress the particular human perspective, and claim that values depend on human beings only. Values are relational and require a rational being, hence animals and non-human nature are not *per se* objects of morality, unless indirectly, by virtue of a surrogate decision maker. According to the anthropocentric view, only (rational) human beings deserve moral protection although one should respect and protect nature either for the sake of human beings (instrumental view) or for the sake of nature itself (non-instrumental view). Anthropocentrism is faced with the objection of speciesism, the view that the mere affiliation to the species of *Homo sapiens* is sufficient to grant a higher moral status to human beings in comparison with animals. Singer has powerfully claimed, however, that the "mere difference of species in itself cannot determine moral status" (Singer 2009:567).

Nonanthropocentrism (or Physiocentrism)

Nonanthropocentrism (or Physiocentrism) mainly consists of three main branches:
1. Pathocentrism
2. Biocentrism
3. Ecocentrism, which can be further divided into an individualistic and holistic version.

All nonanthropocentric approaches share the common claim that there are *objective* or more straightforward naturalistic values which are nonrelational (intrinsic) and do not pre-suppose rational human beings. Nature (including animals) itself is valuable, independently of whether there are any human beings or not (noninstrumental view), even though one has to acknowledge the fact that many arguments about intrinsic value also have instrumental underpinnings. Supporters of pathocentrism argue that all sentient beings deserve moral consideration and protection, equally/egalitarian or nonequally/nonegalitarian with reference to human beings (Singer 1975, Regan 1983, Wolf 1996). Adherents of biocentrism claim that all beings should be part of the moral community. Finally, supporters of ecocentrism argue that the whole of nature deserves moral protection, either according to an individualistic or holistic approach. If individualistically, all *things* in nature are bearers of moral values and are of equal moral worth.

If holistically, there are traditionally at least three main positions:
1. Ecofeminism
2. Deep ecology
3. The land ethics.

Ecofeminists believe that there is a parallel between the systems of domination that affect both women and nature. Therefore, if human beings are willing to change the way they act towards nature, they must understand the real causes of the problem—the idea that nature is rather irrational and passive as well as needing to be controlled by human beings (Plumwood 1986, Warren 1987).

According to deep ecologists, human beings should view themselves as being a part of and not distinct from the natural world by virtue of a refined notion of the self. All living things, according to the founder of deep ecology, Arne Naess, have an equal right to flourish (biospherical egalitarianism).

Proponents of the land ethics argue that one should stop treating the land as a mere resource, but view it as a precious source of energy. Aldo Leopold, the founder of land ethics, famously claims: "A thing is right when it tends to preserve the integrity, stability, and beauty of the biotic community. It is wrong when it tends otherwise." (Leopold 1949/1989:218-25).

From the discussions in this chapter, it can be observed that Bioethics is a vast subject and rightly, shall continue to influence how research is conducted. Most of the literature and thoughts on Bioethics pertains to the moral values, ethics, humaneness, equality of all living organism, environmental impacts, etc. In the literature and references mentioned, the direction is more towards how data is collected as a part of research. With advancements and developments of societies across the world, with time these factors, especially complete transparency in data collection and live-testing techniques would possibly become a norm, without exceptions. But Bioethics does not limit itself in definition on conduct and collection of data (testing on live subjects/organisms, in its basic form is another form of data collection), but also includes how the statistics are analyzed, evaluated, drawn inferences from and reported. A progressive research and science community relies on each other's efforts for mutual advancements. Technological advancements of today form the basis of tomorrow's research. Any incorrect analysis of the data or evidence collected will result in inappropriate conclusions, and if published this can lead to the wrong getting multiplied until the counter-results are published later.

It is said that ends justify the means. The data need to be collected ethically to answer a justifiable research problem. If the foundation is wrong, then the results need to be taken skeptically. Thus the research question should not harm the rights of any of the stakeholder groups involved. In biomedical research or even in clinical interpretation of the results of studies, the practitioner should understand how they were derived and what it means when the findings are applied to subjects in his practice or research. It is important that the handling of data, collected in a bioethical manner, is scientific in nature. Science believes in evidence and logic; this becomes highly imperative in Medical Sciences. Chapter 2 explores the subject of *Evidence-based Medicine* and *Levels of Evidence*.

CHAPTER 2

Evidence-based Medicine and Levels of Evidence

EVIDENCE-BASED MEDICINE

Evidence-based medicine (EBM) is a medical practice approach aimed at optimized decision making by utilizing evidence available from the research. It is a philosophy emphasizing systematic and rigorous assessment of evidence for decision making in all the spheres of healthcare. It integrates the following:
- Evidence available, with
- Expertise of the decision maker, aligned with
- Expectations and values of the patients.

Evidence available refers to observations made with an expressed purpose of generating insight into the issue at hand.

The goal of Evidence-based Medicine (EBM) is to improve peoples' health through decisions, which maximize quality of health and increase the lifespans or reduce morbidity and mortality related to ill-health.

FUNDAMENTAL PRINCIPLES

The fundamental principles of *Evidence-based Medicine (EBM)* are:
- Hierarchy of evidence
- Insufficiency of evidence alone.

Hierarchy of Evidence

The principle of hierarchy of evidence states that evidence available for clinical decision making can be arranged in an hierarchical manner from the most rigorous to the least. This hierarchy reflects the relative authoritativeness of different categories of biomedical research. However, no single, universally-accepted hierarchy of evidence exists. There is a broad consensus on the relative strength of the principal types of research, or epidemiological studies. In 1997, T Greenhalgh stated that, "the relative weight carried by the different types of primary study, when making decisions about clinical interventions (the "hierarchy of evidence"), puts them in the following order":
1. Systematic reviews and meta-analyses of "randomized clinical trials (RCTs) with definitive results" (Confidence intervals that do not overlap the threshold clinically significant effect)

2. RCTs with non-definitive results (a point estimate that suggests a clinically significant effect but with confidence intervals overlapping the threshold for this effect)
3. Cohort studies
4. Case-control studies
5. Cross-sectional surveys
6. Case reports.

However, this is an oversimplified approach as the type of research problem dictates the hierarchy. For example, while usually RCTs are ranked above observational studies, while expert opinion and anecdotal experience are ranked at the bottom; but if the question is about prognosis, then observational studies will be more relevant than randomized controlled trials. Many evidence hierarchies put systematic review, and meta-analysis above RCTs as these combine data from many RCTs and other study types. Some authors differentiate between systematic reviews, which combine the studies using a qualitative method and meta-analysis, which combines the quantitative aspect as well. Systematic reviews can be qualitative or both qualitative and quantitative. Therefore, the line that differentiates them is very fine, and in modern-day practices, they can be used almost interchangeably. Evidence hierarchies are the basis and an integral part of evidence-based medicine.

Insufficiency of Evidence Alone

In an ideal world, everything should be guided by rationale and logic alone. However, biomedical sciences deal with an intricate world where the right of interpretation is distributed. And rightly so, one being should ideally not dictate the terms for another. The change in this paternalistic attitude has been dictated by the removal of ethnocentric and egocentric superiority principle. The other human being also should have and exercise his own free choice. The principle of insufficiency of evidence alone means that evidence alone is not enough in clinical decision making. It has to be integrated with practitioners' clinical expertise as well as patient's expectations and values.

COMPONENTS OF EVIDENCE-BASED PRACTICES

Thus three components of evidence based practices are:
1. Evidence
2. Expertise (clinical and social)
3. Expectations (of the patient and the society)

Therefore, an evidence being available in medical literature becomes useful only when it is suitably employed using expertise in the area of endeavor to meet the expectations of the patient.

History and Etymology

"Evidence-based medicine", can broadly be differentiated into two distinct but interrelated activities:
- Evaluation of evidence of effectiveness in the context of policies like clinical practice guidelines and population-level policies.

- Introduction of epidemiological methods in medical education and management of individual patient-level decisions.

The traditional approach to medical decision-making relied heavily on subjective evaluation by each individual physician of the research evidence in the domain. It was variously called "clinical judgment" and "the art of medicine" and "the experts' opinion". It was flawed insofar as it was highly colored by subjectivity, bias and personal beliefs, in addition to other factors. In 1967, Alvan Feinstein in 'Clinical Judgment' wrote on clinical reasoning and identified biases that could affect it. In 1972, Archie Cochrane's 'Effectiveness and Efficiency' demonstrated that many practices previously assumed to be efficacious did not have the proof of controlled trials. John Wennberg showed physicians' practices had very heterogeneous nature and were extremely variable. Meanwhile research in the US showed that many procedures performed by physicians were considered inappropriate even by the standards of their own experts. This awareness of the weaknesses in medical decision-making at the levels of individual patients and populations has spurred the interest in evidence-based methods.

The term "evidence-based" was first used by David M Eddy, who introduced the Markov Models to healthcare. In 1987, he used the phrase in workshops and one manual for dissemination of formal methods for design of clinical practice guidelines, commissioned by the Council of Medical Specialty Societies. The manual was published by the American College of Medicine, though at a much later date. In Journal of the American Medical Association, Eddy first used this term in 1990 in an article on principles of evidence-based guidelines and population-level policies.

At the McMaster's University, investigators such as Prof Guyatt, began using the term during the 1990s. "Evidence-based medicine" as a term, was used with reference to medical education. This branch of evidence-based medicine has its roots in clinical epidemiology. Gordon Guyatt's description of a new approach to teaching the practice of medicine at McMaster University for prospective or new medical students used the term in 1992. David Sackett and colleagues formally defined this term later as follows, "The conscientious, explicit and judicious use of current best evidence in making decisions about the care of individual patients ... integrating individual clinical expertise with the best available external clinical evidence from systematic research." Its objective was to improve individual decision by making it more structured and objective, by using the evidence from research. It applied the principles derived from population-based data to the care of an individual patient, "while respecting the fact that practitioners have clinical expertise reflected in effective and efficient diagnosis and thoughtful identification and compassionate use of individual patients' predicaments, rights, and preferences." It has been described as "a systemic approach to analyze published research as the basis of clinical decision making."

By 2000, "evidence-based medicine" had become an umbrella term for using evidence in both population-level and individual-level policy making. The term EBM is now used to describe the programs designing evidence-based guidelines and those teaching evidence-based medicine to practitioners. In fact, now it is further expanded to include "evidence-based health services", wherein health service decision makers at the organizational or institutional level are encouraged to use objective data in decision making.

Evidence-based guidelines and policies usually always have consensus with experience-based practices aimed at ethical clinical judgment. This can cause contradictions and 'unintended crises'. Knowledge users, especially the knowledge leaders, must use a broad range of management knowledge in their decision making in addition to mere formal evidence. It is usually a contingency approach at the *event level*. Evidence-based guidelines provide the basis for administration of health care systems and have a big role in the governance and control of healthcare institutions and systems.

Scope of the Use of Evidence

The EBM may be touted as a gold standard of clinical practice but it is unlikely that it would be used in every situation. The meta-analysis and systematic reviews of multiple RCTs answers narrow clinical questions. Randomized controlled trials are expensive and there exist many broad clinical questions. Results from randomized controlled trials (RCTs) may not be relevant in all the treatment situations. The research topics are often influenced by the sponsors' interests and negative results may not be reported. Often a time lapse between the conduct of RCT and publication of its results is there. Historically, racial minorities and people with co-morbidities are under-represented in trials thereby restricting generalizability of results. The EBM applies to groups of people and the probabilities are calculated on the principle of random throws while in case of individual decisions, the element of chance dictates that confidence intervals only can be used with any degree of certainty. "The knowledge gained from clinical research does not directly answer the primary clinical question of what is best for the patient at hand". However, it should still not discourage a clinician or a medical student from using this knowledge to make verifiable and valid decisions in individual cases. It is just a scientific method of clinical judgment.

There is no universal system of hierarchy of evidence. However, a broad consensus exists between the various attempts. In 1979, the Canadian Task Force on the Periodic Health Examination described 'Levels of Evidence' for periodic health examination based on the evidence in the medical literature. The authors developed evidence rating system and grading rating method. Grade A recommendation was given if there was good evidence to support a recommendation that a condition be included in the periodic health exam and lower ratings developed for other levels.

*Canadian Task Force on the Periodic Health Examination's Levels of Evidence**

Level	Type of evidence
I	At least 1 RCT with proper randomization
II.1	Well-designed cohort or case-control study
II.2	Time series comparisons or dramatic results from uncontrolled studies
III	Expert opinions

*Adapted from Canadian Task Force on the Periodic Health Examination. The Periodic Health Examination. Can Med Assoc J. 1979;121:1193-254.

Other societies and professional bodies also came up with their rating methods. In clinical medicine, David Sackett's recommendations constitute a notable landmark and are quoted below:

Levels of Evidence from Sackett[*]

Level	Type of evidence
I	Large RCTs with clear-cut results
II	Small RCTs with unclear results
III	Cohort and case-control studies
IV	Historical cohort or case-control studies
V	Case series, studies with no controls

*Adapted from Sackett DL. Rules of evidence and clinical recommendations on the use of antithrombotic agents. Chest. 1989;95:2S–4S.

Type and level of evidence needs to be modified according to the questions asked and the answers demanded. Research questions are divided into the categories: Treatment, Prognosis, Diagnosis, and Economic/ Decision analysis. The levels of evidence developed by the American Society of Plastic Surgeons (ASPS) for Prognosis differ from the levels developed by the Center for Evidence-based Medicine (CEBM) for Treatment, showing the heterogeneity in the rating schemes, which is scientifically correct as different questions are being asked.

Levels of Evidence for Prognostic Studies[*]

Level	Type of evidence
I	High quality prospective cohort study with adequate power or systematic review of these studies
II	Lesser quality prospective cohort, retrospective cohort study, untreated controls from an RCT, or systematic review of these studies
III	Case-control study or systematic review of these studies
IV	Case series
V	Expert opinion; case report or clinical example; or evidence-based on physiology, bench research or "first principles"

*Adapted from the American Society of Plastic Surgeons (http://www.plasticsurgery.org/Medical_Professionals/Health_Policy_and_Advocacy/Health_Policy_Resources/Evidence-based_GuidelinesPractice_Parameters/Description_and_Development_of_Evidence-based_Practice_Guidelines/ASPS_Evidence_Rating_Scales.html)

Levels of Evidence for Therapeutic Studies[*]

Level	Type of evidence
1A	Systematic review (with homogeneity) of RCTs
1B	Individual RCT (with narrow confidence intervals)
1C	All or none study
2A	Systematic review (with homogeneity) of cohort studies
2B	Individual Cohort study (including low quality RCT, e.g. <80% follow-up)

Contd...

Contd...

Level	Type of evidence
2C	"Outcomes" research; Ecological studies
3A	Systematic review (with homogeneity) of case-control studies
3B	Individual case-control study
4	Case series (and poor quality cohort and case-control study)
5	Expert opinion without explicit critical appraisal or based on physiology bench research or "first principles"

*From the Center for Evidence-based Medicine, http://www.cebm.net.

Most evidence ranking schemes grade evidence for therapy and prevention only. They do not cover diagnostic tests, prognostic markers or treatment harm. Several organizations have developed grading systems for assessing the quality of evidence. The Oxford CEBM Levels of Evidence is one such example. It provides 'Levels of evidence, for claims of prognosis, diagnosis, treatment benefits, harms and screening. Originally introduced in 2000, the Oxford CEBM Levels were redesigned to improve comprehensibility and comprehensiveness in 2011. The Oxford CEBM Levels of Evidence can be used by patients, doctors and policy makers. Reproduced below are the levels as an example of the many efforts. These efforts over many years were further refined and presented in 2011 by CEBM. (The 2011 revision document of Oxford CEBM Levels of Evidence is reproduced in Appendix).

Nature of problem	Therapy/ prevention, etiology/ harm	Prognosis	Diagnosis	Differential diagnosis/symptom prevalence study	Economic and decision analysis
Level					
1a	SR (with homogeneity*) of RCTs	SR (with homogeneity*) of inception cohort studies; CDR" validated in different populations	SR (with homogeneity*) of Level 1 diagnostic studies; CDR" with 1b studies from different clinical centers	SR (with homogeneity*) of prospective cohort studies	SR (with homogeneity*) of Level 1 economic studies
1b	Individual RCT (with narrow confidence interval"¡)	Individual inception cohort study with > 80% follow-up; CDR" validated in a single population	Validating** cohort study with good" " reference standards; or CDR" tested within one clinical center	Prospective cohort study with good follow-up****	Analysis based on clinically sensible costs or alternatives; systematic review(s) of the evidence; and including multi-way sensitivity analyses

Contd...

Contd...

Nature of problem	Therapy/ prevention, etiology/harm	Prognosis	Diagnosis	Differential diagnosis/symptom prevalence study	Economic and decision analysis
1c	All or none§	All or none case-series	Absolute SpPins and SnNouts""	All or none case-series	Absolute better-value or worse-value analyses """"
2a	SR (with homogeneity*) of cohort studies	SR (with homogeneity*) of either retrospective cohort studies or untreated control groups in RCTs	SR (with homogeneity*) of Level >2 diagnostic studies	SR (with homogeneity*) of 2b and better studies	SR (with homogeneity*) of Level >2 economic studies
2b	Individual cohort study (including low quality RCT; e.g. <80% follow-up)	Retrospective cohort study or follow-up of untreated control patients in an RCT; Derivation of CDR" or validated on split-sample§§§ only	Exploratory** cohort study with good"" " reference standards; CDR" after derivation, or validated only on split-sample§§§ or databases	Retrospective cohort study, or poor follow-up	Analysis based on clinically sensible costs or alternatives; limited review(s) of the evidence, or single studies; and including multi-way sensitivity analyses
2c	"Outcomes" Research; Ecological studies	"Outcomes" research		Ecological studies	Audit or outcomes research
3a	SR (with homogeneity*) of case-control studies		SR (with homogeneity*) of 3b and better studies	SR (with homogeneity*) of 3b and better studies	SR (with homogeneity*) of 3b and better studies
3b	Individual Case-Control Study		Non-consecutive study; or without consistently applied reference standards	Non-consecutive cohort study, or very limited population	Analysis based on limited alternatives or costs, poor quality estimates of data, but including sensitivity analyses incorporating clinically sensible variations

Contd...

Contd...

Nature of problem	Therapy/ prevention, etiology/ harm	Prognosis	Diagnosis	Differential diagnosis/symptom prevalence study	Economic and decision analysis
4	Case-series (and poor quality cohort and case-control studies§§)	Case-series (and poor quality prognostic cohort studies***)	Case-control study, poor or non-independent reference standard	Case-series or superseded reference standards	Analysis with no sensitivity analysis
5	Expert opinion without explicit critical appraisal, or based on physiology, bench research or "first principles"	Expert opinion without explicit critical appraisal, or based on physiology, bench research or "first principles"	Expert opinion without explicit critical appraisal, or based on physiology, bench research or "first principles"	Expert opinion without explicit critical appraisal, or based on physiology, bench research or "first principles"	Expert opinion without explicit critical appraisal, or based on economic theory or "first principles"

Reproduced from Bob Phillips, Chris Ball, Dave Sackett, Doug Badenoch, Sharon Straus, Brian Haynes, Martin Dawes since November 1998. Updated by Jeremy Howick March 2009 for CEBM

- * By homogeneity we mean a systematic review that is free of worrisome variations (heterogeneity) in the directions and degrees of results between individual studies. Not all systematic reviews with statistically significant heterogeneity need be worrisome, and not all worrisome heterogeneity need be statistically significant. As noted above, studies displaying worrisome heterogeneity should be tagged with a "-" at the end of their designated level.
- " Clinical Decision Rule. (These are algorithms or scoring systems that lead to a prognostic estimation or a diagnostic category)
- "¡ See note above for advice on how to understand, rate and use trials or other studies with wide confidence intervals.
- § Met when all patients died before the Rx became available, but some now survive on it; or when some patients died before the Rx became available, but none now die on it.
- §§ By poor quality cohort study, we mean one that failed to clearly define comparison groups and/or failed to measure exposures and outcomes in the same (preferably blinded), objective way in both exposed and non-exposed individuals and/or failed to identify or appropriately control known confounders and/or failed to carry out a sufficiently long and complete follow-up of patients. By poor quality case-control study we mean one that failed to clearly define comparison groups and/or failed to measure exposures and outcomes in the same (preferably blinded), objective way in both cases and controls and/or failed to identify or appropriately control known confounders.
- §§§ Split-sample validation is achieved by collecting all the information in a single tranche, then artificially dividing this into "derivation" and "validation" samples.
- " " An "Absolute SpPin" is a diagnostic finding whose Specificity is so high that a Positive result rules-in the diagnosis. An "Absolute SnNout" is a diagnostic finding whose Sensitivity is so high that a Negative result rules out the diagnosis.
- " " " Good reference standards are independent of the test, and applied blindly or objectively to applied to all patients. Poor reference standards are haphazardly applied, but still independent of the test. Use of a non-independent reference standard (where the 'test' is included in the 'reference', or where the 'testing' affects the 'reference') implies a level 4 study.
- " " " " Better-value treatments are clearly as good but cheaper, or better at the same or reduced cost. Worse-value treatments are as good and more expensive, or worse and the equally or more expensive.

** Validating studies test the quality of a specific diagnostic test, based on prior evidence. An exploratory study collects information and trawls the data (e.g. using a regression analysis) to find which factors are 'significant'.

*** By poor quality prognostic cohort study we mean one in which sampling was biased in favor of patients who already had the target outcome, or the measurement of outcomes was accomplished in <80% of study patients, or outcomes were determined in an unblinded, non-objective way, or there was no correction for confounding factors.

**** Good follow-up in a differential diagnosis study is >80%, with adequate time for alternative diagnoses to emerge (for example, 1–6 months acute, 1–5 years chronic)

CEBM Grades of Recommendation

A	Consistent level 1 studies
B	Consistent level 2 or 3 studies or extrapolations from level 1 studies
C	Level 4 studies or extrapolations from level 2 or 3 studies
D	Level 5 evidence or troublingly inconsistent or inconclusive studies of any level

The US Preventive Services Task Force (USPSTF) is another hierarchy which is quoted pari passu as an example below:

Level I	Evidence obtained from at least one properly designed randomized controlled trial.
Level II-1	Evidence obtained from well-designed controlled trials without randomization.
Level II-2	Evidence obtained from well-designed cohort or case-control analytic studies, preferably from more than one center or research group.
Level II-3	Evidence obtained from multiple time series designs with or without the intervention. Dramatic results in uncontrolled trials might also be regarded as this type of evidence.
Level III	Opinions of respected authorities, based on clinical experience, descriptive studies, or reports of expert committees.

Levels of recommendations	The US Preventive Services Task Force Suggestion on action
Level A	Good scientific evidence suggests that the benefits of the clinical service substantially outweigh the potential risks. Clinicians should discuss the service with eligible patients.
Level B	At least fair scientific evidence suggests that the benefits of the clinical service outweigh the potential risks. Clinicians should discuss the service with eligible patients.
Level C	At least fair scientific evidence suggests that there are benefits provided by the clinical service, but the balance between benefits and risks are too close for making general recommendations. Clinicians need not offer it unless there are individual considerations.
Level D	At least fair scientific evidence suggests that the risks of the clinical service outweighs potential benefits. Clinicians should not routinely offer the service to asymptomatic patients.

Similarly other professional groups have also given their hierarchies. The idea of enumerating all these here is to understand that hierarchies are not constant or universal, even though largely a consensus exists. This is an evolving field and there is a lot of activity which will further define the paradigms of engagement in future.

A notable unification of levels of proof with action to follow in non-technical jargon is GRADE (Grading of Recommendations, Assessment, Development and Evaluations). It is "a systematic and explicit approach for making judgements about the quality of evidence and strength of the recommendations." Developed by the Grading of Recommendations, Assessment, Development and Evaluations (GRADE) Working Group, it assesses and reports:
- Methodological flaws within the component studies
- Consistency of results across different studies
- Generalizability of research results to the wider patient base
- Effectiveness or efficacy of the treatments shown.

It is now felt to be the most effective method of linking evidence-quality evaluations to clinical recommendations. Treatment comparisons are rated as 'high-, moderate-, low-, or very low-quality evidence' depending on the quality of the evidence. This system grades the quality of evidence into four levels based on their confidence in the observed effect being close to what the true effect is on a numerical basis. The confidence value is based on judgments assigned in five different domains in a structured manner.

Type of evidence		
Initial score based on type of evidence	+4	RCTs/ SR of RCTs, +/– other types of evidence
	+2	Observational evidence (e.g. cohort, case-control)
Quality		
Based on		Blinding and allocation process
		Follow-up and withdrawals
		Sparse data
		Other methodological concerns (e.g. incomplete reporting, subjective outcomes)
Score	0	No problems
	–1	Problem with 1 element
	–2	Problem with 2 elements
	–3	Problem with 3 or more elements
Consistency		
Based on		Degree of consistency of effect between or within studies
Score	+1	Evidence of dose response across or within studies (or inconsistency across studies is explained by a dose response); also 1 point added if adjustment for confounders would have increased the effect size
	0	All/most studies show similar results
	–1	Lack of agreement between studies (e.g. statistical heterogeneity between RCTs, conflicting results)
Directness		
Based on		The generalizability of population and outcomes from each study to our population of interest
Score	0	Population and outcomes broadly generalizable
	–1	Problem with 1 element
	–2	Problem with 2 or more elements

Contd...

Contd...

Effect Size		
Based on	The reported OR/RR/HR for comparison	
Score	0	Not all effect sizes >2 or <0.5 and significant; or if OR/RR/HR not significant
	+1	Effect size >2 or <0.5 for all studies/meta-analyses included in comparison and significant
	+2	Effect size >5 or <0.2 for all studies/meta-analyses included in comparison and significant

The final GRADE score uses 4 categories of evidence quality based on the overall GRADE scores for each comparison: high (at least 4 points overall), moderate (3 points), low (2 points), and very low (one or less). This score does not comment or reflect on the methodology of the individual RCTs but the actual quality of evidence on a specific outcome in the population of interest. The GRADE working group has looked upon 'quality of evidence' and 'strength of recommendations' based on the quality as two different but related concepts. Since these are usually confused with each other, this separation made use of GRADE easier for people, without an advanced training in these techniques, especially so since it focused on the specific outcomes in specific populations.

Some journals have started assigning a level of evidence to the papers they publish and societies are encouraging their authors to assign a level in their abstract for conference proceedings. This allows the reader to know the level of evidence of the research, though it does not guarantee the quality of the research submitted. This still has to pass the acid test of a meta-analysis. However, it is a good beginning.

In the discussion in this chapter, it may be noted that many different *levels of evidence* have been promulgated by different organizations, but one factor common to all is the 'scientific approach to the research study'. Scientific studies involve a proper and systematic approach to define, study, collect and analyze data, draw relevant inferences, conclude and report the findings. In basic terms, the field of science which deals with data in such a manner is known as *Statistics*. Therefore, appropriate knowledge of statistics allows the researcher to conduct his research in a scientific manner, thereby enhancing its quality and reliability.

CHAPTER 3

Introduction to Statistics

Statistics is "the branch of science that deals with the collection, organization, analysis, and interpretation of numerical data." It is also defined as 'the study of the collection, analysis, interpretation, presentation, and organization of data.' It studies the methods of collection, summarizing and drawing of conclusion from the data. Statistics is widely applicable in different academic and applied functional fields such as physical and social sciences, humanities, business, government and industry. The sum total of attributes studied are of no use if some information regarding their characteristics cannot be derived or some actionable intelligence cannot be gained from them. The aim of collection of data is to convert it into information which may be used for action or decision. This justifies the resources expended on it.

ETYMOLOGY

The word statistics is derived from modern Latin phrase *statisticum collegium* meaning 'lecture about state affairs' or *status* meaning a "Political State" or a Government. Shakespeare used the word *Statist* in Hamlet in 1602. The Latin word gave rise to the Italian word 'statista' meaning statesman or politician and the German word 'Statistik' was used to describe the analysis of data about the state. In those days, statistics was used by rulers who needed information about land, agriculture, commerce, state population, etc. to assess their military potential, wealth and taxation. W Hooper used the word statistics in the translation of Baron BF Bieford's "Elements of Universal Erudition" in 1771. In this treatise, statistics refers to the science that teaches us the political arrangement of all the modern states of the known world.

HISTORY

In the early 19th century, the word statistics was formally used to mean collection and classification of data. In India, administrative data collection was seen around 300 BC, during the reign of Chandragupta Maurya. Abul Fazl recounts of surveys in *Ain-i-Akbari* in the late 16th Century.

The theoretical modern statistics came around the mid-17th century with the introduction of *Theory of Probability* and *Theory of Games and Chance*. These introductions were supposedly driven by the mathematicians and gamblers in France, Germany and England. Theory of Probability is one of the most critical concepts and can be considered as the backbone of modern statistics.

Before this, around early 17th century, 'Vital Statistics', dealing with the demographics, originated. Captain John Grant (1620–1674) in England was the first one to study the birth and death statistics and is called as the Father of Vital Statistics. His works and subsequent contributions from Casper Newman, Sir W Petty, Dr Price, etc. culminated in introduction of life insurance.

NOTABLE CONTRIBUTORS

Few main contributors in the development of statistics have been (In random order, does not signify importance of the contribution/ contributor and is not in any chronological order):

Contributor	Contribution Field
John Graunt	Vital statistics
B Pascal	'Problem of points' coefficients of binomial expansion
James Bernoulli	First treatise on 'Theory of Probability', Binomial distribution
Abraham De'Moivre	'The Doctrine of Chances', probability and annuities, pioneer work on normal distribution
Laplace	Probability (Laplace and Gauss rediscovered normal distribution while working on their respective theories on law of errors)
Gauss	Gave 'Principle of Least Squares' and 'Normal Law of Errors'
Thomas Bayes	Inverse probability
Lagrange	Interpolation
Sir Francis Galton	Regression, pioneered use of statistics in biometry
Karl Pearson	Standard Deviation, Skewness, Correlation, Introduced 1st test of significance- 'Chi- square Test'
WS Gosset (used pseudonym 'Student')	Student's t-test, introduced exact sample test for small samples
Sir Ronald Fisher	Known as 'Father of Statistics', applied statistics theory to diverse scientific fields such as agriculture, genetics, etc. Introduced concepts of point estimates, exact sampling distributions, variance, analysis of variance (ANOVA) and design of experiments, amongst others.
Jacques Quetelet	'Consistency of Great Numbers' which is basis of sampling theory
Laspeyres	Index numbers
Irving Fisher	Index numbers

AN Kolmogorov	'Foundations of Probability'—Axiomatic approach to probability
Simeon Poisson	Poisson distribution
WA Shewhart	Statistical Quality Control
Wilcoxon	Non-parametric test

INDIA IN MODERN CONTEXT

In India, PC Mahalanobis, eminent scientist and applied statistician, is regarded as 'Father of Indian Statistics.' Apart from his pioneering studies in anthropometry, he founded the 'Indian Statistical Institute, Kolkata' in 1932. In 1933, the acclaimed journal *"Sankhya"* (the Hindi word for 'Number') was founded by him, after getting inspired by Karl Pearson's *'Biometrika.'* His other important contributions relate to his works on large scale sample surveys and introduction of concept of pilot surveys. These were acknowledged by international academia including Sir RA Fisher.

APPLICATIONS

Statistics has application in practically every field, and which is probably the reason why it does not have a clear-cut definition. Due to the vastness of its application, no definition is able to capture its true essence. Therefore, the simplest definition of statistics, "the science which deals with the collection, analysis and interpretation of numerical data" provided by 'Croxton and Cowden', appears to be the most suitable as it describes statistics in the simplistic form and does not limit its scope of application. Statistics does not have the true and exact characteristics like other pure sciences, as it has to deal with multiplicity of factors, cause and effect, etc. which cannot be captured in data always. It is more like a systematic technique and scientific method for obtaining meaningful information and knowledge, thereby providing assistance to all diverse fields or subjects, for their own technical advancements.

Statistics through its application, would possibly be classified as one of the most used subjects, irrespective of it being used- knowingly or unknowingly; by literates or illiterates, and in this lies its biggest disadvantage. Due to so much common use, statistics are quoted at every instance, without understanding the basis and its implication. It is generally assumed that quoting statistics tends to lend weight to the arguments. Possibly this is the reason of use and misuse of statistics.

Any data, by itself is neither correct nor incorrect. It is just that a datum or a 'data' set—an observation or set of observations 'present' or 'available' at a given point of time. How it is analyzed and inferences drawn from it, is not in the control of 'data' itself, but the entity who is handling this data. So much data is projected, used, available in print and non-print mediums, that it is difficult to trust all of it and more so, the validity of the inferences drawn, unless the source has high consistency and reliability. In general public life, percentage and inferences drawn from it are commonly misused. For example: a statement—highest number of travelers die, (say 61%) in road accidents, hence travelling by other modes should be preferred. Such a statement is incomplete and may have been wrongly inferred. It is true that 61% of deaths have been in road accidents, but considering roads are used by (say) 90% of the population, then as a ratio of users and deaths the inference may not hold true.

Misrepresentation and misinterpretation of data is seen and is not appropriate, but manipulation of data is another issue which leads to misuse of statistics. Misuse can happen through the use of inappropriate measures, altering definitions to suit outcomes, insufficient or inappropriate selection of samples, incomplete data or using specific period data and use of improper statistical significance tests, etc. Irrespective of whether these occur due to ignorance or deliberateness, these surely cast doubt.

In scientific community, such misquoting of statistics is not prevalent, but the challenges differ. The reason being that at research level, advanced statistics is used for analysis. Therefore, some of the key issues faced here include defining research problem, sampling design and methods utilized, establishing proper hypothesis, using appropriate test of significance, and reporting appropriately the results, including the limitations of the research study.

Types of Statistical Analysis

Statistical analysis is generally divided in two parts, namely:
1. Descriptive statistics, and
2. Inferential statistics

This demarcation is appropriate from research perspective, but cannot be termed as 'two broad divisions of statistics as a subject,' the reason being that such division ignores many aspects of statistics like vital statistics, index numbers, etc. Irrespective of suitability of such a division, these are commonly used in the terminology. In basic terms, these can be stated as:
- Descriptive statistics summarizes and describes data using indexes such as mean, standard deviation, skewness and kurtosis.
- Inferential statistics draws conclusion from data taking into account random variation and probability.

SUMMING UP

Statistics is a collection of methods and techniques to analyze numerical data or non-numerical data, which can be converted into numerical form without loss of interpretation. Therefore, it finds use in diverse fields like economics, biology, psychology, business management, etc. Most of these specialized fields do not require indepth knowledge of all aspects of statistics, but require thorough understanding of certain aspects, whereas for other aspects only basic familiarity and awareness would suffice. For example, for business management knowledge of concepts like Index number, modeling, simulation, regression, etc. is essential, but thorough knowledge of actuarial statistics would not be required, unless the person works in the insurance field.

BIOSTATISTICS

Biostatistics is the branch of statistics which deals with presentation and interpretation of scientific data in biology, public health and other biomedical sciences. This is a special branch because subjects (patients, mice, cells, etc.) exhibit variation in their response to various factors due to constitutional make-up/ genotype or the physical factors interacting with it/phenotype. Therefore, differences observed could be due to different treatments or they may be attributable to chance, measurement error, or other characteristics of the individual subjects. Biostatistics studies enunciate and elucidate different sources of variation. It is an interdisciplinary collaboration to advance knowledge by methodological use of statistical methods for biomedical applications. It applies statistical theory to real-world problems; helps design and conduct biomedical experiments and clinical trials; study computational algorithms; display clinical data. Biostatistics is an integral element in advancements of knowledge in disciplines of biology, public health policy, clinical medicine, proteomics, genomics, healthcare economics, etc.

Broadly stating, the concepts in statistics do not change, but only the terminology to which concepts are being applied, change. Method of testing the significance or calculating the Arithmetic Mean will remain same, irrespective of whether the data represents the monthly sales or weight of the person. Therefore, the critical aspect is—which is the best tool of statistics for the given field and the problem under the study, so that actionable information can be derived from it.

Section 2

Research Methodology and 'Back to School'

- Research Methodology
- Inside Research Methodology: The Nuts and Bolts
- Back to School

CHAPTER 4

Research Methodology

Research is "an inquiry into the nature of, the reasons for, and the consequences of any particular set of circumstances, whether these circumstances are experimentally controlled or recorded just as they occur".
—*Bernard Ostle and Richard W Mensing in "Statistics in Research"*

Research is a systematic and scientific search for relevant information on a given specified topic. It is a "systematized effort to gain new knowledge". It is the art of scientific investigation. It is a mental adventure to satisfy intellectual curiosity. It is a way of life and a method of thought. It is an approach to question rationally the validity of thought at every stage of progress.

Research must define the condition as well as analyse cause and effect. It should be free from bias or prior inclination towards the decision that emerges. A researcher should patiently assess his work, take feedback and make course correction if necessary. Attention to minute details is essential. It requires persistence and sustained effort. It may not succeed in the first attempt. But these failures should act as stepping stones to greater success. Without dogged determination no research is possible. Socrates almost drowned disciple Plato and released him just before the final gasp repeatedly to teach him the value of a burning desire to succeed. Socrates wanted to develop a burning desire for knowledge in Plato that would be as great as the desire for a breath of air at the verge of suffocation. A burning desire coming from within, is the greatest motivation for research and truth. This cannot be imposed from without by someone. The seeker must have the self-driven motivation to find the truth. And this should not be a mere curiosity. It should be almost an obsession as great as the compulsion to eat to remain alive.

OBJECTIVES OF RESEARCH

The main objectives of research can be broadly outlined as follows:
- To become familiar with a phenomenon or gain new insight into a subject. These studies are called *exploratory* or *formulative* research studies.
- To describe the attributes and characteristics of study group or situation or condition. Such studies are called *descriptive* research studies.
- To find frequency of occurrence and association of variables.

- To test the hypothesis of association and judge causality relation between variables. These are called *hypothesis testing* or *analytical* or *diagnostic* research studies.

TYPES OF RESEARCH

Many different ways have been used to classify research efforts by various authors. The basic types of research can be broadly discussed under following headings. (These are not exhaustive or mutually exclusive and an overlap exists).

- *Conceptual and empirical*: Former is related to expression of abstract ideas or theories. Empirical research is based on data and relies on experience or observation of facts/ phenomenon. Evidence from empirical studies and experiments is the strongest support for a hypothesis.
- *Applied and fundamental*: Applied research attempts to find the solution for a well-defined problem faced by a branch of learning or by the society today. Fundamental research is concerned with generating knowledge or gathering it purely for the sake of formulation of theory. It refers to gathering knowledge for knowledge's sake, for example, research in mathematics.
- *Descriptive and analytical*: Descriptive research deals with distribution and description of attributes or variables in a population. The variables in a population are measured to report what is happening or what has happened but the occurrence of the condition is not controlled or regulated.

    Analytical research uses facts or information already available to make a critical evaluation of the information on relationship between variables thereby testing hypothesis.
- *Conclusion and decision orientation*: Decision oriented research is aimed at guiding decisions and the researcher is not free to redesign the query or its import. In conclusion oriented research, the researcher is free to pick up a problem, redesign his query and conceptualize as per his wishes.
- *Laboratory and field setting research*: Latter is conducted in places where the subjects exist in their natural surroundings (field setting) with no control over environment. The former is conducted in controlled environment of laboratories. The factors under study can be controlled and/or modified.
- *Experimental and observational research*: In the former, variables are controlled and altered. The effects of these alterations are measured and studied. In observational research, the variables can be measured but they cannot be controlled or manipulated. Experimental research gives more reliable results scientifically and better quality of evidence. However, such results may not be generalizable in field setting where observational research would be better suited. Hence the nature of question or research problem decides which study design is better suited.
- *Animal and human research*: Animals are often used as experimental models for diseases and conditions to test drugs, interventions, etc. before these can be tried on humans.

    Among humans there are different phases of research which are done in an ethical manner. There are studies on healthy human volunteers and on disease or condition specific volunteers. These are called clinical trials, e.g. Phase I Clinical

Trials, Phase II Clinical Trials and Phase III Clinical Trials. For example, after the preclinical testing in Phase I the drug is tested on healthy volunteers for dose-ranging from the minimal subtherapeutic dose to ascending therapeutic doses. This evaluates the safety, determines a safe dosage range and identifies side effects in healthy volunteers. In Phase II the drug is tested on patients to assess efficacy and safety giving the therapeutic dose. In Phase III the drug is tested on patients and its efficacy, effectiveness, safety and therapeutic dose are assessed.
- *Historical research*: Historical remains and documents are used to learn about ideas and events of the past.

APPROACHES TO RESEARCH

The basic approaches to research are:
- Quantitative
    - Inferential
    - Experimental
    - Simulation
- Qualitative

Qualitative Approach

Qualitative approach refers to the subjective assessment of attitudes, opinions and behavior. Generally interviews and projective techniques are used. The results are either non-quantitative or semi-quantitative. They cannot be subject to rigorous quantitative analysis.

Quantitative Approach

Quantitative approach refers to use of measurement techniques and methods which allow use of rigorous quantitative analysis of variables.

Inferential Approach

In inferential approach, a database is formed of variables from which characteristics and relationships among variables of population can be judged or inferred. This database is like a masterchart and it allows tabulation and comparisons to be made.

Experimental Approach

In experimental approach, some given variables are manipulated in a controlled environment to observe and record their effect on other variables. This may follow rigorous double cross over or may simply give the results as associations.

Simulation Approach

In simulation approach, an artificial environment or simulation model is created, within which the relevant information or data can be generated based on the rules defined therein.

Scientific progress comes from inquiry. As Hudson Maxim, the famous chemist and inventor, says, "Doubt is often better than overconfidence, for it leads to inquiry and inquiry leads to invention." Research promotes scientific and inductive thinking.

It develops logical and organized thinking. It is of use in:
- Government work for policy making
- Business and industry for solving various problems
- Education and applied sciences for developing new concept to overcome bigger challenges.

RESEARCH METHODS

Research methods refer to all techniques used by researchers in studying a problem or performing research operations. They can be broadly divided into three main categories:
1. Data collection methods
2. Statistical techniques used for elucidating relationships between data and the unknowns
3. Methods used to evaluate the accuracy of results obtained.

It is important to understand various techniques, the basic assumptions underlying each of them and be aware of their strengths and limitations. Research methods are the behaviors and instruments used for selecting and constructing research technique. Research techniques are the behavior and instruments used for performing research operations like observations, data recording, data processing, etc.

RESEARCH METHODOLOGY

Research methodology is the study of systematic solution of research problems. It studies how research is done scientifically. The research decisions must be exposed to evaluation before they are implemented. They must be stated very clearly and explicitly along with the reasons why these were taken. Research methodology not only concerns the research methods, but also the logic behind the methods used- in the context of the research question. It is pertinent to state if other methods are available, the reason why they are not being used, and describing in a logical manner why the chosen one is being preferred over others. Broadly it means describing:
- Why the research study was undertaken?
- How the problem was defined?
- What was the hypothesis and why it was formulated that way?
- What data would be captured and by using which method?
- How it would be analyzed and why it would be analyzed in this manner?
- Strengths and limitations of the current course of actions
- Any alternative approaches and why they were not used or recommendations for using them.

In 'The Grammar of Science', Karl Pearson wrote, "The unity of all sciences consists in its methods, not its material. The man who classifies facts of any kind whatsoever, who sees their mutual relation and describes their sequences, is applying the scientific method and is a Man of Science".

Basic Postulates of Scientific Method

The basic postulates of scientific method are:
- Empirical evidence
- Ethical neutrality
- Utilization of empirical evidence
- Objective consideration
- Probability prediction
- Propagation of methodology among all concerned for careful scrutiny before embarking on action
- Formulation of general axioms or theories through use of generalizable and repeatable results.

Research Process

Research process refers to the series of steps or actions along with their desirable sequence used to carry out research on a given problem **(Flow chart 4.1)**. The steps are as follows:
- Formulation of the research question
- Doing an extensive and exhaustive literature search
- Development of a hypothesis
- Preparation of a research design
- Collection of data
- Analysis of data
- Testing of the hypothesis
- Generalization and interpretation of findings
- Monitoring and evaluation of the project
- Preparations of a report with formal write up of conclusions.

Flow chart 4.1: Research process

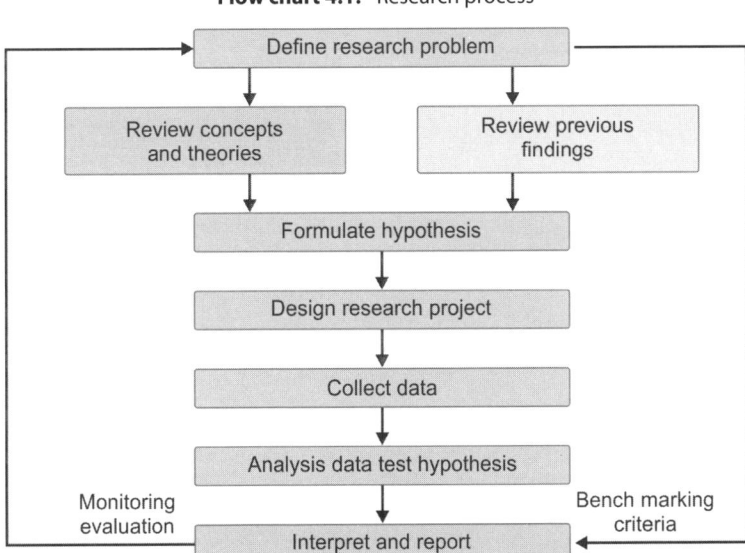

These steps may overlap and may not be mutually exclusive or be separate and distinct.

Broad Fields of Biomedical Research

According to T Greenlaugh the broad fields of biomedical research can be differentiated into the following:

- *Therapy*: Testing the efficacy of drug treatments, surgical procedures, alternative methods of service delivery, or other interventions. Preferred study design is randomised controlled trial (RCT).
- *Diagnosis*: Demonstrating whether a new diagnostic test is valid (can we trust it?) and reliable (would we get the same results every time?). Preferred study design is cross sectional survey in which both the new test and the gold standard are performed.
- *Screening*: Demonstrating the value of tests which can be applied to large populations and which pick up disease at a presymptomatic stage. Preferred study design is cross sectional survey.
- *Prognosis*: Determining what is likely to happen to someone whose disease is picked up at an early stage. Preferred study design is longitudinal cohort study.
- *Causation*: Determining whether a putative harmful agent, such as environmental pollution, is related to the development of illness. Preferred study design is Cohort or Case-control Study, depending on how rare the disease is, but case reports may also provide crucial information.

CHAPTER 5

Inside Research Methodology: The Nuts and Bolts

In the previous chapter, apart from research objectives, types of research and research methodology were discussed, as also the outline of research process was stated. Before discussing in detail the research process, it would be prudent to understand what constitutes a research and what the components of a good research are. Research design and research plan are important components of the research process, therefore these are detailed after the research process. All these are being discussed in our run up to measurement, descriptive and analytical statistics. This is just a recapitulation of the processes that almost all practitioners are conversant with. The purpose of covering these topics here, even at the risk of repetition, is that this broadly defines the methods we work with and leaves no ambiguity in what we aim to study using the statistical methods. This discussion is not exhaustive. It is a synopsis for the purpose of this book and the readers are advised to refer to appropriate texts. Here this is just the first peg to hang the further concepts on. In our quest 'from known to unknown', this is the universal known first step.

WHAT CONSTITUTES A RESEARCH?

Broadly a research problem should satisfy the following conditions. It must be:
- Attributed to an individual/group/organization who does not know what the best course of action is, i.e. the individual/ organization must have some difficulty or problem the solution of which is not known to it even after exhaustive search.
- Defined by independent uncontrolled variables.
- There must be two or more alternative courses of action ($C_1, C_2, C_3, C_4 \ldots C_n$) defined and one or more values of dependent or controlled variables.
- There must be at least two outcomes, one of which is more desired. If the outcomes are proven to be equally good or bad then the effort is wasted. The choice does not matter in such a situation. Some objectives must be required to be achieved.
- The courses of action must provide a chance of attaining the objective (This chance should not be equal because then the choice would not matter) e.g. Toss of coin for heads or tails. This is called "principle of unequal efficiencies of the desired outcomes".
- There should be some doubt about alternatives.

Thus the researcher is attempting to find an optimum solution to a given problem, in the given environment in a manner that similar results would be found by other researchers in given environment following the same methodology.

This implies that a gap in knowledge or a gray area is sought to be filled through research.

Practical Considerations

Selecting a research problem is both a science and an art. In selecting the problem or solving it, creativity should not be sacrificed. However, the topic should not be such that, one is not able to do justice to it in the time available. Also bear in mind the reason why the research is being conducted:
- Working towards a research degree
- To get respectability
- Service to scientific community
- Intellectual joy derived from the creative work.

If it is for a degree, the primary objective is to get to your results in time, therefore objective needs to be decided accordingly.

The topic should be carefully selected.

A topic which has been answered repeatedly may not have too many areas to shed light on.

Controversial topic should generally be avoided for degree aspiring students.

The subject should be familiar for the researchers and facilities should be available. The definition of scope should allow for creative and innovative ideas to be incorporated (flexible and familiar subject).

Problems which are too narrow or too vague may result in poor analysis and low repeatability.

The subject, qualifications, training of the researcher, budget available, ability to get cooperation of the study subjects, access to funding agencies all have a bearing on selection of the problem.

If conditions are favorable any study can be undertaken, but when constraints are faced creative approach can go a long way in coming up trumps.

Good Research

What constitutes a good research? A good research should be:
- *Systematic*: Specified steps must be taken in a logical predefined sequence according to a well-defined set of rules with knowledge of all concerned.
- *Logical*: It should be guided by logic and rationale. Processes used are induction and deduction. Induction is the process of reasoning from a part to the whole and deduction is the process of reasoning from some conclusion which follows from that particular premise in a verifiable manner.
- *Empirical*: It should be based on experience of a real situation from which concrete data is analyzed to provide external validity to the research results.
- *Replicable*: Other researchers can get similar results by following the methodology described by the current researchers. This, therefore, can be used for making valid decisions in similar situations.

Research Process

As outlined in the previous chapter, the research process includes:

Formulation of Research Question

The problem to be studied should be selected. There are broadly two categories:
1. Problem relating to state of condition or nature.
2. Problems regarding relationship between variables.

The research question needs to be stated in a broad general way first, and then the ambiguities regarding it, if any, are removed. Problem has to be understood thoroughly. Then it should be phrased in terms of analysis for operational purposes in, as specific a manner as possible. The statement of the objective determines what and how data is collected, explored, analyzed and reported. Being specific in above steps ensures that relevant characteristics of data are picked up for understanding the problem thoroughly and answers sought using specified appropriate techniques.

Extensive and Exhaustive Literature Search

Extensive literature search is necessary before writing the synopsis or executive summary. The problem must focus on gap in knowledge. Sometimes the question has already been answered and the gray area may not exist at all, making the project redundant. By extensive search, it allows the researcher to know early and ensures no regret at a later date.

Types of Literature

Two types of literature can be explored.
1. Conceptual literature and
2. Related or empirical literature

Conceptual literature refers to the articles or books written by authorities giving their opinions, experience, theories and ideas on the subject. Experts use this concept to outline alternative courses of action or present a chosen approach to an issue. Conceptual literature deals with studies done earlier on the subject. In specific terms, conceptual literature's materials, data, and other related topics originate from the books, journals, etc.

Related or empirical literature helps the researcher to know and understand the subject better. In specific terms, it usually comes from the modern technology like internet and other researches from students undertaking graduate thesis or dissertations. The quality of this evidence is not always fully verifiable. Researcher needs to familiarize with both and use each appropriately.

Often empirical literature is important when relationship between variables is being tested using some hypothesis, while conceptual literature is of use where state of a condition is being studied. But both go hand in hand and only the weightage given changes, not the direction of the two while a synopsis is made. PUBMED, MEDLINE, INDEX MEDICUS, various libraries such as National Medical Library, Gopher sites, web crawlers and search engines are used. The abstracting and indexing journals, published and unpublished bibliographies, academic journals, conference proceedings, books, government reports, international bodies reports are good initial *go to* places. Good index cards were earlier indispensable for organizing this information. Nowadays it is probably easier to make a database or a spreadsheet.

Development of a Working Hypothesis

A working hypothesis should be defined and it must be very specific, precise, clearly defined and limited to the research in hand. This sharpens thinking and focuses attention on the important components or aspects of the research question. It is very pertinent to utilize the following to develop a hypothesis:
- Discussions with colleagues and experts on the subject matter.
- Review of similar studies
- Examining data and records
- Conducting pilot studies and exploratory research on the subject.

Preparation of Research Design

Research design states the conceptual structure of the boundaries within which research will be conducted using specified tools and techniques **(Flow chart 5.1)**. The objective is to collect relevant information with minimum efforts, expense or delay. Research design depends on, among other things, 'research purpose' which may be in nature:
- Exploratory
- Descriptive

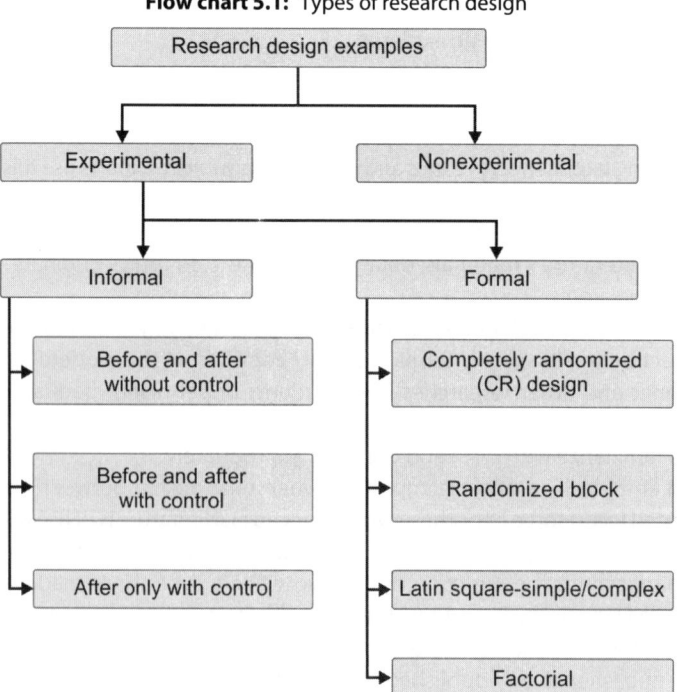

Flow chart 5.1: Types of research design

- Diagnostic
- Experimental.

Important considerations for research design are what would be the 'means of information' and explanation of reasons for using selected 'means of information'. Means of information may depend on availability of time, budget and skills. Sampling is an important tool for means of information.

Sample Design

Sample design is the definite plan for obtaining a representative sample from a given population which is determined before any data is collected. Sample design considerations include defining the population, whether the study would be based on census or sample. If based on sample, then how would be the samples be drawn. A brief of these is given below:

- *Universe/population*: All items in a field of enquiry.
- *Census inquiry*: Complete enumeration of all items in *universe*. Often this entails so much expense and time that it is only used if the research relevant population is restricted or small in size. Otherwise sampling is undertaken.
- *Sample:* A selection from population, which is considered as a fair representative of the population and its characteristics. This is further divided as listed below.
- *Probability sample:* Each element/ item of the population has a definite and known probability of being induced in the sample.
- *Nonprobability sample:* No information of probability of being included in sample can be determined.

The commonly used sampling methods include the following:
- Deliberate/purposive nonprobability sampling
- Simple random sampling
- Systematic sampling
- Stratified sampling
- Cluster sampling
- Complex sample designs.

Sampling theory and probability has been covered in detail in their respective chapters in this book.

Data Collection

Data collection methods could be either primary or secondary. Primary methods would include:
- Personal observations
- Personal interviews\telephone interviews
- Questionnaire
  - By mail
  - Directly administered
  - Self-administered
  - By telephone
- Checklists
- Indirect sources.

Secondary data may be taken from government records, census, economic surveys, industry reports, disease registries or other published sources.

Data being collected can be numerical (Quantitative) or categorical (Qualitative).

Quantitative data can be discrete (number of patients, number of books) or continuous (height, weight, etc., i.e. can take any value even decimals/ fraction within the given range). Depending upon the measurement scale, quantitative data can be measured on a numerical scale where zero value too represents a state. For example temperature scale, where zero degree means a specific state of temperature but not absence of temperature. Such quantitative data is also known as 'Interval data'. Contrary to this, a numerical data where zero signifies nil or absence of the object/ subject being measured, such as weight, length, etc. is called as ratio scale based data.

Qualitative data can be ordinal or nominal. Nominal data is on nominal scale, i.e. yes or no, present or absent, urban or rural, etc. Primarily the data is divided in any number of groups and not necessarily two. The nominal data does not have any order within the group. In case an order is introduced within the group (say given ranks), such a qualitative data becomes ordinal data.

Data can also be stated as cross-sectional, temporal or spatial. Cross-sectional data refers to measurement of variable at same period/point of time like patients treated in a particular year by different hospitals, number of deaths in the last month due to different causes, etc. Temporal data refers to a specific variable but for many periods/ points of time. For example, number of patients treated by a hospital each year in the last (say) seven years, number of deaths caused by road accidents each month in the last one year, etc. Further, data based on geographical location is called spatial data. For example, number of patients treated by hospital in a given period in different cities. Type of data required primarily depends upon the nature, objective and subject of the study.

Analysis of Data and Testing of Hypothesis

Data from records, performance or patient sheets should be condensed into manageable groups or tables. This leads to the creation of a database or chart called the master chart.

Categories of data may be transformed into symbols to be tabulated and counted. This is called coding or extraction.

Improvement of quality of data by filling up the losses or correcting errors is called editing. Tabulation is the process of putting this data in tables. Using algorithms and normalization the data may be electronically corrected by a process called transformation, e.g. continuous data like blood pressure may be converted into binomial data like hypertension or interval data like grades of hypertension. Using the data, a hypothesis stated earlier may be tested using statistical tests such as Chi-square, t-test, f-test, Wilcoxon Rank Sum test, etc.

Generalization based on data may be made if no specific hypothesis had been stated earlier, along with recommendation for future research.

Generalization and Interpretation

Generalization can be arrived at, if a hypothesis has been tested and upheld many times. Thus a theory can be built, whereby the results can be generalized over a larger population. If the researcher had no theory or hypothesis to begin with, he/she may attempt to explain his/ her findings on the basis of some rationale/theory. This process is known as interpretation.

Monitoring and Evaluation of the Project

During the course of the research project there may be reasons to stop the project, for example, there may be serious adverse effects in treated group or the beneficial effects in treated group show that withholding such therapy from the control group would be unethical. This requires continuous oversight and monitoring of results during the execution of the project.

Evaluation refers to the matching of the execution with the design. Set at the start of the project, compared with what has been achieved at the end of the research, including time taken for enrolment, cost of enrolment, overall time taken, total cost, etc. This is of value to the funding agencies, and has a bearing on future grants to same group of researchers or institution or research problem.

Preparation of Report and Reporting of Conclusions

The report depends on what the purpose of the research was and is guided by who funded the project, while thesis submitted for award of degrees should adhere to the norms laid down by the individual university/accreditation system. Stating generically, the report should have the following:
- The Preliminary pages should carry:
  - Title and date
  - Acknowledgements
  - Forward
  - Table of contents
  - List of tables and graphs.
- Main text should preferably be divided into:
  - Introduction and background
  - Objectives
  - Methods/materials and methods
  - Main report with results in logical sequence
  - Discussion of results in logical sequence and identifiable sections
  - Conclusions stated clearly and precisely
  - Summary.
- Ending pages should be devoted to the following:
  - Bibliography of references cited in text
  - Appendices of technical data
  - Master charts or supplementary material, if any
  - Index is required especially in Government Report.

Research Design

Research design is the blueprint for collection, measurement and analysis of data. It has been defined by C. Selltiz et al. as, the arrangement of conditions for collections and analysis of data in a manner that aims to combine relevance to the researchers purpose with economy in procedure.

It answers questions about:
- What it is being done?
- By whom?
- Where?
- On whom?
- Which data will answer what aspect of the problem?
- From when till when will it be carried out?
- How will sampling data collection and analysis be done?
- Style of report and further actions.

It must always contain the following information:
- Clear operational statement of the research problem.
- Information gathering procedure and techniques alongwith logic behind the choice
- Target population and attributes under study
- Data processing and analysis tools to be used along with their rationale.

A good research design minimizes bias and maximizes the reliability of the data collected and analyzed, in the context of the research question at hand. The gold standards are smallest experimental error and the largest range of information. Since both cannot be achieved simultaneously the focus shifts depending on study type for example, in exploratory research we may want to collect a large range of information to study the different aspects of the problem. On the other hand, where we want to study the distribution and relation between variables we would go for the design with smallest experimental error which would minimize bias and maximize reliability.

Practical Considerations in Defining Research Problems and Research Design

- Start by stating the problem in a broad general manner
- Remove ambiguities in the problem statement
- Do an experience survey by discussing with experts and colleagues who have knowledge about the subject.
- Do an exhaustive survey of the available literature. Knowledge of the existing literature helps to identify the gaps to be covered and helps in furtherance of knowledge.
- Again discuss the problem with added knowledge and narrow down the research problem to specifics, and come up with an operational project problem.
- Rephrasing of problem in operational and analytic terms is the final step. It must be done after doing due diligence and considering inputs from a wide variety of sources especially those with insight into the problem and research dynamics. Eventually only the researcher has the unique perspective on the project in hand.

Types of Research Design

Older literature broadly divided research design on the basis of study objectives for ease of discussion. Now there are cases of overlap which may not strictly stick to a

single domain. For sake of simplicity and ease of discussion we use the older method of discussion bearing in mind the fact that these are not water tight compartments today. Research designs subserve the following:
- Exploratory research studies
- Descriptive and diagnostic research studies
- Hypothesis testing studies.

Exploratory Research Studies

The main objective of these projects is to get insight into the problem with discovery of ideas and issues. It leads to more precise investigation with good working hypothesis operationally. Broadly we can use:
- Survey of existing literature
- Experience survey
- Using questionnaires, interviews and insight stimulating case studies.

It is important to keep the overall design flexible so that a wide range of insights can be picked up to understand the contours of the problem. Generally 'purposive' or 'judgment' sampling suffices. Usually, there is no preplanned statistical design and it is formulated on the way.

Descriptive or Diagnostic Research Studies

In descriptive research studies more rigid designs are used. The objective is to minimize bias and maximize the reliability of the data collected. Probability sampling is best for this and often random sampling is used. The instruments for data collection are well researched and enunciated. The statistical analysis proceeds along specific predetermined lines. Survey design and comparative design are used operationally quite frequently.

Hypothesis Testing or Analytical Research Studies

They can be nonexperimental or experimental. In the former no manipulation of the variable is made. In the latter, under controlled conditions a variable is modified and the results of this modification are studied. It is one of the most robust study designs for many types of research questions. Prof RA Fisher while working at Center for Agricultural Research at Rothamsted Experimental Station laid the foundations of this type of designs. He enumerated the following principles of experimental design:
- Principle of randomization
- Principle of local control
- Principle of replication.

Randomization

The elements are selected randomly to be studied for attributes. As a result of this the extraneous factors can be assumed to be randomly distributed in the population and its representative sample. They will, therefore, have no effect on the outcome and all variations due to these factors can then be attributed to chance alone.

Local Control

The extraneous factor which induces variability is made to vary over as wide a range as possible and its effect is recorded. This can be removed from experimental error.

Replication

The experiment or treatment needs to be repeated more than once. It increases the accuracy with which main effects, interactions and relationships of variables can be measured. It allows the statistical tests to be used.

RESEARCH PLAN

The synopsis or formal write up of experimental or research project constitutes a research plan. It should essentially contain the following in addition to local academic/funding agency requirement.
- Research objectives in crisp noting, outlining primary and secondary outcome measures.
- Explicitly stated research problem
- Each major concept to be measured defined in operational terms.
- Method of solving problem
- Population to be studied and sampling design
- Details of measurement plan
- Details of data processing methods
- Result of pilot study, if any
- Sources of funding or budget plan, if required
- Bibliography.

Uses of Research Plan

- Helps to organize and crystallize ideas or action plans
- Provides inventory for materials and methods

Flow chart 5.2: Research plan

```
Planning
(Research design, sample design, etc.)
          ↓
Execution
(Performing experiment, data collection, etc. as
defined in the research design
          ↓
Analysis
(As outlined in research design)
          ↓
Report writing
Guided by research design, and based on
results as derived on execution and analysis
```

- As a formal document, it can be given to others for discussion and comments
- Forms the final blueprint for conducting the research **(Flow chart 5.2)**

Proper establishment of research design and sample design, ensures that the research would not slip due to poor planning. After this, the focus shifts to execution of the research, which is followed by analysis and report writing.

The discussion above depicts the importance of the research design, in the overall scheme of research. Therefore, it is suggested that one should give the planning stage due importance for successful conduct of research projects. The next chapters would be devoted to core statistical concepts of research projects, tools, techniques and methods, with a brief note on elementary mathematics.

CHAPTER 6

Back to School

It has been particularly noted that individuals from non-mathematical education in higher classes find it relatively difficult to understand the concepts of statistics. Though the interest to learn the concepts is of high order, but due to inadequate exposure to some elementary knowledge about operators, symbols, rules, etc, the interest is lost. To avoid such instances, the following pages would touch upon the basics of some mathematical concepts, which will help in understanding statistical concepts, where such mathematical concepts are involved.

GREEK ALPHABETS

Pronunciation	Capital/small letters	
Alpha	A	α
Beta	B	β
Gamma	Γ	γ
Delta	Δ	δ
Epsilon	E	ε
Zeta	Z	ζ
Eta	H	η
Theta	Θ	θ
Iota	I	ι
Kappa	K	κ
Lamba	Λ	λ
Mu	M	μ
Nu	N	ν
Xi	Ξ	ξ
Omnicron	O	o
Pi	Π	π
Rho	P	ρ
Sigma	Σ	σ
Tau	T	τ

Contd...

Contd...

Pronunciation	Capital/small letters	
Upsilon	Y	υ
Phi	Φ	φ
Chi	X	χ
Psi	Ψ	ψ
Omega	Ω	ω

COMMON MATHEMATICAL OPERATORS/SYMBOLS

Operator/symbol	Denotes	Remarks
∞	Infinity	Not to be confused with alpha (α)
$\propto$	Proportional to	
$\sim$	Approximately	
!	Factorial	Used in permutation and combination
$\cong$	Approximately equal to	
<	Less than	5 < 7 (5 less than 7)
>	Greater than	7 > 5 (7 greater than 5)
$\leq$	Less than or equal to	
$\geq$	Greater than or equal to	
$\pm$	Plus minus	7 ± 5 = 12 or 2; if specified as Range then 2 – 12
$\approx$	Almost Equal to	Asymptotic to
$\equiv$	Identical	
$\therefore$	Therefore	
/	Forward Slash (Division)	
Σ	Summation	(Refer note below)
$\int$	Integral	
Π	Product	
$\because$	Because	
$\cap$	Intersection	Frequently used in Set Theory and Venn Diagrams
$\cup$	Union	
$\in$	Element of	
$\ni$	Member of	
$\subset$	Subset of	
$\supset$	Superset of	
:	Ratio	
::	Proportion	

Note: The symbol Σ denotes Summation, and is very frequently used is statistics. It implies sum of all constituent values. It can be written as Σ or $\Sigma_{t=0}^{n} x_i$ or

$$\sum_{i=0}^{n} x_i = x_0 + x_1 + x_2 + x_3 + \ldots + x_{n-1} + x_n$$

and implies that the value of ranges from zero (*i* = lower value = start) to, (at top = end value) and all these values are to be added up.

For example, $\Sigma_{i=9}^{20} x_i$ where it is given that is an even number, then the result is as follows:

$$\sum_{i=9}^{20} x_i = 10 + 12 + 14 + 16 + 18 + 20 = 90$$

MODULUS

Denoted by $|x|$, two vertical bars on each side, Modulus represents the Absolute Value of the number, without regard to its positive or negative sign, i.e. $|-2| = 2$.. In other words, it gives the distance of the value from zero and can be positive number or zero, but never negative. In mathematical terms it is defined as:

$$|x| = \begin{cases} x, & \text{if } x \geq 0 \\ -x, & \text{if } x < 0 \end{cases}$$

Further, few basic properties of Modulus are as follows:

if $|x| = 0$, then $x = 0$
if $|x - y| = 0$, then $x = y$
$|x \times y| = |x| \, |y|$

similarly $\left|\dfrac{x}{y}\right| = \dfrac{|x|}{|y|}$ provided $y \neq 0$

$|x + y| \leq |x| + |y|$

BASICS OF SET THEORY AND VENN DIAGRAMS

In the study of probability, distributions, sampling and many others concepts of statistics, terms like 'mutually exclusive' are used. Though these concepts can be grasped without the knowledge of set theory, but reviewing basic terms and properties of set theory allows better and faster understanding of these concepts. Therefore a basic treatment, without getting into too much details of the subject is being considered here.

A well defined collection of objects is called a Set. The objects could be anything like planets in solar systems, natural numbers, equipment in office, books in a library, but the collection has to be well defined. This 'well defined' clause ensures clear cut understanding as to who/what can be member of that set and who/what cannot.

If an object 'x' belongs of Set D, i.e. 'x' is an element of Set D, then it is denoted as $x \in D$ and vice versa $x \notin D$ signifies x is not an element of Set D.

A set is always denoted by Capital letters like A, B, C..., but certain capital letters are very commonly used for specific sets, like N for all natural numbers, W for all whole numbers, Z for all integers, Q for all rational numbers and R for all real numbers.

Sets can be expressed in either 'Roster Method' or 'Set-builder Form'. Assume a set A of all natural numbers between 1 and 9. In 'Roster Method' will this will be depicted as "Set A = {2, 3, 4, 5, 6, 7, 8}" i.e. all elements are written within the specific type of brackets shown. It is a convenient method, but has problems when the number of elements in a set are too many. To overcome this, 'Set-builder Form' mentions the property, on basis of which the elements have become members. Set A as stated earlier in Roster Method, can be stated in Set-builder Form as: Set A = Set A = {x|x ∈ N,

$1 < x < 9$} Set and read as "Set A consists of x, where x is an element of set of natural numbers, and where x is greater than 1 but less than 9." (In a generic manner, the statement can also be stated as 'Set A consists of, where has the property of being a natural number between 1 and 9').

A set having no elements is called 'Null Set' or 'Empty Set'. It is denoted by ϕ (phi). A set having one or more than one element is called 'Non-empty Set', whereas a set containing only one element is called 'Singleton Set'. Mentioning a set with zero elements (null set) as {0} would be incorrect, as this notion indicates a singleton set whose element is '0'. A set whose elements can be counted is called 'Finite Set', and similarly a set whose elements are infinite is known as 'Infinite Set'. Every finite set will have its 'cardinal number'. The number (count) of distinct elements in a finite set is called its cardinal number and denoted as $n(A)$. For the above example, the cardinal number is $n(A) = 7$.

Two sets A and B would be considered as 'Equal Sets', if all the elements of Set A are present in Set B and all elements contained in Set B are present in Set A. Equal Set differs from Equivalent Sets, because two sets are considered 'Equivalent Sets' if their cardinal number is equal. Hence Set A = {10, 15, 20} and Set B = {1, 2, 3} are Equivalent sets but not Equal sets. Equal Sets will always be equivalent, but two equivalent sets may or may not be equal. The order of elements and repetition is not considered. For example: A set of alphabets used in word 'example' would be written as Set X = {e, x, a, m, p, l}.

'Subsets': If A and B are two sets, with every element of set A being present in Set B, then Set A is called the 'Subset' of Set B (denoted as $A \subset B$), while Set B is called the 'Superset' of Set A (denoted as $B \supset A$). It is inferred that the superset will contain all elements of the subset plus some more elements. Further every set is a subset of itself, and empty set ϕ is a subset of every set. If $A \subset B$ and $B \subset A$, then it follows that Set A = Set B and vice versa. Every non-empty sets having number of elements will have:
  2^x number of total subsets and
  ($2^x - 1$) number of proper subsets. (Null set is not a proper subset).

Universal Set: Denoted by ξ or U, Universal Set, is a set which contains all the other sets under consideration. From this arises the term of 'Complementary Sets'. If A = {1, 2, 3, 4} and Universel Set ξ = {1, 2, 3, 4, 5, 6, 7}, then Complementary Set of A, denoted as A' = {5, 6, 7} i.e. all elements which are not present in Set A, but present in the Universal Set.

Operation of Sets

Union of Sets A and B (Denoted by $A \cup B$ and read as 'A Union B') is the set of all those elements contained in sets A and B, i.e. if A = {2, 3, 4, 5} and B = {5, 1, 2, 6, 8}, then $A \cup B$ = {2, 3, 4, 5, 1, 6, 8}. Properties of Union of Sets:
  i. Both sets A and B are subsets of set $(A \cup B)$.
  ii. $A \cup B = B \cup A$.
  iii. If A is a subset of B, then $A \cup B = B$, as repetitions are not considered.
  iv. Union of a Set A and its Complementary Set A' (pronounced as 'A dashed'), i.e. $A \cup A' = ξ$ (Universal Set). For ease of explanation only two sets A and B are considered, but Union can be of multiple sets like $((A \cup B) \cup C) \cup D$, etc.

Intersection of Sets: (Denoted as $A \cap B$) is the set of all those elements which are common to both the sets. For example, consider the previous values, if $A = \{2, 3, 4, 5\}$ and $B = \{5, 1, 2, 6, 8\}$ then $A \cap B = \{2, 5\}$. From this it follows that two sets are called Disjoint sets if $A \cap B = \phi$. Properties of Intersection of Sets:
  i. $A \cap B = B \cap A$
  ii. If $A \subset B$ then $A \cap B = A$.

Venn Diagrams

The pictorial representation of the sets and their relationship is called the Venn Diagrams **(Fig. 6.1)**. In Venn Diagrams, the Universal Set ξ is represented as rectangle, which contains all the sets. Each set is represented by circle or oval any other closed shape. The constituents of a set (say set A) are marked inside its representative circle and therefore the area outside the A set's area is its complementary set (A'). Union of sets will include all the constituents of the two or more sets, while intersection of two or more sets would be represented by the overlap of the representative circles of the sets whose intersection is desired. If the universal set (rectangle) contains two representative circles which are not overlapping (no intersection), then these sets are disjoint sets (no element in common).

Consider the above Venn Diagram. The rectangle denotes the Universal Set ξ. The three representative closed oval/circles are three sets A, B and C. From the above Venn Diagram, the following can be inferred:

Set $A = \{2, 8, 4, 6, 11, 1\}$, Set $B = \{4, 13, 1, 5, 7, 9, 16\}$ and
Set $C = \{6, 11, 1, 3, 7, 9, 12\}$
$A \cup B = \{2, 8, 4, 6, 11, 1, 13, 5, 7, 9, 16\}$ and so on
$A \cap B = \{4, 1\}$ (Overlap area of Circle A and Circle B);
$(A \cap B) \cap C = \{1\}$ (Central Portion where all three circles overlap)

Value '18' does not lie in any of the circle but is an element of Universal Set. Similarly lot of information about Union, Intersection, Complementary etc can be gathered from this.

'Mutually Exclusive'—a term very commonly used in Probability and other concepts of Statistics, fundamentally denotes sets which do not have common element, i.e.

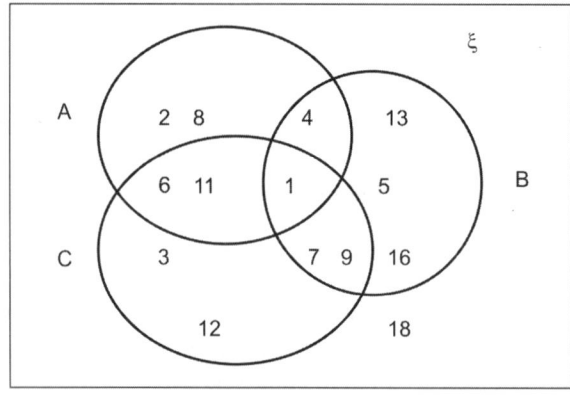

Fig. 6.1: Venn diagram

two disjoint sets and therefore Intersection of mutually exclusive sets (or events in Probability), is a null set $(A \cap B) = \phi$.

INDICES

If a number is multiplied by itself, the resultant number can be written in a shorter notation as below:
$$7 \times 7 \times 7 \times 7 \times 7 \times 7 = 7^6$$
and called '7 to the power of 6'. The upper digit (superscript) is called the Index or Exponent. Indices as a topic, covers the study of the property of these Exponents/Index.

Some of the important properties of Indices are as given below:
$$x^m \times x^n = x^{m+n}$$
i.e. can be extended to $x^m \times x^n \times x^p \ldots = x^{m+n+p}$
$$(x^m)^n = x^{m \times n}$$
$$\frac{x^m}{x^n} = x^{m-n}$$
$$\frac{1}{x^m} = x^{-m}$$
$$x^{1/m} = \sqrt[m]{x} \ (m^{th} \text{ root of } x)$$
$$x^0 = 1 \ (\text{any no. to the power of } 0 = 1)$$
$$x^m y^m = (xy)^m$$

PROGRESSIONS: ARITHMETIC AND GEOMETRIC PROGRESSIONS

Arithmetic Progression

A series in which the difference between any two numbers placed next to each other is constant, for example 2, 5, 8, 11, 14, 17… is called an Arithmetic Progression. The constant difference is called as Common Difference. If 'a' is the first number of any such series and 'd' is the common difference, then such a series can be represented as:
$$a, a + d, a + 2d, \ldots, a + (n-1)d$$
The nth term will be $(a + (n-1)d)$ as first term was 'a'. The sum of any such series having 'n' terms is given as:
$$a + (a+d) + (a+2d) + \ldots + (a + (n-1)d) = \frac{n}{2}(2a + (n-1)d)$$

Geometric Progression

In 'Geometric Progression' the ratio deduced on dividing any number by its preceding term remains constant. This constant ratio is called 'Common Ratio.' Series like 4, 16, 64, 256, 1024…., is an example of Geometric Progression, where the common ratio is 4. Such a series can be written as:
$$a, ar, ar^2, ar^3 \ldots ar^{n-1}$$
where 'a' is the first number and 'r' is the common ratio. Sum of such a series having 'n' terms (n^{th} term would be ar^{n-1}) is given as:

$$a + ar + ar^2 + ar^3 = \ldots + ar^{n-1} = \frac{a(r^n - 1)}{r - 1} \text{ (for } r > 1\text{) OR}$$

$$= \frac{a(1 - r^n)}{1 - r} \text{ (for } r < 1\text{)}$$

Other Sum of Series:

a. $1 + 2 + 3 + 4 + \ldots + n = \dfrac{n(n+1)}{2}$

b. $1^2 + 2^2 + 3^2 + 4^2 + \ldots + n^2 = \dfrac{n(n+1)(2n+1)}{6}$

c. $1^3 + 2^3 + 3^3\, 4^3 + \ldots + n^3 = \left[\dfrac{n(n+1)}{2}\right]^2$

LOGARITHMS AND ANTILOGARITHMS

Also referred to as 'Log' and 'Antilog.'
Consider the following equation: $3^4 = 81$, i.e. 3 raised to the power of 4 = 81.

Let 3 be called as the base, 4 as the power and 81 as the result, or in Log terminology, 4 is the Logarithm of 81 to the base 3. (mathematically $Log_3 81 = 4$; in generic terms if $x^y = N$, then $Log_x N = y$).

Therefore Logarithm of any number to a given base, is the index of the power to which the base must be raised to equal to that number.

Logs to the base 10 are called common logarithms and to the base 'e' are called natural logs. The logarithm tables are generally to the base 10, unless otherwise specified.

Some properties of logarithms:
1. The Log of 1 to any base is 0
2. Log of any number to the same base is 1, i.e. $Log_x x = 1$
3. $Log_x (mnp) = Log_x m + Log_x n + Log_x p$
4. $Log_x \left(\dfrac{m}{n}\right) = Log_x m - Log_x n$
5. $Log_x (m)^r = r\, Log_x m$
6. $Log_x y \times Log_y x = 1$
7. Log of 0 or negative numbers are not defined.

Before the computing devices became popular, logarithm were used to simplify the calculations. In statistics, Log to the base 10 is generally considered.

Reading log tables:

X	Log X
0.5	$\bar{1}.6990$
8	0.9031
57	1.7559
128	2.1072
1361	3.1338

Sample reading of table: value 1361

Leaving aside the Unit place, the number of remaining places is counted and written (i.e. 1361 has Tens, Hundreds and Thousand place, hence 3 is written). The value of 1361 is checked in the log table—first two digits are searched in the first column containing series starting from 10. After finding the first two digit, i.e. 13, the row is checked for next digit, i.e. 6 in the first set of columns marked 0–9. The value shown there is 1335. In the same row of 13, thereafter last digit, i.e. 1 is marked on the second set of column's marked 1–9 and its value noted, i.e. 3 in this case which is added to the earlier value, i.e. 1335, making it 1338. Hence Log 1361 = 3.1338. In case of values between 0 and 1, the notion of putting a bar over the digits is used, as shown above for 0.5. In case the value had two decimal values, say 0.25 the digit would have been 2 instead of 1 with a bar over it.

Antilog: Antilog is reverse of Log. For example, for Antilog of 2.1072, the values after decimal point are found in the table. Similar to log, the digit 0.10 is found first on the first column, then in its row the value under 7 is tracked (1279 in this case) which is added to the value under 2 (1 in this case) in the second set of column values. This gives the number as 1279 +1 = 1280. Considering the value before decimal was 2, therefore in the value 1280, the decimal is put after 3 places, i.e. 128.0 = 128.

THE EXPONENTIAL FUNCTION 'e'

The number

$$e = 1 + \frac{1}{1!} + \frac{1}{2!} + \frac{1}{3!} + \frac{1}{4!} .... = 2 < e < 3$$

$$= \text{Approximated to } 2.718281828$$

And,

$$e^x = 1 + \frac{x}{1!} + \frac{x^2}{2!} + \frac{x^3}{3!} + \frac{x^4}{4!} ....$$

is called the exponential function 'e'.

RATIO

Ratio is the relation one quantity has with other of the same kind and is measured by the fraction which the first unit is of the second. Assume two values and, then their Ratio is denoted by '$x:y$' (read as x is to y) and calculated as $\frac{x}{y}$ or $x \div y$. Therefore ratio signifies what multiple or parts of multiple, is the first quantity in terms of second quantity.

The numerator is called the 'antecedent' and denominator as 'consequent'.

From the above, it can be noticed that the Ratio does not change, if both the numbers are divided or multiplied by a common number, i.e. Dividing $x : y$ by a constant number c.

$$= \frac{x}{c} : \frac{y}{c} = \frac{\frac{x}{c}}{\frac{y}{c}} = \frac{x}{c} \times \frac{c}{y} = \frac{x}{y} = x : y$$

Similarly it can be done for multiplication.

As from the definition, 'same kind' implies, unit of measurement has to be same for arriving at proper ratios. If two different ratios are stated in two different units, then these can be combined by compounding the same. Assume two ratio's, $x:y$ and $m:n$, these can be combined by compounding, i.e. $\frac{x \times m}{y \times n}$ and arriving at the new compound ratio.

Important properties of ratios:

1. Invertendo if $\frac{x}{y} = \frac{m}{n}$ then $\frac{y}{x} = \frac{n}{m}$
2. Alterendo if $\frac{x}{y} = \frac{m}{n}$ then $\frac{x}{m} = \frac{y}{n}$
3. Componendo if $\frac{x}{y} = \frac{m}{n}$ then $\frac{x+y}{y} = \frac{m+n}{n}$
4. Dividendo if $\frac{x}{y} = \frac{m}{n}$ then $\frac{x-y}{y} = \frac{m-n}{n}$
5. Componendo Dividendo if $\frac{x}{y} = \frac{m}{n}$ then $\frac{x+y}{x-y} = \frac{m+n}{m-n}$

PROPORTION

When the ratio of two numbers is equal to the ratio of two other numbers, then all these four numbers are stated to be in proportion, i.e. if $x:y = m:n$ (primarily denoted as $x:y::m:n$). When such numbers are in proportion then:

'Product of extremes' $(x \times n)$ = 'Product of means' $(y \times m)$.

Continued Proportion: Instead of four different terms, if there are three numbers such that ratio of 1st and 2nd, is equal to the ratio of 2nd and 3rd, then these numbers are said to be in continued proportion, i.e. if $x:y::y:z$, then x, y, z, are in continued proportion.

FACTORIAL

The symbol, ! denotes a mathematical operator called Factorial. Factorial of the number 'n' is given as the product of all positive integers less than or equal to 'n,' where n is a non-negative integar. From definition:

$$n! = n \times (n-1) \times (n-2) \times (n-3) \times \ldots \times 3 \times 2 \times 1$$

For example, $4! = 4 \times 3 \times 2 \times 1 = 24$

Factorial of 0 is 1 i.e. $0! = 1$

Calculations of factorial, while calculating combination can be simplified, as stated below:

$x!$ can also be stated as $n(n-1)(n-2)!$ or $n(n-1)(n-2)(n-3)!$ or $n(n-1)!$ or

Therefore in combination equation, if $n = 7$ and $r = 5$, then

$$^nC_r = \frac{n!}{r!(n-r)!} \Rightarrow {}^7C_5 = \frac{7!}{5!(7-5)!} = \frac{7 \times 6 \times 5}{5! \times (2)!} = \frac{7 \times 6}{2} = 21$$

i.e. $n!$ should be deduced to the level of either $r!$ or $(n-r)!$ (whichever is greater), and then cancelled out.

PERMUTATIONS AND COMBINATIONS

This topic is highly relevant in Probability and Probability Distributions like Binomial, etc and therefore would require slightly more detailed approach.

Suppose Set A = (Four different chairs marked y, m, c, k). Out of these 4 chairs, two chairs are to be selected to be placed on the dais, at locations marked A and B. It can be deduced that these two chairs can be a combination from any of these 6 possible options - (y, m), (y, c), (y, k), (m, c), (m, k), (c, k) - without any restriction about the order of selection. Selection of chair 'y' first followed by 'm' i.e. (y, m) is considered same as selecting 'm' first followed by 'y' i.e. (m, y).

Now, consider again the situation but with the sequence or arrangement is also a criteria, i.e. combination (y, m) ≠ (m, y). In such a situation the possible options are (y, m), (m, y), (y, c), (c, y), (y, k), (k, y), (m, c), (c, m), (m, k), (k, m), (c, k), (k, c) = 12 different options.

The above is a simple example, but such counting of options is not possible when the number of items in the set increases, number of rules for arrangement increase, etc. For example, if the number of chairs is increased to 200 divided into 10 different types of chairs and the selection is to be done for 10 slots, then it is not easy to count the number of options possible. If there are further restrictions or conditions like if the first chair if a specific type, then what are different options etc, complicate the counting of options further. 'Permutation' and 'Combinations' are counting rules which help simplify these.

Permutation

All possible arrangements which can be achieved for a set of things and where the order of arrangement is important too, i.e. ordered arrangement of a collection of objects is called Permutation.

If 'n' is the total number of objects, out of which 'r' object needs to be arranged/ selected, then the total number of options possible (permutations possible) for arranging/ selecting these is given as:

$$^nP_r = \frac{n!}{(n-r)!}$$

Where ^nP_r denotes permutation of n objects taken r at a time
Using this formula in the earlier example, n = 4, r = 2, i.e.

$$^4P_2 = \frac{4!}{(4-2)!} \quad \frac{4 \times 3 \times 2 \times 1}{2 \times 1} = 12$$

From above it follows that if there are 'n' objects and all of them have to be arranged/selected, then it is possible to arrange/select them in (using the above permutation formula), i.e a case where 'r' equals 'n',

$$^nP_n = \frac{n!}{(n-n)!} = \frac{n!}{0!} = \frac{n!}{1} = n!$$

The number of permutations of 'n' objects, out of which 'r' objects occur/selected and with the freedom of repetition being permitted, is given as n^r. This is a case where 'n' remains same even after selection (selection is being used as a generic term and represents occurrence of an event). Consider a box of 3 balls marked A, B, C out of which two need to be randomly picked, but with replacement. The first pick can be any of these, i.e. 3 possibilities. After first pick, the ball is placed back in the box and the experiment is repeated. Again there are 3 possibilities. Therefore total possibilities are $3 \times 3 = 9$, and from the formula $n^r = 3^2 = 9$.

If one thing can be done in 'x' different ways and other in 'y' different ways, then both the things can be done in "$x \times y$" ways. For example, if a dish can be made by 3 different recipes and 3 different types of heating mechanism, then it is deduced that the dish can be prepared in $3 \times 3 = 9$ different ways.

Combination

A Combination is the possible number of ways in which 'r' objects can be selected from total number of objects 'n' and where the order of arrangement or selection or occurrence is not important.

The number of combinations that are possible for 'r' objects selected out of total objects 'n' is given as:

$$^nC_r = \frac{n!}{r!(n-r)!}$$

Combination ^nC_r can also be stated as $\frac{^nP_r}{r!}$

From this formula and the concept, the following can be inferred:

$$^nC_r = {^nC_{n-r}}$$

The number of combinations, of 'n' objects out of which 'r' are to be selected, subject to the condition that 'p' particular objects will always get selected, will be given as $^{n-p}C_{r-p}$ i.e. the selection process becomes applicable to $(n - p)$ objects for selecting the remaining $(r - p)$ objects.

The number of combinations of 'n' objects out of which 'r' are to be selected, subject to the condition that 'p' particular objects will not get selected/occur, is given as $^{n-p}C_r$.

Section 3

Describing the Data

- Frequency Distribution
- Measures of Central Tendency
- Measures of Dispersion (or Variation)
- Skewness, Moments and Kurtosis

Section 3

Describing the Data

CHAPTER 7

Frequency Distribution

Statistics, broadly speaking, is a tool to guide collection of data and its analysis to derive meaningful information which can aid in decision making. Most research studies generate data, which needs to be properly classified, tabulated and analyzed. It is possible to sort data, like arranging it in an increasing or decreasing order of magnitude, called arrays, but sorting only organizes and does not reduce the volume of data. Hence, classification of data assumes significance in analysis of data. The discussions in this chapter must have been used in different forms by the readers without really having a formal introduction to the basic theories involved. The applications are more important than the background.

CLASSIFICATION OF DATA

Classification of data, after its collection and editing involves 'grouping' of related entries into 'classes'. The process of classification summarizes the large amount of data into a small number of groups, which helps in further processing. The classification can be on any parameter as required by research objectives. Common classifications are based on time, geography, attributes, defects, etc. Some of the advantages of classifying the data are:
- It provides initial summarization and compresses the data volume.
- It allows preliminary comparison among the various groups or classes.
- It helps to take a bird's view of the collected data.
- It forms the basis for further statistical analysis of the data.

Consider the 'Raw Data' of the weights (in kg) of adult male patients admitted in a hospital, at a given point of time **(Table 7.1)**.

This given raw data contains values, but does not provide any valuable information. Hence, it needs to be condensed for further analysis. As a first step, to facilitate counting of data, 'Tally Marks' are used. Tally Marks are the vertical bars (|) with each bar signifying an occurrence. To simplify the counting of the tally marks, after every four occurrences, instead of putting a fifth bar as a vertical bar, the fifth bar is drawn as a diagonal line over the first four bars drawn previously. One such bunch, which is visually read easily, signifies five occurrences. Frequency or 'Number of Occurrences' is then tabulated as shown in **Table 7.2**.

Section 3: Describing the Data

Table 7.1 Raw data (Admitted patients' weight in kg)

68	52	57	69	74	73	85	49	57	78
71	81	98	73	68	67	78	45	54	59
74	73	77	66	49	50	54	66	73	80
58	61	60	70	66	67	74	73	80	95
78	73	62	64	69	59	58	74	75	77
82	84	69	67	76	74	79	83	79	62
57	60	61	71	69	68	75	77	71	69
70	72	40	56	71	66	49	53	59	64

Table 7.2 Data given in table 1 converted in a frequency table by use of tally marks

Weight	Tally marks	No. of patients
40	|	1
45	|	1
49	|||	3
50	|	1
52	|	1
53	|	1
54	||	2
56	|	1
57	|||	3
58	||	2
59	|||	3
60	||	2
61	||	2
62	||	2
64	||	2
66	||||	4
67	|||	3
68	|||	3
69						5
70	||	2
71	||||	4
72	|	1
73	|||| |	6
74						5
75	||	2
76	|	1
77	|||	3
78	|||	3
79	||	2
80	||	2
81	|	1
82	|	1
83	|	1
84	|	1
85	|	1
95	|	1
98	|	1
Total		**80**

Through the use of Tally Marks, the raw data representation has improved and is better organized, but it is still large and cannot be easily utilized. To achieve simplification, the range of data is divided into suitable number of Class Intervals and their frequencies are written alongside, as shown in **Table 7.3**.

In this operation, the weights have been grouped together and their corresponding frequencies also added. In **Table 7.3**, the weight is called the 'variable' (x), and the number of patients is the frequency (f). A Variable is one which 'varies' and a frequency is 'how frequently one occurs.' *This representation of the variable along with its frequency is called "Frequency Distribution".*

Defining, a 'Frequency Distribution' or a 'Frequency Table' is a form of representing a data, which has been classified on the basis of a variable and its associated frequency. The 'Variable' refers to the attribute that varies in quantum or magnitude in a frequency distribution. Frequency distribution can also be described as 'how the frequencies are distributed over the various values of the variable'.

VARIABLES: DISCRETE AND CONTINUOUS

The variable (x) can either be 'Discrete' or 'Continuous'.

Discrete Variable

A 'Discrete Variable' is a variable, which varies in 'steps' and does not take fractional values in between. For example, number of patients in a hospital, number of books in library. It is not possible to have number of people in a fraction count.

Continuous Variable

A 'Continuous Variable' is capable of taking fractional values within the range of possibility. For example, height of a person which is written as 172.2 cm.

FREQUENCY DISTRIBUTIONS: DISCRETE AND CONTINUOUS

On basis of the above, a 'Discrete Frequency Distribution', is defined as the Frequency table of a discrete variable (as shown in **Table 7.4**), and similarly 'Continuous Frequency Distribution' is the frequency table of a continuous variable **(Table 7.5)**.

The frequency table is read as: 40 clinics have 1 doctor, 29 clinics have 2 doctors and so on.

Table 7.3 Frequency table with classes (Grouping)

Weight in kg (Variable x)	No. of patients (Frequency f)
40–50	5
50–60	14
60–70	23
70–80	29
80–90	7
90–100	2
Total	80

Table 7.4 Discrete frequency distribution

No. of doctors in a clinic (x)	No. of clinics in the region (f)
1	40
2	29
3	33
4	15
5	3
Total	**120**

Table 7.5 Continuous frequency distribution

No. of doctors in a hospital (x)	No. of hospitals (f)
5–15	3
15–25	6
25–35	8
35–45	12
45–55	6
Total	**35**

This is read as: There are 3 hospitals which have between 5 and 15 doctors, 6 hospitals where the number of doctors is between 15 and 25, and so on.

Discrete frequency distributions are simpler in form, as they have variable and its associated frequencies, but continuous frequency distributions require more elaboration.

Continuous Frequency Distribution

These distributions are most popular and extensively used. The notations used in continuous frequency distributions are explained through **Table 7.6**.

Class: The grouped values of variable x are called the class. 40–50, 50–60,..90–100 are all classes.

Range: It is the difference between the smallest and the largest observation in the data.

Class Limits: The lowest and the highest value that can be included in a class is called its 'Class Limits'. In **Table 7.6**, for Class 40–50, the class limits are $40 \leq x < 50$; for Class 50–60 the class limits are $50 \leq x < 60$. Note that variable $x = 50$, is getting included in the Class 50–60, i.e. towards the lower limit.

Class interval: (Denoted as '*i*'): The difference between Upper and Lower Limits of the Class is known as Class Interval. For example, the class intervals in **Table 7.6** are 10 for all classes (50 minus 40, 60 minus 50, etc.).

Class frequency: The frequency associated with the corresponding class is known as the Class Frequency for that class. For example, the Class frequency for Class 80–90 is 7 and for Class 90–100 is 2 and so on.

Class Mid Point: (Also called Class Mark)—The mid-point of a Class is called Class Mark. It is calculated as:

Chapter 7: Frequency Distribution

Table 7.6 Continuous frequency distribution-notations

Weight in kg (Variable 'x')	Class marks (Class mid point)	No. of patients (Frequency 'f')
40–50	45	5
50–60	55	14
60–70	65	23
70–80	75	29
80–90	85	7
90–100	95	2
Total		**80**

$$\text{Class Mark} = \frac{(\text{Class upper limit} + \text{Class lower limit})}{2}$$

For example, Class Mark for Class 70–80 = (70 + 80)/2 = 75

Class Mark is important as for future data analysis and operations, it is considered as the representative value of its class.

Exclusive and Inclusive Method of Class Intervals

Class intervals can be designed in two ways:
1. Exclusive method,
2. Inclusive method.

In Exclusive Method, the class intervals are so established that the upper limit of a class becomes the lower limit of the succeeding class (as represented in **Table 7.5** or **Table 7.6**). In Inclusive Method, the upper limit of the class interval is included in the class itself. For example, Classes like 40–49, 50–59, 60–69, etc. would include an observation value of 49 in the class 40–49. Such Class Intervals are suitable for discrete variables, but in case of continuous variables, inclusive method would not be applicable because it would not be possible to include a value like 49.5 in any class interval, as it does not fall in either of the classes 40–49 and 50–59. Even while using the inclusive method for discrete variable, it has to be ensured that none of the frequency lies in between the lower limit of a class and the upper limit of the preceding class.

In spite of the significance of data classification, there do not exist any standard rules or procedures on classifying the data. Albeit the following points should be duly considered while classifying the data:
- The classes need to be clearly defined, be exhaustive and mutually exclusive so that all given values of observations (frequencies) get included in one and only one of the classes.
- Unless absolutely strong reasons are present, the classes should have equal width, i.e. class interval should be equal. Equal width of the class allows comparison of frequencies, whereas in case of unequal widths, the frequencies are not truly comparable. For situations where classification has unequal class width, comparison among the classes can be done by dividing the frequency value by

its corresponding width of the class-interval. The ratios, so generated are called 'Frequency Densities,' which can then be compared.
- Exclusive Method is preferred as it provides continuity of data. In case inclusive method has been used for Classes, a Correction Factor is applied to the class intervals to make them continuous. The Correction Factor is calculated as:

$$\text{Correction Factor} = \frac{\text{(Lower limit of 2nd class − upper limit of 1st class)}}{2}$$

The Correction factor so calculated is then added to all upper limits and subtracted from lower limits of the Class, to get a Continuous Frequency Distribution.

For Example: If the Classes are mentioned as 40–49, 50–59, 60–69… these can be converted to Continuous Classes by applying the Correction Factor.

Correction Factor = (50 − 49)/2 = ½, i.e. 0.5

Hence, the Corrected Classes would become as 39.5–49.5, 49.5–59.5, 59.5–69.5…
- Open ended classes should be avoided as indeterminate classes like, say, less than 40 (<40) or greater than 100 (>100) will not allow calculation of Class Mid Points (Class Marks) and thereby affect further computations.
- The number of classes to be formed should neither be very less nor too many. It is suggested to have number of classes between 5 and 25. Though more classes can still be used, depending upon the number of observations/frequencies, but having less than 5 classes may not reveal sufficient details and information about the population. Ultimately, the number of classes depends on the nature of data and objectives of the research. Guidance about the number of classes can be taken from Sturges' rule.

$$\text{Sturges' rule } k = 1 + 3.322 \log_{10} N$$
(where, k = No. of classes, and
N = No. of observations or Total Frequency)

For example, for $N = 10$, as per this rule $k = 1+3.322 = 4.322$, i.e. 4 Classes, (log 10 = 1, from Log tables) and whereas for $N = 100,000$, $k = 17.61$ ($k = 1 + 3.322 * \log 100,000 = 1 + 3.322 * 5 = 17.61$), i.e. 18 classes. The derived value of 'k' should be used as a suggested number and not a sacrosanct number for determining number of classes. With advent of computers having immense computing power, any number of classes can be easily handled, hence it is suggested that number of classes should be guided by the study objectives, where 'k' value can be considered as the suggested minimum number of classes.
- *Size of class interval*: Having established the number of classes for the given data set, the size of Class Interval i can be calculated from the formula given further.

$$i = \frac{L-S}{k}$$

Where,
i = Class Interval, L = Largest Observation
S = Smallest Observation and k = Number of Classes

As (L minus S) is called the 'Range' of data, and formula for 'k' already known, the above formula, for calculating the magnitude or size of the class interval, can be rewritten as:

$$\text{Size of Interval} = i \frac{\text{Range}}{1 + 3.322 \log N}$$

While determining the class interval and number of classes, using these formulas, the resultant value may be a fraction or odd intervals. Suitable approximation and adjustments should be carried out for such cases.

RELATIVE FREQUENCY DISTRIBUTION

A Relative Frequency Distribution is a frequency distribution table, where the class frequencies have been converted to Relative Class Frequencies, to show the percentage of the total number of observations in each class. To convert the frequency to relative frequency, divide it by Total Frequency (N) and multiply by 100 to get the percentage value. For example, reproducing the previous Frequency Distribution Table **(Table 7.5)** and converting it to a Relative Frequency Distribution Table, as shown in **Table 7.7**.

The sum of all the relative frequencies would be equal to 100, if the relative frequencies are shown as a percentage (i.e. for class 40–50, 5÷80 * 100 = 6.25%) and 1, if they are shown as proportion value (i.e. 5÷80 = 0.0625).

BIVARIATE OR TWO WAY FREQUENCY DISTRIBUTION

In the previous examples so far only one variable along with its associated frequency has been considered. Such frequency Distributions are also called univariate frequency distribution. Contrary to this, when two variables are involved in a frequency distribution, it is known as Bivariate Frequency Distribution. Consider the Raw Data having two variable i.e. Height (in cm) and Weight (in kg) of the players of a team **(Table 7.8)**.

The two-way frequency distribution table for such a data would be as shown in **Table 7.9**.

The considerations applicable for construction of a univariate frequency distribution are also applicable to the bivariate frequency distribution. The two variables can have, and generally have different number of classes and class intervals.

Table 7.7 Relative frequency distribution table

Weight in kg (Variable 'x')	Class marks (Class mid point i)	No. of patients (Frequency 'f')	Relative frequency (%)
40–50	45	5	6.25
50–60	55	14	17.50
60–70	65	23	28.75
70–80	75	29	36.25
80–90	85	7	8.75
90–100	95	2	2.50
Total		N = Σf = 80	Σ = 100%

Table 7.8 Two variable data sample

Player no.	Height (cm)	Weight (kg)
1.	171	59
2.	180	75
3.	165	63
4.	166	69
5.	155	72
6.	159	66
7.	165	71
8.	170	69
9.	177	72
10.	184	72
11.	159	67
12.	162	62
13.	164	70
14.	166	73
15.	170	69

Table 7.9 Bivariate frequency table

Weight (kg)	Height (cm)				Total
	150–160	160–170	170–180	180–190	
55–60	0	0	1	0	1
60–65	0	2	0	0	2
65–70	2	1	1	0	4
70–75	1	3	2	1	7
75–80	0	0	0	1	1
Total	3	6	4	2	N = 15

MARGINAL DISTRIBUTION

The frequency distribution of the values of one variable, along with its frequency totals is called the Marginal Distribution of that variable. From **Table 7.9**, the last column giving the total along with variable—Weight is the marginal distribution of weight in kg, and similarly the first row along with last row depict the marginal distribution of height in cm.

Understanding of frequency distribution is critical and important as it has relevance in all statistical analysis. Classification of data prepares the observations for further analysis through the various tools and techniques of Statistics.

Once the data has been classified, the next step is to understand and know if there is any single value which can be taken as a representative of the entire frequency distribution. What are the different ways to calculate this single entity and their relative advantages and disadvantages? These questions become the subject of the next chapter.

CHAPTER 8

Measures of Central Tendency

Measures of central tendency are also referred to as measures of central value or measures of location. All these terms are interchangeably used without any loss of meaning.

CENTRAL VALUE CONCEPT

A central value of the data set refers to the 'single value' which is considered as a representative of the entire data set; and the value around which all the observations or values are supposedly located. For easier understanding, it can taken alike the concept of *center of gravity* where though the mass of an object is spread throughout, but it is assumed that the entire mass of the object lies at it single point known as its centre of gravity. Similarly, the data values are spread and located at various points, but a single value is calculated, which is considered as a representative of all the other values. This is useful when comparisons have to be made and communicated.

MEASURES OF CENTRAL TENDENCY

The statistical tools which calculate or locate this central value are collectively called as *Measures of Central Tendency*.

In statistical terms, central value is also known as *Average*, (a term not to be confused with the generic term 'average' used in daily lives).

Averages, as defined by Professor Bowley "are statistical constants which enable us to comprehend in a single effort the significance of whole". According to AE Waugh, "an average is a single value selected from a group of values to represent them in some way—a value which is supposed to stand for whole group, of which it is a part, as typical of all the values in the group".

Statistically, the word 'Average' is the central value of the data, which has been derived by using any of the various different methods of measuring the central tendency. These various methods include Arithmetic Mean, Median, Mode, etc. The generic term 'average' used very commonly is statistically called *arithmetic mean* or simply mean. This distinction between *arithmetic mean* and *averages*, is statistically necessary to be maintained in all terms and discussions.

Significance

The measures of central tendency apart from providing a central value, also facilitate comparison between two or more relevant data sets, either at a point of time or in a period of time. For example, average increase in salaries of the employees in different years, average interest rates in selected years, etc. While comparing and drawing inference from the data, the multiplicity of forces affecting the data at various points of time should be considered and also same measure of central value should be compared. It is not appropriate to measure the arithmetic mean of one set of values with median or mode or any other measure of the other set of values, as these are not comparable. Each measure of central tendency has its own significance, different method of calculation and applicability. Ideally, one data set should have one central value, but none of the various measures are able to explain the distribution in completeness.

Ideal Measure of Central Tendency

Theoretically, an ideal measure of central tendency should have the following characteristics:
- Easily understandable.
- Simplistic computation.
- Defined rigidly through a formula and have only one interpretation.
- Should take into account all observations of the data set.
- Should be suitable for further statistical treatment, i.e. it should be possible to calculate composite averages of multiple data sets by using their respective averages and sizes.
- Sampling stability should be there. The value should be affected very little or not at all due to fluctuations of sampling. The averages calculated by drawing different samples from the same population should have values close to the each other and also close to the central value of the whole population or data set.
- The ideal measure of central tendency should not be largely affected by extreme values in the data, as extreme values tend to distort the data.

TYPES OF MEASURES OF CENTRAL TENDENCY

The different methods of measuring the central tendency are:
- Arithmetic mean
- Median, quartile and percentile
- Mode
- Geometric mean
- Harmonic mean
- Ogive.

Arithmetic Mean ($\bar{x}$)

Denoted as $\bar{x}$, it is calculated by dividing the sum of all the observations by the number of the observations. Though it is usually called as average, but due to differences mentioned earlier, in statistical parlance, it is called 'Arithmetic Mean' or simply the *Mean*. Arithmetic Mean can be either "Simple Arithmetic Mean" or "Weighted Arithmetic Mean."

Simple arithmetic mean is used when only the individual observations are provided and there are no corresponding frequencies for the variable. Weighted arithmetic mean is used, if a variable and its associated frequencies are provided. These corresponding frequencies are called the 'weights'. More the occurrence of the variable, higher is its frequency and higher would be its influence on the central value. If weights are equal for all the variables, then the calculated values of simple arithmetic mean and weighted arithmetic mean would be equal. If greater weights are assigned to greater values, then simple arithmetic mean would be less than weighted arithmetic mean, and if greater weights are assigned to smaller values, then simple arithmetic mean would be greater than weighted arithmetic mean.

Weighted mean versus simple mean: Wherever frequencies or weights are provided, weighted mean is a necessity as otherwise the mean derived from simple mean formula for this type of data would be erroneous and misleading.

The arithmetic mean of data values is calculated as per given formulas:

Simple Arithmetic Mean

Simple arithmetic mean $\bar{x} = \dfrac{x_1 + x_2 + x_3 + \ldots + x_n}{n} = \dfrac{\Sigma x}{n}$

Where, $\bar{x}$ (Read as x bar) = Arithmetic mean
$x_1, x_2, x_3, \ldots, x_n$ = Various values of the variable 'x'
Σx (Read as Sigma x) = Sum of all the observations or values of 'x'
n = Total number of observations

Weighted Arithmetic Mean

Weighted arithmetic mean $\bar{x} = \dfrac{f_1 x_1 + f_2 x_2 + f_3 x_3 + \ldots + f_n x_n}{f_1 + f_2 + f_3 + \ldots + f_n} = \dfrac{\Sigma fx}{\Sigma f}$

Where
$x_1, x_2, x_3, \ldots, x_n$ = Various values of variable x
$f_1, f_2, f_3, \ldots, f_n$ = Corresponding frequencies of the variable x
$f_1 x_1, f_2 x_2, f_3 x_3, \ldots$ = Product of variable value x and its associated frequency f
Σx = Sum of all observation or values of x
Σfx = Sum of all the products of variables x and their frequencies f i.e. $f_1 x_1$.
Σf = Sum of frequencies = N (Total number of observation).

For continuous or grouped frequency distributions, x is taken as the class mid point/class mark of the corresponding class.

Example: Simple arithmetic mean

The number of medical students, specializing in different branches of medicine are: 26, 29, 30, 35 and 30. Calculate arithmetic mean (AM) to infer the mean number of students in each branch.

Solution: The variable is the number of students and number of observations are $n = 5$, hence applying the formula for simple arithmetic mean.

$$\bar{x} = \frac{26+29+30+35+30}{5} = \frac{150}{5} = 30$$

The arithmetic mean of the number of students in each branch of medicine is 30.

Example: Weighted arithmetic mean of a discrete frequency distribution.

Calculate arithmetic mean for the following data:

x: 1 2 3 4 5
f: 10 11 12 13 14

Solution: The variable (x) and their associated frequencies (f) are given. The calculation table for this data is:

Variable (x)	Frequency (f)	Variable (x) frequency (fx)
1	10	10
2	11	22
3	12	36
4	13	52
5	14	70
Total	$\Sigma f = 60$	$\Sigma fx = 190$

$$\text{Mean } \bar{x} = \frac{\Sigma fx}{\Sigma f} = \frac{190}{60} = 3.17$$

Hence, the arithmetic mean of the above discrete frequency distribution is 3.17

Example: Weighted arithmetic mean of a continuous frequency distribution.

Calculate the arithmetic mean of the following sample data pertaining to age of the patients when they were first admitted in a hospital for treatment.

Age of the patient (in years)	0–10	10–20	20–30	30–40	40–50
Number of patients	6	10	12	24	4

Solution: The calculation table for the data is:

Patient's age (Years)	Class mid point (x)	Number of patients (f)	fx
0–10	5	6	30
10–20	15	10	150
20–30	25	12	300
30–40	35	24	840
40–50	45	4	180
Total		$\Sigma f = 56$	$\Sigma fx = 1500$

$$\text{Mean } \bar{x} = \frac{\Sigma fx}{\Sigma f} = \frac{1500}{56} = 26.79$$

Hence, the mean age of the patient when he/she was admitted to the hospital for the first time is 26.79 years.

Note: There are short cut methods of calculating mean too. These methods intend to make the calculations easier by calculating the deviations from an 'assumed mean' and then adjusting in the final stage the assumption. The same are not considered here as these short cuts methods were commonly used before the computers became popular. As today, the entire computing of data is done on computers, knowing these shortcuts methods are not necessary.

Important Properties of Mean

1. The sum of deviations of the observations from their mean is zero. Mathematically, Sum of deviations form

$$\bar{x} = (x_1 - \bar{x}) + (x_2 - \bar{x}) + \ldots + (x_n - \bar{x}) = \Sigma(x_i - \bar{x}) = 0$$

2. The sum of squares of deviations of a set of values is at minimum, when deviations are considered from their mean. This property forms the basis of the regression analysis dealt later.
3. Combined mean of two or more sets of data series or values can be computed from the means of the individual data series or values, by using the formula:

$$\text{Combined mean } \bar{x} = \frac{n_1\bar{x}_1 + n_2\bar{x}_2 + n_3\bar{x}_3 + \ldots + n_n\bar{x}_n}{n_1 + n_2 + n_3 + \ldots + n_n} = \frac{\Sigma n_i \bar{x}_i}{\Sigma n_i}$$

Where, $n_1, n_2, \ldots n_n$ = Number of observations or total frequency in each set of values, and $\bar{x}_1, \bar{x}_2, \bar{x}_3 \ldots \bar{x}_n$ = the corresponding mean of the each set of values.

For example, if three data series have sizes (number of observations) of 12, 20 and 30, and their respective means are 20, 32 and 44, then the combined mean of the three series collectively would be:

$$\text{Combined mean } \bar{x} = \frac{(12 \times 0) + (20 \times 32) + (30 \times 44)}{12 + 20 + 30} = \frac{2200}{62} = 35.48$$

Merits and Demerits of Mean

In relation with the requirements of an ideal measure of central tendency, mean fulfills certain criteria which are its merits.

The notable merits of mean include its rigid definition, relative ease to understand and compute, being based on all observations of the data set, suitability to apply further statistical tools to it like combined mean, and being relatively less impacted by sampling fluctuations.

Despite the merits, mean has many demerits which limit its usage under certain conditions. The main drawback of mean is that its gets significantly affected by extreme values and in such situations its results are distorted to that extent. Arithmetic mean provides a good measure of central value when the distribution is symmetrical with a normal curve or is evenly distributed. For skewed distributions or asymmetrical distributions, the mean value does not prove to a suitable measure of location.

Other demerits of mean include the inability to calculate it in case any observation is missing, or in a grouped frequency distribution, the starting class or last class is open ended.

MEDIAN (M_d)

Median is denoted as M_d and, as a measure of central tendency splits the distribution or data into two equal halves, i.e. median is the value which has equal number of observations on both sides (higher and lower) of it. Thus it is a *positional average* and also referred to as 50th 'percentile.'

Calculation of Median

a. *For data consisting of individual values or items (without frequencies)*
   Consider the following observations: 10, 11, 15, 25, 12, 17, 16
   Step 1: Arrange the values in ascending or descending order.
   Step 2: Median M_d is the central value, i.e. value which lies in middle of the series in terms of its position.
   Solving the above example: *Step 1*: 10, 11, 12, 15, 16, 17, 25
   Total number of observations are seven, hence central value would the 4th observation, i.e. median $M_d = 15$.
   The above example has odd number of observations, therefore, the middle value is clearly defined and known. In case, the number of observations is even, then there exist two middle values. In such cases, median is obtained by calculating mean of the two middle terms, i.e.

$$\text{Median } M_d = \text{Mean } (\bar{x}) \text{ of the values above}$$
$$\text{and below the } \left(\frac{N+1}{2}\right) \text{th term}$$

Theoretically, in even numbered series any value between the two middle points can be taken as the median, but conveniently we take the arithmetic mean of the two middle values.

Example: Calculating the median for data having even number of entries.
   Calculate median for the following data: 10, 12, 17, 18, 15, 20, 14, 19.
   Step 1: Arranging in ascending order the values: 10, 12, 14, 15, 17, 18, 19, 20.
   Step 2: Count the number of entries, i.e. 8 which is even.
   Step 3: Calculating the middle location, i.e. (8+1)/2 = 4.5; hence mean of 4th and 5th value needs to be calculated. Therefore

$$\text{Median } M_d = \frac{(15+17)}{2} = 16$$

b. *Calculation of median for discrete frequency distribution.*
   For discrete frequency distributions, the median is calculated by considering the cumulative frequencies. If N = Total number of observations or total frequency, find $\frac{N}{2}$. Track the cumulative frequency just greater than the $\frac{N}{2}$; the value of the variable corresponding to this value of cumulative frequency would be the median of the discrete frequency distribution.

Example: Calculate the median for the following hypothetical data:

| Height of newborn child (cm) | : | 45 | 46 | 42 | 51 | 54 | 56 |
| Number of newborns | : | 110 | 225 | 185 | 180 | 235 | 265 |

Solution: Arranging the data in ascending order and presenting it in a tabular form.

Height of newborn child (cm) (x)	Number of newborns frequency (f)	Cumulative frequency (cf)
42	185	185
45	110	295
46	225	520
51	180	700
54	235	935
56	265	1200
	$\Sigma f = 1200$	

The middle cumulative frequency = $\dfrac{N}{2} = \dfrac{1200}{2} = 600$

Just higher cumulative frequency = 700 and variable value corresponding to 700 is 51. Hence, median = 51.

c. *Calculating median of a continuous frequency distribution*

In continuous frequency distribution, the individual observations lose their individuality, and become a part of a class, but the objective is not to find a specific item but to find a point which divides the distribution into two halves. Hence, for continuous distribution instead of $\left(\dfrac{N+1}{2}\right)$ th term, $\left(\dfrac{N}{2}\right)$ th term is used. The following steps detail the procedure for calculation of median of a continuous frequency distribution.

Step 1: Arrange the continuous frequency distribution in ascending order.

Step 2: Calculate $\left(\dfrac{N}{2}\right)$, where N = Total frequency, and find the class corresponding to the cumulative frequency just greater than $\left(\dfrac{N}{2}\right)$. This identified class is called the median class.

Step 3: Apply the formula for calculating median which is given here:

$$\text{Median } M_d = l + \dfrac{i\left(\dfrac{n}{2} - c.f.\right)}{f}$$

Where l = Lower limit of the median class
 i = Class interval of the median class
 f = Frequency of the median class
 cf = Cumulative frequency of the class preceding the median class
 n = Total number of observations (Σf)

Example: Find the median age of the data here.

Age : 20–30 30–40 40–50 50–60 80–90 70–80 60–70
Frequency : 7 21 39 40 2 17 27

Solution: Arranging the data in ascending order of continuous distribution and representing in tabular form, as here:

Age (x)	Frequency (f)	Cumulative frequency (cf)
20–30	7	7
30–40	21	28
40–50	39	67
50–60	40	107
60–70	27	134
70–80	17	151
80–90	2	153
	$\Sigma f = 153$	

Calculating $\dfrac{n}{2} = \dfrac{153}{2} = 76.5$

Cumulative frequency just greater than 76.5 is 107 and the class corresponding to 107 is '50–60'. Hence, median class is class '50–60, i.e.

- l = Lower limit of the median class = 50
- i = Class interval of the median class = 60–50 = 10
- f = Frequency of the median class = 40
- cf = Cumulative frequency of class preceding the median class = 67
- n = Total number of observations (Σf) = 153

Applying the formula

$$\text{Median } M_d = 50 + \dfrac{10\left(\dfrac{153}{2} - 67\right)}{40}$$

$$= 50 + \dfrac{10(76.5 - 67)}{40} = 50 + 2.38 = 52.38$$

Hence, the median age is 52.38 years.

The same formula is applied even in the case of unequal class intervals being present in the distribution.

Advantages

The main advantages of using median as a measure of central value are:
- Effect of extreme values is not there as it is 'locational' in nature.
- Can be applied to distributions having open ended classes at start and at end.
- Can be applied to distributions having unequal class intervals. These first three points allow it to overcome the main demerits of mean.
- Like mean, it is rigidly defined and easy to comprehend and compute
- Can be calculated graphically through cumulative frequency curves

Due to its listed advantages, it is considered most suitable measure of central tendency for qualitative data, where measurement is not possible but ranking or scores are available.

Disadvantages

Median has disadvantages also, as listed here:
- It does not consider all observations, and considers only the location of the value and not the value itself.
- It cannot be determined exactly in cases of even numbered data or discrete distributions are to be assessed.
- Median is not suitable for further algebraic treatments as it is not possible to calculate the combined median of multiple series based on the sizes and median of those series.
- In relation with mean, median is more affected by sampling fluctuations.

QUARTILES, DECILES AND PERCENTILES

Quartiles, deciles and percentile, are closely related to median, and similar to median, these are other positional measures. Like the median, which splits the data into two halves, the values of the variable which divide the total frequency into four equal parts are called the quartile (denoted as Q). Similarly, deciles (denoted as D) divide the data in 10 equal parts, and percentile (denoted as P) divide it in 100 equal parts.

The terms percentile is mentioned regularly in the scores of entrance test or examinations. Suppose a candidate scores 490 in an examination (5 tests each of 100 marks) out of maximum 500 and his percentile is mentioned as 99 percentile. The following inferences can be drawn from this statement:
- The mean score of the candidate is 98 (490÷5).
- The percentage score of the candidate is 98% ((490 ÷ 500) × 100).
- Relative to other candidates, his score is higher than 99% of the candidates and lower than 1% of the candidates. This inference comes from his percentile score, as these positional measures split the data.

The three quartiles are denoted as Q_1, Q_2, and Q_3. The nine deciles are D_1, D_2, D_3,.... and D_9 and 99 percentiles are denoted as P_1, P_2, P_3.... P_{99}.

Quartiles split the data in four equal parts, i.e. four quarters hence the name quartile. Q_1 is the first quartile and implies that 1/4th of the total frequency is less than it and 3/4th values are greater than this value. Similarly at second quartile, i.e. two quarters on either side, or one half on each side. Hence, second quartile is same as the median value of the data. For a given data, series the following points or value would be common but may be referred by different measure, depending upon the measure under consideration.

First decile D_1 = 10th percentile P_{10}
First quartile Q_1 = 25th percentile P_{25}
Second quartile Q_2 = 50th percentile P_{50} = 5th Decile D_5 = Median M_d
Third quartile = 75th percentile P_{75}
This can be shown pictorially on a percentage line as in **Figure 8.1**.

Computing Quartile/Decile/Percentile (Fig. 8.1)

To compute these, similar methods are employed as used in median. The first step is to find the class containing the required quartile/ decile or percentile. This can be done as follows:

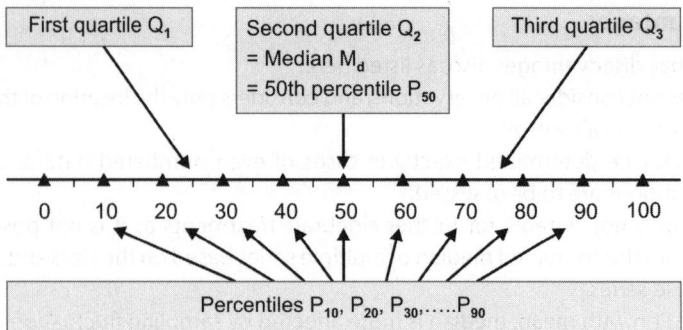

Fig. 8.1: Computation of quartile/decile/percentile

For $Q_1 = \dfrac{(n+1)}{4}$ th item for individual and discrete observations, and $\dfrac{n}{4}$ th item for continuous distribution.

For Q_2: As $Q_2 = M_d$, hence it is to be calculated exactly such as median and as discussed earlier.

For $Q_3 = \dfrac{3(n+1)}{4}$ th item for individual and discrete observations, and $\dfrac{3n}{4}$ th item for continuous distribution.

For $D_1 = \dfrac{(n+1)}{10}$ th item for individual and discrete observations, and $\dfrac{n}{10}$ th item for continuous distribution.

For $D_8 = \dfrac{8(n+1)}{10}$ th item for individual and discrete observations, and $\dfrac{8n}{10}$ th item for continuous distribution.

For $P_1 = \dfrac{(n+1)}{100}$ th item for individual and discrete observations, and $\dfrac{n}{100}$ th item for continuous distribution.

For $P_{90} = \dfrac{90(n+1)}{100}$ th item for individual and discrete observations, and $\dfrac{90n}{100}$ th item for continuous distribution.

From the above formulas, similarly the relevant 'class' for other deciles and percentile can also be computed.

Once the class containing the desired quartile/decile or percentile has been identified, the next step is to use the formula utilized for median, but with slight adjustment. The value for the $\dfrac{n}{2}$ is replaced suitably with the calculation used to find the median class, to derive the required formula, as shown here:

$$\text{Median } M_d = l + \dfrac{i\left(\dfrac{n}{2} - cf\right)}{f}$$

$$\text{Quartile } Q_1 = l + \frac{i\left(\frac{n}{4} - cf\right)}{f}$$

Where l = Lower limit of the median or quartile class
  i = Class interval of the median or quartile class
  f = Frequency of the median or quartile class
  cf = Cumulative frequency of the class preceding the median/quartile class
  n = Total number of observations (Σf)

MODE (M_0)

Denoted as M_0, mode of the series is the observation which occurs most frequently in the data, i.e. value which has the highest frequency or greatest frequency distribution in its surrounding values. The mode values is also referred to as modal value.

Calculating Mode

- *For individual values*: Mode is largest number of values among the given values.
- *For discrete frequency distribution*:

Mode of such a frequency distribution is the value of corresponding to the maximum frequency. For example, in the following discrete frequency distribution.

x	f
5	10
6	12
7	24
8	17
9	12

The maximum frequency is 24 which correspond to variable value = 7.
Hence, Mode $M_0 = 7$ (Sometimes also written as $\bar{x}_{mod}$)
In case, the maximum frequency is repeated or maximum frequency lies at extreme ends of the data or data has some irregularity, then it not possible to compute mode from the above method. In such cases, mode is calculated by 'Grouping Method,' which allows the greatest frequency density area of the data to show up.

For Continuous Frequency Distribution

$$\text{Mode } M_0 = l + \frac{i(f_1 - f_0)}{(2f_1 - f_0 - f_2)}$$

Where l = Lower limit of the modal class
  i = Class interval of the modal class
  f_1 = Frequency of the modal class
  f_0 = Frequency of the class preceding the modal class.
  f_2 = Frequency of the class succeeding the modal class.
(The modal class is the class, which corresponds to maximum frequency)

It is necessary to check that class intervals are equal before applying the mode formula, as otherwise for unequal class intervals the formula may provide erroneous results. Further the formula is not applicable, if the distribution is bimodal (having two modes) or multimodal (having multiple modes).

Relationship Among Mean, Median and Mode

For a continuous data distribution, the mode, mean and median can be summarized as: The mode is the value with highest frequency, the median splits the data in two equal halves each containing equal number of observation (positional), and mean splits the data in two equal halves based on the values or weights **(Fig. 8.2)**.

The empirical relationship among these three measures of central tendency, given by Karl Pearson, is:

$$\text{Mean} - \text{Median} = \frac{\text{Mean} - \text{Mode}}{3}$$

Or,

$$\text{Mode} = 3\,\text{Median} - 2\,\text{mean, i.e. } M_o = 3\,M_d - 2\bar{x}$$

Example: Find the mode of the following frequency distribution:

Salary/ week (in '00 $): 0–10   10–20   20–30   30–40
Number of persons: 20   47   30   13

Solution: The maximum frequency in the data is 47, which corresponds to the class 10–20, hence the class 10–20 is the modal class. Thus the other required values are:
l = Lower limit of the modal class = 10
i = Class interval of the modal class = 10
f_1 = Frequency of the modal class = 47
f_0 = Frequency of the class preceding the modal class = 20
f_2 = Frequency of the class succeeding the modal class = 30
Applying the mode formula:

$$\text{Mode } M_o = l + \frac{i(f_1 - f_0)}{(2f_1 - f_0 - f_2)}$$

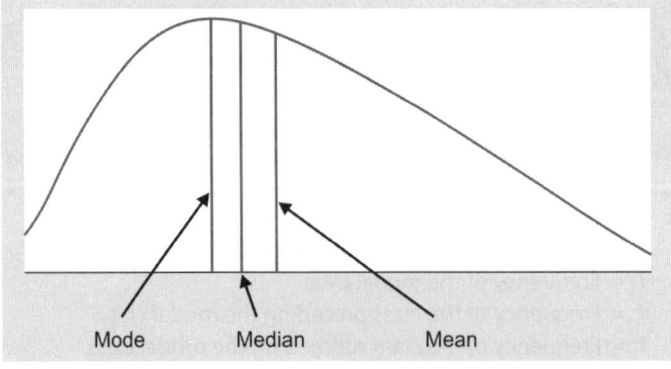

Fig. 8.2: Relationship among mean, median and mode

$$\text{Mode } M_o = 10 + \frac{10(47-20)}{(2 \times 47 - 20 - 30)}$$

$$\text{Mode } M_o = 10 + \frac{270}{44} = 16.14$$

The mode of the data is 16.14, which implies that that $1,614 ($ 16.14 × '00), is the salary amount which will have maximum frequency or in simpler terms, maximum number of people would be earning a salary of $ 1,614/ week.

Example: Given the mean and median of a sample are 50 and 53 respectively, find the mode.

Solution : Mode = 3 Median − 2 Mean, i.e. mode = 3 × 53 − 2 × 50 = 59.

Merits of Mode

Mode is easiest to calculate and understand. It is not affected by extreme values or open-ended classes present in the data. Modal value can be calculated even when the data presents itself with unequal class widths, subject to the condition that modal class and its immediate neighboring (preceding and succeeding classes) have equal class interval.

Demerits of Mode

Demerits of mode lie in its definition which is not very clearly defined, especially in cases where the frequency distribution is bimodal or multimodal. It is based on the frequencies of the modal class and its immediate neighboring classes, while ignoring all the other class frequencies. Further it is not possible to derive composite values of multiple frequency distributions by using the size and modes of the individual distributions, unlike mean. Lastly, it is largely impacted by sampling fluctuations. These limitations restrict the usage of mode to specific situations only.

Geometric Mean

Geometric mean (abbreviated as GM) of a set of values consisting of n observations is the nth root of their product, as shown here mathematically:

$$\text{Geometric mean (GM)} = \sqrt[n]{x_1 \, x_2 \, x_3 \, x_{4...} \, x_n}$$

Where $x_1 \, x_2 \, x_3 \, x_4 \ldots x_n$ are the observations and n is the total number of observations.
This implies that if there are 8 observations is a data set, then the 8th root of their product is calculated to arrive at their geometric mean. As the number of observations become very large, their root can be calculated with the help of computing devices like computers, etc. Alternatively the logarithms (or log) can be used. If logarithms are used, then the formula for geometric mean (for individual observations) is given as:

$$\text{Geometric mean (GM)} = \text{Antilog} \left(\frac{\Sigma(\log x)}{N} \right)$$

Section 3: Describing the Data

For discrete frequency distribution, the formula is given as:

$$\text{Geometric mean (GM)} = \text{Antilog}\left(\frac{\Sigma(f \log x)}{N}\right)$$

Where, x = The discrete variable
f = Corresponding frequency
N = Total number of observations

For continuous frequency distribution, 'x' is the midpoint of the class intervals of the classes, and the formula to calculate remains the same.

Example for individual observations: Find the Geometric Mean of: 6, 10, 14, 18, 22

Solution: Calculating the logarithmic values for $N = 5$ observations

x	log x
6	0.7782
10	1.0000
14	1.1461
18	1.2553
22	1.3424
	$\Sigma \log x = 5.522$

$$\text{Geometric mean (GM)} = \text{Antilog}\left(\frac{\Sigma(\log x)}{N}\right)$$

$$\text{Geometric mean (GM)} = \text{Antilog}\left(\frac{5.522}{5}\right) = \text{Antilog } 1.1044 = 12.72$$

Hence, geometric mean of the given values is 12.72

Example for continuous frequency distribution: Find the geometric mean for the following distribution:

Variable:	5–15	15–25	25–35	35–45
Frequency:	10	15	21	4

Solution: Representing the distribution in tabular form and calculating the log values.

Variable	Mid point (x)	Frequency (f)	Log (f)	f × logx
5–15	10	10	1.0000	10.0000
15–25	20	15	1.3010	19.5150
25–35	30	21	1.4771	31.0191
35–45	40	4	1.6021	6.4084
		$\Sigma = N = 50$		$\Sigma = 66.9425$

$$\text{Geometric mean (GM)} = \text{Antilog}\left(\frac{\Sigma(f \log x)}{N}\right)$$

$$\text{Geometric mean} = \text{Antilog}\left(\frac{66.9425}{50}\right) = \text{Antilog } 1.3389 = 21.82$$

Hence the geometric mean of the given continuous distribution is 21.82.

Merits of Geometric Mean

The merits of geometric mean are its clear-cut definition and calculation process, its consideration for all observations and the fact that it is not much affected by sampling fluctuations. Further it allows subsequent mathematical treatment to be performed on it.

Composite geometric mean of multiple series can be calculated from the geometric means and size of the various series, as per equation given here:

Composite geometric mean

$$= \text{Antilog} \left(\frac{n_1 \log G_1 + n_2 \log G_2 + \ldots + n_n \log G_n}{n_1 + n_2 - \ldots + n_n} \right)$$

Where $G_1, G_2, \ldots G_n$ are the GMs of the multiple series and $n_1, n_2 \ldots n_n$ are their respective sizes.

Compared to arithmetic mean, geometric mean provides more weights to small items and less weights to large items. Hence, it is generally lower or equal to, but never greater than the arithmetic mean for a given distribution.

Demerits of Geometric Mean

It is based on the concept of logarithm, and therefore, not easy to comprehend for some users. Another major drawback is that the formula contains 'product of the values', and if in the data, one or more values are zero, geometric mean cannot be calculated. On the same lines, if a frequency distribution contains odd number of negative observations, the resultant product is negative, and root of negative is an imaginary number, thus making the geometric mean calculations invalid.

Significance of geometric mean lies in the fact that it can be used accurately in calculating the 'Average (as explained earlier, there is difference in word average and mean) Percentage increase' in data series such as increase in interest rates, sales, patients handled, etc. i.e where the data given is in percentages. In such situations, arithmetic mean provides misleading measure. If some series, increase has been to the tune of 9% and 13%, the average increase per year over two year period would not be equal to 11% (average of 9 and 13), because in such a scenario the base on which percentage is being calculated changes every year. Geometric Mean would be suitable in such data cases.

HARMONIC MEAN

Harmonic mean of a data is the reciprocal of the arithmetic mean of the reciprocals' of the given data. In mathematical form, the formula for harmonic mean is given as:

$$\text{Harmonic mean (HM)} = \frac{N}{\left(\frac{1}{x_1} + \frac{1}{x_2} + \frac{1}{x_3} + \ldots + \frac{1}{x_n} \right)} = \frac{N}{\sum \left(\frac{1}{x_i} \right)}$$

For discrete frequency distributions, the formula after considering the associated frequencies is:

$$\text{Harmonic mean (HM)} = \cfrac{N}{\cfrac{f_1}{x_1} + \cfrac{f_2}{x_2} + \cfrac{f_3}{x_3} + \ldots + \cfrac{f_n}{x_n}} = \cfrac{N}{\sum\left(\cfrac{f_i}{x_i}\right)}$$

For continuous frequency distribution, harmonic mean is calculated in accordance with the formula shown above for discrete frequency distribution, with Class Mid Point being taken as values.

Example: Calculate harmonic mean for the following data:

Weight (Kg):	40	50	60
Number of Persons:	20	15	45

Solution: Representing the data in tabular form:

Weight in Kg (Marks) (x)	Number of person (f)	f/x
40	20	0.50
50	15	0.15
60	45	0.75
	$\Sigma = 80 = N$	$\Sigma = 1.40$

Applying the harmonic mean formula

$$\text{Harmonic mean (HM)} = \frac{80}{1.40} = 57.14$$

The harmonic mean of the weights given is 57.14 Kg.

Harmonic mean, though very infrequently used, finds relevance in cases where average speeds or average per unit rates, per unit kilometer, etc. need to be calculated. The connotation per unit implies two units of measurement which can expressed reciprocally. For example, grams per square meter, kilometer per hour, etc. To elaborate, consider a doctor travels from his house to hospital at an average speed of 30 km/hr, and returns at an average speed of 40 km/hr. Find the average speed over the entire travel of 120 km. The arithmetic mean can be calculated as (30+40)/2 = 35 km/hr. Verifying this with the actual values: As the doctor has travelled 60 km (Total travel 120 km, therefore one side journey is 60 km) at a speed of 30 km/hr, he completed the journey in 2 hours (60 km÷30 km/hr). While returning his average speed was 40 km/hr which allowed him to complete the return journey in 1 hour 30 minutes (60 ÷ 40 = 1.5). Total time taken by the doctor to cover 120 km = 3 hours 30 minutes, hence his actual average speed is 120÷3.5 =34.29 km/hr, which is different from the calculated arithmetic mean.

Now for same values, calculating the harmonic mean to check its appropriateness:

$$= \cfrac{N}{\sum\left(\cfrac{f_i}{x_i}\right)} = \cfrac{2}{\left(\cfrac{1}{30} + \cfrac{1}{40}\right)} = \cfrac{2}{\left(\cfrac{7}{120}\right)} = \cfrac{240}{7} = 34.29$$

The harmonic mean is reflecting the actual average speed, while the arithmetic mean is showing a higher value. Thus, it can be said that for cases having 'per unit,' harmonic mean gives appropriate results, instead of any other measure of central tendency.

Salient Points of Harmonic Mean
- It is based on all observations of the data.
- Gives better results as compared to other measures in problems of time and rates.
- Gives large weights to small items, therefore used in economic analysis.
- Not suitable for series containing an observation as zero. Harmonic Mean is used in very limited cases, but cases where small items need to be given higher weight or per unit measurements are involved.
It is worthnoting that for any distribution

Arithmetic Mean ≥ Geometric Mean ≥ Harmonic Mean

$$\text{Geometric Mean} = \sqrt{\text{Arithmetic Mean} \times \text{Harmonic Mean}}$$

Ogive

In the discussion on median, it had been mentioned that median can be calculated through graphical method. Such a graph which allows calculation of median and other quartiles/percentile is called an 'Ogive'.

Ogive is the graphical method to measure the central tendency. Ogive is a cumulative frequency curve for a frequency distribution. The x-axis of the graph represents the Class Interval, while on y-axis the cumulative frequency is plotted by drawing a freehand curve. The resulting curve is called an 'Ogive'. By drawing lines from the axis to the curve, values of the variable as well as median, quartile, deciles, etc. can be inferred.

An illustration of Ogive is as given here **(Fig. 8.3)**. Normally, Ogive is drawn for increasing cumulative frequency (Less than Ogive Curve). In case, the reverse cumulative frequencies are also plotted, it is called as 'More than Ogive curve'. The point where these two curves intersect, is the median of the data series.

Alternatively, the cumulative percentage frequency values can also be plotted for Ogive. These Ogives, based on 'Cumulative Percentage Frequencies', help to deduce other quartiles and percentiles by drawing the lines to the x-axis. The same is illustrated in the **(Fig. 8.4)**.

WHICH MEASURE TO USE?

As detailed in this chapter, there is no ideal measure of central tendency. The usage of the measure depends upon the objectives, the type of data available, requirements of post-calculations, etc. Collectively, all the various measures provide good amount of information and can be used simultaneously to describe a distribution, but when this is not possible, the following broad guidelines for selecting the appropriate measure of central value are helpful.
- Median is best suited for open-ended grouped distributions.
- Mode can be used to evaluate qualitative data or where the objective is to find most common value or value with highest frequency.
- Geometric mean is useful in averaging ratios, percentages or average rates of increase or decrease.

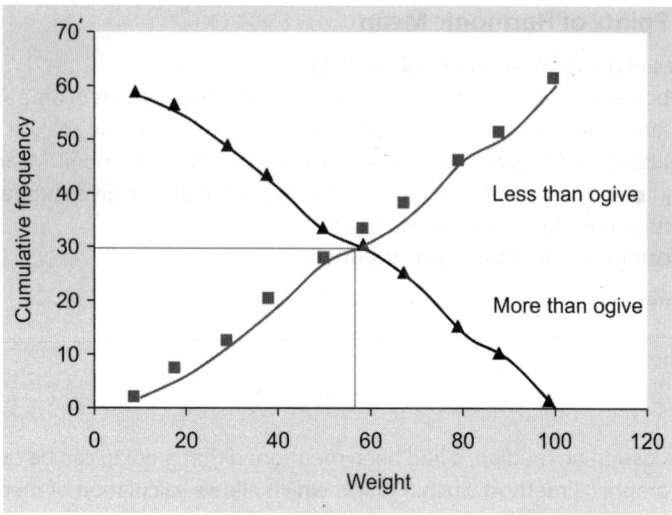

Fig. 8.3: Ogive for increasing cumulative frequency

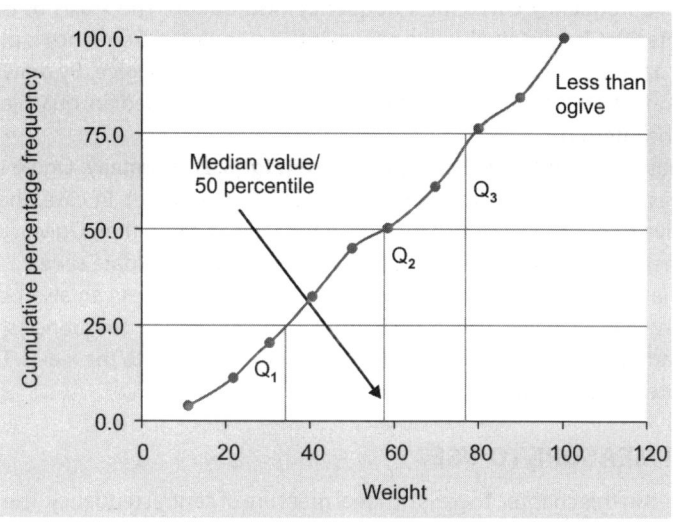

Fig. 8.4: Cumulative percentage frequency values

- Harmonic mean should be preferred for 'rates per unit' cases.
- Arithmetic mean should be generally avoided in highly skewed distributions, or distribution with open-ended classes, or having presence of extreme values or contains ratios.
- Apart from above specific about arithmetic mean, it should be used a first choice due to its universal understanding, and the facility of application of subsequent statistical techniques. In general sense, the user will not go wrong by use of arithmetic mean, unless the data is in ratios, percentages or per units.

Measure of central value provides a single value about the distribution, but for understanding the distribution this single value is insufficient. It is like a situation where the average depth of river is known to be five feet, but which does not mean the river is having a constant depth of five feet. There are bound to be areas where the depth could be even 20 feet, and may be some stretches where the depth is just 3 feet. It points, toward a need of a measure which provides more information about the variable's distribution. Such information can be studied through dispersion.

Therefore *Measure of Dispersion* is essentially required, so that these along with measures of central value, collectively can allow superior inferences about the data.

CHAPTER 9

Measures of Dispersion (or Variation)

UNDERSTANDING DISPERSION

The objective of studying the variation or dispersion of the frequency distribution arises because the measures of central tendency are not able to adequately describe the frequency distribution. The concept of dispersion can be explained by taking an example about the population data of a state. Assume a state has a population of 12 million people and its area is 0.5 million sq kms. Therefore, the population density (Mean number of people per sq. km.) of this state would be 24 people/sq km. But in reality, it is not expected that there would 24 people living within every sq. km, instead there would be high concentration of population in big cities, reasonable number in towns and very less concentration in villages. In addition to these there possibly would exist, large tracts of land having no population at all, due to forests, inhospitable terrain like steep mountains/deserts, etc. The population density of all these sub-areas of the state (say districts) can be construed as a frequency distribution. The population density, as a central measure, provides a single number, representative of the entire, but does not convey whether the population is living in a concentrated areas or spread throughout the state.

The measures of central tendency, though provide a single number as representative of the distribution, but does not give an idea about the 'scatteredness or concentration' of the distribution.

'Dispersion' or 'spread' is the measurement of the variation (scatteredness) of the variable around its central value. Collectively, the tools to measure this variability of the distribution from its central value are called "measures of dispersion." The central value is not restricted to mean, but may have been derived by any of the methods discussed in the chapter on 'measures of central tendency.'

Elaborating further on the concept, assuming three sets of data as follows:

Set A—50, 50, 50, 50 (Total 200)

Set B—20, 80, 60, 40 (Total 200)

Set C—10, 10, 10 170 (Total 200)

Even though the central value of all these sets is equal [Arithmetic mean of all these sets is 50 (200÷4)], these sets are completely different, with not even a single value common across the sets. Set A does not have any have dispersion at all, and the arithmetic mean completely represents the data. Set B has some variability but it is small as compared to Set C, where the dispersion is quite large. Thus, the need arises to study the dispersion characteristics of a distribution in order to supplement the information derived from measures of central tendency.

The various measures of dispersion allow understanding of the 'degree of variability' or 'degree of dispersion' from the central value, but do not give the 'direction of the variation.' Measures of dispersion indicate reliability of the central value—a large variation may indicate the central value is not a typical representative of the distribution, subject to the sample size being not too large.

Another advantage of the dispersion value is that it allows comparison of different series on their uniformity and consistency. A low value of dispersion implies low degree of variation, i.e. greater uniformity of the distribution. Many significant statistical tools like regression, hypothesis testing, analysis of variance, etc. draw a lot from the basic concept of dispersion and thus clarity about this subject is very important.

TYPES OF MEASURES OF DISPERSION

The different types of measures of dispersion are:
- Range
- Quartile deviation or interquartile range
- Percentile deviation
- Mean deviation
- Standard deviation
- Lorenz curve

These different methods require different calculation processes and provide different information about the state of dispersion.

Range

Range (R), is defined as the difference between the two extreme values of the distribution.

$$Range\ (R) = L - S,$$
where L = Value of the largest observation, and
S = Value of the smallest observation

Range is the simplest measure of dispersion, providing positional spread of the distribution, but does not provide any information about the behavior of the distribution within these two extreme points. It cannot be applied to open-ended distributions. Despite its drawbacks, range is useful in situations where there is not much interest about the intervening information, but only values desired are the maximum and minimum. For example, the various parameter in weather reports like temperature, humidity, wind speed, etc. mention only maximum and minimum values. Similarly, quality parameters applicable to most products mention minimum and maximum limits for acceptance.

QUARTILE DEVIATION

Quartile deviation is also called interquartile or semiquartile range.

Range was the difference between two extremes, while quartile deviation is the difference between the two quartiles Q_3 and Q_1, divided by 2.

$$\text{Quartile deviation } (Q) = \frac{(Q_3 - Q_1)}{2}$$

where Q_3 = Third quartile and Q_1 = First quartile of the distribution.

From the discussion in the earlier chapters, each quartile consists of 25% of the observations, therefore difference between Q_3 and Q_1 will consist of central 50% of the observations, because from the first three quarters included in Q_3, the first quarter Q_1 has been subtracted which resulted in only two quarters remaining. The mean of these two remaining central quarters is the quartile deviation of the distribution. Relatively, a small value of quartile deviation indicates small variation in the central 50% of the observations and vice versa. For computation of quartile deviation, the first and second quartiles need to be calculated, as explained in the chapter on central tendency.

Quartile is a positional measure and based on quartile/median, therefore has advantages similar to median like, it can be calculated for open ended distributions or can be well employed where large extreme values cause undesired fluctuations. Ironically, this advantage is its drawback too, as it ignores 50% of the values and is based on the central 50% of the values only.

PERCENTILE DEVIATION

Similar to quartile deviation, percentile deviation is computed from the formula:

$$\text{Percentile deviation} = \frac{(P_{90} - P_{10})}{2}$$

Where P_{90} = 90th percentile of the distribution.
P_{10} = 10th percentile of the distribution.

MEAN DEVIATION

If $x_1, x_2, x_3, \ldots x_n$ is a frequency distribution consisting of 'N' observations and 'A' is its average (any central value and not only mean), then mean deviation of the distribution is given as:

$$\text{Mean deviation (for individual observations data)} = \frac{1}{N}\sum_{i=0}^{n}|x_i - A|$$

$$\text{Mean deviation (for discrete distributions)} = \frac{1}{N}\sum_{i=0}^{n}(f_i|x_i - A|)$$

where N = Total number of observations, i.e. $\sum f_i$

$|x_i - A|$ = Modulus of difference between the variable and the average.

('Modulus' abbreviated as 'Mod', denoted as two vertical lines before and after the value, is the absolute value while ignoring the –ve sign, i.e. Mod –3 = |–3| = 3).

It should be clearly understood that mean deviation is the 'mean value of deviation from the any central value' and not necessarily 'the deviation from only mean as a central value'.

Mean deviation is based on all the observations; hence, it is better than range or quartile deviation. Relative to standard deviation it is less affected by the extreme values. For any given distribution, its mean deviation value is least if measured from median, as compared to its value from mode or mean, as measures of central value.

STANDARD DEVIATION

Standard deviation is the most popular measure of dispersion and is denoted by the Greek symbol sigma (σ).

As against a mean deviation which can be calculated from any central tendency measure like the median, mode or mean, the standard deviation of a distribution is always calculated from its mean. This is because the sum of 'squares of the deviations' from mean is minimum. This property is very important and will appear frequently in many advanced statistical tools. A low value of standard deviation indicates low dispersion which implies high degree of uniformity in the distribution. Similarly, a high value of standard deviation indicates high dispersion implying less uniformity.

Standard deviation of the distribution is calculated by the given formula:

For individual observations

$$\text{Standard deviation}(\sigma) = \sqrt{\frac{\sum_i^n (x_i - \bar{x})^2}{N}}$$

For discrete and continuous distributions

$$\text{Standard deviation}(\sigma) = \sqrt{\frac{\left(\sum_i^n f_i (x_i - \bar{x})^2\right)}{N}}$$

Where

$N = \Sigma f_i$

= Sum of all frequencies or observations (in case of individual observations)
x_i = Variable value (class mid point in case of continuous distribution)
f_i = Corresponding frequencies
$\bar{x}$ = Mean of the distribution.

Root Mean Square Deviation

Sometimes standard deviation is also referred to as root mean square deviation, but there exists a very fine difference between the two. The difference being, when standard deviation is calculated from an 'arbitrary' number instead of mean of the distribution, it is referred to as root mean square deviation. In the formula also, the mean is replaced by 'A' the arbitrary number, as shown below:

$$\text{Root mean square deviation (RMS)} = \sqrt{\frac{\sum_i^n (x_i - A)^2}{N}}$$

Naturally it follows that for any given distribution, multiple 'root mean square' can be calculated, but of all the root mean squares so calculated, the least value of the root mean square would be equal to the standard deviation. Thus, standard deviation is also the least value of the root mean square deviation of the distribution.

Variance

The square of standard deviation is called the variance of the distribution.

$$\text{Variance} = \sigma^2$$

$$\text{Standard deviation } (\sigma) = \sqrt{\text{Variance}}$$

Similarly the square of 'root mean square deviation' is called 'mean square deviation' i.e.

Mean square deviation = (Root mean square deviation)^2

Combined Standard Deviation

Composite standard deviation of two distributions can be calculated by using the following formula:

Combined standard deviation

$$\sigma_{combined} = \sqrt{\frac{N_1\sigma_1^2 + N_2\sigma_2^2 + N_1 d_1^2 + N_2 d_2^2}{N_1 + N_2}}$$

Where

N_1, N_2 = Sizes of the two distributions,

σ_1, σ_2 = Standard deviations of the two distributions,

$d_1 = d_1 = (\bar{x}_1 - \bar{x}_{combined}); d_2 = (\bar{x}_2 - \bar{x}_{combined})$ and $\bar{x}$ is the respective mean

Standard deviation allows more precise understanding of the distribution. For a symmetrical distribution, the value of its 'mean ± 1σ' covers 68.27% of the observations, 'mean ± 2σ' covers 95.45% of the observations and 'mean ± 3σ' covers 99.73%.

Tchebysheff's Theorem

As per *Tchebysheff's theorem*, no matter what spread the distribution has, 75% of its values will fall within 'mean ± 2σ' and 88.89% will lie within 'mean ± 3σ' limits. The difference in these two statements on percentage of values covered is the character of the distribution. In the first statement, the values are subject to the distribution being 'symmetrical', whereas in second statement, there is no such restriction; hence the percentage of values lying within the stated limits is lower. This concept is the covered in detail under normal distribution, normal curves and is related to the test of significance which are of paramount importance in any research data analysis, and would be taken up in detail later.

Though standard deviation gives more weight to the extreme items, it is still the most widely used and best measure of dispersion.

LORENZ CURVE

Lorenz curve is the graphical method of studying dispersion, and is a cumulative percentage curve. The variables as well as the frequencies are cumulated and their respective percentages are calculated.

Plotting: Being percentages, scale is 0–100 on both X-axis as well as Y-axis. The X-axis, pertains to the 'cumulative frequencies percentage', while Y-axis is allotted for 'cumulative variable percentage'. As a standard measure points (0, 0) and (100,100) are joined by a straight line, known as 'line of equal distribution.' Thereafter, the values of the distribution are plotted and joined through a free hand drawn curve. This resultant curve is a graphical measure of dispersion. As it is a percentage curve, more than one distribution curve can be plotted on the graph and their dispersion can be compared.

Graph of Lorenz Curve

The following graph depicts two hypothetical data series A and B, pertaining to age of the staff employed by two hospitals **(Fig. 9.1)**. The age group alongwith the cumulative frequencies of both the series, are expressed in their percentage terms. These are plotted and the resultant curves provide an indication of the dispersion of the two series. From the Lorenz curve given below, it can be inferred that hospital A has lesser age dispersion than hospital B. The series closer to the line of equal distribution would have less dispersion. It needs to be noted that hospital A may have less age dispersion, but it does not imply that it has lower mean age than hospital B. Dispersion provides the measure of spread around the central value and does not comment about the numerical value of the 'central value'.

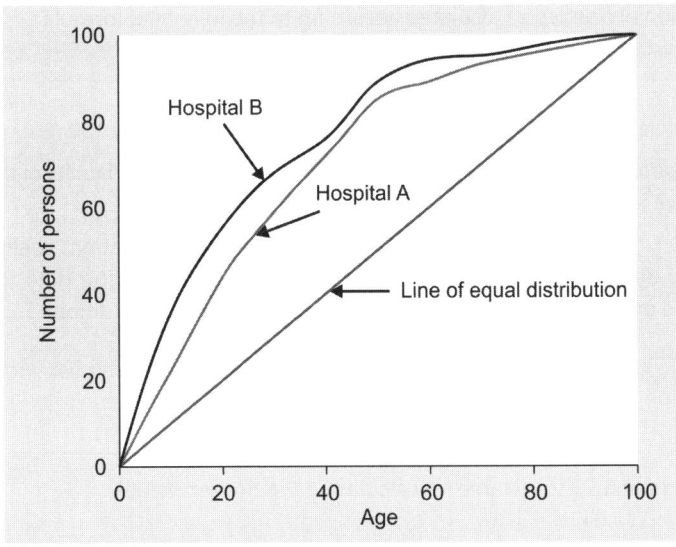

Fig. 9.1: Lorenz curve

COEFFICIENT OF DISPERSION AND ITS MEASURES (ALSO REFERRED AS A RELATIVE MEASURE OF VARIATION)

The measures of dispersion discussed in the previous sections, provide absolute value which can be compared across two different sets of data, but at times the two sets of data may not have same but have different units of measurements. In such cases, these measures of dispersion are no more comparable across the two sets of data. Therefore, for such situations, apart from calculating the dispersion measures, coefficients of dispersion are also calculated. These co-efficients being ratios are independent of units of measurement. The relevant formula for calculating these are: (Notations are same as given earlier)

$$\text{Coefficient of range} = \frac{L-S}{L+S}$$

$$\text{Coefficient of range} = \frac{Q_3 - Q_1}{Q_3 + Q_1}$$

Coefficient of mean deviation =

$$\frac{\text{Mean deviation}}{\text{Central value used to calculate above mean deviation}}$$

(In case of coefficient of mean deviation, if mean deviation has been calculated on basis of arithmetic mean, then the denominator would be the arithmetic mean. If mean deviation has been calculated on the basis of median, the denominator would be median and so on).

$$\text{Co-efficient of standard deviation} = \frac{\text{Standard deviation}}{\text{Arithmetic mean}} = \frac{\sigma}{\bar{x}}$$

Coefficient of variation: It is the most common measure of relative variation. Coefficient of variation of a series is calculated according to the following formula:

$$\text{Co-efficient of variation (CV)} = \frac{\sigma}{\bar{x}} \times 100$$

Or, Coefficient of variation = Coefficient of standard deviation × 100

Interpretation: Higher the coefficient of variation, higher is the variability of the series, and hence lower is its uniformity, and vice- versa.

Example: Find the mean deviation from mean, standard deviation, coefficient of standard deviation and coefficient of variation for the following data reflecting incidence of nonaccidental death (due to illness or medical condition):

Age group:	10–20	20–30	30–40	40–50	50–60	60–70
No. of persons:	33	31	41	55	72	68

Solution
Preparing the necessary table for calculating the desired values.

Age group	Mid point (x_i)	No. of person (f)	fx_i	Deviation from mean $(x-\bar{x})$ $=(x-45.2)$	$(x-\bar{x})^2$	$f(x-\bar{x})^2$	Modulus of deviation $|x-\bar{x}|$	$f|x-\bar{x}|$
10–20	15	33	495	–30.2	912.04	30097.32	30.2	996.6
20–30	25	31	775	–20.2	408.04	12649.24	20.2	626.2
30–40	35	41	1435	–10.2	104.04	4265.64	20.2	828.2
40–50	45	55	2475	–0.2	0.04	2.2	0.2	11
50–60	55	72	3960	9.8	96.04	6914.88	9.8	705.6
60–70	65	68	4420	19.8	392.04	26658.72	19.8	1346.4
Total		N=300	13560			80588	100.4	4514

The required results can now calculated as follows:

$$\text{Mean } \bar{x} = \frac{\Sigma fx_i}{N} = \frac{13560}{300} = 45.2$$

Mean deviation from mean $\frac{1}{N}\sum_i^n (f_i |x_i - \bar{x}|) = \frac{4514}{300} = 15.05$

Standard deviation

$$= \sqrt{\frac{\sum_i^n f_i(x_i - \bar{x})^2}{N}} = \sqrt{\frac{80588}{300}} = \sqrt{268.63} = 16.39$$

Coefficient of standard deviation $= \frac{\sigma}{\bar{x}} = \frac{16.39}{45.2} = 0.36$

Coefficient of variation $= \frac{\sigma}{\bar{x}} \times 100 = 0.36 \times 100 = 36$

WHICH MEASURE OF DISPERSION TO USE?

Just like the measures of central tendency, there is no ideal measure for dispersion also. Range is commonly used where the interest in only in the maximum and minimum value or the overall spread of the distribution. Though less preferred, range is helpful in a supplementary role to other measure of dispersion like standard deviation.

Quartile deviation is sparingly used, while mean deviation has some applicability. Unless otherwise warranted, standard deviation is the first choice for measuring the dispersion, alongwith arithmetic mean as the measure of central value. This combination permits further analysis and comprehension of the frequency distribution through calculation of variance and the coefficient of variation.

The measures of dispersion are important tools for description of data and are fully utilized when two or more data-sets are being compared. These add to the information derived from measures of central tendency, but more information about the data can still be derived, which is the subject of next chapter.

10

Skewness, Moments and Kurtosis

INTRODUCTION

Next step in analysis of the data values is to know the characteristics of its skewness, moments and kurtosis. Measures of central tendency provide a single value representative of the distribution, whereas the measures of dispersion provides information on degree of dispersion or spread of the values. With these tools, the central value and variability of the distribution becomes known, but there is no information about the direction in which this variability is occurring. Consider an example, of two sets of values A and B:

	Observation Values						Sum	Mean	Std. dev. σ	Median	Skewness
Set A	37	36	38	41	38	110	300	50	29.4	38	Positive
Set B	8.2	79	74	72.8	27.5	38.5	300	50	29.4	55.6	Negative

In the two sets of values above the respective sums, means and the standard deviations are equal, but with visual inspection it can be inferred that these two sets vastly differ in their 'scatteredness'. In the Set A, except one observation, all other observations are concentrated about 10–12 units lower than the mean; and out of 6 observations, five are lower than mean. In Set B, no value is very distant from the central value and 3 values are higher and 3 values are lower than the mean value. Visual inspection indicates that the 'direction of dispersion' in the two sets should differ. The same gets measured by the Value of Skewness as shown in the last column. As one of the skewness values is negative, while other being positive; which implies that the direction of dispersion in both the series are opposite to each other.

Therefore to understand the frequency distribution more accurately, other tools are additionally required. Concepts like skewness, moments and kurtosis fill this gap to provide further understanding about the distribution.

SKEWNESS

Skewness means 'Lack of Symmetry'. Denoted as S_k, Measure of Skewness provides the direction and extent of asymmetry in the distribution (**Fig. 10.1**). A symmetrical

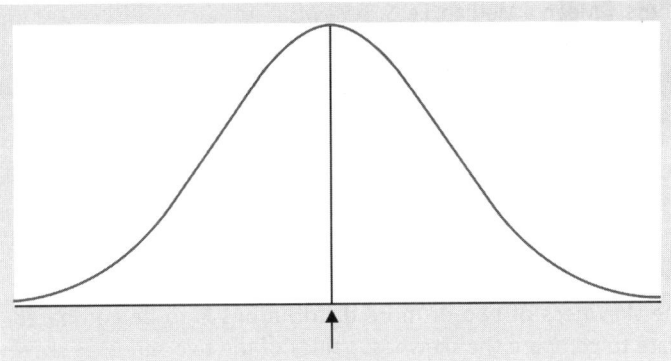

Fig. 10.1: A normal curve (its mean = median = mode)

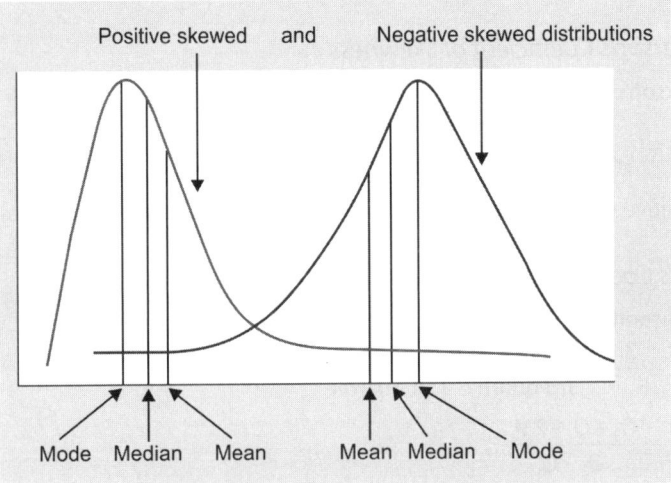

Fig. 10.2: Skewed curve for the distribution

distribution is one which is evenly spread around its central values. A distribution will have skewness or can be called skewed, if
i. Mean ≠ Median ≠ Mode
ii. Quartiles Q_2 and Q_3 are not equidistant from median i.e. $Q_2 - Q_1$ $Q_3 - Q_{2M}$

In symmetrical distributions, the mean, median and mode of the distribution are equal and the curve of such a distribution is called the normal curve, as shown below. (Normal distribution would be dealt in detail later pages).

The direction of skewness can be positive or negative, as shown in the positively and negatively skewed distributions **(Fig. 10.2)**.

Measures of Skewness

For standalone distributions, skewness can be measured in absolute terms, by any of the following methods:

Skewness = Mean − Median, i.e. $S_k = \bar{x} - M_d$
Skewness = Mean − Mode, i.e. $S_k = \bar{x} - M_o$
Skewness = 2 Median − 1st Quartile − 3rd Quartile
i.e. $S_K = 2M_d - Q_1 - Q_3$

From the above equations, if the value calculated is positive i.e. $(\bar{x} - M_d) > 0$ or $(\bar{x} - M_o) > 0$, the distribution is called 'positively skewed distribution.' Similarly if the values are < 0, the distribution is called "negatively skewed distribution".

Coefficient of Skewness

When the skewness of two or more distributions is to be compared, it is not appropriate to compare the skewness values of the two series, as skewness is an absolute measure. Hence for comparing skewness of different distributions, a relative measure is used, called coefficient of skewness. Coefficient of skewness can be calculated by any of the below listed methods:

Karl Pearson's Coefficient of Skewness

Karl Pearson's coefficient of skewness

$$= \frac{\text{Mean} - \text{mode}}{\text{Standard deviation}} = \frac{\bar{x} - M_o}{\sigma}$$

A positive value of the coefficient will imply positive skewness and vice versa.

Bowley's Coefficient of Skewness

Bowley's coefficient of skewness

$$= \frac{\text{3rd quartile} + \text{1st quartile} - 2 \text{ median}}{\text{3rd quartile} - \text{1st quartile}}$$

$$= \frac{Q_3 + Q_1 - 2M_d}{Q_3 - Q_1}$$

Bowley's coefficient is based on quartiles and is also known as 'quartile coefficient of skewness'. It is derived from the normal curve notion, that the first and third quartiles are equidistant from Median. Value of Bowley's coefficient of skewness would range between +1 and −1. A positively skewed distribution would give a positive coefficient and a negatively skewed distribution would result in a negative value of the Bowleys's coefficient.

Coefficients of skewness calculated from Karl Pearson's method and Bowley's method are not comparable and in rare situations can give contrary results. This is due to the basis of calculation being very different. The main drawback of Bowley's coefficient, is its dependence on just 50% central values while ignoring the observations lying on the ends of the distribution.

Kelly's Coefficient of Distribution

As a measure of distribution, Kelly's coefficient of skewness improves upon the drawback encountered in the Bowley's method, by extending the coverage of distribution from central 50% values to central 80% values. This is achieved by using the 10th percentile and 90th percentile instead of 1st and 3rd quartiles. Hence the formula is as below:

Kelly's coefficient of skewness

$$= \frac{(90\text{th percentile} + 10\text{th percentile} - 2 \text{ median})}{90\text{th percentile} - 10\text{th percentile}}$$

$$= \frac{P_{90} + P_{10} - 2M_d}{P_{90} - P_{10}}$$

Coefficient of Skewness based on Moments

Coefficient of skewness based of moments is given as:

$$S_K = \frac{\sqrt{\beta_1}\,(\beta_2 + 3)}{2\,(\beta_2 - 6\beta_1 - 9)}$$

Where β_1 and β_2 are values calculated from 'moments about mean'. Further details on moments are covered later in this chapter. For a symmetrical distribution $\beta_1 = 0$.

KURTOSIS

Kurtosis refers to the degree of 'peakedness' of the curve of the distribution. In addition to the central value, and degree of dispersion, skewness allows information to be deduced about the direction of the symmetry of the distribution (**Fig. 10.3**). Another characteristic which provides more information about the frequency distribution is the 'peakedness' or 'flatness' of the curve. This property allows comparison of two curves of different frequency distributions. The significance of kurtosis arises, as two or more frequency distributions can be symmetrical around mean, with similar range, yet have different curves. Distinctly, two distributions can have same central value, dispersion and be either positive or negative skewed, but still they may have difference in terms of their 'curve peakedness'.

If a curve is more peaked than the Normal curve, it is called 'leptokurtic curve' and in case it is more flatter then it is known as 'Platykurtic curve'. Accordingly 'Mesokurtic curve' is the one which is neither peaked nor flat, i.e. normal curve.

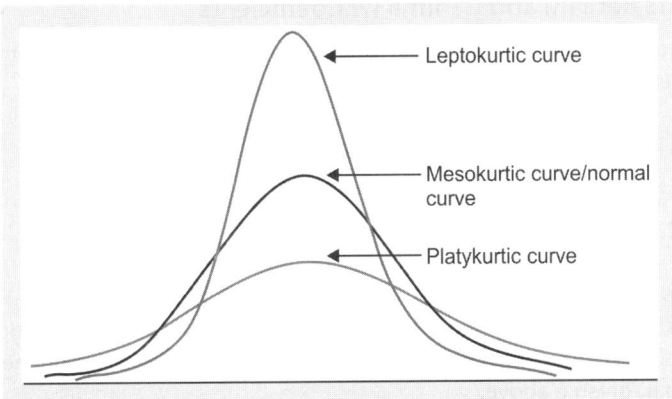

Fig. 10.3: Kurtosis—types of curves

Measure of Kurtosis

Kurtosis is measured as a coefficient and denoted as β_2.

$$\text{Kurtosis } (\beta_2) = \frac{\mu_4}{\mu_2^2}$$

Where 'μ_4' is the 4th Moment and 'μ_2' is the 2nd Moment.

MOMENTS

Moments of a distribution are defined as the "various powers (like 'x to the power two $= x^2$, x to the power three $= x^3$, x to the power four $= x^4$ ... and so on) of the deviation of the variable from any arbitrary point A" divided by the total number of observations or frequency. Moments are denoted by μ (mu). In maximum calculations, instead of calculating moments from any arbitrary point, the Mean of the distribution is used, and these Moments are called 'Moments about Mean'. Mathematically, the equation for Moments about mean of rth order is as given below:

$$\text{Moments } (\mu_r) = \frac{1}{N} \sum_{i=1}^{n} f_i (x_i - \bar{x})^r$$

Where μ_r is the rth moment, x_i is the variable, f_i its corresponding frequency, N is sum of all frequencies Σf_i, and $\bar{x}$ is the mean of the distribution.

Generally the moments are calculated upto fourth power only, as beyond that even the small deviations become highly significant (because of higher power). By giving the equation, the powers 0, 1, and 2, by replacing 'r', we get the following:

$$\text{Moments } (\mu_0) = \frac{1}{N} \sum_{i=1}^{n} f_i (x_i - \bar{x})^0 = 1, \text{ i.e. unity}$$

$$\text{First moment } (\mu_1) = \frac{1}{N} \sum f_i (x_i - \bar{x})^1$$

$= 0$ (as sum of deviation from arithmetic mean is zero)

$$\text{Second moment } (\mu_2) = \frac{1}{N} \sum f_i (x_i - \bar{x})^2 = \sigma^2$$

i.e. square of standard deviation = variance.

Similarly third and fourth moments about mean can be calculated.

Pearson's Beta (β) and Gamma (γ) Coefficients

Pearson defined the following four coefficients, based on various "Moments (μ) about mean"

$$\beta_1 = \frac{\mu_3^2}{\mu_2^3} \text{ i.e. } \frac{\text{Square of 3rd moment about mean}}{\text{Cube of 2nd moment about mean}}$$

= Measure of skewness of the distribution

$\gamma_1 = +\sqrt{\beta_1}$, or $\gamma_1 = \sqrt{\frac{\mu_3^2}{\mu_2^3}} = \frac{\mu_3}{\sigma^3}$ hence β_1 and γ_1 are measures of skewness.

But as β_1 does not take negative values, it is not able to provide direction of the skewness. Therefore γ_1 is a better measure as it can be negative based on the formula involving μ_3 given σ above.

If $\gamma_1 = 0$, Distribution is symmetrical;

if $\gamma_1 > 0$, the distribution is positively skewed
if $\gamma_1 < 0$, the distribution is negatively skewed

$$\beta_2 = \frac{\mu_4}{\mu_2^2} \text{ i.e. } \frac{\text{Fourth moment about mean}}{\text{Square of 2nd moment about mean}}$$

= Measure of kurtosis of distribution curve

$$\gamma_2 = \beta_2 - 3,$$

if $\beta_2 > 3$ → Leptokurtic curve;
$\beta_2 = 3$ → Normal curve; and
$\beta_2 < 3$ → Platykurtic curve

SIGNIFICANCE

Significance of moments lies in the ease of calculations for arriving at mean (moment about origin), standard deviation and variance (from second moment), skewness (from 2nd and 3rd moment) and kurtosis (from 2nd and 4th moment).

Therefore, the concepts and measures of central tendency, dispersion, skewness and kurtosis allow detailed study and understanding of the frequency distribution. The coefficients, wherever applicable, also allow comparison of two or more frequency distributions. With these fundamentals in place, further understanding of statistical concepts becomes easier. The main concepts which will be used very frequently in advanced methods of statistics are frequency distribution (especially continuous), arithmetic mean, median, standard deviation and skewness.

Section 4

Probability, Probability Distribution and Sampling Theory

- Theory of Probability
- Theoretical Discrete Probability Distributions
- Theoretical Continuous Probability Distributions
- Sampling Theory

CHAPTER 11

Theory of Probability

INTRODUCTION

Probability theory is the foundation of statistical inference. All the tests of significance, Chi-square, F test, etc. rely on probability for acceptance or rejection of a hypothesis. This is because the level of significance, against which the hypothesis is evaluated, is actually a probability. The details of this would be dealt in their respective chapters, but on a minimum scale, this indicates that knowledge of the 'concept of probability' is very important for proper understanding of probability distribution including normal distribution, and subsequently inferential statistics.

Probability, in a generic sense, is widely used by common people without being aware of it. 'I am sure the train will arrive on time', 'The chances of survival postsurgery are 50–50', 'No chance that there would be rain today', 'there is 50% chance that in a toss of a coin, a head will show up'—these and similar statements are stated and heard very often, but except last one all the others statements are more of subjective estimates about the uncertainty and not probability. Theory of probability provides mathematical and logical basis to establish numerically valid statements about uncertainty. (Use of word 'uncertainty' in this statement does not imply exclusion of 'certainty' but 'certainty' is considered as a situation where 'uncertainty is not present at all').

Many principles and laws of sciences have been able to establish relationships between various factors like Pythagoras' Theorem, Newton's Laws, Ohm's Law, etc. where, based on values of some components, the other component's value can be determined accurately. Such 'deterministic' models, where input values allow determination of outcomes, are not possible for uncertain or 'probabilistic' events. The study of such probabilistic events for their occurrence and nonoccurrences in a mathematical and numerical manner becomes the subject matter of probability.

Probability is the defined as the 'measure of the likeliness of occurrence of an event or success' or 'the extent to which an event is likely to occur'. It is expressed as the ratio of the favorable cases to the total number of possible cases. As the number of favorable cases cannot be a negative number and being a ratio, where numerator cannot be greater than the denominator, probability takes form, as a number 'equal to' or 'between' 0 and 1, where 0 indicates nil possibility or impossibility, while 1 indicates certainty or 100 percent possibility. Probability is also the name given to the branch of statistics which studies the possible outcomes of given events, the relative likelihood of different outcomes and the distributions of these probabilities. Sometimes it is expressed as a percentage wherein '1' refers to 100% possibility.

TERMINOLOGY IN PROBABILITY THEORY

Consider the following simplistic case: A dice is rolled say, 5 times and the number that rolls up is recorded.

1. *Experiment*: The rolling of dice many number of times is called the experiment. Experiment, consisting of one or many trials, is performed basically under identical conditions, without resulting in a unique result, but results in many possible results.
2. *Trial:* The rolling of a dice is the trial.
3. *Event:* The outcome of the trial is known as 'event'. In this case the possible events are 1, 2, 3, 4, 5 or 6. It is also referred as 'cases'.
4. *Exhaustive events:* The total number of possible outcomes, of any trial is called as its 'exhaustive events or exhaustive cases.' In this example the exhaustive events are numbers 1, 2, 3, 4, 5 and 6 as there no other possible outcome except these.
5. *Success or favorable events:* It is the number of events or outcome from the exhaustive events, which are desirable as an outcome. In this example if success is defined as 'dice showing an odd number', then the favorable events would be those where the dice shows up numbers 1, 3 and 5.
6. *Equally likely event:* The event would be considered 'equally likely' if all outcomes are equally possible; and there is no logical/ mathematical evidence to support that any one or more of possible outcomes, may be preferred over others. In the case being considered, outcomes from the roll of dice are an equally likely event. To explain this concept further, consider a basket containing 2 balls each of four colors yellow, magenta, cyan and black. One ball is drawn from the basket. The possible outcomes are that the drawn ball is of yellow or magenta or cyan or black color and all being equally likely. Assume that the first ball drawn is of yellow color. Now, if a second ball was drawn from the basket, without placing back the first drawn ball, the event is no more 'equally likely' because the probability of drawing a yellow ball is less than the probability of drawing any other colored ball. The probability has reduced because the number of possible favorable cases has reduced from 2 to 1. Had the yellow ball been replaced after the first trial, the second trial would then have been, again an 'equally likely' event.
7. *Mutually exclusive event:* If in a trial, the occurrence of an outcome eliminates or precludes occurrence of all other possible outcomes, it is termed as mutually exclusive event. This implies that in mutually exclusive events, two or more than two outcomes are not possible in the same trial. In the roll of dice trial, all the events are mutually exclusive, as occurrence of any one of them (either of 1, 2, 3, 4, 5, 6 possibilities), eliminates the possible occurrence of others in the single dice roll.

8. *Independent and dependent events:* The outcomes or events are termed as independent, if in an experiment;
   - The 'outcome of trial' is not affected by the outcome of the preceding trials and
   - Also in turn, does not affect the outcome of the subsequent trials.

   In the dice example, the outcome of roll of dice does not get influenced by the result (outcome) of the preceding rolls of dice (trial), and therefore is an independent event. In the previously described experiment of basket containing colored balls, if the ball is placed back after each draw, then it becomes an independent trial. In case the ball is not replaced, then outcome of second draw is no more independent but becomes dependent on the outcome of the first trial and is said to be 'dependent event.'

9. *Success and failures:* In probability, the occurrence of the desired outcome is generically termed as 'success' and its nonoccurrence as a 'failure'. In the dice roll example, if the desired outcome is say, number 2, then the 'probability of success' is probability of occurrence of number 2 in the roll of dice. Probability of failure would imply nonoccurrence of number 2 or in other words occurrence of any number except 2. As a standard notation, probability of success is denoted as 'p' and that of failure as 'q.'

APPROACHES TO PROBABILITY

With this basic terminology in place, probability can now be defined as 'the mathematical expression of the likelihood of occurrence of an event'. The theory of probability has been approached from different aspects and these are:

'*a priori*' Probability or Classical Approach

Consider a trial having 'n' number of all possible outcomes. These outcomes are equally likely, mutually exclusive and exhaustive in nature. Out of these 'n' possible outcomes, 'x' number of outcomes are considered as 'desired outcomes' (or favorable or success). The ratio of the 'desired outcome/success' to the total number possible outcomes 'n' is called the probability. If probability of occurrence of 'x' is denoted as $P(X)$, then its probability is given as:

$$\text{Probability of 'x' } P(X) = \frac{\text{Total favorable or desired outcomes}}{\text{Total number of possible outcomes}}$$

In the dice example, if success is defined as rolling of an even number, then probability can be calculated as below:

Total number of possible outcome on roll of dice = 1,2,3,4,5,6, i.e. 6 possible outcomes.

Total number of favorable outcomes, i.e. even number= 2,4,6, i.e. 3 favorable outcomes.

∴ Probability of occurrence of even number in roll of dice = 3/6 = 0.5.

Extending the example: The probability of failure, i.e. non-occurrence of an even number can also be calculated similarly. The favorable cases for 'nonoccurrence of even number' = 1,3,5, i.e. 3 possibilities, while the total number of possible outcomes remain same. Therefore, probability of failure (nonoccurrence of even number) = 3/6 = 0.5.

From the above measure of probability, the following can be deduced:
- The value of P(X) would always be between 0 and 1, i.e. $0 \leq P(X) \leq 1$.
- A probability value of '0' denotes 'impossible' event, while '1' denotes 'a certain event'.
- If 'p' is the probability of success and 'q' is the probability of failure, then
$$p + q = 1 \text{ or}$$
$$q = 1 - p$$

The classical probability is referred as 'a priori' because the probability can be calculated in advance like in examples of dice, cards, balls, coins, etc. For example in a two times toss of coin, the probability of heads showing up in both the tosses is 0.25 (Possible outcomes = HH, HT, TH and TT, i.e. 4 and success = HH, i.e. 1 outcome and where H = Head and T = Tail).

The classical approach of probability has limitations as it mandates 'equally likely' and 'exhaustive' nature of total possible outcomes. Real life situations do not allow these natures. What is the probability of a success of a, say, glaucoma surgical procedure? Had it been 50% or 0.50 (equally likely), it would have resulted in nearly half of surgeries failing over a long-term. On the other hand, many situations present themselves where the total number of possible outcomes may not be exhaustive. In such situations, the possible outcomes would be too large to be enumerated or may tend towards infinity. For example, what is the probability that the MRI machine would malfunction after 10000 hours. Such situations are considered through statistical or empirical probability.

'a posteriori' or Empirical Probability (Empirical Approach)

Empirical Probability is based on the relative frequency of occurrences and overcomes the limitations of classical probability. Empirical probability can be defined as 'relative frequency of desirable outcomes in an indefinitely large number of trials'. If out of 'n' total occurrences, 'x' is the number of favorable outcomes, then the probability of 'x' is stated as:

$$\text{Probability of } x, P(x) = \lim_{n \to \infty} \frac{x}{n}$$

In practice, as it not possible to obtain the limit, hence the probability is considered as $\frac{x}{n}$, with the objective to increase the 'n' so that the value so derived reflects the real probability very closely. With $n \to \infty$, this concept considers probability to be a long-term phenomenon and was developed in response to the problems posed by insurance and survival rates. For example, what is the probability that a person will survive until the age of 75?

Consider the case of classical approach. In an experiment of roll of dice, all numbers are equally likely to appear (probability being equal = 1/6 for all numbers). If the dice is rolled 6 times, then as per the classical probability each number is expected to appear once (1/6 x 6 = 1). If we actually perform this experiment, then it is most likely that all the number would not appear once, but would occur with different frequencies. Classical approach cannot explain this, but such situations are explained by the empirical approach. From the definition of empirical, the situation can be rephrased as 'the probability of each number showing up will tend to 1/6 as number of trials tends to infinity.'

The limitation of empirical probability lies is the fact that it cannot be accurately determined when 'n' tends to infinity, even though it is assumed that the limit is finite and unique. The concept of empirical probability is very useful in practical situations and cannot be eliminated from use.

Subjective Probability

This concept goes against the grain of initial discussion about imparting mathematical function to uncertainty, but it is still being stated here. The reason for this is that it allows the reader, to understand that such a possibility exists, and can be used by the practitioner or the researcher, if strongly and absolutely warranted by the situation. In the classical approach, the probabilities are calculated in advance, while in empirical they are based on the relative frequency of occurrence. In the subjective approach—the probability, between 0 and 1, is assigned on subjective basis, which may at best be partly based on mathematical basis, but not fully. Consider the case of cataract surgery. The empirical probability of the surgery resulting in vision loss, based on 100,000 recorded surgeries over six months at 6 hospitals is 0.0005 (50 failures in 100,000 operations). Without further details about why these surgeries failed, one would assume that the failure rate would possibly hold in future too. But in case there was a subjective knowledge about the patients' eye condition before surgery, camp condition in terms of facilities, postoperative care as well as expertise of the staff carrying out the surgeries, there is a possibility of adjusting the failure rate for future accordingly. Because it is based on the subjective assessment, it may sometimes find some relevance in forecasting techniques and in social sciences, but should be avoided in clinical sciences and research of basic sciences.

Axiomatic Approach

This is the approach which is currently in-use. Due to advantages and limitations of classical and empirical, an axiomatic approach to probability was developed, which included both these approaches. Technically an axiom is defined as 'a self-evident fact requiring no proofs' or mathematically 'a proposition which is accepted without proof and from which further theorems can be developed.' The axiomatic approach to probability relies completely on the logic of deduction and provides no exact definition of probability but provides postulates on basis of which probability can be calculated and other mathematical functions can be performed. This approach uses set theory, with the universal set representing the total possible outcomes 'N' or finite sample space.

The three fundamental axioms are:

Axiom of Positiveness

The probability will always be a positive number, i.e. Probability of an event A, $P(A) \geq 0$.

Axiom of Certainty

The probability of entire sample space, $P(N) = N/N = 1$ [combining the two statements (this simply means that the probability can range from 0 to 100%)] translates into— probability of any event would lie between $0 \leq P(A) \leq 1$).

Axiom of Union

If event A and B are mutually exclusive events (Disjoint), then the probability of occurrence of either of them (i.e. A or B) would be equal to sum of their individual probabilities, i.e.

$$P(A \cup B) = P(A) + P(B)$$

i.e. *Probability of A or B = Probability of A+ Probability of B*
This can generically stated as:

$$P\left(\bigcup_{i=1}^{n} A_i\right) = \sum_{i=1}^{n} P(A_i)$$

As the axiomatic approach draws from set theory, it is worthwhile to mention certain relevant probability terms and their corresponding set theory nomenclature, which will facilitate understanding of other laws of probability. In the chapter 'back to school' basics of set theory have been explained which may be referred.

Set theory terminology	Implication in probability terms
$A \cap B = \phi$ (Null set)	No common elements, the events A and B are mutually exclusive
$A \cap B$	Occurrence of both events A and B
$\bar{A} \cap \bar{B}$	Nonoccurrence of both event A and B
$A \cup B$	From A and B, atleast one of the events occur
$A \subset B$ (A is subset of B)	If event A occurs, then B automatically occurs
$\bar{A} \cap B$	Event A does not occur but event B occurs

Other Important Theorems and Laws of Probability

- Probability of impossible event is Zero, i.e. $P(\phi) = 0$
- Probability of complementary event of A, given as $\bar{A}$ is stated as $P(\bar{A}) = 1 - P(A)$
- Law of addition of probabilities, where A and B are not mutually exclusive (not disjoint), probability of occurrence of event A or B is given as:

$$P(A \cup B) = P(A) + P(B) - P(A \cap B)$$

i.e. for *non-mutually* exclusive events.
 Probability of A or B = (Sum of their respective probabilities) less (probability of combined occurrence of A and B).

It may be noted that this is different from axiom of union which provided the equation for mutually exclusive cases, whereas Law of Addition covers non-mutually exclusive events.

- Law of multiplication of probabilities: For two independent events A and B, the probability that will both occur (combined occurrence) is equal to the product of their respective probabilities, i.e.
$$P(A \cap B) = P(A) \times P(B)$$
Probability of A and B = Product of their respective probabilities
- If A, $\bar{A}$ and B are all mutually exclusive events, then $P(\bar{A} \cap B) = P(B) - P(A \cap B)$
- If event B is a subset of event A, then $P(\bar{A} \cap B) = P(A) - P(B)$ and $P(B) \leq P(A)$
- For any two events A and B, $P(A \cap B) \leq P(A) \leq P(A \cup B) \leq P(A) + P(B)$

CONDITIONAL PROBABILITY

Unlike the independent events considered in multiplicative law, if the two events A and B are dependent events, then multiplicative law does not hold good. For such dependent events, their probability is called conditional probability (occurrence of one event is dependent or conditional to the occurrence of the other event). Conditional probability is denoted as P (A/B) which implies, probability of A, given event B has occurred. Conditional probability can be calculated as:

$$\text{Conditional probability } P(A \mid B) = \frac{P(A \cap B)}{P(B)} = \frac{\text{Probability of (A and B)}}{\text{Probability of B}}$$

Or vice versa

$$\text{Conditional Probability } P(B \mid A) = \frac{P(A \cap B)}{P(A)} = \frac{\text{Probability of (A and B)}}{\text{Probability of A}}$$

Subject to the condition that probability of the first event on whose occurrence second event is dependent, is not equal to Zero.

From above mathematical equation, the Multiplicative Law of Probability can be restated as:

$P(A \text{ and } B) = P(A \cap B) = P(B) \times P(B|A)$, $P(B) > 0$ or
$P(A \text{ and } B) = P(A \cap B) = P(A) \times P(A|B)$, $P(B) > 0$

Example: Conditional Probability:
A basket contains different coloured balls as stated: Yellow-4, Magenta- 3, Cyan- 3, Green - 3, Red-2 and Blue-1. Two balls are to be randomly drawn for play. What is the probability that both the balls would be red in color.

Solution: The total number of balls: 4+3+3 +3 +2+1 = 16
Total number of Red Balls : 2
Probability that first selected ball is red in colour = 2/16 i.e. 0.125
Considering first event has occurred, calculating probability of second event
Total number of balls remaining: 16−1 = 15
Total number of red balls remaining: 2−1 = 1
Probability second selected ball is also red in color =1/15 = 0.067
Hence, probability that both balls are red in color (from conditional probability)
$$= 0.125 \times 0.067 = 0.008$$
Example: A death causing strain of virus manifests itself in multiple ways, making diagnosis difficult. Depending on the stage of infection, even on proper diagnosis the patient may not survive. One patient died for which the concerned doctor was

charged with medical negligence by patient's kin. The available statistics (Number of cases) pertaining to this virus/infection is as below:

Number of patients treated for virus : 1000
Number of patients correctly diagnosed : 800
Correct diagnosis but still patient died : 320
Incorrectly diagnosed patient who died : 140

Evaluate statistically, the probability that the patient was diagnosed correctly.

Solution: Converting the above data as probabilities, and restating the same:

Probability of correct diagnosis $= \dfrac{800}{1000} = 0.80$

Probability of incorrect diagnosis $= \dfrac{(1000-800)}{1000} = \dfrac{200}{1000} = 0.20$

Probability of death despite correct diagnosis $= \dfrac{320}{800} = 0.40$

Probability of death on incorrect diagnosis $= \dfrac{140}{200} = 0.70$

Assume Event A_1 = The infection was correctly diagnosed and
Event A_2 = The patient who had infection died.

Therefore, from above $P(A_1) = 0.80$, $P(\bar{A}_1) = 0.20$, $P(A_2|A_1) = 0.40$ and $P(A_2|\bar{A}_1) = 0.70$.

The probability whether diagnosis was correct or not for a patient who died can be stated as a case of finding $P(A_1|A_2)$, i.e. probability of correct diagnosis with patient having died, i.e.:

$$P(A_1|A_2) = \dfrac{P(A_1 \cap A_2)}{P(A_2)}$$

$P(A_2)$, i.e. event of death is possible in both situation, i.e. correct as well as incorrect diagnosis, hence:

$$P(A_2) = P(A_1 \cap A_2) \cup P(\bar{A}_1 \cap A_2)$$

Therefore,

$$P(A_1|A_2) = \dfrac{P(A_1 \cap A_2)}{P(A_1 \cap A_2) \cup P(\bar{A}_1 \cap A_2)}$$

Using conditional probability equations

$$P(A_1|A_2) = \dfrac{P(A_1) P(A_2|A_1)}{P(A_1) P(A_2|A_1) + P(\bar{A}_1)\left(P(A_2|\bar{A}_1)\right)}$$

Replacing values:

$$P(A_1|A_2) = \dfrac{0.80 \times 0.40}{(0.80 \times 0.40) + (0.20 \times 0.70)} = \dfrac{0.32}{0.32 + 0.14} = 0.70$$

With the available statistics, the conditional probability that the dead patient was correctly diagnosed is 0.70 or 70%. Purely from statistical perspective, there is high chance that the patient died due to manifestation of the infection instead of doctor's negligence.

BAYES THEOREM

Bayes theorem is an extension and unique method of calculating conditional probabilities in the sense it takes into consideration 'revised probabilities' based on additional information. In Bayesian approach, probability is a statistical procedure to estimate parameters of an underlying distribution based on the observed distribution. It starts with a 'prior distribution' including an assessment of the relative likelihoods of parameters. Data is collected and this gives the observed distribution. The likelihood of the observed distribution with given parameter values, is multiplied with the prior distribution and normalized to obtain a unit probability over all possible values. This is the posterior distribution. The 'probability intervals' are therefore calculated using the standard procedure. The limitations imposed by this method are inherent to the procedure as the validity of the result depends on validity of the prior distribution which may not be assessed statistically.

If $A_1, A_2, A_3 ... A_n$, are mutually exclusive (disjoint) and collectively exhaustive events with individual probabilities), $P(A_i) \neq 0$ ($i=1, 2, 3, ... n$), then for any other event B, such that event B belongs to same sample space and is a subset of all A_i i.e. $\cup_{(i=1)}^{n} A_i$, such that $P(B) > 0$, then the probability of A_i conditional to event B is given as:

$$P(A_i | B) = \frac{P(A_i) P(B | A_i)}{\sum_{i=1}^{n} P(A_i) P(B | A_i)}$$

This can be explained by considering a simple situation consisting of two sets of events A_1 and A_2, which are mutually exclusive and completely exhaustive. Another event B is such that it is subset of both the events A_1 and A_2, as represented in the Venn diagram given in **Figure 11.1**.

For such a case, the conditional probability of event A_1, given event B is:

$$P(A_1 | B) = \frac{P(A_1 \text{ and } B)}{P(B)} \qquad \ldots\ldots 1$$

and conditional probability of event A_2, given event B is:

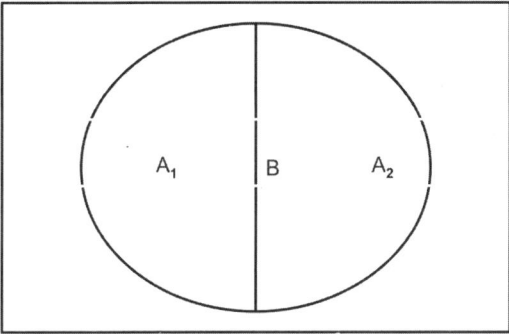

Fig. 11.1: Venn diagram

$$P(A_2|B) = \frac{P(A_2 \text{ and } B)}{P(B)} \qquad \ldots\ldots 2$$

Where $P(B)$ (which intersects A_1 and A_2) is given as:
$P(B) = P(A_1 \text{ and } B) + P(A_2 \text{ and } B)$ 3
As, $P(A_1 \text{ and } B) = P(B|A_1) \times P(A_1)$ 4
and $P(A_2 \text{ and } B) = P(B|A_2) \times P(A_2)$
Using equations (4) in equation (3),
$P(B) = P(B|A_1) P(A_1) + P(B|A_2) P(A_2)$ 5
Now using equation (4) and (5) to rewrite equation (1), i.e.

$$P(A_1|B) = \frac{P(B|A_1) \times P(A_1)}{P(B|A_1)P(A_1) + P(B|A_2)P(A_2)}$$

Or

$$P(A_1|B) = \frac{P(B|A_1)P(A_1)}{\sum_{i=1}^{2} P(A_i)P(B|A_i)}$$

Similarly for:

$$P(A_2|B) = \frac{P(B|A_2)P(A_2)}{\sum_{i=1}^{2} P(A_i) P(B|A_i)}$$

This is simple form of the equation stated earlier for Bayes theorem.

The probabilities $P(A_1), P(A_2),\ldots, P(A_n)$ are 'a priori' probabilities, because they are present before the start of experiment. The probabilities $P(B|A_1)$, $P(B|A_2)\ldots P(B|A_n)$ are the 'likelihoods' as they reflect the probabilities of occurrence of event B against each 'a priori' probability. The probabilities $P(A_1|B), P(A_2|B),\ldots, P(A_n|B)$ are termed as 'posterior' probabilities because they are calculated after the probability results are known.

As can be noticed, the 'a priori' probabilities are unconditional, whereas posterior probabilities are conditional. Bayesian approach, though seemingly similar to conditional probability, actually differs. In conditional probability, the statement generally is 'probability of an event subject to prior occurrence of another event which has a specific probability value'. Against this in Bayesian approach, the statement will be 'Probability of an event subject to another event whose probability will be result of sample or experiment'.

Example: A hospital has two emergency OT's, OT1 and OT2 for critical and emergency cases. Records of last year indicate that 60% of the surgeries were performed in OT1. Out of the total surgeries conducted in OT1, in 3% cases the patients could not be revived (failure). Similarly in OT2, 40% of the surgeries were conducted and out of all the surgeries conducted in OT2, 5% patients could not be revived (failure). If one of these failure cases is randomly drawn, what is the probability that surgery was performed in OT1 or OT2.

Solution:
Assume A_1 = Event of selecting a case pertaining to OT1
 A_2 = Event of selecting a case pertaining to OT2
 B = Event of selecting a failure pertaining to either OT1 or OT2

Then $P(A_1) = 60\% = 0.60$
$P(A_2) = 40\% = 0.40$
$P(B|A_1) = 3\% = 0.03$
$P(B|A_2) = 5\% = 0.05$

The posterior probabilities can be easily calculated by using the tabular format, as shown below:

Event	Prior probabilities $P(A_i)$	Conditional probability $P(B\|A_i)$	Joint probability $P(A_i \text{ and } B) = P(B\|A_i) \times P(A_i)$	Posterior probability $P(A_1\|B) = \dfrac{P(B\|A_1)\,P(A_1)}{\sum_{i=1}^{2} P(A_i)P(B\|i)}$
A1	0.60	0.03	0.018	$= \dfrac{0.018}{0.038} = 0.47$
A2	0.40	0.05	0.020	$= \dfrac{0.020}{0.038} = 0.53$
Total	1.00		$P(B) = 0.038$	1.00*

* The sum of posterior probabilities will always be equal to unity and this acts as a check on the calculations.

Therefore, it can be concluded that the probability that a randomly selected failure case belonged to OT1 = 0.47 (47%) and similarly for OT2 = 0.53 (53%).

CONCLUSION

The basic concept of theory of probability assists in understanding of the probability distributions and the test of significance on which analysis of all research relies upon.

In research we do not prove the hypothesis to be correct or incorrect. What is calculated is the probability of observing the given values if the null hypothesis is not correct. By default we accept the null hypothesis if the probability of observing such condition is not too extreme. We will discuss more about this in the subsequent chapters.

From probability arises the concept of 'distribution of probability'. A properly normalized function assigning a probability 'density' to each possible outcome within given interval is called a 'probability density function' or 'probability distribution function' or in short 'pdf'. The cumulative value of the function (integral for continuous and sum for discrete) is called a 'cumulative distribution function' or cdf. There are many different probability distributions and these are considered in detail over the next two chapters.

CHAPTER 12

Theoretical Discrete Probability Distributions

Some information about discrete and continuous frequency distribution was covered earlier in chapter on frequency distribution. Prior to studying the theoretical probability distributions, whether discrete or continuous, few additional underlying concepts and terminology need to be considered. Usually these concepts may not be directly used by practitioners in applied biostatistics, but they form the bedrock of important assumptions. The reader may find the reading a little tedious at places due to lack of mathematics background but for the sake of comprehensiveness these concepts have been placed here. And for the reason mentioned in the previous sentence, this discussion needs to find a mention in any book on the subject for biosciences graduates or postgraduates.

UNDERLYING CONCEPTS AND TERMINOLOGY

Probability Distribution

In the earlier sections of the book, measures of central value, dispersion and frequency distribution were discussed. Now, assume the further given cases.

In a hall, there are 100 chairs of 5 different colors—Red, Blue, Green, White and Yellow. The counting of these chairs resulted in following values:

Red	–	20
Blue	–	15
Green	–	30
White	–	10
Yellow	–	25 Total = N = 100

This is a simple case of a frequency distribution, where the color of the chair is a *variable* and the number count of chairs is the frequency. More specifically, a case of discrete frequency distribution as the variable is discrete and not continuous. (Though an example of continuous could have been considered, but to explain the terminology, a simpler discrete variable is preferred).

After counting the chairs, if a chair is randomly selected, the probability of it being of a particular color is as follows:

Red	–	0.20	
Blue	–	0.15	
Green	–	0.30	
White	–	0.10	
Yellow	–	0.25	Total Probability = 1.00

The above chart or for that matter any chart, depicting the distribution of probabilities of the variable is called the *Probability Distribution* of that variable.

Random Variable

Random variable is also known as *'chance variable'* or *'stochastic variable.'*

A random variable refers to a real number whose value is determined by the outcome of a random experiment.

For example, consider a random experiment consisting of toss of a coin three times. The outcome of the trial is showing up of tail or head. Assume the random variable to be *Number of tails*. The possible outcome of the experiment is, along with the value of random variable is shown below: (H = Head and T = Tail, HHH means coin showing up 'heads' three times).

Possible outcome	Value of random variable
HHH	0 (as Random variable is the numbers of tails)
HHT	1
HTH	1
THH	1
HTT	2
TTH	2
THT	2
TTT	3

The total number of possible outcomes is called the sample space (S), whereas the outcome of the random experiment corresponding to the real number is denoted as $\chi(\omega)$ and called the Random Variable. If the outcomes are considered a combination of two real numbers, it is called two-dimensional random variable.

If such a random variable takes on integer values such as 0, 1, 2, 3, ... etc., it is called discrete random variable. For example, number of patients visiting a hospital. Alternatively if such a random variable can take all the values within the specified space or interval or range, it is called a Continuous Random Variable. For example, heights of students of a school, which can not only be values such as 140, 141, 142 cms .. but also 140.5, 140.8, 141.9..cms etc.

If $X_1, X_2, X_3....X_n$ are the discrete values of a random variable, and $P_1, P_2, P_3,....P_n$ are their respective probabilities, where sum of all these probabilities = 1, the resulting distribution $P(X)$ is called the *discrete probability distribution* as shown in the initial table. It is also referred as *Probability Mass Function* of the *Discrete Random Variable*.

Similarly, a random variable, if it takes all possible values within a range is called a Continuous Random Variable. With its probabilities, its P(X) is called *probability density function* and its distribution as *continuous probability distribution*. Based on the infinitesimal intervals, if the values of the probability density function are plotted on a graph, resulting into the curve, the resultant curve is called the *Probability Density Curve* or simply the *Probability Curve*.

Considering the initial table once again, two possibilities exist:
1. It is assumed that these 100 chairs signify the total population and 10 chairs are drawn from it. From the probability distribution, the probability of red chair being selected is 0.2, hence out the 10 chairs picked up at random, 2 chairs (0.2 × 10) are expected to be of red color.
2. Other possibility is that these 100 chairs are a subset or themselves a sample drawn from a normal population consisting of 1000 chairs. In that case it is expected that the normal population consists of 200 chairs (0.2 × 1000) of red color. Based on this example is the next concept.

Relative Frequency Distribution

If probability distribution relates to the distribution of the populations, then relative frequency distribution are the distributions drawn from that population. As the sample size increases and tends towards (getting equal to) the total population, the difference between *relative frequency distribution* and the *probability distribution* becomes narrower and narrower.

Mathematical Expectation

The word *expectation* is regularly used in every walk of life. *Rainfall is expected today*. 'It is expected, the cases of the disease will reduce in future owing to vaccination', etc. are some of examples. Though, in generic sense even the word *probability* is used in describing similar situation, like Probably there would be rain today. On closer examination, we find that there is bias towards some firmness in the word expectation, whereas the word *probably* seems to slightly less firm in its connotation. These words have been interchangeably used, but mathematically they imply different aspects. Mathematical expectation is the probability of the variable multiplied by the variable. For example, if there is a probability of 0.5 for heads to show up in a toss of a coin, then the expectation that it will show up in 10 tosses = 10 × 0.5, i.e. 5 times. It is to be noted that the probability of outcome has remained same because the every toss is independent of previous tosses.

Expectation of Discrete Random Variable

Mathematically, if X is a discrete random variable, consisting of values $x_1, x_2, x_3, ..., x_n$; having corresponding probabilities $p_1, p_2, p_3, ..., p_n$, i.e. $P(X = x_i)$, then the Expectation (E) of variable X is given as:

$$E(X) = p_1 x_1 + p_2 x_2 + p_3 x_3 + ... + p_n x_n = \sum_{i=1}^{n} p_i x_i \text{ where } \sum_{i=1}^{n} p_i = 1$$

Expectation of Continuous Random Variable

If 'X' is a continuous random variable with probability density function $f_X(x)$, then its mathematical expectation is given as:

$$E(X) = \int_{-\infty}^{\infty} x f_X(x)\, dx$$

provided that the integral is absolutely convergent. *Absolutely convergent* implies that even if the equation states limits as $-\infty$ to ∞, but the continuous random variable exists for a finite limit and it's expectation too exists for this finite limit.

Additive Theorem of Expectation

The addition theorem of expectation states that the *mathematical expectation of the sum of many random variables* is equal to the sum of their individual expectations, subject to the existence of all the individual expectations, as shown below.

$$E(X + Y + \ldots + Z) = E(X) + E(Y) + \ldots + E(Z);$$

where $X, Y, \ldots Z$ are 'n' random variables and $E(X), E(Y) \ldots E(Z)$ being their respective expectations.

Multiplicative Theorem of Expectation

The multiplicative theorem of expectation states that the 'mathematical expectation of the product of many random variables' is equal to the product of their individual expectations, subject to the existence of all the individual expectations, as shown below.

$$E(XY \ldots Z) = E(X)\, E(Y) \ldots E(Z);$$

where $X, Y, \ldots Z$ are 'n' random variables and $E(X), E(Y) \ldots E(Z)$ their respective expectations.

Example: A diagnostic center is evaluating to open a new branch in one of the area, out of three possibilities—Areas A, B and C. The risk possibilities depending upon various factors for each of the locations is given below:

Location	Profit (₹)	Probability of profit	Loss (₹)	Probability of loss
Area A	20,000	0.8	10,000	0.2
Area B	40,000	0.5	20,000	0.5
Area C	50,000	0.6	30,000	0.4

If maximization of profit is the sole objective, then in which location should the new branch be opened?

Solution: Calculating mathematical expectation of all the three options:
  Area A = 20,000 × 0.8 − 10,000 × 0.2 = 14,000
  Area B = 40,000 × 0.5 − 20,000 × 0.5 = 10,000
  Area C = 50,000 × 0.6 − 30,000 × 0.4 = 18,000
  As area C has highest expected value at ₹ 18,000, hence the new branch should be located in area C.

Theoretical Distribution

Consider the grouped frequency distribution below, based on *actual data* collected (observations) during the year, on the amount of rainfall in mm/day, and number of such days during the year.

Rainfall in 'mm/day'	Number of days
0 – 50	280
50 – 100	70
100 – 150	11
150 – 200	04 ($\Sigma N = 365$)

Contrary to this, if the values of the frequency distribution had been derived from established theory or historical data, it would have been called *Theoretical Distribution*. For example, against the actual rainfall data available, if from historical data the probabilities had been known, then these probabilities could be classified as theoretical for current year estimation. From these theoretical probabilities, the theoretical frequencies can be calculated. The difference between the theoretical frequencies so calculated, and the actual frequencies in normal, as shown below:

Rainfall in mm/day	Theoretical probability based on past years data	Theoretical frequency	Actual observations
0–50	0.76	274 (0.76 × 365)	280
50 – 100	0.20	73	70
100 – 150	0.03	11	11
150 – 200	0.01	4	4
		$\Sigma f = 365$	$\Sigma f = 365$

The variable along with its theoretical frequencies, so calculated, is called the *Theoretical Distribution* of the variable.

Consider another example, an experiment wherein dice is rolled 60 times. The Probability of any number appearing on the throw of dice is equal for all numbers and equals $\frac{1}{6}$. With this information, the expectation of each numbering showing up during the 60 throws of dice would be equal to $\frac{1}{6} \times 60 = 10$, i.e. it is expected that in a throw of dice 60 times, each number will show up 10 times (because the probability of each number is equal to one another). Representing these values in the below table, along with actual observations, which were observed when the dice was actually thrown 60 times.

Number showing up on throw of Dice →	1	2	3	4	5	6
Probability of the number showing up	$\frac{1}{6}$	$\frac{1}{6}$	$\frac{1}{6}$	$\frac{1}{6}$	$\frac{1}{6}$	$\frac{1}{6}$
Theoretical values	$\frac{1}{6} \times 60 = 10$	10	10	10	10	10
Actual observations	14	15	8	9	7	7

In both the above examples, there exists deviation between the theoretical frequencies and the observed frequencies, which is a very usual phenomenon. This deviation can be attributed to:
- *Sampling fluctuation:* As the number of experiments is increased, the observed frequency distributions tend to get closer to the theoretical distributions.
- Error in probability calculation (seldom a cause of deviation)
- The dice may be biased. (Statistics is an investigative tool and no possibilities are ruled out).

Another important aspect in the above examples is the fact that the theoretical frequencies constituting the theoretical distribution are dependent on the probability of the outcome. If the probabilities were based on different definite laws of probability, then the expected values will also undergo a change; and so will the frequencies too, leading to a different theoretical distribution.

These deviations between theoretical and actual are separately studied for their significance in later chapters, but the concepts discussed so far, form the basis of the understanding of the various theoretical distributions.

TYPES OF THEORETICAL DISCRETE DISTRIBUTIONS

Distributions of the discrete random variable are distributed based on definite probability laws and are called the Theoretical Discrete Distributions. Broadly, this includes the following distributions:
1. Binomial distribution
2. Multinomial distribution
3. Negative binomial distribution
4. Geometric distribution
5. Hypergeometric distribution
6. Poisson distribution.

Binomial Distribution

Consider a Bernoullian process, wherein:
- An experiment is performed repetitively (i.e. has '*n*' number of trials)
- Each trial has only two possible outcomes. These two outcomes, for simplification and common understanding, are termed as either *success* or *failure*.
- Success denotes occurrence of an event, whereas failure signifies nonoccurrence of the event.
- Further the outcome of the trial, i.e. success or failure, does not affect the outcome of any of the subsequent trials, i.e. all the trials are independent.
- The probabilities of outcome remain constant during all the trials.

The probability distribution for such '*n*' independent Bernoullian trials, with probability '*p*' of success and '*q*' for failure (and $1 - p = q$), which remain constant in each trial, can be expressed as:

$$P(r) = \binom{n}{r} p^r q^{(n-r)}$$

where n = number of trails, r = number of success in 'n' trials, p = probability of success, q = probability of failure (q = 1– p) and $\binom{n}{r} = \frac{n!}{r!(n-r)!}$ (Combination ^nC_r – refer chapter 'Back to School')

The probability distribution, of number of success obtained from the above equation, is called Binomial Probability Distribution. The two independent constants 'n' and 'p' in the above equation are called the Parameters of the Binomial Distribution. The distribution is discrete as 'r' can only take discrete integral values, and finally a *Binomial Variate* is a variable which follows binomial distribution.

Example: In a city, 40% of its population is undernourished. If 10 people are randomly selected, what is the possibility that 5 of them would be undernourished?
Solution: From the given data:
　Probability of being under-nourished (p) = 40% = 0.4
　Therefore probability of not being under-nourished (q) = 1 – p = 1 – 0.4 = 0.6
　Number of persons selected (n) = 10
　Number of success (selection of under-nourished persons) (r) = 5

Applying binomial probability equation:

$$P(5) = \binom{10}{5} 0.4^5 \, 0.6^{(10-5)} = 252 \times 0.01024 \times 0.07776 = 0.20$$

i.e. probability that out of 10 randomly selected persons, 5 would be under nourished = 0.2 or 20%.

Binomial Distribution: Graphical Representation

Figure 12.1 depicts a graphical representation of binomial distribution with n = 10 and probability of success = 0.5. The x-axis indicates the number of successes and y-axis the probability. Therefore the value point is read as 'probability of having 4 successes in 10 trials, where probability of success in each trial is 0.5 = 0.20.' The graph also portrays the discrete nature of the distribution.

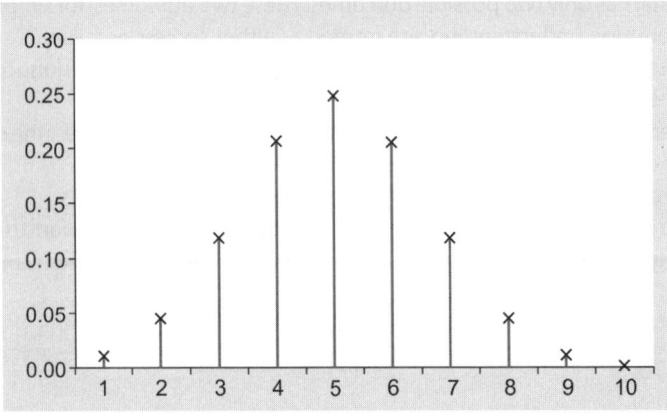

Fig. 12.1: Binomial distribution with parameters n = 10 and p = 0.5

Properties of Binomial Distribution

- For $p = 0.5$ and $q = 0.5$, the Binomial Distribution is symmetrical as can be seen in the below graph. It tends to become right skewed for $p < 0.5$ and left skewed for $p > 0.5$. Drawing the binomial distribution graph with same number of trials $n = 10$, but with different probabilities, i.e. $p = 0.2$ and $p = 0.8$, instead of $p = 0.5$ as shown previously, the direction of skewness can be seen, as shown in **Figures 12.2A and B**.
- The binomial distribution tends to normal distribution, as 'n' increases and tends to ∞. This can be seen from the following graph, having same parameters '$p = 0.5$' while 'n' increases from 50 to 100 **(Figs 12.3A and B)**.

As 'n' the number of trials increases, the points come so close to each other that they appear as a continuous distribution instead of a discrete. These close dots can

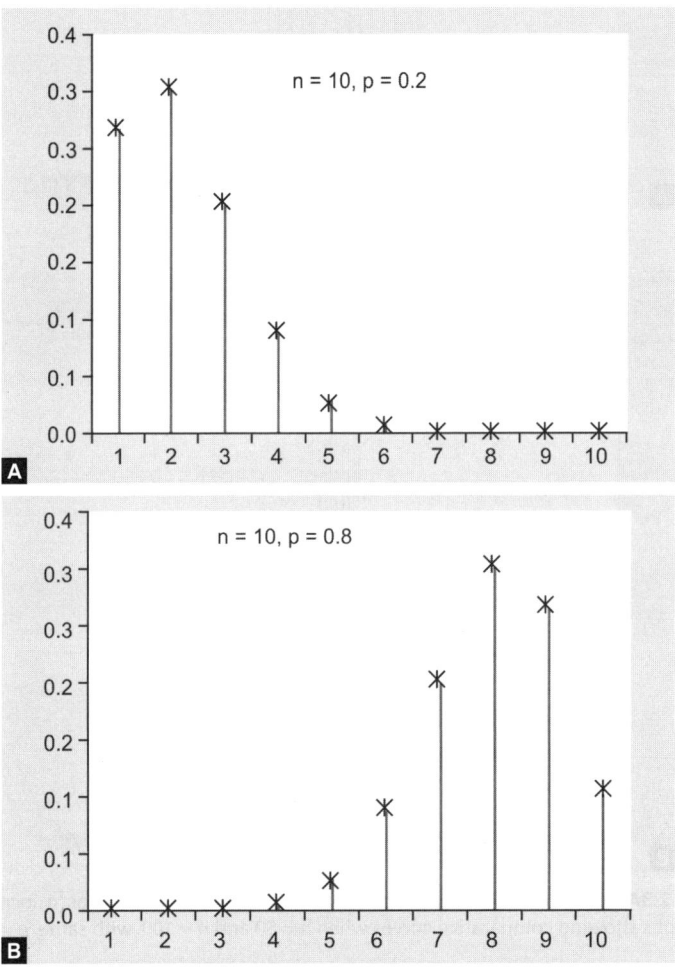

Figs 12.2A and B: Skewness in binomial distribution when probability of success 'p' changes, while 'n' remains same

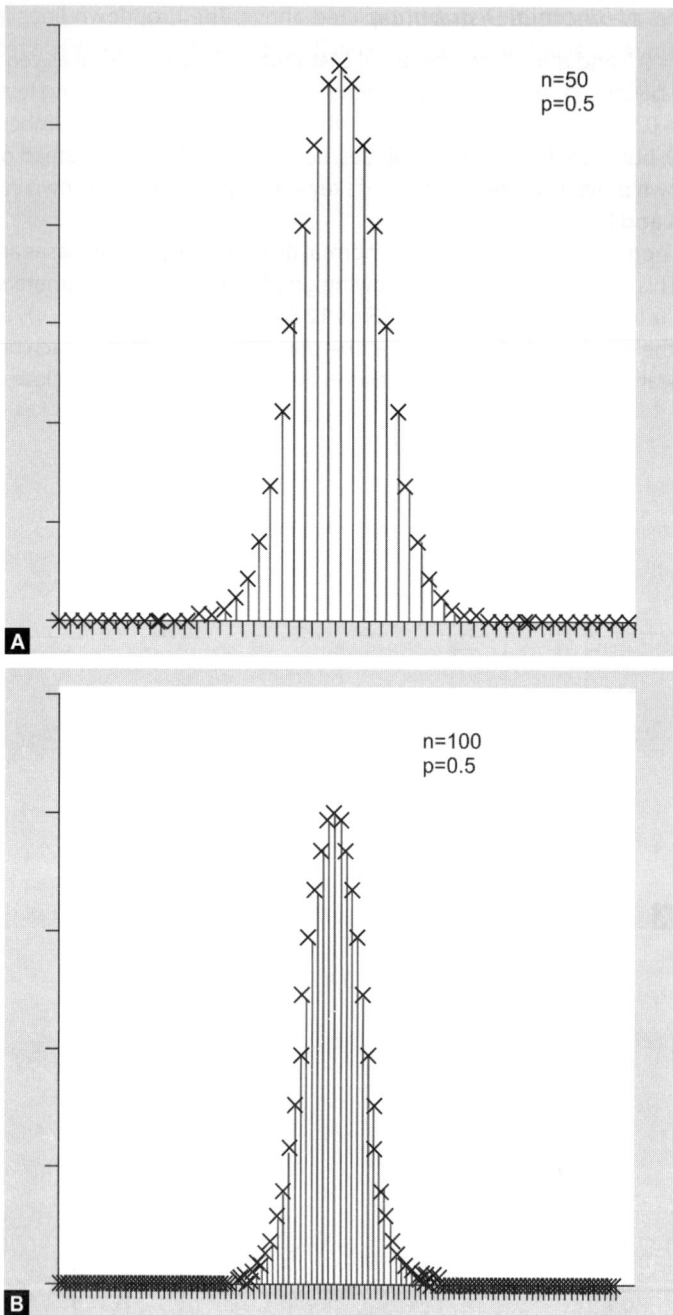

Figs 12.3A and B: Binomial distribution tends to normal distribution as 'n' increases. Graphs showing comparative curves when $n = 50$ and $n = 100$, with same 'p' = 0.5

be seen in the distribution graph depicted above. The dropdown lines from the point seem to appear tightly closed to each other, and the curve shape resembles a normal curve.

Constants of Binomial Distribution

Arithmetic mean $\bar{x} = np$

Standard deviation $\sigma = \sqrt{npq}$

Variance $\sigma^2 = npq$

Measure of skewness $\beta_1 = \dfrac{(1-2p)^2}{npq}$

Measure of kurtosis $\beta_2 = 3 + \dfrac{1-6pq}{npq}$

Mode of the binomial distribution is the value of 'r' for which P(r) is maximum. Hence for a fixed 'n', both mean and mode shift towards right (indicating increase of r) as 'p' increases.

Theoretical frequencies of a binomial distribution are obtained by the following equation:

$$f(x) = N\,P(r) = N\left[\binom{n}{r} p^r q^{(n-r)}\right]$$ where $r = 0, 1, 2 \ldots n$ and $N = \Sigma f_i$

Example: Four coins are tossed simultaneously 120 times, with appearance of head being considered a success. The observations recorded are:

Number of heads appearing in a throw of 4 coins	Number of times of success (Frequency)
0	20
1	30
2	40
3	18
4	12

Calculate the expected frequency distribution. Also calculate the mean and standard deviation of the theoretical and observed frequency distributions.

Solution: Probability of head showing up $'p' = \tfrac{1}{2} = 0.5$, therefore $q = 1 - p = 0.5$

$$N = \Sigma f_i = 120$$

Expected frequency distribution can be calculated by fitting a binomial distribution using the formula (stated above) for values of $'r' = 0, 1, 2, 3$ and 4.

$$f(x) = N\,P(r) = N\left[\binom{n}{r} p^r q^{(n-r)}\right]$$ where $r = 0, 1, 2 \ldots n$

Number of appearance of heads (r)	Applying the formula	Expected frequency f(x)
0	$120 \binom{4}{0} 0.5^0 \, 0.5^4$	7.5
1	$120 \binom{4}{1} 0.5^1 \, 0.5^3$	30
2	$120 \binom{4}{2} 0.5^2 \, 0.5^2$	45
3	$120 \binom{4}{3} 0.5^3 \, 0.5^1$	30
4	$120 \binom{4}{4} 0.5^4 \, 0.5^0$	7.5

Mean and standard deviation of the theoretical binomial distribution:

Mean $= np = 4 \times 0.5 = 2$

Standard deviation $\sigma = \sqrt{npq} = \sqrt{4(0.5)(0.5)} = 1$

Calculating mean and standard deviation of the observed frequency distribution:

Variable (x)	Frequency (f)	f(x)	$(x - \bar{x})$	$(x - \bar{x})^2$	$f(x - \bar{x})^2$
0	20	0	−1.77	3.13	62.6
1	30	30	−0.77	0.59	17.7
2	40	80	0.23	0.05	2.0
3	18	54	1.23	1.51	27.18
4	12	48	2.23	4.97	59.64
	$\Sigma f = N = 120$	$\Sigma fx = 212$			$\Sigma f(x - \bar{x})^2 = 169.12$

Arithmetic mean $(\bar{x}) = \dfrac{\Sigma fx}{N} = \dfrac{212}{120} = 1.77$

Standard deviation $\sigma = \sqrt{\dfrac{\Sigma f(x - \bar{x})^2}{N}} = \sqrt{\dfrac{169.12}{120}} = \sqrt{1.41} = 1.19$

The mean and standard deviation of the theoretical values are 2 and 1 respectively, whereas the mean and standard deviation of the observed values are 1.77 and 1.19 respectively.

From this example it can be seen that there is difference between the observed and theoretical frequencies, and as stated earlier this difference → 0 as the sample size or N → ∞ (dx the difference tends towards zero as the sample size N tends towards infinity or becomes very large).

Multinomial Distribution

Multinomial distribution is generalized case of binomial distribution, where the underlying assumptions are same, but contrary to the two mutually exclusive outcome in a binomial distribution, the multinomial distribution has more than two mutually exclusive outcomes.

Mathematically, if $p_1, p_2, p_3 \ldots p_n$ are the corresponding probabilities to mutually exclusive outcomes $E_1, E_2, E_3 \ldots E_n$, the probability that E_1 occur r_1 times, E_2 occurs r_2 times and so on is given by:

$$P(r_1, r_2, r_3, \ldots r_n) = \frac{n!}{r_1! \, r_2! \, r_3! \, \ldots r_n!} \, p_1^{r_1} \, p_2^{r_2} \, p_3^{r_3} \ldots p_n^{r_n}$$

where, $(r_1 + r_2 + r_3 + \ldots + r_n) = n$, total frequency and $(p_1 + p_2 + p_3 + \ldots + p_n) = 1$

The example below would allow better understanding of the multinomial distribution.

Example: There are 3 hospitals A, B and C in a town, treating 60%, 30% and 10% patients of the town respectively. If 10 patients are randomly selected, what is the probability that:

i. Six of them have been treated by hospital A, 3 by hospital B and 1 by hospital C.
ii. Three have been treated by hospital A, 3 by hospital B and 4 by hospital C.

Solution: As there are three possible outcomes, i.e patient has either been treated at hospital A or B or C, we use multinomial distribution. In this case.

Total sample size = $n = 10$
Probability of having been treated by hospital A = 60% = 60/100 = 0.6
Probability of having been treated by hospital B = 30% = 30/100 = 0.3
Probability of having been treated by hospital C = 10% = 10/100 = 0.1

To find probability for case (i) where $r_1 = 6$, $r_2 = 3$ and $r_3 = 1$, (r_n is the event occurrence). Applying multinomial formula

$$\text{Probability} = \frac{10!}{6! \, 3! \, 1!} (0.6)^6 (0.3)^3 (0.1)^1$$

i.e. Probability = 0.106

For case (ii) also, applying the same formula,

$$\text{Probability} = \frac{10!}{3! \, 3! \, 4!} (0.6)^3 (0.3)^3 (0.1)^4 = 0.0024$$

The probabilities calculated for both cases are different, because in case (i), the occurrence being tested is in similar proportion to the actual probabilities, whereas in case (ii) it specifies an high level of occurrence for an event, which has low probability, i.e. finding higher number of people having been treated by hospital C, which generally treats the least number of patients. This results in a very low probability value of 0.0024.

Negative Binomial Distribution

If for a random variable x, its probability function is given as:

$$p(x) = P(X = x) = P(r) = \binom{x + r - 1}{r - 1} p^r q^n \quad \text{where } x = 0, 1, 2, \ldots$$

Then it is said to follow negative binomial distribution. (The description of characters in the formula remains same as that used in Binomial Distribution especially ( ) signifies *Combination* and not *Brackets*).

Negative binomial distribution is applicable in following conditions which are similar to Binomial Distribution, except condition (iv) stated below:

i. There are 'n' Bernoullian trials, which are independent.
ii. There are only two possible outcomes and which are mutually exclusive and independent.
iii. The probability of success 'p' and failure 'q' remain constant throughout the trial.
iv. The experiment is performed only until a predetermined or prespecified number of successes are achieved. Thereafter the experiment is considered complete.

Explanation

In negative binomial distribution, the situation is different from binomial distribution, in terms of number of trials conducted. In binomial distribution, the number of trials 'n' is fixed, and the number and probabilities of success and failure are deduced for these 'n' trials. In negative binomial distribution the number of trials is not pre-specified, but is a function of the number of success, and the trials are stopped, once the specified number of success has been achieved. If 'r' is the number of successes prespecified, then there should also be, say 'x' failures, before the outcome is 'rth' success, i.e. the total number of trials needed would be $(x+r)$. Imagine the situation, one trial before the $(x+r)$th trials. It is natural that in these $(x+r-1)$ trials, number of success should be $(r-1)$, where r is the total number of pre-specified successes.

Therefore, if 'x' is the number of failed trials preceding the pre-specified rth success, then total number of trials in the experiment would be $(x+r)$. Hence till the trial $(x+r-1)$, there must be $(r-1)$ success. The probability of such cases can be calculated using negative binomial distribution denoted as $f(x: r, p)$ through the following equation:

$$\text{Negative binomial distribution } f(x: r, p) = \binom{x+r-1}{r-1} p^r q^x$$

Where x = Number of failures
r = Prespecified number of success
p = Probability of success
q = Probability of failure $(1-p)$

$$\binom{x+r-1}{r-1} = \frac{(x+r-1)!}{(r-1)!\,(x+r-1)-(r-1)!} = \frac{(x+r-1)!}{(r-1)!\,x!}$$

In negative binomial distribution the variance of the distribution is greater than its mean. This is unlike binomial distribution where the variance is always lower than its mean and Poisson Distribution where mean is equal to variance. Due to this behavior, Negative Binomial Distribution is very suitable for calculating probability in inverse cases like number of shots before the target falls, number of samples required before 10 samples fail, etc.

Constants of Negative Binomial Distribution

$$\text{Arithmetic mean } \bar{x} = \frac{rq}{p}$$

$$\text{Variance } \sigma^2 = \frac{rq}{p^2}$$

$$\beta_1 = \frac{(1+q)^2}{rq}$$

$$\beta_2 = \frac{p^2 + 3q(r+2)}{rq}$$

$$\gamma_1 = \sqrt{\beta_1} = \frac{1+q}{\sqrt{rq}}$$

$$\gamma_2 = \frac{(p^2 + rq)}{rq}$$

Example: In a situation of euthanasia, the probability of an injection of a commonly used 'hit-and-run' sedative in a specified dose causing death to an animal is 10%. What is the probability that an animal would have been unsuccessfully administered 14 doses before it dies assuming that the drug is completely cleared from the body after every injection?

Solution: In the statement, probability of animal dying from injection is 10%, i.e. $p = 0.1$, hence $q = (1 − p) = 0.9$. Further the 15th dose would be considered a success, implying 14 failures, i.e. $x = 14$, and $r = 1$, as this dose would prove to be lethal. With these parameters, applying negative binomial distribution, for calculating the probability:

$$\binom{x+r-1}{r-1} p^r q^x = \binom{14+1-1}{1-1} (0.1)^1 (0.9)^{14}$$

$$= \binom{14}{0} (0.1)(0.228) = 0.0228$$

i.e. the probability that the 15th dose and not before or after, would be the lethal one is 0.0228.

Using this example to elaborate further, from theory of probability, the probability that the first injection is lethal is 0.1. For second injection to be lethal, the first one has to nonlethal, i.e. the probability is $(0.9) \times (0.1) = 0.09$. Accordingly, for the third injection to be lethal, first two have to nonlethal while the third one has to be lethal. Probability of such a case would be $(0.9)(0.9)(0.1) = 0.081$. For 10th injection to be lethal the probability is 0.038 and for 15th it further reduces to 0.0228, as shown above. The probability is reducing with increase in number of injections, due to fact that it is highly unlikely that one lethal dose would not have been administered for so long. Hence negative binomial distribution is a case of *compound probabilities*. Compare this with binomial distribution, where the question will have to reframed as "what is the probability that the animal dies in 15 doses." With all parameters remaining same, the probability that first dose is lethal is 0.1, and the probability that the animal dies by second dose, would be alike *any of the first or second dose is lethal…* and so on. Hence for binomial distribution the probabilities would increase initially to reach a maximum and then decrease. This explains the difference between negative binomial and binomial distributions.

The example considered to explain, negative binomial distribution and its difference with binomial distribution was elementary, as it discussed about first success, whereas negative binomial distribution is used, more often than not, in cases where the interest lies in number of successes being more than one.

Geometric Distribution

A random variable will have geometric distribution, if it assumes non-negative values and its probability mass function is as follows:

$$P(X = x) = q^x p,$$

where $x = 0, 1, 2, \ldots$ and denotes the number of failures preceding first success; p = Probability of success and $q = (1 - p)$ is probability of failure; and probabilities remain constant in all trials.

From above definition, it appears to be a negative binomial distribution, but the fine line dividing these two distributions is the number of successes. In geometric distribution, the consideration is for 'probability of achieving first success preceded by 'x' number of failures', whereas in negative binomial distribution the consideration is *continuation of the experiment until specified number of success is achieved*. Hence geometric distribution is a limiting case of negative binomial distribution, when $r = 1$, i.e. first success. Other assumptions in geometric distribution are similar to negative binomial distribution. Geometric distribution is considered *memoryless* as in repetitive trials, the observed failures do not alter the conditional probabilities of success in following trials.

Further, the above stated probability function ($q^x p$) (where 'x' is the number of failed trials until first success), implies that the $(x + 1)$th trial is a success. In case the probability is desired for xth trial to be successful, then the probability function needs to restated as:

$$P(X = x) = q^{x-1} p, \text{ where } x \text{ is the first successful trial.}$$

Constants of Geometric Distribution

$$\text{Arithmetic mean } \bar{x} = \frac{q}{p}$$

$$\text{Variance } \sigma^2 = \frac{q}{p^2}$$

Example: If the probability that a chemical treatment can kill a pathogen is 0.40, then what is the:
i. Probability that pathogen would be killed in the 5th Chemical treatment.
ii. Probability that there would five treatments before the pathogen is killed.

Solution

- The probability that the pathogen would be killed in the 5th treatment implies fifth attempt is a success (i.e. four failed treatments, before success in fifth), therefore 'x' = 5, p = 0.40 and $q = 1 - 0.4 = 0.6$
  $P(X = x) = q^{x-1} p$, where x is the first successful trial
  $P(X = 5) = (0.6)^{5-1} (0.4) = 0.052$

- In this case, there are number of failures before first success has been given as trial number 5 (the statement implies 5 failure before, i.e. the pathogen gets killed in the sixth trial) therefore:
$P(X = x) = q^x p$, where x = number of failures preceding first success
$P(X = 5) = (0.60)^5 (0.4) = 0.03$.

Hypergeometric Distribution

Theoretical discrete distributions discussed so far, have one of the underlying assumptions as *the probability of success or failure does not change throughout the trial space*. Consider a finite population, in which a random sample drawn is not placed back in the population, i.e. *simple random sampling without replacement*. In such a situation, the probability of subsequent draw also undergoes change. Even though the subsequent trials would become stochastically dependent, but they remain random. Such cases are dealt in hypergeometric distributions.

Hypergeometric distribution is based on following conditions:
- The experiment conducted is finite.
- There are two possible outcomes for each trial viz. success or failure
- The probabilities of these changes with each draw/ trial/event/ random sample.
- Subsequent draws/samples/events are dependent.

Subject to these conditions, the probability mass function of a hypergeometric distribution is given as:

$$P(X = k) = h(k; N, M, n) = P(k) = \frac{\binom{N-M}{n-k}\binom{M}{k}}{\binom{N}{n}}$$

where $P(k)$ = Probability of event k,
N = Total population size,
M = Population of success,
$N - M$ = Population of non-success,
n = Sample size (without replacement) and
k = Number of success in the same.

It is worth observing that hypergeometric distribution tends to binomial distribution when $n \to \infty$ and $\frac{M}{N} = p$.

The following example would help to understand the equation parameters.

Example: A box contains 10 balls, consisting of 3 orange and 7 green. A sample of 4 balls is drawn without replacement. Calculate the probability that the sample would have 2 Green Balls.

Solution: Considering draw of a green ball is success, then from the given details:
N = Total size = 10
M = Population of success (green balls) = 7
n = Sample size = 4
k = Success events in the sample = 2

Being a case of hypergeometric distribution, the probability is calculated by using the formula:

$$P(k) = \frac{\binom{N-M}{n-k}\binom{M}{k}}{\binom{N}{n}}$$

i.e. $P(2) = \dfrac{\binom{10-7}{4-2}\binom{7}{2}}{\binom{10}{4}} = \dfrac{\binom{7}{2}\binom{3}{2}}{\binom{10}{4}} = \dfrac{63}{210} = 0.30$

Hence the probability of getting 2 green balls in a sample of 4 balls drawn without replacement from the box is 0.30.

Uniform Discrete Distribution

This is the simplest form of distribution among all the discrete distributions. Consider a population of size 'n,' distributed uniformly and probability of selecting any sample from the population is uniform and equal for all the samples constituting the total population. Probability for such a random variable is given by:

$$P(r) = \frac{1}{n}, \text{ where } r = 1, 2, 3 \ldots n$$

Uniform discrete distribution find usefulness in basic probability calculations, like probability of any number showing up in a roll of dice (all number have equal probability = 1/6) or drawing a card from pack of cards (1/52), etc. Due to its elementary nature and restrictive condition, it supports other distribution in calculation of the probability of the outcome, subject to its condition being fulfilled.

Poisson Distribution

Poisson distribution, along with binomial distributions is most frequently used discrete distributions.

Poisson distribution is a limiting form of binomial distribution, wherein if, in a binomial distribution, where $P(r) = {}^nC_r p^r q^{n-r}$, the number of trials 'n' become indefinitely large, i.e. $n \to \infty$, and Probability of success is very small, i.e. $p \to 0$, and its mean 'np' is denoted 'λ' (Lambda λ being finite and a positive real number), then poisson distribution can be defined as:

$$P(r) = \frac{e^{-\lambda} \lambda^r}{r!}$$

where $e = 2.7183$ (base of natural logarithm) and λ = mean of poisson distribution.

In poisson distribution, as the probability is very small and population size very large, the success events occur at very random points, as against binomial distribution where events occur as outcome of definite number of trials. Therefore, poisson distribution helps in understanding these very low probability occurrences and not concerned about probabilities of nonoccurrences. Situations which can be evaluated using poisson distributions are number of accidents in a city, calculating count of bacteria, quality control in production, number of deaths due a specific rare ailment, number of machine failures in a high quality phacoemulsification machine, etc.

Constants of a Poisson Distribution

Mean $\bar{x} = \lambda$
Variance $\sigma^2 = \lambda$; (Standard deviation $\sigma = \sqrt{\lambda}$)

$$\beta_1 = \frac{1}{\lambda}$$

$$\beta_2 = 3 + \frac{1}{\lambda}$$

From these constants, it can be inferred that poisson distribution is a skewed distribution as also shown in **Figure 12.4**. If mean of poisson distribution tends to infinity, then its reciprocal, would tend to zero. As it's reciprocal is a measure of skewness, the skewness will also tend to zero. Further its measure of Kurtosis would tend to 3, which implies a mesokurtic curve. These attributes—no skewness and mesokurtic curve is a property of normal curve. Therefore, in these conditions a poisson distribution tends to a normal distribution. As can be seen in the curves for poisson distribution in **Figure 12.4**, if its mean λ is increasing, the curve is losing its skewness, and is tending towards normality. Further it can be noticed that for less common or rare occurrences, the probability is higher. Therefore, poisson distribution is also referred to as *distribution of rare events*.

In poisson distribution if $\lambda \to \infty$ then $\beta_1 = \frac{1}{\infty} \to 0$ and $\beta_2 = 3 + \frac{1}{0} = 3$, i.e. tends towards normal distribution.

As stated earlier, poisson distribution being a limiting factor of binomial distribution, for practical purpose, if size of population is larger than 30 and probability of success ≤ 0.05, use of poisson distribution should be preferred instead of binomial distribution.

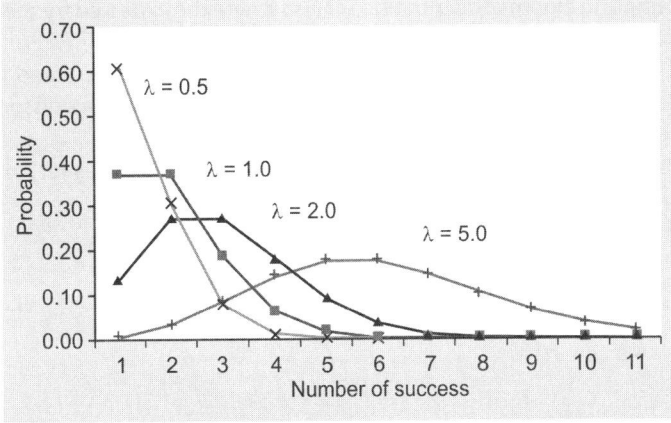

Fig. 12.4: Poisson distribution curves

Example: As a mean rate 1 person falls sick in a month among 10,000 persons living in the area. What is the probability that 5 people will fall sick during a particular month in a village having population of 25,000 people.

Solution: Given—Population Size $(n) = 25{,}000$, Probability of falling sick $(p) = 1/10000 = 0.0001$.

Therefore Mean $= \lambda = np = (25{,}000)(0.0001) = 2.5$

Being a poisson distribution variate, Probability of 5 people falling sick in a month (for $r = 5$)

$$= P(r) = \frac{e^{-\lambda}\lambda^r}{r!} \text{ i.e. } P(5) = \frac{e^{-2.5} 2.5^5}{5!}$$

= The equation can be calculated taking log and antilog (Using log and antilog tables)

i.e. $P(5) = \dfrac{\text{Reciprocal \{antilog }(2.5 \times \log 2.7183)\} \times 97.66}{120}$

$= \dfrac{\text{Reciprocal \{antilog }(2.5 \times 0.4343)\}\, 97.66}{120}$

$= \dfrac{\{\text{Reciprocal (antilog } 1.08575)\}\, 97.66}{120}$

$= \dfrac{\{\text{Reciprocal (12.18)}\}\, 97.66}{120} = \dfrac{1}{12.18} \times \dfrac{97.66}{120}$

= 0.067 i.e. Probability that 5 persons will fall sick in a month is 0.067.

(The detailed calculation shown above, are for ease of understanding the calculation, alternatively the value of $e^{-\lambda}$ can be directly taken from the tables of $e^{-\lambda}$ or can be calculated in excel or through other relevant softwares).

Fitting a Poisson Distribution

Deduction of the theoretical frequencies or the frequency distribution, when select parameters of the distribution are known, is called Fitting of the Distribution. (Refer to the example in binomial distribution section where the expected frequencies, i.e. theoretical frequencies were calculated using the formula).

For a poisson distribution if mean, λ is given or can be calculated from the observations, then the frequency distribution, considering it be a poisson variate, can be derived using the below equations:

If the sample size is N, then the poisson distributed frequencies for $r = 0, 1, 2, 3 \ldots N$ would be,

$$N(P_0) = Ne^{-\lambda}$$
$$N(P_1) = N(P_0)\frac{\lambda}{1}$$
$$N(P_2) = N(P_1)\frac{\lambda}{2}$$

$$N(P_3) = N(P_2)\frac{\lambda}{3}$$

i.e. $$N(P_n) = N(P_{n-1})\frac{\lambda}{n}$$

This formula can be used to find subsequent probabilities too, by cancelling the N from both sides of the equation. The below examples shall make the concept more clear.

Example: A surgical procedure has mean number of failures of 2 cases in a sample of 100. Calculate the probabilities for up to 4 failures and fit a poisson distribution for the same.

Solution: From the given data mean $\lambda = 2$; $N = 100$.

Up to 4 defects, includes number of defects being 0, 1, 2, 3 and 4. Calculating probabilities for $r = 0, 1, 2, 3, 4$.

Using the poisson distribution equation $P(r) = (e^{-\lambda}\lambda^r)/r!$

$P(0) = (e^{-2}2^0)/0! = e^{-2} = 0.135$
$P(1) = (P_0 \times \lambda)/1 = (0.13543 \times 2)/1 = 0.271$
$P(2) = (P_1 \times \lambda)/2 = (0.271 \times 2)/2 = 0.271$
$P(3) = (P_2 \times \lambda)/3 = (0.271 \times 2)/3 = 0.180$
$P(4) = (P_3 \times \lambda)/4 = (0.180 \times 2)/4 = 0.090$

Probability for more than 4 defects would be given as (1−probability up to 4 defects) = 0.053.

Theoretical/Expected frequencies for the above data:

Number of defects (r)	P(r)	Expected frequency = N P(r); (N = 100)
0	0.135	13.5
1	0.271	27
2	0.271	27
3	0.180	18
4	0.090	9
	$\Sigma = 0.947$	$\Sigma = 94.5$

The Σ of P(r) or NP(r) is ≠1 or 100, respectively. This is because the probabilities and expected frequencies have not been calculated for number of defects more than 4. If we add the same, i.e. 0.053 and 5.3, respectively the sum of probabilities becomes 1 and expected frequencies N = 100 (slight difference due to rounding off errors).

Example: A survey was conducted for number of visits by a person from the population, to a hospital for treatment during the month. The survey elicited the following responses:

Number of visits: 0   1   2   3   4
Number of persons: 40   27   15   10   8

Fit a poisson distribution and find expected frequencies.

Solution: Calculating the mean of the distribution:

Number of visits (x)	Number of visits (Observed frequency, f)	fx
0	40	0
1	27	27
2	15	30
3	10	30
4	8	32
Total	$\Sigma f = 100$	$\Sigma fx = 119$

Mean = $\Sigma fx/\Sigma f$ = 119/100 = 1.19, i.e. λ of the distribution = 1.19.

Based on the value of $\lambda = 1.19$, the following probabilities and expected frequencies can be derived.

Number of visits (x)	Number of visits (observed frequency) (f)	Probability function P(r)	Probability calculated by: $P(r) = (e^{-\lambda}\lambda^r)/r!$	Expected frequencies NP(r)
0	40	P(0)	0.304	30
1	27	P(1)	0.362	36
2	15	P(2)	0.215	22
3	10	P(3)	0.085	9
4	8	P(4)	0.025	3
Total	$\Sigma f = 100$			$\Sigma Np(r) = 100$

There exists a difference between the observed frequencies and theoretical expected frequencies. Further statistical analysis by like 'Goodness of Fit' are used to confirm whether the fit is close enough to be considered or not, i.e. whether the frequency distribution is good enough to be called a Poisson Distribution or not.

There are still more discrete distributions but they are rarely used. Binomial and poisson distributions are mostly used along with negative binomial and geometric distribution.

CONCLUSION

In conclusion, it can be stated that, through the concept of probability distribution, a random variable is the subject of study on two aspects, namely:

i. The collection of all possible values that the random variable can manifest in and

ii. The probability for each of these values which the random variable takes.

Depending upon the characteristics of the random variable, the appropriate probability distribution is selected for its study. The first differentiation is whether the random variable is discrete or not. If it is discrete, then one of probability distributions discussed in this chapter can be used.

Contrarily, more often than not, random variable is continuous in nature. For such random variables, different types of probability distributions are required for its analysis. For continuous random variable, no formula is there to provide the probability at specific point, but the focus lies in finding the probabilities within a stated interval. The discussion on important continuous probability distributions then becomes subject of the next chapter.

CHAPTER 13

Theoretical Continuous Probability Distributions

Frequency distribution, depending on the variable, can be either discrete or continuous. Basics of frequency distribution-discrete and continuous, theoretical and observed have already been dealt in the earlier chapters on frequency distribution and theoretical discrete distributions. These remain the same and, therefore, are not being covered again in this chapter.

The following types of distributions pertain to the probability function of a continuous random variable and are known as Continuous Distributions. Other continuous distributions though exist, but they are not being considered here as they are used in a very limited manner. Normal distribution is most widely used and forms the basis of the various tests of significance which are vital in inferential statistics. Therefore, the following pages are devoted to it to deal with it in reasonable detail. Broadly, the following continuous distributions are of use in biostatistics:
- Normal distribution
- Rectangular or uniform distribution
- Lognormal distribution
- Exponential distribution.

NORMAL DISTRIBUTION

The nomenclature does not signify and should not be construed that all other distributions are *not* normal or *abnormal*. A normal distribution is a continuous theoretical distribution.

Consider a variable which is continuous in nature, i.e. instead of taking up distinct values it can take any intermediate value and for such variables a continuous distribution is required for its study. Normal distribution or sometimes called as normal probability distribution is one such distribution which helps in study of such variables.

Normal distribution forms the basis of most of the research data analysis as the hypothesis testing, tests of significance, etc. are all based on normal population assumption.

The probability of a random variable 'x', under normal distribution is given by:

$$P(x) = \frac{e^{\frac{-(x-\mu)^2}{2\sigma^2}}}{\sigma\sqrt{2\pi}}$$

Where P(x) = Probability of the continuous Random Variable x (–∞ < x < ∞)
μ = Mean of the normal distribution
e = Constant approximated at 2.7183
σ = Standard deviation of the normal distribution, σ^2 = Variance

Fitting of the Normal Distribution

The expected frequencies for the normal distribution can be calculated by the following formula, where N is the population size.

$$P(x) = \frac{Ne^{\frac{-(x-\mu)^2}{2\sigma^2}}}{\sigma\sqrt{2\pi}}$$

The frequency distribution curve for a normal distribution is ubiquitous in medicine texts and is given in **Figure 13.1**.

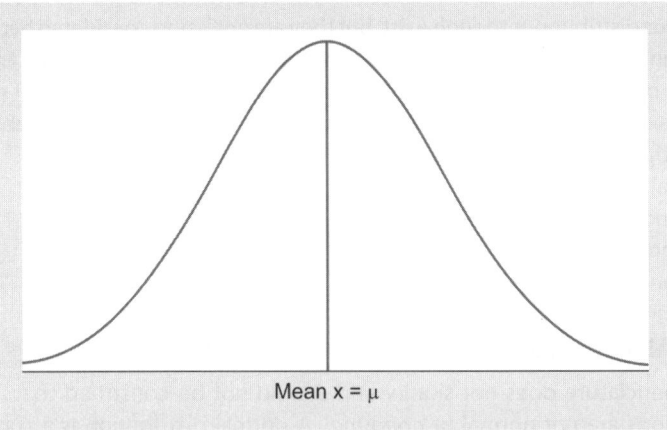

Fig. 13.1: A normal curve—(Bell-shaped and therefore also called Bell-curve)

Features and Properties of Normal Distribution

- Normal distribution is an approximation to Binomial distribution, when n → ∞, and probability 'p' or 'q' may or may not be equal, but are not very small.
- Normal distribution is also a limiting case of Poisson distribution, when λ → ∞, i.e. mean is very large.
- The graph of a normal distribution is known as *normal curve* or *normal frequency curve*. It is also popularly called the 'Bell-shaped curve.'
- The normal curve is symmetrical around its mean i.e. $\bar{x} = \mu$.

Chapter 13: Theoretical Continuous Probability Distributions

- For a given pair of mean μ and standard deviation σ, there is one and only one normal distribution and hence such a combination will have only one unique curve.
- The mean, median and mode coincide in a normal distribution, i.e.
  Mean = Median = Mode
- The maximum probability lies at the center point, i.e. when $x = \mu$ (mean)
- The constants β_1 and β_2 for normal distribution are 0 and 3 respectively. Hence the normal curve does not have any skewness, and also it is mesokurtic (no peakedness and no flatness).
- The normal distribution and so its Curve is asymptotic to x-axis, i.e. the two-tails of the curve tend to ∞, and never touch the x-axis.
- The quartiles Q_1 and Q_3 are equidistant from the Median (Q_2) or mean.
- For ease in calculation and representation, the curve is considered to have a *unit area under it*, by equating the total frequency to 1.
- Inflexion point of the curve (point where the curvature of the curve changes- concave to convex or vice versa) occurs at $\mu \pm \sigma$.
- For a normal distribution, the mean deviation about Mean is (4σ)/5 (approximately) and quartile deviation at (2σ)/3. Collectively the ratio's between its quartile deviation (QD), mean deviation (MD) and standard deviation (SD) is given by:
  QD:MD:SD :: (2σ)/3:(4σ)/5:σ
- Normal distribution is a unimodal distribution, i.e. it has only one mode and whose value = μ.
- The sum and difference of two independent normal variates is also a normal variate. If x_1 and x_2 are two normal variates, then for combined ($x_1 + x_2$), the $\mu_{combined} = (\mu_1 + \mu_2)$ and $\sigma_{combined} = \sigma_1 + \sigma_2$. Same holds true for difference also.
- Normal distribution follows the property of *central limit theorem*, which states that the mean of the distribution $\bar{x}$, belonging to any population, would tend towards normal distribution as the sample size n increases. This implies that even for non-normal distributed populations, distribution of the mean of the samples drawn from the population will exhibit normal distribution properties, as n → increases. This property of normal distribution forms its basis for extensive usage is the research studies.

Property of the Normal Curve

In a normal curve, the probability and the area under the curve is distributed as follows:

x-value defined range	Probability for the defined range	Corresponding area under curve within the range	Remarks (also called as)
$\mu - \sigma < x < \mu + \sigma$	0.6826	68.26%	1 Sigma Limits
$\mu - 1.96\sigma < x < \mu + 1.96\sigma$	0.9500	95.00%	5% Significance
$\mu - 2\sigma < x < \mu + 2\sigma$	0.9544	95.44%	2 Sigma Limits
$\mu - 2.58\sigma < x < \mu + 2.58\sigma$	0.9900	99.00%	1% Significance
$\mu - 3\sigma < x < \mu + 3\sigma$	0.9973	99.73%	3 Sigma Limits

The 1, 2 or 3 Sigma limits are extensively used in statistical quality control methods, while the 1% and 5% significance limits are the most common limits used in the statistical tests of significance.

Standard Normal Curve

A normal curve with mean $\mu = 0$ and standard deviation $\sigma = 1$, is known as standard normal curve, and the distribution is called *standard normal distribution* **(Fig. 13.2)**. The values of *Area under a curve* of such a *standard normal curve* are the ones which are calculated and tabulated in the tables of *area under normal curve*. Hence, it should be noted that the area under normal curve tables are for a normal distribution having $\mu = 0$ and $\sigma = 1$.

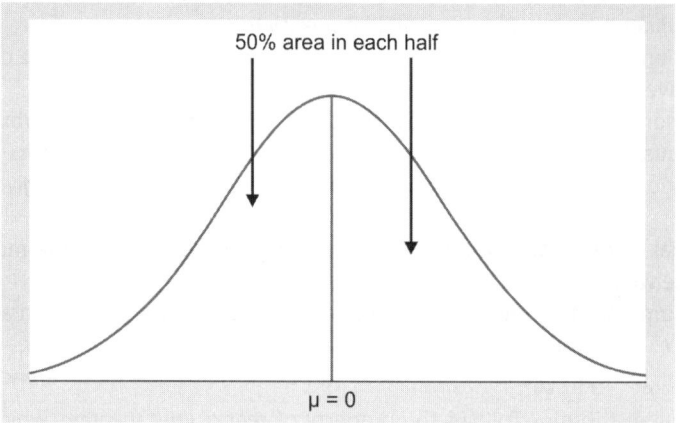

Fig. 13.2: A standard normal curve has mean = 0 and standard deviation = 1. Area under the standard normal curve is taken as *unity*

Area Property of a Normal Curve

The area between any two points on x-axis provides the proportion or percentage of cases or probability for the cases to fall between the two stated points. Refer to the table providing the percentage or probability given in the table of area under the curve in previous section. As the normal distribution is symmetrical about its mean, the area under the curve is split in two equal halves on either side of its mean. Therefore, in table the value is considered '±' from its mean. As the total area under the curve is considered unity, each half constitutes 0.5 of the probability or 50% of the area.

For explanation consider the case of $\mu - 2\sigma < x < \mu + 2\sigma$: In this case the value of variable x lies between $\mu \pm 2\sigma$, i.e. with mean being at center of the curve and limits being 2σ on either side of the central point. The area of the curve that would get covered within these limits is 95.44% and the probability that a normal distribution variate will fall within these limits would be 0.9544. This area would be divided equally on both the sides of the normal curve's mean value, i.e. 47.72% on either side.

Chapter 13: Theoretical Continuous Probability Distributions

In the earlier points, it has been specifically mentioned that for a given combination of mean and standard deviation, there is one and only one normal curve.

Consider the following case, where the curves for values: Curve A ($\mu = 50$, $\sigma = 10$), Curve B ($\mu = 50$, $\sigma = 15$), Curve C ($\mu = 30$, $\sigma = 20$) and Curve D ($\mu = 40$, $\sigma = 10$) are plotted. **Figure 13.3** of normal curves shows that the normal curve for each different value combination of mean and standard deviation gives rise to different normal curve.

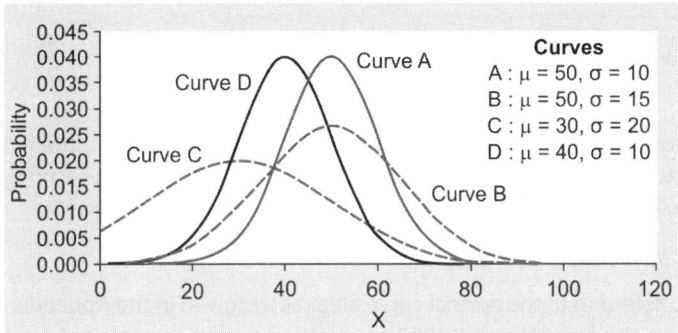

Fig. 13.3: Normal curves for different mean and standard deviation

For arriving at the *Area under the curve* values for each of these individual curves is not easy. Considering that there are infinite combination-possibilities of the values of mean and standard deviation, it is impossible to have infinite number of tables for *area under curve values*. This is resolved by converting the normal curve of any normal distribution into *standard normal curve* by equating the $\mu = 0$ and $\sigma = 1$. This is achieved by changing the scale and origin of the curve. It is like super imposing the normal distribution curve over a standard normal curve, i.e the mean of the normal distribution is taken as 0, and the scale is taken as units of standard deviation from the mean. The new scale so developed is called the *z*-Scale, and is equated as:

$$z = \frac{(x - \mu)}{\sigma}$$

The change of scale and origin is shown above in the graphical format.

As the origin is transformed to zero and with the curve being symmetrical about its mean, the values on the left of the mean would then have negative deviation from mean. This is similar to the *XY*-axis graphs, where the values to the left of *XY* (0,0) are negative for *X*-axis. This eases the broad understanding of the normal curve, as it can be visualized as curve on Graph having two quadrants viz (*XY*) and (–*XY*) with the *Y*-axis coinciding with the mean of the normal distribution, and scale on *X*-axis being expressed in terms of units of standard deviation as shown in **Figure 13.4**. This transformation also eliminates the need to have *area under the curve* tables for negative values of *z*, as they can be equated with +*z*, because of symmetrical behavior of normal distribution. Hence the tables are for 50% values. For negative values of *z*,

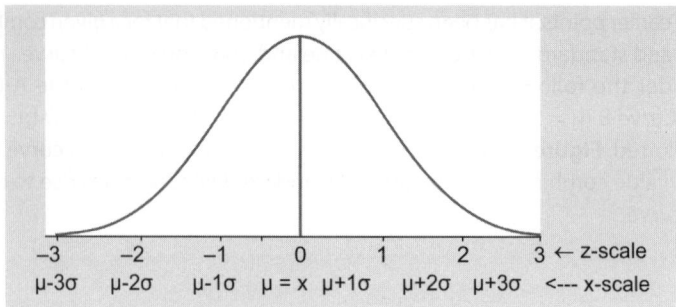

Fig. 13.4: Change of scale and origin of normal curve to convert any normal curve to standard normal curve

the area under the cover would be same as that +z. Further as the probability of any variable to exist beyond 3σ limits is 0.0027, most of the tables have values till $z = 3$. With availability of computing software, this is no more a restriction.

Example: What is the area under the normal curve for $0 < z \leq 2.59$?

Solution: Referring to the normal curve area tables (given in the Appendix: Table 5), for $z = 2.59$, the area given is 0.4952, i.e. probability of the variate to be in this zone is 0.4952 and area under the curve for this limit is 49.52%. This area being measured is shown as shaded portion in the normal curve **(Fig. 13.5)**.

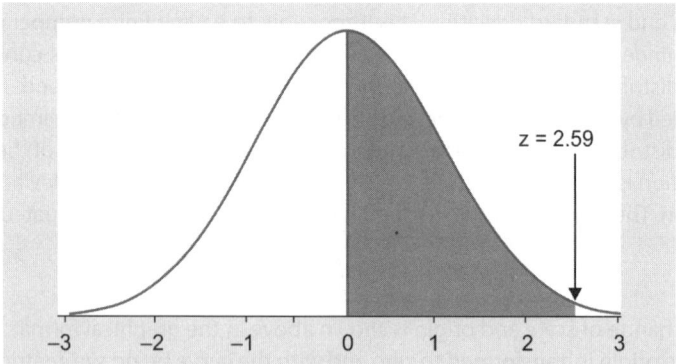

Fig. 13.5: Area under the curve $0 < z \leq 2.59$

Example: Find the area under the curve for $z > 0.40$.

Solution: Representing graphically $z > 0.40$.

Area for $z > 0.40$ will not be directly available in the tables **(Fig. 13.6)** because the table has the values from mean $\mu = 0$, with the upper end being open-ended, therefore for calculating for $z > 0.40$, it has to be considered as:

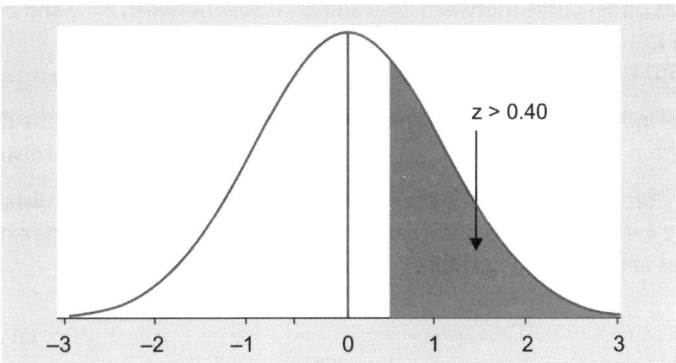

Fig. 13.6: Area under the curve z > 0.40

Area for $z > 0.40$
  = 1 − [(Area on left of the mean μ or $z = 0$) + (Area between point $z = 0$ and $z = 0.40$)]
  = 1 − (0.5000 + 0.1554) (*Note*: Area under the total curve = 1 with 0.50 in each half)
  = 1 − (0.6554)
  = 0.3446 i.e. 34.46%

Alternatively it could have been solved by only considering the half of the curve in i.e.
  = 0.5 − 'Area between point $z = 0$ and $z = 0.40$'
  = 0.5 − 0.1554 = 0.3446

Example: Find the probability of normal variate lying between two points $z = 0.40$ and $z = 0.50$ **(Fig. 13.7)**.

Solution: Representing graphically the required area.
Therefore this can be calculated as follows:
Area between ($z = 0.40$ and $z = 0.50$)

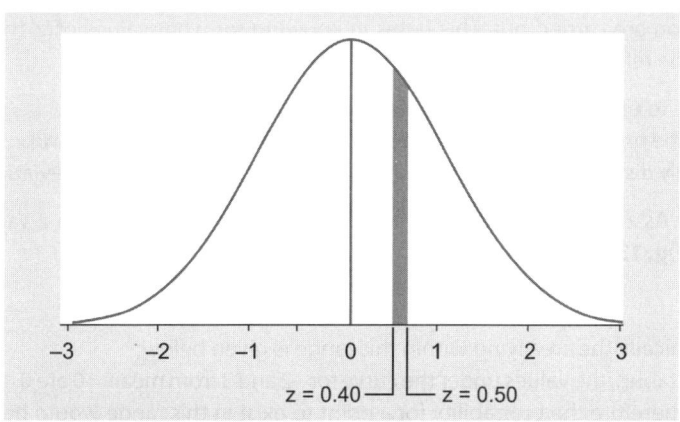

Fig. 13.7: Area under the curve for 0.40 < z < 0.50

= Areas under curve (between $z = 0$ and $z = 0.50$)−(between $z = 0$ and $z = 0.40$)
= 0.1915 − 0.1554 (from normal curve tables)
= 0.0361 or 3.61%.

Example: Find the probability that a normal variate x would lie within the deviation ranging from $(\mu - 0.5\sigma)$ to $(\mu + 0.7\sigma)$, where μ is the mean of the normal distribution.

Solution: Considering $\mu = 0$, the area under the curve needs to be calculated for $z = -0.5$ and $z = +0.7$, which would be sum of these areas, because the range extends on both sides of the mean **(Fig. 13.8)**.

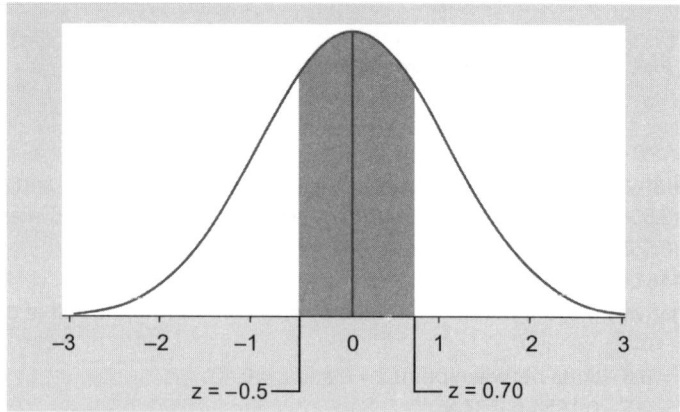

Fig. 13.8: Area under the curve $-0.50 \leq z \leq 0.70$

From area under normal curve table, the area between '$z = 0$' and '$z = -0.5$' = 0.1915; and area between '$z = 0$' and '$z = 0.7$' = 0.2580, therefore area under the normal curve for the given range = 0.1915 + 0.2580 = 0.4495.

Hence the probability of the variate to exist within this range is 0.4495 (44.95%).

It is strongly suggested that for calculation of normal distribution probabilities/areas under the curve, appropriate graph should be constructed before the calculation are carried out. This helps in knowing for what values of 'z' the tables have to be referred to.

Example: To understand calculation of z variate itself.

Find the probability for a point to exist between two points $x_1 = 20$ and $x_2 = 45$, for a normally distributed population having mean $\mu = 30$ and standard deviation $\sigma = 5$.

Solution: As z variate = $(x - \mu)/\sigma$, the two points are transformed to z variates as below **(Fig. 13.9)**:

$$z_1 = (20 - 30)/5 = -2$$
$$z_2 = (45 - 30)/5 = +3$$

Graphically the area lying within this range is given below:

From tables, the values under the curve for −2 and 3 from mean = 0 are 0.4772 and 0.4987; therefore the probability for a point to exist in this range would be (0.4772 + 0.4987) = 0.9759 or 97.59%.

Chapter 13: Theoretical Continuous Probability Distributions

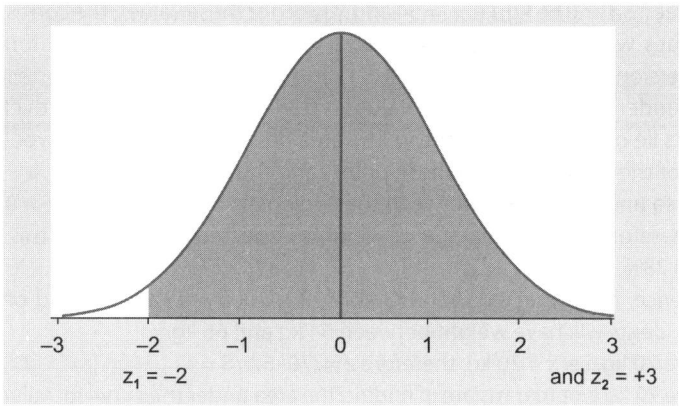

Fig. 13.9: Area under the curve for $-2 \leq z \leq 3$

Example: The mean weight of 1000 students is 62 kgs with standard deviation of 8 kg. Find how many students weight is expected to be:
- Less than 40 kg
- Between 54 and 66 kg
- Above 70 kg

Provided the weights follow normal distribution.

Solution: Given mean $\mu = 62$ kg and standard deviation $\sigma = 8$
- Less than 40 kg.
  - As shown in graph **(Fig. 13.10)**, the value can be calculated by: (0.5 − z value at 40 kgs). Calculating z; $z = (x - \mu)/\sigma = (40-62)/8 = -22/8 = -2.75$
  - At $z = -2.75$, the area under the curve is 0.4970, hence for $z < -2.75 = (0.50 - 0.4970) = 0.003$.
  - Therefore, the probability for any weight to be below 40 kg is 0.003 or 0.3%.
  - Given the number of students is 1000, therefore the expected number of students whose weights would be less than 40 kg = 1000 × 0.003 = 3.

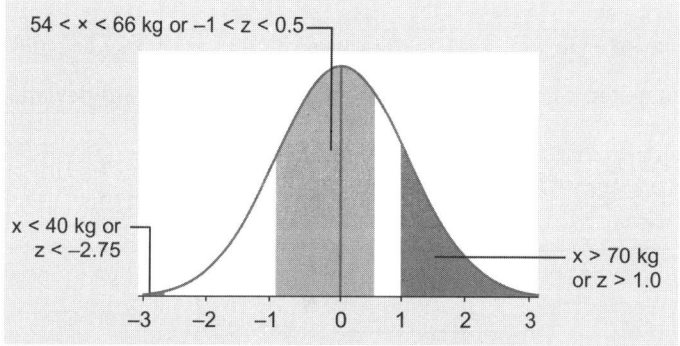

Fig. 13.10: Area under the normal curve for various values

- Between 54 and 66 kg, i.e. $x_1 = 54$ and $x_2 = 66$ for these values, the corresponding z values would be $z_1 = -1.00$ and $z_2 = 0.5$. (By using the same formula for conversion into z as used in point (1) above). From tables, the corresponding area under the curve for these values is 0.3413 and 0.1915, respectively. As the points lie on either side of mean, the area between these points, would be the sum of the areas between these points and the mean $z = 0$.
    - Area under the curve between the two points = 0.3413 + 0.1915 = 0.5328.
    - Therefore, the percentage of weights expected to be within this range is 53.28%.
    - Hence, it is expected that 533 (0.5328 × 1000 = 532.8 rounded off to 533) students will have weight between 54 kg and 66 kg.
- Above 70 kg, i.e. $x = 70$ kg, therefore $z = (70-62)/8 = 1$
  Above 70 kg would translate in finding the area under the curve for values $> z = 1$
    - Area under the curve at $z = 1$ is 0.3413.
    - Therefore, area above $> (z = 1) = 0.5 - 0.3413 = 0.1587$, i.e. 15.87%.
    - Hence, the expected number of students to weigh more than 70 kgs = 159 (0.1587 × 1000 = 158.7, rounded off to 159).

Fitting a Normal Curve

Fitting of a normal curve can be done as below:

To understand these methods, consider a case of continuous distribution having the below values:

Class: 0–10   10–20   20–30   30–40   40–50   50–60
Frequency: 5   10   27   35   15   8

By Method of Ordinates

Class	Frequency (f)	Mid point (x)	fx	Deviation $d = x - \bar{x}$	fd	fd²
0–10	5	5	25	−26.9	−134.5	3618.05
10–20	10	15	150	−16.9	−169.0	2856.10
20–30	27	25	675	−6.9	−186.3	1285.47
30–40	35	35	1225	+3.1	+108.5	336.35
40–50	15	45	675	+13.1	+196.5	2574.15
50–60	8	55	440	+23.1	+184.8	4268.88
	$N = \Sigma f = 100$		$\Sigma fx = 3190$		$\Sigma fd = 0$	$\Sigma fd^2 = 14939$

As a first step calculate the arithmetic mean and standard deviation of the distribution:

$$\text{Arithmetic Mean } \bar{x} = \frac{\Sigma fx}{\Sigma f} = \frac{3190}{100} = 31.90$$

$$\text{Std Deviation } \sigma = \sqrt{\frac{\Sigma fd^2}{N} - \left(\frac{\Sigma fd}{N}\right)^2}$$

$$= \sqrt{\frac{14939}{100} - \left(\frac{0}{100}\right)^2} = \sqrt{149.39} = 12.22$$

Chapter 13: Theoretical Continuous Probability Distributions

Next step is to find the height of the curve: For a normal curve the maximum height (frequency), is at its mean value. Hence using the normal distribution formula and multiplying it by N (the total population) and '*i*' the class interval (because while calculating the mean and standard deviation the mid point was considered).

$$P(31.90) = (iN)\frac{e^{\frac{-(x-\mu)^2}{2\sigma^2}}}{\sigma\sqrt{2\pi}} = \frac{iNe^{\frac{-(31.90-31.90)^2}{2\sigma^2}}}{\sigma\sqrt{2\pi}} = \frac{(10)(100)e^0}{(12.22)(2.5066)} = 32.64$$

The (maximum frequency) height of the curve at its mean = 32.64.
Finding the heights (y-axis) at values of 1σ, 2σ and 3σ:

$$P(1\sigma) = (iN)\frac{e^{\frac{-(\sigma)^2}{2\sigma^2}}}{\sigma\sqrt{2\pi}} = \frac{iNe^{\frac{-1}{2}}}{\sigma\sqrt{2\pi}} = \frac{(10)(100)(0.6065)}{(12.22)(2.5066)} = 19.80$$

$$P(2\sigma) = (iN)\frac{e^{\frac{-(2\sigma)^2}{2\sigma^2}}}{\sigma\sqrt{2\pi}} = \frac{iNe^{-2}}{\sigma\sqrt{2\pi}} = \frac{(10)(100)(0.1353)}{(12.22)(2.5066)} = 4.42$$

$$P(3\sigma) = (iN)\frac{e^{\frac{-(3\sigma)^2}{2\sigma^2}}}{\sigma\sqrt{2\pi}} = \frac{iNe^{\frac{-9}{2}}}{\sigma\sqrt{2\pi}} = \frac{(10)(100)(0.0111)}{(12.22)(2.5066)} = 0.36$$

The 1σ, 2σ and 3σ points on x-axis are already known, as the mean and standard deviation is known. Along with the y-points, the plotting can be done and the normal curve be drawn by free-hand draw.

By Using Area under the Curve

- As a first step calculate the arithmetic mean and standard deviation as explained in the previous part. Taking the values of Mean (= 31.90) and standard deviation (=12.22) from the previous part of this example, and proceeding with the calculations as below:

Class	Frequency (f)	Lower limit of class (x)	$(-x-\bar{x})$ $= (x-31.90)$	$z = \dfrac{x}{\sigma}$	#	$	Expected frequency	Normalized frequency
0–10	5	0	−31.9	−2.61	0.4953	0.032	3.20	3
10–20	10	10	−21.9	−1.79	0.4633	0.1293	12.93	13
20–30	27	20	−11.9	−0.97	0.3340	0.2704	27.04	27
30–40	35	30	−1.9	−0.16	0.0636	0.309	30.90	31
40–50	15	40	8.1	0.66	0.2454	0.1852	18.52	19
50–60	8	50	18.1	1.48	0.4306	0.0587	5.87	6
		60	28.1	2.30	0.4893			

$N = \Sigma f = 100$

Table explained:
Area under normal curve between 0 and z (z value as shown in previous column)— From the normal curve tables (for z = 2.61 the area is 0.4953 and so on).

$. Area under each class: (or percentage of values that would be located in the class). The previous column generated the area between '0 and z' for the respective class. From the area of a class, its next class area is subtracted, and then the remaining area pertains to that class. (0.4953 – 0.4633 = 0.32). The changeover from negative side of the curve to the positive side (z value sign) should be kept in mind, as in that case the area under the curves need to be added up to derive the difference, i.e. (0.2454 + 0.0636 = 0.309).

Expected frequency = Probability × Σf = 0.032 × 100 = 3.2

Normalized frequency = Rounded off the expected frequency. Sometimes the total may not add up to N, as the area under the normal curve is considered up to four decimal points, which can lead to such a situation. The same does not have a material effect as the N is large.

Representing the curved based on observed values and the normalized value, in a graphical manner to see the normal curve fitted vis-à-vis original values, would visually signify the points that have been shifted and the quantum of shift **(Fig. 13.11)**.

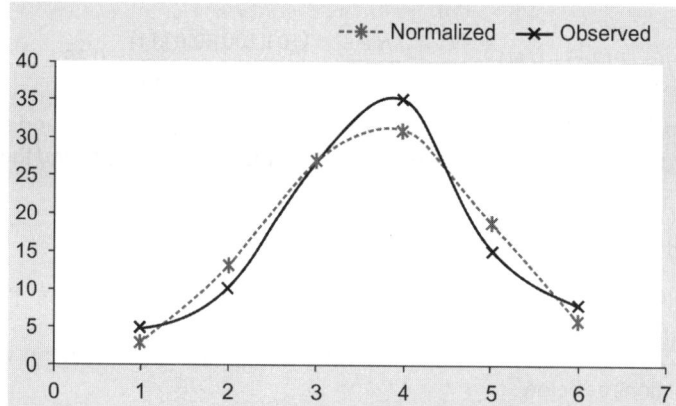

Fig. 13.11: Fitting of normal curve

RECTANGULAR DISTRIBUTION

Also referred to as uniform continuous distribution

A random variable x will have continuous uniform distribution, if over an interval (a, b), its probability remains constant.

$$f(x) = k, \quad \text{for } a < x < b$$

Using mathematical operator, as total probability is 1,

$$\int_a^b f(x)\,dx = 1$$

$$k \int_a^b dx = 1$$

i.e. mathematically f(x)

$$= \frac{1}{b-a}; \text{ where } a, b \text{ are two parameters of the distribution } (a, b)$$

Graphically the distribution is represented (as shown in **Fig. 13.12**):

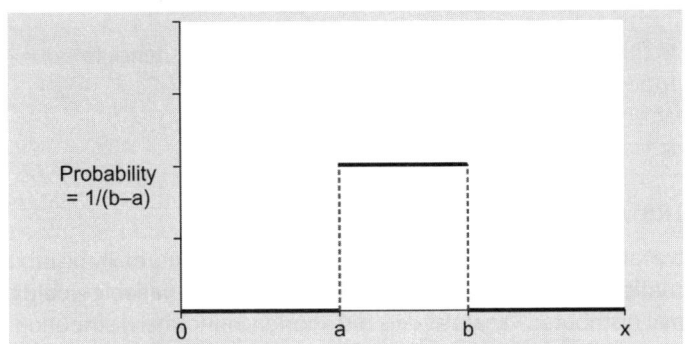

Fig. 13.12: Rectangular/uniform continuous distribution

For such a distribution,

$$f(x) = 0 \text{ if } -\infty < x < a$$

$$= \frac{(x-a)}{(b-a)} \text{ if } a \leq x \leq b$$

$$= 1 \text{ if } b < x < \infty$$

Since f(x) is not continuous at point a and b, therefore f(x) = 1/(b − a) ≠ 0, at any point in distribution, except points a and b. The constants of uniform distribution are:

$$\text{Mean} = \frac{a+b}{2}; \text{ and Variance} = \frac{(b-a)^2}{12}$$

A uniform distribution having a = 0 and b = 1, is called a standard uniform distribution.

Example: In a small hospital, nurse visits the ward every 40 minutes. Find the probability that in case a patient needs the nurse at a random time, (a) he will have to wait for maximum 20 minutes and (b) he will have to wait atleast 30 minutes for the nurse to arrive.

Solution: Let the random variable be the waiting time 'x', which is uniformly distributed from 0 to 40 minutes, i.e. a rectangular distribution with parameter (0, 40).
The probability density function for this would be
f(x) = 1/(b − a) = 1/(40 − 0) = 1/40 for 0 < x < 40
a. Therefore, probability that waiting time would be maximum 20 minutes =

$$\int_0^{20} f(x)dx = \frac{1}{40}\int_0^{20} dx = \frac{1(20-0)}{40} = \frac{20}{40} = 0.5$$

b. *Atleast 30 minutes:* As calculated in previous point f(x) = 1/40, therefore for waiting atleast 30 minutes, the probability can be calculated as:

$$\int_{30}^{40} f(x)dx = \frac{1}{40}\int_{30}^{40} dx = \frac{1(40-30)}{40} = \frac{10}{40} = 0.25$$

Example: Find the mean and variance of the following rectangular distribution.

Variable x:	x_1	x_2	x_3	x_4	x_5
Frequency f:	20	22	19	21	18

Solution: In this rectangular distribution $a = 18$ and $b = 22$, hence $(b - a) = 4$
$f(x) = 1/(b - a) = ¼$ for $18 \leq x \leq 22$
As Mean $= (a + b)/2 = 40/2 = 20$
As Variance $= (b - a)^2/12 = (22 - 18)^2/12 = 16/12 = 1.33$

LOGNORMAL DISTRIBUTION

If a positive random variable x, which by itself may not be normally distributed, but its logarithmic values are normally distributed, then such a variable would be called a lognormal distributed variate and its distribution lognormal distribution.

Lognormal distribution is right skewed distribution, with many small values and few large values. Therefore, the mean of a lognormal distribution is generally greater than its Mode. The lognormal distribution operates in the range $0 \to \infty$.

The probability of lognormal distribution is given as:

$$P(x) = \frac{e^{\frac{-(\log x - \mu)^2}{2\sigma^2}}}{\sigma\sqrt{2\pi}}$$

Where $P(x)$ = Probability of the continuous random variable x $(0 < x < \infty)$
μ = Mean of the distribution
e = Constant approximated at 2.7183
σ = Standard deviation of the distribution.

EXPONENTIAL DISTRIBUTION

Exponential distribution is useful in expressing probabilities for situations consisting of waiting times before an event occurs and where this waiting time is unknown. For example, how much time before an electronic item malfunctions, waiting time before the next customer arrives, time before the next earthquake, etc.

A continuous random variable x, having positive values, will be said to have exponential distribution if its probability density function is given as:

$f(x) = \lambda e^{-\lambda x}$; where λ (Rate parameter) > 0 and $x \geq 0$

The rate parameter λ is equal to $1/m$, where m is the mean rate like failure rate, service rate, occurrence rate, etc. **(Fig. 13.13)**.

The cumulative distribution function of exponential distribution is:

$f(x) = 1 - e^{-\lambda x}$

From this it can be used to derive the probability for finding occurrence of an event between two time intervals.

Consider a point of time 'a', with $a \leq x$, then $f(a \leq x) = 1 - e^{-a\lambda}$. Conversely for point of time 'b' where $b \geq x$, the probability equation would be $(1 - f(a \leq x)) = 1 - (1 - e^{-a\lambda}) = e^{-a\lambda}$

Hence probability $p(a \leq t \leq b) = e^{-a\lambda} - e^{-b\lambda}$

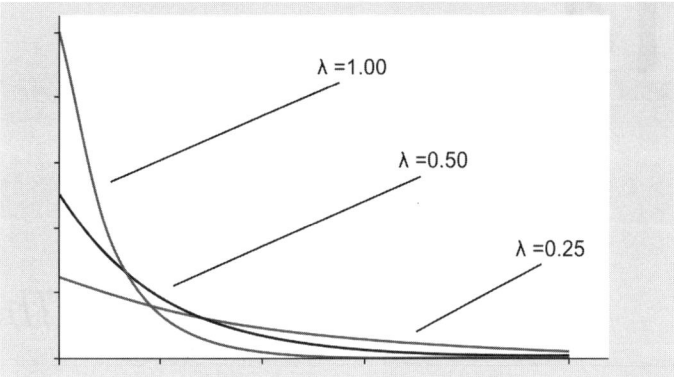

Fig. 13.13: Exponential distribution curves for different rate parameter λ

Constants of exponential distribution:
    Mean = m; Median = m, Variance = m^2
The exponential distribution is positively skewed distribution and lacks memory.

Example: On an average the waiting time at the reception of a hospital is 3 minutes. What is the probability that a customer has to wait for more than 5 minutes before her turn comes.

Solution: Assuming the case to be an exponential distribution, with mean $m = 3$, the distribution equation can be stated as:

$$f(x) = \lambda e^{-\lambda x} \text{ where } \lambda = \frac{1}{m} = \frac{1}{3},$$

$$\text{therefore } f(t) = \frac{1}{3} e^{-\frac{t}{3}}, \text{ where } t \text{ time} \geq 0$$

Probability the customer has to wait for more than 5 minutes, inversely means $(1 - P(t \leq 5))$, therefore:
$P(t \geq 5) = 1 - P(t \leq 5)$
$\quad = 1 - (1 - e^{(-5/3)}) = e^{-1.167}$
$\quad = 0.189$

CONCLUSION

Continuous distributions have maximum applications is research analysis. Most of the subjects under study in medical science tend to have continuous characteristics. The application of normal distribution in hypothesis testing has already been stated. Apart from normal distribution, other distributions play important role in survival analysis, which shall be taken up in subsequent chapters. For some readers these discussions will become clearer after reading the subsequent chapters. Hence please refer to these discussions again after finishing the book for even sharper insight than the impression first had.

CHAPTER 14

Sampling Theory

INTRODUCTION

Statistics is most commonly defined as the "field of study that deals with the collection, organization, analysis, and interpretation of numerical data." This and probably all other definitions too, imply essentiality of the numerical data on which the various statistical tools and techniques can be applied. Such data may already be available or may need to be collected. In the event of it being unavailable, it has to collected, collated or generated through surveys/ studies of the *population*. Many times, it is not feasible to study the entire population. Therefore, samples from population are selected in order to study them and thereafter, if required, to estimate the population characteristics.

As individuals, in general life, the first tryst with sampling occurs immediately on birth, when the blood *sample* is collected for testing and for determining various parameters including blood group. A student will leaf through book or magazine pages while reading one or two (sample) pages before deciding to buy it or not. Trailers are shown before the release of the movies or even through the soap operas, which itself is a form of sample trying to influence a particular action. A new product is launched with free samples being given out, trying to induce usage of the product. People sampling perfumes, visually inspecting pulses/cereals before making a decision, are all forms and examples of sampling. Tasting of the food before eating is again a form of sampling. Medical professionals may be unaware, but they are one of the many heavy users of sampling results. All blood, urine, stool, sputum tests, etc. and advanced procedures like biopsy, etc. are nothing but samples.

Sampling is required to study the sample characteristics or to estimate the population characteristics based on the sample results. The tests of significance are then performed to check the reliability of these results, or to check whether the values so derived about the population parameters could be true or not with a certain degree of confidence. Testing of hypothesis, tests of significance, etc. are part of inferential statistics and would be dealt in subsequent chapters, while the focus in this chapter would be:

- What is sampling and why is it required?
- How can samples be obtained—*Types of Sampling*?

- The advantages and disadvantages of sampling.
- What should be the sample size?
- What are sampling errors and standard error?
- What are underlying basis of theory of sampling?

Prior to initiating the discussions, it is important to understand the term *Population* and *Sample*. In statistical terms, the *group* or *aggregate* (of individuals, objects, service, product, and attribute… possibly anything) under study is called *Population* or *Universe*. This group can be finite or infinite, it can be a quantified or qualified (an attributes presence or absence, which may or can be quantified indirectly).

Sample is a *constituent* or *finite subset* of the population under study. A Sample may contain one unit or many units of the population, depending on the need of the research. If required, many samples can be drawn from the population, with each sample itself comprising of many units. The size of the sample extracted or to be selected from the population is called the *Sample size*.

WHAT IS A SAMPLE AND WHY IT IS REQUIRED?

A sample is a finite subset drawn from the *population* under study, under the presumption that it will be representative of the characteristics of the population and by analyzing it, estimates about the population can be inferred.

In many research subjects, there arises a need to collect primary data about the population, either because of unavailability or unreliability of the secondary data. This primary data can be collected through complete enumeration of the *population* (census method) or by taking a *sample* from the population (sampling method). Census method, wherein complete enumeration has been done, will certainly provide accurate results about the population, but in certain situations, like the ones given below, it is not possible to carry out census. In such situations sampling method is the method opted for:

- Where the population size in infinite
- Where the population size is finite but extremely large.
- Where the population undergoes changes before census can be completed for example, counting the number of leaves in a forest. Before the census can be completed, it is possible that old leaves die out and new ones arrive.
- Where the study objective requires destruction of the population object. For example, quality check in a crackers manufacturing unit, strength testing of materials through crash or drop tests, etc.
- Where the population does not exhibit any variation at all within itself or has parameter values which are consistent throughout the population. In such cases, sample study takes the form of *Standard Tests* with specification on all aspects' based on samples, for example, standards in electronic devices and monitors. (This chapter does not consider such cases/tests while discussing the sampling theory, but this point is being mentioned here to convey the scope).
- In situations where census is possible, but is a very expensive affair. In such situations, either sampling is followed or a combination of census and sampling, i.e. census is done occasionally, while sampling method is used at regular intervals. The above listed situations are indicative and not exhaustive in nature.

TYPES OF SAMPLING METHODS

Sampling methods are the different ways in which samples can be drawn from the population. These methods could be (a) Probability based or (b) Nonprobability based.

Probability Based Random Sampling Methods

Simple Random Sampling

In simple random sampling (SRS), the samples are drawn randomly from the population and each unit of the population has equal chance of being selected in the sample. If the total population size is N then the probability of each unit getting selected is $\frac{1}{N}$, and if from it, a sample size of units is to be selected then, the number of such different samples which are possible are $^{N}C_{n}$. For example, if the population contains A, B, C, D as its elements and a sample containing 2 letters is to be selected, then $^{4}C_{2} = 6$ number of different samples are possible, which are AB, AC, AD, BC, BD, CD.

Simple random sampling can take two forms—(i) With replacement and (ii) Without replacement. In case of 'with replacement' (SRSWR), the selected unit is not removed from the population, but is placed back in the population and is available for selection as a sample unit again. For example, replacing the drawn card back into the pack of playing cards, before drawing the next card.

In 'without replacement' (SRSWOR), the selected unit of the population becomes a part of the sample and is not placed back into the population, i.e. no unit can get selected twice. For example, Lottery system.

Systematic Sampling

In systematic sampling, the first unit selection is done on random basis, either random selection or through the use of random number table (provided in appendix). The other sample units are selected on the basis of a pre-decided system. The population data is organized in some order before conducting the systematic sampling. For example, if the population consists of 100 items, organized and numbered, and 5% of it is required as sample. Hence 5 items need to be selected. The sample interval in such a case = 100 ÷ 5 = 20. Therefore, the first unit is selected from the population's first 20 units, on basis of random table and each subsequent unit is the 20th unit from this selected unit. For example, if from the random table the first unit selected = 11th unit, then the second unit to be selected would be 11+20 = 31st unit, and accordingly the other units would be 51st, 71st, and 91st unit of the population.

The researcher can decide on any other system to be followed, but the system has to be established before selection and the first unit has to be selected through a random process. The example considered in the above paragraph, where the increment was in linear basis is called as linear systematic sampling. It puts a restriction in terms of the first unit being closer to the ends, therefore to avoid such a restriction another variant circular systematic sampling is also used. In circular systematic sampling, the first unit can be any unit from the entire population and thereafter all other unit are as per pre-decided system. In case, the population sequence is exhausted before all samples could be selected, the selection sequence continues from the first unit of the population. For example, if in the above case, the first selected unit had been 35th, the next units would have been 55th, 75th, 95th, 5th and 15th unit of the population, i.e. after reaching the last unit of the population, the count is maintained by counting from start of population (so as to form a circular sequence of the population).

Stratified Sampling

Stratified sampling involves classifying the population into distinct groups called *Strata*. These strata should be as homogeneous as possible within the strata; and as heterogeneous and different as possible, compared to other strata. Thereafter, simple random sampling is carried from each stratum. For example, population can be divided on various parameters such as income (low income, mid income, high income), or place of residence (rural/urban), or age-groups. This form of sampling offers more detailed analysis, less influence of undesirable samples and is particularly useful in skewed distributions. Another advantage offered by this sampling method is the scope to use different sampling technique in each different stratum.

Cluster Sampling

In cluster sampling, the entire population is divided into different groups called clusters. Such one cluster is considered as a *sampling unit* and the selection of sample is done among these clusters or 'sampling units'. On selection as a sample, the entire cluster along with its constituents becomes a part of the sample. It is different from stratified sampling as there is no condition imposed upon clusters in terms of homogeneity or heterogeneity. It can be further classified as single stage or multi stage cluster sampling.

In single stage cluster sampling, the population is divided into clusters, and the desired number of clusters is randomly selected as sample. For example, if there are 200 persons and 40 are to be selected as samples; then, one feasibility is to divide the entire population in 20 clusters with each containing 10 units. From these 20 clusters, 4 clusters are randomly selected to make available 40 sample units (4 × 10 units in each cluster). The disadvantage of this method lies in it becoming too narrow by allowing complete enumeration of the cluster, but completely missing out on other clusters and their constituent units.

Multi stage cluster sampling overcomes the disadvantage of single stage cluster sampling. In this process, at the first stage more numbers of clusters are randomly selected, followed by randomly selecting specified number of units from each cluster. In the example considered for single stage cluster sampling, a multi stage cluster sampling would be like: (for example), from the 20 clusters, first randomly selecting 8 clusters and thereafter randomly selecting 5 units from each selected cluster. The number of samples collected this way remains 40 (8 × 5), which is similar to single stage cluster sampling, but represents larger base of clusters, and thereby expected to provide relatively better representation and possibly improved results.

Nonprobability Sampling Methods

Purposive Sampling

In purposive sampling, the sample units are selected with a definite purpose. Accordingly the results are also biased and may not be properly representative of the population parameters. For example, to prove that health facilities are available to majority of population, survey is conducted only in areas which have high numbers of hospitals.

Judgment Sampling

In this type of sampling, the sample constituents get included in the sample space based on the judgment of the researcher. The selection in sample is based on the researcher's judgment as to which units when included in the sample would constitute a proper representative of the population. This method has the bias, but in cases where small sample sizes are required, it may ensure inclusion of important units.

Quota Sampling

In quota sampling, a pre-determined number of samples are taken from each group, after duly segregating the population into various groups. The number of samples may differ from group to group. For example, if 20 people are to be selected to study heart condition and job stress, for men aged between 50–60, researcher may decide to provide quota for selecting 5 men from government jobs, 4 men from private sector, 6 from worker jobs, 3 from unorganized workforce and 2 from self-employed category.

Convenience Sampling

It is a generic term and includes many ad hoc methods of selecting units to form the sample. In basic form any unit of population gets included in the sample based on the convenience of the researcher. For example, a medical professional conducting research, decides to randomly select the sample units from his patients who visit him/her for consultation.

Snowball Sampling

This method of sampling is very unique and used only where the qualifying population is limited and not easily identifiable. In this the first set of units are randomly selected from the limited population available and then referrals are taken from the first set of respondents, for few more respondents to get included in the sample. At times this is used in medical research, but more so used in customer surveys. The disadvantage of this method is that the referrals most of the time, belong to same class, income group, etc. as that of the main sample unit who referred them.

ADVANTAGES AND DISADVANTAGES OF SAMPLING

The alternate to sampling is to study the population completely (or census). Earlier in the chapter, few points were mentioned as to why sampling may be necessary,

which indirectly are the advantages of sampling. Apart from above, the benefits of sampling, in comparison to complete population study include:
- Saves time
- Is economical
- Can be done with limited resources
- Quality of measurement is more reliable, as personnel required for data collection is less, increasing the scope of employing qualified and competent people.
- In research, in technical and scientific areas, where research is generally conducted by individuals or small groups, it is not possible for such a small group to cover the entire population for study. Therefore, sampling becomes a necessity.

The main disadvantage of sampling is its accuracy and reliability level in comparison with the study of the entire population. Even if the samples considered are true representative of the population, theoretically some amount of difference is expected to remain. This is discussed in the section *sampling errors* in the chapter.

Inspite of these disadvantages, sampling is unavoidable when the population size is large, testing involves destruction of the subject or cost and time are a constraint. Complete population study is preferred in certain situations like, where population size is small, the variance in the values is high or cost of sampling error is very large. Beyond these guidelines, there is no fixed format as to when to use sampling or when to go for complete population study. Accordingly the researcher needs to take an informed decision based on the area/topic of research, etc.

WHAT SHOULD BE THE SAMPLE SIZE?

Deciding on the sample size is never easy as every increase in the number of samples may increase the cost, time and resources. While on the other hand, having restricted sample size may not look representative of the population. There exists no rule on about the sample size, but the following guidelines help in arriving at an approximate number.
- *Population size*: The sample size is in direct proportion of the population. Larger is the population size, larger would be the sample size required.
- *Accuracy*: As the sample size increases, the accuracy of estimation of population parameters increases, i.e. for more accuracy increase the sample size.
- *Population's heterogeneity level*: Higher heterogeneity levels in the population will require a relatively larger sample size, so that the heterogeneity gets sufficiently represented in the sample.
- *Resources available*: Quantum of available time, funds, manpower directly impacts the sample size. Any constraints on account of these should not negatively affect the research and coming out with report having insufficient supporting data will lead to rejection.

In biomedical research there is a broad consensus on the level of significance and power of the studies. However, these are much less stringent than the terms for physical sciences. In the absence of any specific rule to decide the sample size, it is left to the choice of the researcher to decide on the sample size, depending on the topic of research.

It is also a fact that most of the tests of significance, assume normality of the population. From the formulae and understanding of significance tests, an indicative reverse calculation can be done to approximate the sample size. The precaution for using these methods is, that if they are carried out, they should be done at the research planning stage or during development of sampling plan and not to be calculated once the data collection has been initiated or is in progress.

Sample Size for Approximating Population Mean

$$n \geq \left(\frac{Z_\alpha \times \sigma}{\text{Margin of error}}\right)^2$$

where, n = minimum suggested sample size,
Z_a = Confidence limits (z value) for the desired confidence level,
σ = (Estimated) standard deviation of the sample, and
Margin of error = The estimated gap between population and sample means
The above formula would be easier to understand from the below example:

Example: A research is being planned on finding the mean weight of men in a population. The researcher feels that the sample mean should be within 3 kgs of the population mean ($\bar{X} = \bar{x} \pm 3$) and weights of the population would have a standard deviation of 15 Kgs. What should be the minimum sample size, the researcher should look for, if the level of confidence at 95%.

Solution: From the given details, Margin of error = 3,
Standard deviation = 15, Level of confidence = 95%

As the 95% (i.e. 0.95) area would be equally distributed in both the tails, under the premise that the population is normally distributed, area in each tail = 0.95÷2 = 0.475. From the tables given in appendix for *area under the normal curve*, a value of 0.475 corresponds to the confidence limit as z_α = 1.96. Therefore with the data now available, applying the formula:

$$n \geq \left(\frac{Z_\alpha \times \sigma}{\text{Margin of error}}\right)^2$$

$$n \geq \left(\frac{1.96 \times 15}{3}\right)^2$$

$$n \geq 96.04$$

i.e. the sample size should be a minimum of 97 units.

Sample Size for Estimating Population Proportion

The formula when the parameter of interest is the population proportion, is given as:

$$n \geq \left(\frac{Z_\alpha \times \sqrt{p(1-p)}}{\text{Margin of error}}\right)^2$$

where, n = minimum suggested sample size,
Za = confidence limits (z value) for the desired confidence level

p = value of unknown proportion of the population and
Margin of error = The estimated gap between population and sample proportion

As the objective being to estimate the minimum sample size, the numerator on the right hand side of the equation needs to be maximized. The proportion 'p' and $(1-p)$ would be fractions, with sum of these being 1. Hence the maximum value of p and $(1-p)$ would be when both of them are equal, i.e. equal to 0.5.

Example: In continuation of the previous example, instead of mean, the researcher expects the proportion of weights in sample should be within 3% range of the true weight, at 95% confidence level. Determine the suggested sample size.

Solution: From the given data,
$$Z_a = 1.96, \text{ and}$$
Margin of error = 3% = 0.03 and 'p' = 0.5

Applying the formula:
$$n \geq \left(\frac{Z_\alpha \times \sqrt{p(1-p)}}{\text{Margin of error}}\right)^2$$

$$n \geq \left(\frac{1.96 \times \sqrt{0.5(0.5)}}{0.03}\right)^2$$

$$n \geq 1067.33$$

i.e. the minimum sample size should be 1068 units.

This method of estimating the sample size provides a broad guideline and its result is not a sacrosanct value. The researcher's judgment, considering the field of research and other relevant factors should be accorded first priority in deciding the sample size.

Fixed Size Sampling and Sequential Sampling

As stated earlier, such estimations should be done before commencement of data collection. At times, in certain specific instances, researchers do not fix the sample size before the data collection/survey but instead decide on a rule which allows them to keep on collecting samples or stop the sample collection, once the evidence is available. In such a situation the collection and analysis of sample data is working in tandem. For example, if the research topic is what is the distribution of other weights among a population, if the weight of 27% people is 71 Kgs. In such a situation the researcher may stop the study, once his data reflects 27% people having the weight of 71 Kgs. On stop, the weights of the other sample units can be used to define the percentage population lying in different weight group. Such a method is called *sequential sampling*, as against the previous method of fixed number pre-determined sample size which is known as fixed size sampling.

SAMPLING ERRORS

The results derived from samples may or may not be true with respect to the population, due to the simple reason that the estimates are based on the small part of the population and not the whole. This small part or samples are rarely perfect miniatures of the population, which leads to the deviations between the population's true parameters and the estimated parameter derived from samples. These deviations are generally referred to as *sampling errors* or *sampling fluctuations*, and are inherent to any sampling process.

The sampling errors could be either *biased* or *unbiased* in nature. The biased errors are noncompensating in nature as they do not decrease even with increase in sample size, whereas the unbiased errors (compensating) or random sampling errors decrease with increase in sample size. The causes of these errors could be wrong sampling methods, errors in data collection, improper data analysis and many more. The best way to minimize sampling errors is to develop a robust sampling plan, adequate sample size, proper implementation and use appropriate analytical methods to analyze and draw inferences.

STANDARD ERROR

Assume from a population A, five random samples A1, A2, A3, A4 and A5 are drawn. Each sample consists of multiple units of the randomly selected from the population. From each sample its mean can be calculated. Let these sample means be denoted as $\bar{x}_1, \bar{x}_2, \bar{x}_3, \bar{x}_4$ and $\bar{x}_5$.

The sample means will vary from sample to sample. The extent of this variation in the sample means is measured through the standard error of the sample mean and is given as:

$$\text{Standard error of sample mean } (\sigma_{\bar{x}}) = \frac{\sigma}{\sqrt{n}}$$

Where σ = Standard deviation of the population and n = sample size

$$\text{Standard error of proportion } (\sigma_p) = \sqrt{\frac{PQ}{n}}$$

Where P = Population proportion; $Q = (1-P)$ and n = sample size

The significance of standard error lies in its inverse proportion to the sample size n, i.e. as 'n' increase, the standard error decreases, which implies the reliability of the sample increases. (*Note:* Standard error has been dealt with in detail in the chapter 'Hypothesis').

UNDERLYING BASIS OF SAMPLING THEORY

Sampling theory is based on two important laws of statistics, which are:

Law of Statistical Regularity

Based on the theory of probability, Law of Statistical Regularity, conveys that if a sample of reasonable size is drawn randomly from a large group (population), then the sample, on an average, is bound to exhibit the same characteristics as that of the large group (population).

Law of Inertia of Large Numbers

Law of inertia of large numbers states that, larger the sample size, greater is the accuracy with which population parameters can be estimated from the sample characteristics.

Sampling theory is a critical aspect involved in research studies. In case, complete enumeration of large populations had been possible and that too in a cost effective and timely manner, then sampling would not have been required. But such an ideal condition does not exist and therefore sampling is a tool of immense importance. With sampling methods, appropriate samples can be selected, which can then be studied and tested through various inferential statistical techniques.

Section 5

Correlation, Regression, Partial and Multiple Correlation and Regression, Theory of Attributes

- Correlation
- Regression Analysis
- Partial and Multiple Correlation and Regression
- Theory of Attributes

CHAPTER 15

Correlation

The aim of most medical researchers is to find out whether two or more variables are associated or not. If they are associated, whether this association is significant or not; and if significant, then is there a proximate relationship (equation) in quantitative terms between or among them. These three aims can be stated as correlation, test of significance and regression. This chapter will explain the concept of correlation and the various measures of correlation.

WHAT IS CORRELATION?

Correlation deals with the study of relationship between two or more variables. The correlation measure provides the 'degree and direction of association' between the variables. The 'degree' of association refers to the strength (how strongly the variables are related?), and the 'direction' indicates whether the relationship move is tandem in same direction or opposite directions. Consider two related variables X and Y, taking two different sets of values as given here:

SET A		SET B	
Variable X	Variable Y	Variable X	Variable Y
10	45	15	50
13	57	18	45
17	71	21	40
21	85	24	35

In set A, the variable X is continuously increasing and so is variable Y. On visual inspection there appears to be a similar level of change of values of both variables. This series indicate same directional and strong relationship between the two variables.

In set B, while variable X is increasing, variable Y is decreasing. Visually it can be seen that the amount of increase is constantly 3 units in variable X and a decrease in Y of 5 units, i.e. the two variables have strong relationship but in opposite and counter directions.

Therefore, both the sets A and B are said to be correlated, even though direction and strength varies.

Thus, 'The study of two or more variables, to assess their degree and direction of association is known as correlation.' Correlation is said to be 'positive' if the two variables increase or decrease simultaneously, and 'negative' when one variable is increasing (or decreasing) while the other variable is decreasing (or increasing), i.e. movement is in opposite direction. Further correlation between two variables is termed as 'simple correlation,' while between more than two variables is called 'multiple correlation.'

Though the concept of correlation is straight forward, the chances of misinterpretation are high. Therefore, before proceeding with correlation analysis, the following should be considered:

a. The two variables should appear to have some association and or, are related. Though this runs contrary to the definition of correlation, this statement is to safeguard against faulty use of correlation method. The statistical methods for measuring correlation are objective in nature and consider solely the values of two variables. These methods do not identify or check the variables for their appropriateness and subjectivity. For example, consider two variables, number of students pursuing medical course and number of accidents in a city. On application of the correlation measure on such a data, it may throw up fair level of correlation, as both may reflect increase over the years- one, due to increase in medical colleges or student intake and other, due to significant increase in number of vehicles on roads. But it is clearly known that such variables are not related and hence should not be subjected to the correlation measures. Therefore, subjective assessment is necessary to eliminate use of correlation on such variables. We discuss this in inferential issues further. For the time being, suffice it to say that the correlation between such two completely independent variables is termed as 'spurious correlation' or sometimes even called 'nonsense correlation.'

b. Results from small samples should be viewed with appropriate criticality. Though an ideal number of sample size is not defined, but wherever possible it is better to have as many values as possible and reasonable, depending upon the variables under study. A small sample may throw widely different results as there is a high possibility that, the values of variables which got included in samples, belong to a string of freak values which run contrary to the true behavior of the population.

c. Existence of other variables, which are not part of the variables under study, but influence the variables in such a manner, directly or even indirectly, that the variables under study exhibit correlation. Such cases need to be evaluated before applying the measures of correlation, so that the inference drawn from the results are placed in proper context.

CORRELATION MEASURES

Correlation between two variables can be measured or studied in following manner:
1. Scatter diagram
2. Karl Pearson's coefficient of correlation
3. Spearman's rank correlation
4. Kendall's rank correlation.

Scatter Diagram

This method involves plotting of the values of the two variables (x, y) on xy-axis graph. The resulting diagram is called the scatter diagram. The inference about the relationship between the two variables is drawn from visual inspection of the diagram **(Fig. 15.1)**. The following sample scatter diagram would help understand the concept:

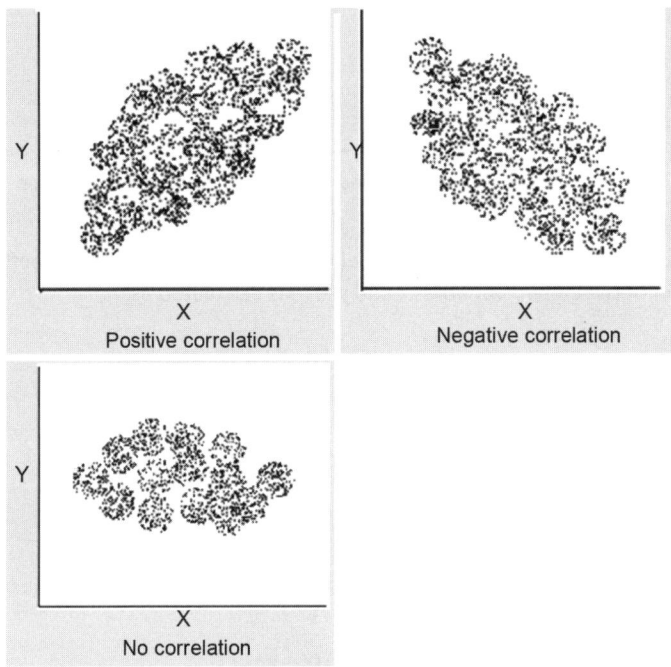

Fig. 15.1: Scatter diagram

The correlation between the two variables is inferred from the pattern of the diagram of dots. The strength of correlation can be inferred from the 'scatteredness' of the dots from the imaginary central line moving diagonally upwards or downwards from left to right. A perfect positive or negative correlation between two variables would result in a straight line as shown here **(Fig. 15.2)**.

Scatter diagram method is an easy way to get an initial and general idea about the variables, but does not provide any accurate measure of their correlation. This limitation gets pronounced in case comparison is to be made about two positively or negatively correlated data series, as by visual inspection, it may not be possible to compare them properly.

Karl Pearson's Coefficient of Correlation

Denoted by r, Pearson's coefficient measures the strength or degree of relationship between the two variables along with the direction of relationship, with an underlying

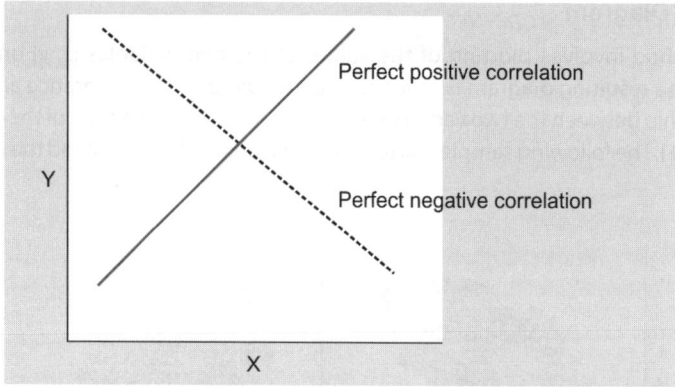

Fig. 15.2: Scatter diagram showing correlation between the variables

assumption that the two variables have a linear relationship. Pearson's coefficient of correlation between two variables x and y, (r_{xy}) is calculated as:

$$r_{XY} = \frac{Cov(X, Y)}{\sigma_X \sigma_Y}$$

This can be simplified as: $= \dfrac{\Sigma xy}{N\sigma_X \sigma_Y}$

$$= \frac{\Sigma(X-\bar{X})(Y-\bar{Y})}{N\sqrt{\dfrac{\Sigma(X-\bar{X})^2}{N}}\sqrt{\dfrac{\Sigma(Y-\bar{Y})^2}{N}}}$$

$$= \frac{\Sigma(X-\bar{X})(Y-\bar{Y})}{\sqrt{\Sigma(X-\bar{X})^2 \times \Sigma(Y-\bar{Y})^2}}$$

i.e. Pearson's coefficient of correlation 'r' = $\dfrac{\Sigma xy}{\sqrt{\Sigma x^2 \times \Sigma y^2}}$

Where

$x = (X - \bar{X})$ and $y = (Y - \bar{Y})$,
and N = Number of pair of values,
σ = standard deviation,
Cov (X, Y) = Covariance (X, Y)

The above formula can be applied directly to discrete variable data, but in case of grouped data which is classified in a two-way frequency distribution, the formula and the calculations are slightly modified. The complete formula application process for grouped data is being explained through an example in the following pages, while the formula is as given below:

Pearsons coefficient of correlation r for grouped data

$$= \frac{N\Sigma fd_x d_y - \Sigma fd_x \Sigma fd_y}{\sqrt{N\Sigma fd_x^2 - (\Sigma fd_x)^2}\sqrt{N fd_y^2 - (\Sigma fd_y)^2}}$$

Assumptions and Properties

a. The value of 'r_{xy}' lies between −1 to +1, i.e. $-1 \le r(XY) \le +1$ and with the value signifying as below:
   −1 = Perfectly negative correlation
   +1 = Perfectly positive correlation
   0 = No correlation
b. The strength and direction of correlation of two variables is same irrespective of which variable series is considered as 'X' or 'Y', i.e. $r_{xy} = r_{yx}$
c. If the Pearson's coefficient of correlation between two variable is '0', it does not indicate that the two variables are 'independent', but simply implies, that the there does not exist a linear relationship between the two variables. It is possible that the two variables may have different type of polynomial relationship (quadratic/cubic, etc). In contrast, if two variables are independent then their coefficient of correlations would be '0'.
d. It is very rare that two variables would be only related to each other and none other. In practical situation, the variables apart from impacting each other are faintly or strongly impacted by large number of other contributory variables. Hence, any inference of the value of 'r' should be appropriately considered in view of presence or absence of a strong co-variate which can influence the results.
e. Pearson's correlation coefficient 'r' is equal to the geometric mean of the two coefficients of regression, i.e. $r_{XY} = \sqrt{b_{XY} \times b_{YX}}$ Concept of regression is detailed in subsequent chapter.
f. Value of 'r' = 0.30 cannot be considered as having double the correlation as compared to a series having 'r' = 0.15. In interpretation, it is generally stated by using words like 'strongly', 'fairly', or 'weakly' correlated alongwith positive or negative sign, as the correlation direction may possibly be. At times in addition to the correlation coefficient, the coefficient of determination is also calculated and is the given as r^2 Being a square, with value being always between 0 and 1, it is not able to provide the direction, hence it is used alongwith correlation coefficient. Coefficient of determination tones down the correlation coefficient.

Probable Error

How reliable is the value of correlation coefficient of a sample in representing its population? Probable error is a historical method of evaluating the 'reliability' of the correlation coefficient of the sample, as a representative of the population, and is given as:

Probable error (PE) (r) = 0.6745 × Standard error = $0.6745 \times \dfrac{(1-r^2)}{\sqrt{N}}$

Where N = Number of pair of observations and r = correlation coefficient of the sample.

The constant 0.6745 relates to the 0.6745σ (0.6745 × Standard Deviation) limits of normal curve which translates into 50% of the area under the normal curve. It is assumed that the distribution from which the sample has been drawn has normal distribution behavior.

Section 5: Correlation, Regression, Partial and Multiple Correlation ...

Interpretation

If value of 'r' < Probable Error (r): It implies correlation coefficient is not significant, while a value of 'r' > 6 × Probable Error (r) signifies correlation coefficient to be highly significant. The range $r \pm PE(r)$ provides the limits within which the correlation coefficient is expected to lie in.

This above method is to estimate at best 'approximately' the reliability, but wherever possible, 't-test' for testing the significance of correlation coefficient should be given preference and should be primarily used.

Example: A study was conducted for 12 days on the pollution measure—PM_{10} (Particulate Matter 10 µg/m³) and the number of patients reporting asthma attacks during this period. Evaluate correlation between these by methods: (a) Scatter diagram and (b) Pearson's coefficient of correlation for the survey data given below.

Day	1	2	3	4	5	6	7	8	9	10	11	12
PM_{10}	176	149	179	417	379	365	297	399	410	261	352	190
Number of patients reporting asthma attacks	11	19	21	31	33	34	14	19	24	8	11	9

Solution: Consider the variable PM_{10} as 'X' and number of patients reporting asthma attacks as 'Y'. Plotting these pairs of on xy-axis graph, we get the following scatter diagram.

From the above scatter diagram, it can be inferred that there appears to be a positive correlation between the two variables **(Fig. 15.3)**.

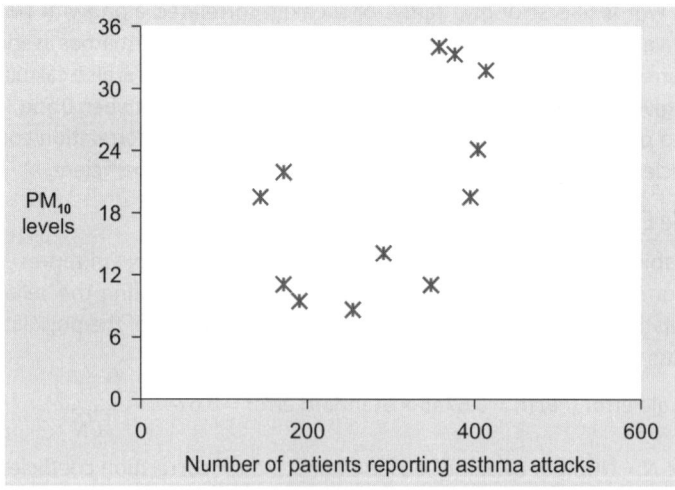

Fig. 15.3: Scatter diagram showing correlation

Calculating Karl Pearson's coefficient of correlation:

PM_{10} (X)	Number of patients (Y)	$x = X - \bar{X}$	x^2	$y = Y - \bar{Y}$	y^2	xy
176	11	−121.83	14842.55	−8.50	72.25	1035.56
149	19	−148.83	22150.37	−0.50	0.25	74.42
179	21	−118.83	14120.57	1.50	2.25	−178.25
417	31	119.17	14201.49	11.50	132.25	1370.46
379	33	81.17	6588.57	13.50	182.25	1095.80
365	34	67.17	4511.81	14.50	210.25	973.97
297	14	−0.83	0.69	−5.50	30.25	4.56
399	19	101.17	10235.37	−0.50	0.25	−50.59
410	24	112.17	12582.11	4.50	20.25	504.77
261	8	−36.83	1356.45	−11.50	132.25	423.55
352	11	54.17	2934.39	−8.50	72.25	−460.45
190	9	−107.83	11627.31	−10.50	110.25	1132.22
$\Sigma X = 3574$	$\Sigma Y = 234$	$\Sigma X = 0.04$	$\Sigma x^2 = 115151.67$	$\Sigma Y = 0$	$\Sigma y^2 = 965$	$\Sigma xy = 5926$

N = Number of paired observations = 12

Mean of variable $X = (\bar{X}) = \dfrac{3574}{12} = 297.83$

Mean of variable $Y = (\bar{Y}) = \dfrac{234}{12} = 19.5$

As, Pearson's correlation coefficient 'r'

$$= \dfrac{\Sigma xy}{\sqrt{\Sigma x^2 \times \Sigma y^2}} \text{ where } x = (X - \bar{X});\ y = (Y - \bar{Y})$$

Therefore, applying the formula on the values in the above table:

$$r = \dfrac{5926}{\sqrt{115151.67 \times 965}} = \dfrac{5926}{10541.4} = +0.56$$

Hence, there is a reasonably good positive correlation between pollution measure PM_{10} levels and asthma attacks reported.

Example: Assume that the variable X (Level of PM_{10}) was measured as a continuous frequency distribution (Grouped Data), and further details of the age group of the patient were also collected. The data so collected was represented as a Two-way classification table, as shown below. For such a data, calculation of Pearson's coefficient of correlation between 'pollution level PM_{10}' and ' age of patient suffering an asthma attack', is detailed below. The example being considered here contains limited sample values for ease of understanding of calculations.

Section 5: Correlation, Regression, Partial and Multiple Correlation ...

Age group of patient reporting asthma attack	Pollution level PM$_{10}$			
	100–200	200–300	300–400	400–500
0–20	14	4	21	18
20–40	19	6	26	12
40–60	27	12	50	25
Total	60	22	97	55

Solution: As the Pearson's coefficient of correlation is not impacted by change in scale and origin, we use the deviation from assumed mean for simplified calculations.

Creating the matrix for calculations:

Y Age group		X Pollution level PM$_{10}$								f	fdy	fd²y	Σ (fd$_x$d$_y$)
		100–200		200–300		300–400		400–500					
	d$_y$ \ d$_x$	−2		−1		0		1					
0–20	−1	14	28	4	4	21	0	18	−18	57	−57	57	14
20–40	0	19	0	6	0	26	0	12	0	63	0	0	0
40–60	+1	27	−54	12	−12	50	0	25	25	114	114	114	−41
	f	60		22		97		55		N = 234	Σ = 57	Σ = 171	Σ = −27
	fd$_x$	−120		−22		0		55		Σ = −87			
	fd²$_x$	240		22		0		55		Σ = 317			
	Σ (fd$_x$d$_y$)	−26		−8		0		7		Σ = −27			

Explanation: Values in 'bold' are the data given. The groups '300–400' and '20–40' are considered as assumed mean (origin) for variable X and Y respectively. The values dx and dy are deviations from the respective assumed mean, i.e. for group '100–200' it is '−2' (taken from group '300–400') and on Y variable age group '0–20' it is '−1' and so on.

Value *28* (highlighted in first row containing data) is calculated as = 14 × −2 × −1, i.e. cell value × its respective $d_x \times d_y$; Similarly, value 4 in same row, next column = 4 × −1 × −1 and so on.

Value f = 57 is calculated as row total =14 + 4 + 21 + 18 and f = 60 as column total (14 + 19 + 27) and so on.

For grouped data, the formula is given as:
  Pearsons coefficient of correlation

$$r = \frac{N\Sigma fd_x d_y - \Sigma fd_x \Sigma fd_y}{\sqrt{N\Sigma fd_x^2 - (\Sigma fd_x)^2} \sqrt{N fd_y^2 - (\Sigma fd_y)^2}}$$

$$r = \frac{234 \times (-27) - (-87)(57)}{\sqrt{234 \times 317 - (-87)^2} \sqrt{234 \times 171 - (57)^2}}$$

$$r = \frac{-1359}{258.09 \times 191.7} = -0.03$$

The Pearson's coefficient of correlation value of −0.03 (close to '0') implies that there is no correlation between the pollution levels and the various age groups of the patients. This translates in the statement that 'increase in pollution levels does not indicate increase in 'incidence of asthma attacks in the higher aged people' (no positive correlation)' and similarly does not indicate 'increase the incidence among the lower aged people' (no negative correlation). This result does not indicate that there is no correlation between incidence of asthma attacks and pollution level, but specifically that there is no linear correlation between 'age group of patients having asthma attacks' and 'pollution level.' In other words, it implies that pollution level impacts all age groups in similar manner and does not exhibit differing impacts in different age groups. From the same values, in the previous example, the results exhibited a positive correlation between pollution levels and incidence of asthma attacks. Statistically these statements are different, and readers should note that this hypothetical values based example was especially considered to bring out the concept and inferences in their entirety.

Spearman's Rank Correlation

In cases where the variables under study are attributes or a situation where instead of absolute values of the variables, the relative ranks are available, Spearman's rank correlation is used as a measure of the correlation coefficient. Consider a sample of 20 students. They are subjected to various subjective and qualitative tests to assess the relationship between leadership quality as one variable and thinking prowess as the other variable. The assessment covers two aspects- the individual traits in isolation and individual's behavior in a group. The performance is not quantified, but ranks are given to all students for variable category. Such cases are evaluated for correlation by the Spearman's correlation coefficient which is given as:

$$\text{Spearman's correlation coefficient } r_s = 1 - \frac{6\Sigma d^2}{n(n^2-1)}$$

Where d = difference in ranks between the paired values and n = number pairs

In case multiple values are given the same rank, then the above formula is modified as follows:

Spearman's correlation coefficient r_s

$$= 1 - \frac{6\left(\Sigma d^2 + \frac{1}{12}(m^3 - m) + \frac{1}{12}(m^3 - m)...\right)}{n(n^2-1)}$$

Where m = number of items whose ranks are same.

Interpretation: The interpretation of Spearman's rank correlation is same as Karl Pearson's correlation coefficient, as it is derived from it and its values, too ranges from $-1 \leq r_s \leq 1$.

The advantage of Spearman's correlation coefficient lies in the fact that it can also be applied to, instead of Karl Pearson's method, to data where absolute values are given, subject to no absolute value being repeated. Further it does not assume normal distribution of the variables. It has limited application, as it cannot be used for bivariate frequencies.

Section 5: Correlation, Regression, Partial and Multiple Correlation ...

Example: A group of 10 students are evaluated and ranked through various activities, tests and subjective assessments, on the leadership skills and their mathematical ability.

Student:	1	2	3	4	5	6	7	8	9	10
Rank in leadership:	6	8	7	9	10	5	2	1	3	4
Rank in mathematical ability:	3	6	9	7	1	2	8	5	4	10

Find correlation between the two variables.

Solution: Creating the table for performing the Spearman's rank correlation calculations:

Student number	Rank in leadership	Rank in maths ability	Difference in rank (d)	d^2
1	6	3	3	9
2	8	6	2	4
3	7	9	−2	4
4	9	7	2	4
5	10	1	9	81
6	5	2	3	9
7	2	8	−6	36
8	1	5	−4	16
9	3	4	−1	1
10	4	10	−6	36
n = 10			$\Sigma = 0$	$\Sigma = 200$

Applying the Spearman's rank correlation formula:

$$\text{Spearman's correlation coefficient } r_s = 1 - \frac{6\Sigma d^2}{n(n^2 - 1)}$$

$$r_s = 1 - \frac{6 \times 200}{10(100 - 1)}$$

$$= 1 - \frac{1200}{990}$$

$$= 1 - 1.212$$

$$= -0.212$$

This result '−0.212' implies a relatively weak negative correlation between the 'leadership skills' and the 'mathematical ability' of the students.

Example: (with same ranks) Consider the previous example with slightly modified ranking as given here:

Student:	1	2	3	4	5	6	7	8	9	10
Rank in leadership:	5	8	5	9	10	5	2	1	3	4
Rank in mathematical ability:	3	6	9	7	1	2	8	5	3	10

i.e. the rank data contains common ranks—rank five to three students in first variable and rank three to two students in the second variable. Calculation of Spearman's rank correlation for such cases is done as below:

Student number	Rank in leader ship	Rank in maths ability	Modified rank leadership	Modified rank-maths ability	Difference in rank (d)	d^2
1	5*	3^	6*	3.5^	2.5	6.25
2	8	6	8	6	2	4
3	5*	9	6*	9	−3	9
4	9	7	9	7	2	4
5	10	1	10	1	9	81
6	5*	2	6*	2	4	16
7	2	8	2	8	−6	36
8	1	5	1	5	−4	16
9	3	3^	3	3.5^	−0.5	0.25
10	4	10	4	10	−6	36
n = 10					Σ = 0	Σ = 208.5

Note:
* The common rank is five for three students, instead of rank 5, 6 and 7. Average of 5, 6 and 7 = 6, hence rank 5, 6, 7 is replaced with rank 6 and applied to the corresponding three students for calculating Spearman's rank correlation.
^ The common rank is 3 for two students instead of possible ranks 3 and 4. Average of 3 and 4 = 3.5, which is considered as the rank for both the students.

Applying the Formula

Spearman's correlation coefficient r_s

$$= 1 - \frac{6\left(\Sigma d^2 + \frac{1}{12}(m^3 - m) + \frac{1}{12}(m^3 - m)\ldots\right)}{n(n^2 - 1)}$$

$$r_s = 1 - \frac{6\left\{208.5 + \frac{1}{12}(3^3 - 3) + \frac{1}{12}(2^3 - 2)\right\}}{10(10^2 - 1)}$$

$$r_s = -0.28$$

The results indicate relatively weak negative correlation between the two variables.

Significance Test of Spearman's Rank Correlation

It is necessary to understand whether the Spearman's rank correlation is statistically significant or not. The significance test allows statistical understanding, whether the correlation is due to chance or not and depends upon the degrees of freedom of the data. In the significance test of Spearman's correlation coefficient, the degrees of freedom is equal to the n, i.e. number of paired observations considered. The modulus value of calculated Spearman's correlation coefficient is compared with the critical values of the Spearman's rank correlation table for given degrees of freedom and inference drawn.

For example in the above problem, a value of 0.28 (modulus) for $n = 10$ degrees of freedom, falls outside the significance level of 0.20 (for two tailed test - values as per rank correlation table provided in the appendix), which implies that the correlation coefficient result obtained, has a probability of more than 20% being occurring due to chance. Had the value of r_s been 0.65, then the implication would have been that the probability of r_s occurring due to chance is between 0.01 and 0.05 i.e. between 1% and 5% for two-tailed test. A lower probability for chance occurrence indicates a higher reliability of the correlation coefficient.

Kendall's Rank Correlation (τ) and its Significance Test

Kendall's rank correlation uses ranks to measure the strength and direction of the correlation between two variables (paired observations). Its value ranges from $-1 \leq 0 \leq 1$, with -1 representing perfectly negative correlation, whereas $+1$ representing perfectly positive correlation. Kendall's rank correlation represents the probability, as it is the difference between the probabilities of the data being in same order as against probability of it being not in same order.

Kendall rank correlation's calculation is based on the number of concordant and discordant occurrences in the paired observations. The pairs of the observations are considered concordant if the subsequent variable pair ranks move in same direction, i.e. ranks of both variables increases or decrease simultaneously. In case one of the variables rank increases while that of the other decreases in the subsequent pair, then it is termed as discordant. If ranks of the pairs move in perfect tandem either increasing or decreasing, it results in perfect correlation.

Kendall rank correlations requires the two variables to be measured on either ordinal or continuous scale (interval/ ratio scale). It assumes a monotonic relationship between the variables.

Denoted as τ (Greek letter 'Tau') it is also referred as Kendall Tau. Kendall rank correlation has three variants viz. τ_a, τ_b and τ_c. τ_a is the standard Kendall rank correlation used where the data does not contain any tied ranks, whereas τ_b and τ_c represents the methods used for case of tied ranks and differ between themselves on the basis of the how they handle the tied ranks.

Kendall rank correlation is calculated by the following methods:

$$\tau \text{ or } \tau_a = \frac{n_c - n_d}{\frac{n(n-1)}{2}}$$

Where n_c = Number of concordant observations
n_d = Number of discordant observations
n = Total number of paired observations

(Note: $\frac{n(n-1)}{2}$ represents the total number of pairs combinations possible).

$$\tau_b = \frac{n_c - n_d}{\sqrt{\frac{n(n-1)}{2} - t_x} \sqrt{\frac{n(n-1)}{2} - t_y}}$$

Where
t_x and t_y are the number of tied ranks in the two rank series

$$\tau_c = \frac{2(n_c - n_d)}{n^2}$$

Chapter 15: Correlation

The value of Kendall Rank Correlation is generally lower than Spearman's rank correlation. Similar to Spearman's rank correlation, Kendall rank correlation's significance can be tested through any of the following two methods:
a. By referring to the critical values table of τ (given in the appendix)
b. By conducting z-test which provides good approximations for $n \geq 10$.
   The z-test statistic is given as:

$$z = \frac{3\tau\sqrt{n(n-1)}}{\sqrt{2(2n+5)}}$$

The z value can be evaluated similar to the large sample z-test, against the tabulated normal values, under the null hypothesis that the two variables are independent, i.e. expected value of $\tau = 0$.

Example: Given below are the weights and heights of sample. Calculate the correlation between them using Kendall rank correlation method. (Example with less number of observations considered in order to explain the concordant and discordant calculations).

| Height (cm) | : | 152 | 167 | 175 | 182 | 159 | 171 | 169 |
| Weight (kg) | : | 58 | 75 | 70 | 73 | 67 | 77 | 65 |

Solution: Preparing the table and providing the ranks:

Height	Weight	Ranks-height	Ranks-weight	Arranging one of the series ranks in proper order →→→ (Heights in this example)	Ranks-height	Ranks-weight *
152	58	1	1		1	1
167	75	3	6		2	3
175	70	6	4		3	6
182	73	7	5		4	2
159	67	2	3		5	7
171	77	5	7		6	4
169	65	4	2		7	5

* Ensure that while ordering the ranks in one of the series, the corresponding rank from the other series are also carried along, so that the paired observations remain same and are not altered.

The next step is finding the number of discordant and concordant pairs.

Ranks-height	Ranks-weight	Col. 1	Col. 2	Col. 3	Col. 4	Col. 5	Col. 6
1	1	----					
2	3	C	----				
3	6	C	C	----			
4	2	C	D	D	----		
5	7	C	C	C	C	----	
6	4	C	C	D	C	D	----
7	5	C	C	D	C	D	C

(*Explanation:* Col. 1 entries- The first pair is (1,1). It is compared with every subsequent paired ranks and concluded whether it is concordant or discordant. For e.g. the second pair is (2,3), i.e. both rank are higher than (1,1) which implies that it is concordant and therefore marked as 'C'. Accordingly the column 1 is filled by continuing the comparison with the subsequent ranks with (1,1). Thereafter the second pair (2,3) is compared with the subsequent ranks and marked in column 2. For e.g. the pair (2,3) when compared with the next pair (3,6) implies concordance (increase in both ranks) but when compared with the next rank (4,2) implies discordance as one rank is increases whereas the other is decreases. Hence, it is marked 'D'. This process becomes tedious in case the number of pairs is very high as that would require corresponding increase in number of columns. Idea is every pair of ranks is to be compared with each other. The above provided table eases the process of comparison).

Total number of concordant observations = 15

Total number of discordant observations = 6
Using the appropriate formula:

$$\tau \text{ or } \tau_a = \frac{n_c - n_d}{\frac{n(n-1)}{2}}$$

$$\tau \text{ or } \tau_a = \frac{15-6}{\frac{7(7-1)}{2}} = \frac{9}{21} = 0.429$$

The two variables height and weight are fairly well correlated.

Correlation is a very important tool and is used in research studies very frequently. It looks at the linear relationships and is heavily influenced by the terminal observations or the ones lying at the end. Though it provides the degree and direction of relationship between the variables, it does not indicate any thing about the weights of relationships between the variables. The study of cause and effect, linear and nonlinear relationships is done through regression analysis, which is dealt with in detail in the next chapter.

CHAPTER 16

Regression Analysis

Consider two variables X and Y. In correlation, the objective of study is to find the relationship-direction and the relationship-strength between these two variables. Correlation does not comment on, as to how much effect, does variable X has on Y or vice versa, i.e. if variable X changes by 'a' units in a non-linear manner, then how much change will it bring in variable Y or vice versa. Correlation is sensitive to the end points in the scatter diagram. The study of nonlinear relationships and weightage of different variables in explaining the change of the dependent variable forms the basis of regression analysis. Regression technique is used in evaluating the variables, to determine the existence and quantum of the *cause and effect* relationship between them. This tool uses the available variable values, to establish a relationship equation between the variables. This equation can then be used to forecast or determine further values of one variable, provided the value of other variable is known.

The term *regression* was coined by Sir Francis Galton in 1821 to describe a biological phenomenon. The phenomenon was that the heights of descendants of tall ancestors tended to regress downwards toward a normal average while that of short ancestors tended to regress upwards. In other words, the tall people had shorter children and shorter parents tended to have taller children when compared to their parents. This phenomenon was termed as *regression* (stepping-backwards) toward the mean.

The earliest form of regression: The method of least squares was published by Legendre in 1805 and by Gauss in 1809. They used this method to determine the orbits of bodies about the Sun from astronomical observations.

COMPONENTS OF REGRESSION ANALYSIS

- *Independent variable*: The variable which influences and causes change in the other variable is called an independent variable. Independent variable is used to derive the value of other variable through the regression equation. It is also called as *regressor/ explanatory* variable.

- *Dependent variable*: The variable whose value is dependent or is influenced by the other variable is called the dependent variable. It is also known as *regressed/ explained* variable.

- *Regression lines:* Regression line is the line of best fit representing the data and is obtained through principles of least squares. An indication about the line can be made from the scatter diagram. The plotted values would seem to concentrate 'on, around and/ or along' an imaginary line. This line could be straight or curved and provides an initial idea about the variables.
- *Regression equations:* The algebraic forms of regression lines are called regression equation. Linear regression equation is expressed as '$y = a + bx$', where 'y' is the dependent variable and 'x' is the independent variable. Constant 'a' is the 'y-intercept' or the value of 'y' when $x = 0$ and constant 'b' is the slope of the line representing quantum of change in variable 'y' for unitary change in variable 'x'.
- *Regression curve*: For a continuous distribution, the conditional mean $E(Y|X = x)$ is called the regression function of Y on X, and its graph is called the regression curve of Y on X. If the regression curve is a straight line, the regression is considered as a linear. If, for a set of data, one of the regression curves is linear, then it is not necessary that other regression curve has to be linear. On the contrary it can be nonlinear.

TYPES OF REGRESSION

- *Linear regression*: Developed in the nineteenth century this is the oldest type. Computations on small data can be done manually. It can be used for interpolation. Though not very suitable for predictive analytics, it is used as a forecasting tool in business and commerce. There are many drawbacks in this method, e.g. sensitivity to outliers, cross-correlations (both in the variable and observation domains) and over-fitting.
- *Logistic regression*: Used for binary results (chance of succeeding or failing, e.g. for a new tested drug or a credit card transaction which can be coded as 0 and 1 only and both being mutually exclusive). It has found extensive use in clinical trials, scoring and fraud detection and even quality control. Its limitations are that it is not robust. It is model-dependent. Regression coefficients are computed using complex iterative, numerically unstable algorithms. By using logit transform the results can be approximated to linear regression. Poisson or Cox regressions are used for a nonbinary response, categorical data (classification), ordered integer response (age groups) and continuous response (regression trees).
- *Ridge regression*: It is a version of linear regression. It puts constraints on regression coefficients to make them much more natural. It is less subject to over-fitting. It is easier to interpret. Therefore it is more robust than its originator.
- *Lasso regression:* It is a version of linear regression. It is less subject to over-fitting. It can automatically do variable reduction and thereby allow regression coefficients to be zero.
- *Ecologic regression*: Ecologic regression performs a single regression per strata. It is useful when the data is segmented into several large strata, groups or bins. When millions of regressions are done in case of big data some will be totally wrong (probability theory) and the best fits will be overshadowed by noisy ones. This creates an artificial goodness-of-fit. Ecologic regression is very useful in identifying extreme events and causal relationships like rare diseases or natural hazards like big floods.

- *Logic regression:* It is a specialized logistic regression (useful for fraud detection and binary responses in clinical trials where each variable is a 0/1 rule). It is more robust. All variables have to be recoded into binary variables, for example, in scoring algorithms.

- *Bayesian regression:* It is more flexible and stable than traditional linear regression. It is useful as it does not assume that the error has a normal distribution. Remember that the error should still be independent across observations. However, the prior knowledge often translates into artificial (conjugate) priors—a weakness of this technique.

Other regression types mentioned in passing are quantile regression and Jackknife regression. Least absolute deviations (LAD) or least absolute errors (LAE) regression is more robust than linear regression but is unstable and many solutions are possible in this technique. Jackknife regression gives approximate, yet very accurate, robust solution to regression problems. It can use *independent* variables that are correlated and/or non-normal and is useful when the assumptions of traditional regression like noncorrelation among variables, normality and homoscedasticity of data are violated.

In medical practice, we commonly are concerned with linear and logistic (with variants Poisson or Cox) regression. In linear regression the model specified is that the dependent variable, bears a linear relationship with independent variable and represented as $y = a + bx + e$, where e is the error term.

Addition of x^2 to the preceding regression gives a parabola:

$$y = a + bx + b_2 x^2 + e$$

This is still linear regression. The right hand side is quadratic in the independent variable x, but it is linear in the parameters a, b and b_2.

As stated, linear regression is most common and widely used, and therefore the following discussions will pertain to linear regression. Thereafter logistic regression is dealt in the later part of this chapter.

LINEAR REGRESSION ANALYSIS AND LINE OF BEST FIT

The regression analysis is about finding the *line of best fit* which is obtained by *Least Squares Method* for the two variables. As per least square method, the line for which the sum of *square of deviations* is least is the best fit line representative of the data. The deviations are possible from both the axis, hence there are two lines for data set having two variables X and Y. One line representing equation '$y = a + bx$' (also referred as 'Regression Y on X', i.e. y = Dependent variable, x = Independent variable) and other represented as '$x = a + by$' (also referred to as 'Regression X on Y', i.e. x = Dependent variable and y = Independent variable). Therefore, it is critical to understand and know, before carrying out regression analysis as to which variable is being classified as *Dependent* and *Independent*. The following sample graph depicts the concept in pictorial manner **(Fig. 16.1)**.

The two regression lines depicted pertain to same data points (shown as small crosses). The two regression lines are formed depending upon which variable is being considered as dependent and independent. In case of 'Y on X' line the deviations on y-axis are considered, whereas for Regression Line 'X on Y' the deviations are considered on x-axis, as shown for few points in the above chart.

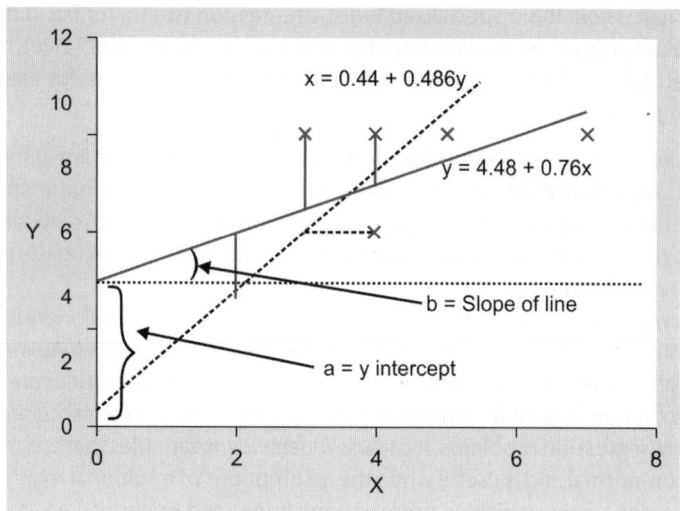

Fig. 16.1: Regression lines: 'Y on X' and 'X on Y'

The line of best fit by least squares method can be derived by solving the following equations for constants 'a' and 'b':

For regression equation Y on X: $\Sigma y = Na + b\Sigma x$
$\Sigma xy = a\Sigma x + b\Sigma x^2$

For regression equation X on Y: $\Sigma x = Na + b\Sigma y$
$\Sigma xy = a\Sigma y + b\Sigma y^2$

Example: Given are the variables X and Y. Establish the regression equation for following:
a. Variable X is dependent on independent variable Y (X on Y)
b. Variable Y is dependent on independent variable X (Y on X)

| Variable X : | 4 | 3 | 4 | 7 | 5 | 2 |
| Variable Y : | 6 | 9 | 9 | 9 | 9 | 4 |

Solution: Preparing the table for ease of calculations:

Variable X	Variable Y	X²	Y²	XY
4	6	16	36	24
3	9	9	81	27
4	9	16	81	36
7	9	49	81	63
5	9	25	81	45
2	4	4	16	8
$\Sigma X = 25$	$\Sigma Y = 46$	$\Sigma X^2 = 119$	$\Sigma Y^2 = 376$	$\Sigma XY = 203$

N = 6

a. Regression equation 'X on Y' using the regression equations:

$\Sigma x = Na + b\Sigma y$ and $\Sigma xy = a\Sigma y + b\Sigma y^2$

i.e.   25 = 6a + 46b and 203 = 46a + 376b

Solving these equations for a and b, we get the values as:
$$a = 0.442 \text{ and } b = 0.486$$
Hence, the regression equation for X on Y is $x = 0.442 + 0.486y$

b. Regression equation 'Y on X' using the regressions equations

$$\Sigma y = Na + b\Sigma x \text{ and } \Sigma xy = a\Sigma x + b\Sigma x^2$$

i.e. $46 = 6a + 25b$ and $203 = 25a + 119b$

Solving these equations for a and b, we get the values as :
$$a = 4.48 \text{ and } b = 0.764$$
Hence the regression equation for Y on X is $y = 4.48 + 0.764x$

Regression Coefficient

The slope of the line i.e. 'b' is called the regression coefficient. For regression 'X on Y' it is denoted as b_{xy} and for regression 'Y on X' it is denoted as b_{yx}. Regression coefficient can be calculated as:

$$b_{XY} = \frac{\Sigma xy}{\Sigma y^2}; \text{ and } b_{YX} = \frac{\Sigma xy}{\Sigma x^2}$$

where $x = (X - \bar{X})$ and $y = (Y - \bar{Y})$

Establishing Regression Equations by using Regression Coefficient

While using the regression coefficient, the regression equations 'X on Y' is stated $(X - \bar{X}) = b_{xy}(Y - \bar{Y})$ and can be directly derived without using the algebraic equations. Consider the variable values, of the previous example. Based on these, we establish the table for solving for regression equation 'X on Y' and 'Y on X'.

Variable X	Variable Y	$x = (X - \bar{X})$	$y = (Y - \bar{Y})$	x^2	y^2	xy
4	6	-0.17	-1.67	0.03	2.78	0.277784
3	9	-1.17	1.33	1.36	1.78	-1.55556
4	9	-0.17	1.33	0.03	1.78	-0.22223
7	9	2.83	1.33	8.03	1.78	3.777764
5	9	0.83	1.33	0.69	1.78	1.111104
2	4	-2.17	-3.67	4.69	13.45	7.944464
$\Sigma X = 25$	$\Sigma Y = 46$			$\Sigma x^2 = 14.83$	$\Sigma y^2 = 23.33$	$\Sigma xy = 11.33$

Mean of X $(\bar{X})$ = 25/6 = 4.17
Mean of Y $(\bar{Y})$ = 46/5 = 7.67
Regression X on Y is given as $(X - \bar{X}) = b_{xy}(Y - \bar{Y})$ where $b_{XY} = \frac{\Sigma xy}{\Sigma y^2}$

$\Rightarrow (X - 4.17) = \frac{11.33}{23.33}(Y - 7.67)$

$\Rightarrow (X - 4.17) = 0.486 (Y - 7.67)$
$\Rightarrow X = 0.44 + 0.486Y$, which is same as calculated in previous example.

Similarly using $(Y - \bar{Y}) = b_{yx}(X - \bar{X})$ where $b_{YX} = \dfrac{\Sigma xy}{\Sigma x^2}$; the regression 'Y on X' can be calculated.

Important Considerations in Linear Regression

- For given variable values X and Y, their regression line may or may not pass through any of the variable values, but the lines will always pass through the mean values of both the variables. Hence, the point of intersection of the regression lines 'X on Y' and 'Y on X' is the mean value point of both the variables.
- The regression coefficient, unlike coefficient of correlation, are not symmetric, i.e. $b_{xy} \neq b_{xy}$.
- The square root of the product of regression coefficients is equal to the correlation coefficient, i.e.

$$\text{Correlation coefficient } r = \sqrt{b_{XY} \times b_{YX}}$$

- The sign of correlation coefficient would be (+) if regression coefficients are positive and (–) if regression coefficients are negative.
- Both regression coefficients will always be of same sign, either positive or negative and cannot simultaneously be > 1.
- Inferences can be drawn about the correlation between the two variables on the basis of their regression lines 'X on Y' and 'Y on X'.
  - Perfect correlation (+/–) will get reflected in the regression lines coinciding with each other.
  - *No correlation exists* can be inferred if the two regression lines are perpendicular to each other.
  - If the angle between the two regression lines is less (acute), then it signifies high correlation, whereas an obtuse angle implies low correlation between the two variables.

Residual Variance or Standard Error of Estimate

The regression equation allows estimation of dependent variable on the basis of the independent variable. Though fairly close, the estimated value of the dependent variable may not be exactly estimated or predicted. Standard error of estimate or residual variance provides a measure of preciseness or inaccurateness of the estimation of the dependent variable. Denoted by S_{xy} *(standard error of estimate for X on Y)* or S_{yx} *(standard error of estimate for Y on X)*, it is calculated by using the formula:

$$s_{xy} = \sigma_x \sqrt{1 - r^2} \text{ and } s_{yx} = \sigma_y \sqrt{1 - r^2}$$

Through mathematical derivations, this formula for *standard error of regression* can be simplified and can be written as:

$$S_{xy} = \sqrt{\dfrac{\Sigma X^2 - a\Sigma X - b\Sigma XY}{N}} \text{ and } s_{yx} = \sqrt{\dfrac{\Sigma Y^2 - a\Sigma Y - b\Sigma XY}{N}}$$

The value of standard error indicates the accuracy of estimation. Smaller the value of this, the closer would be the predicted values to the regression line. It can also be considered the measure of the error in calculating the regression. If the value is '0', then it implies all values to fall on the regression line and hence no error. The effort in regression is to increase the R^2 to explain the variation as much as possible.

$e = y - \hat{y}$, the difference between the value of the 'dependent variable predicted by the model $\hat{y}$, and the *true value of the dependent variable y'* is called *Residual*. This is of importance in diagnostics and should be randomly distributed for a valid regression observation. If it follows any pattern, then the regression cannot be relied upon. To put is simply, the relationship between predictor independent variable '*x*' and the dependent variable '*y*' is defined by the model when all the other predictor variables included in the model are *held fixed*. It is also explained by some, as the change in '*y*' for a one-unit change in '*x*' when the other covariates are held fixed. This is the expected partial derivative of *y* with respect to *x*.

Assumptions in Standard Linear Regression Model

Some important assumptions made by standard linear regression models with ordinary least squares are as follows:

Error-free Predictors

Predictor variables *x* are assumed to be treated as fixed values and be error-free.

Linearity

The response variable is assumed to bear a linear relationship with the predictor variables. Predictor variables can be arbitrarily transformed and this makes linear regression an extremely powerful inference method. Techniques like polynomial regression use linear regression to fit the response variable as an arbitrary polynomial function.

Homoscedasticity or Constant Variance

Response variables are assumed to have the same variance in their errors as they are drawn from the same population regardless of the values of the predictor variables. In practice, this assumption is invalid because the errors are heteroscedastic. This is the reason that researchers need to look for fanning out and do diagnostics using residuals. If the residuals appear clustered and spread apart on their predicted plots for larger and smaller values for points along the linear regression line then the mean squared error for the model will be wrong. Other estimation techniques like weighted least squares and heteroscedasticity-consistent standard errors are available for such situations to handle heteroscedasticity.

Independence of Errors

Independence of errors assumes that the errors of the response variables are not correlated with each other. Methods like generalized least squares can handle correlated errors. They need significantly more data. Bayesian linear regression can also be used in this situation with its usual limitations.

Lack of Multicollinearity

Lack of multicollinearity in the predictors is an important assumption and should be tested in the diagnostics. When two or more perfectly correlated predictor variables come into play then the parameter vector β will have no unique solution in ordinary least squares (OLS) method used as a standard. Methods for fitting linear models with multicollinearity are available but they may need additional assumptions like *effect sparsity*—that a large fraction of the effects are exactly zero or use iterated algorithms for parameter estimation. (e.g. generalized linear models).

Goodness of Fit

Goodness of fit of the model and the statistical significance of the estimated parameters should be confirmed with the model of regression used. Commonly used checks of goodness of fit are:
- R-squared analysis,
- Analyses of the pattern of residuals and
- Hypothesis testing.

Statistical Significance

Statistical significance is checked by F-test of the overall fit and t-tests of individual parameters. The examination of the residuals can be used to invalidate a model if they follow a pattern and are not spread out on a scatterplot. The results of a t-test or F-test are interpreted using the normal methods of the probability values. They may be difficult to interpret if the assumptions of the model are violated. One needs to be careful with small samples as the estimated parameters may not follow normal distributions and complicate inference. For example, the error term may not have a normal distribution. With relatively large samples the central limit theorem makes life much easier and the normality can be assumed.

LOGISTIC REGRESSION

Apart from simple regression, which has been discussed at length, another regression form which is widely used in clinical research is *logistic regression*, and therefore it requires detailing.

In simple regression, the variables are continuous in nature and can take any value from $-\infty$ to $+\infty$. Further, assume the regression equation $y = a + bx$, where y is a dependent variable and x is the independent variable. In this simple regression both y and x could take any values within the continuous infinite range. Now, assume that the dependent variable y is of binomial/ binary nature instead of continuous, i.e. it can take only two possible values which are mutually exclusive. Instead of values, it can be stated that the dependent variable can take two mutually exclusive states, i.e. success/failure, presence/absence, etc. These states can be binary coded as 0 (failure) and 1 (Success). With 0 and 1, simple regression equation can be developed but is of no practical use. Such situations where one of the data is in binary form cannot be resolved through simple regression, as explained below:

Consider the following data on dosage of medicine (x) and result on life of individual (y) (0 = Patient died and 1 = patient survived):

Dosage (mg): 250 275 230 300 210 330 240 290 280 240 310
Patient status: 0 1 0 1 0 1 1 0 1 1 1

For the above data, the simple regression equation is derived as $y = -1.177 + 0.0068x$ and the regression line, along with its scatter diagram is as drawn in **Figure 16.2**.

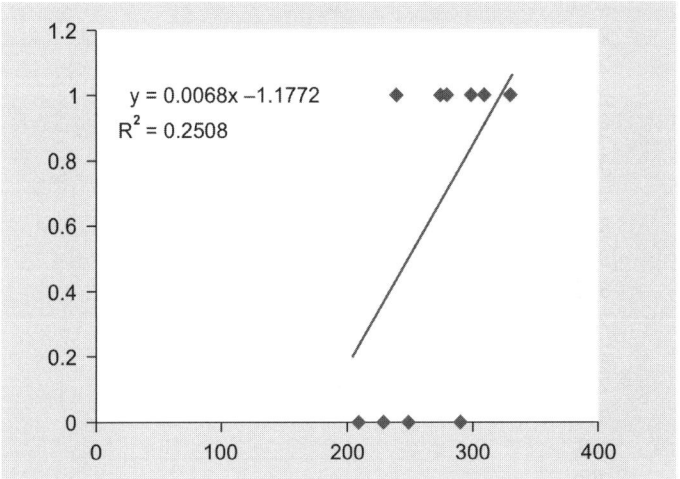

Fig. 16.2: Regression line when dependent variable is in binary form

Now if this regression equation is used to estimate the outcome at dosage of 150 mg, the result is – 0.16, i.e. a negative value which cannot be deciphered. Similarly for dosage of 350 mg, the resulting value of $y = 1.20$, again a value which is not defined. As death and survival are two mutually exclusive categories defined and binary coded as 0 and 1, any other value of y is undefined and cannot be deciphered. Thus, it can be seen that simple regression is not able to handle data where it has been categorized or binary coded. Though discriminant analysis also operates in similar fashion, but in discriminant analysis the independent variable has to be continuous in nature, whereas logistic regression does not impose any such restriction.

The logistic regression equation between the dependent and independent variable is given as:

$$p = \frac{1}{1 + e^{-y}}$$

Where,
p = Probability of success, i.e. dependent binary variable taking the value of 1 $(1 - p)$ = Probability of failure or absence, i.e. binary dependent variable = 0 and, $y = a + bx$, where x is the independent variable.

This logistic equation can be converted into linear form as:

$$y = \log_e \left(\frac{p}{1-p}\right) = a + bx$$

(The Log is to base 'e' and can be converted to the base 10. To convert (for example) $\log_e 5 = 2.303 \times \log_{10} 5$). The concept would be further clarified by considering an example.

Example: Find the regression equation *patient status on specific dosage* between these two data. (Patient status: 1 = Patient survived and 0 = Patient died).

Dosage (mg)	Patient status
250	0
275	0
230	0
240	1
370	0
360	1
285	0
330	1
360	1
315	0
290	1
230	0
260	1
310	1
245	1
280	0
375	1
200	0
225	1
300	1

Solution: The above data is grouped and calculations performed as shown below:

Grouped dosage	Mid point (class mark)	Class frequency	Number of Success	Probability of success (p)	(1-p)	Logit y = Log (p/(1-p))
200–250	225	6	3	0.5	0.5	0.000
250–300	275	6	2	0.33	0.67	−0.301
300–350	325	4	3	0.75	0.25	0.477
350–400	375	4	3	0.75	0.25	0.477

The simple regression equation between grouped dosage class mark (Mid point) and the logit y is derived through simple regression formula and the graph is plotted as shown in **Figure 16.3**:

The simple regression (linear fit) is derived as $y = -1.162 + 0.0044x$
Therefore, the logistic regression equation for the data is stated as:

$$\log_e \left(\frac{p}{1-p}\right) = -1.162 + 0.0044x$$

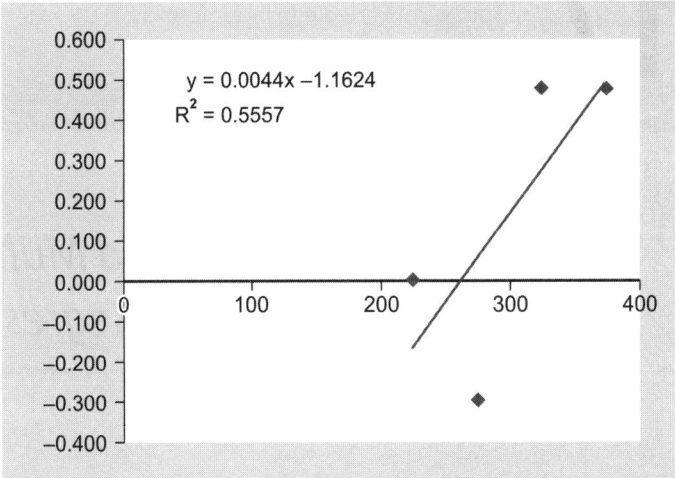

Fig. 16.3: Simple regression equation between grouped data and logit y

$$\text{Or, } p = \frac{e^{-1.162+0.0044x}}{1+e^{-1.162+0.0044x}}$$

Based on these the probabilities of dependent variable y can be derived for given value of independent variable x, as shown here:

Independent variable x	Dependent variable y $y = -1.162 + 0.0044x$	Probability $p = \dfrac{e^{-1.162+0.0044x}}{1+e^{-1.162+0.0044x}}$
100	−0.722	0.33
180	−0.370	0.41
240	−0.106	0.47
300	0.158	0.54
360	0.422	0.60
420	0.686	0.67
500	1.038	0.74

From the table it can be noticed that the probability of the dependent variable remains within 0 and 1, even though the regression equation indicates a negative value. The odds ratio reflects the slope of the regression and hence it is of use even when glancing through data. There are several iterations and renditions of logistic regression in use because of computerized analysis. Suffice it to say that it is a powerful tool in the armamentarium; every researcher and practitioner must be conversant with, or at least be aware of the technique.

Logistic regression finds immense application in clinical research as it is able to handle binomial data (binary), which is beyond the scope of simple regression. Attributes like pain, recovery, treatment effect (treated or untreated), infected or uninfected, etc. are some of the possibilities of logistic regression.

CHAPTER 17

Partial and Multiple Correlation and Regression

In the previous chapters on correlation and regression, the study was restricted to two variables, one dependent and other independent. In practice, it is very seldom that two variables would be influencing only each other in isolation. On the contrary, a dependent variable would be influenced by multiple variables, though in varying magnitude. The study of the combined effect of a group of variables (considered independent) on another variable (dependent variable), to determine the correlation and regression among them, is known as multiple correlation and regression.

Now assume another situation where in a multivariate distribution, and the interest is not to study the combined effect, but to study the effect of only *one* of the independent variables (out of multiple) on the dependent variable. This can be achieved either by selecting only those data fields where the values of noninterest independent variables are same. Such data entries would be limited and at times may not be even available. The other alternative is to mathematically remove the linear effect of all the noninterest independent variables. The correlation and regression between two variables eliminating the linear effect of other independent variables present, is called as partial correlation and partial regression.

PARTIAL CORRELATION

If y, x_1 and x_2 are three variable, the partial correlation coefficient between variable y and x_1, while keeping x_2 as constant, is given as:

$$r_{yx_1 x_2} = \frac{r_{yx_1} - r_{yx_2} r_{x_1 x_2}}{\sqrt{\left(1 - r_{yx_2}^2\right)\left(1 - r_{x_1 x_2}^2\right)}}$$

where,

$r_{yx_1 \cdot x_2}$ = Partial correlation coefficient between y and x_1 while keeping x_2 constant.

r_{yx_1}, r_{yx_2} and $r_{x_1 x_2}$ are simple correlation coefficients between the respective variables.

Similarly the equation for variables y and x_2, and x_1 and x_2, can be stated as under:

$$r_{yx_2.x_1} = \frac{r_{yx_2} - r_{yx_1}r_{x_1x_2}}{\sqrt{(1-r^2_{yx_1})(1-r^2_{x_1x_2})}}$$

$$r_{x_1x_2.y} = \frac{r_{x_1x_2} - r_{yx_1}r_{yx_2}}{\sqrt{(1-r^2_{yx_2})(1-r^2_{yx_2})}}$$

Partial correlation coefficient signifies the linear association between two variables, while holding the other variable(s) as constant, i.e. the impact of the other variables has been removed. For example, the value of $r_y x_1 \cdot x_2$ allows us to know whether the relationship between variables 'y' and 'x_1' is due to their own relationship or is it due to the impact of variable 'x_2' on both of them.

If $r_{yx_1.x_2} = 0$, then $r_{yx_1} = r_{yx_2}r_{x_1x_2}$, which means that $r_{yx_1} \neq 0$, if x_2, is correlated to both y and x_1. This implies that variable y and x_1 may appear to be uncorrelated, when the effect of x_2 is eliminated, but when their total (simple) correlation is calculated it may show a correlation due to effect of variable x_2 on both of them.

Partial correlation coefficient is interpreted through its corresponding coefficient of partial determination, which is the square of the partial coefficient value, i.e. $r^2_{yx_1x_2}$. A value of 0.81 for $r^2_{yx_1x_2}$ implies that the 81% of the variation in 'y' that is not associated with x_2 is associated with x_1.

In partial and multiple correlation/regression the notations are very important hence the placement of dots between the subscripts should be thoroughly understood. Accordingly, in partial coefficient of correlation's subscript, the variables before dot signify the variables between which the partial correlation is being calculated, whereas the subscript after the dot, are the variables which have been held constant. Formulas for partial correlation coefficient for more than three variables (higher order) also exist. In case the number of variable under consideration is two only (a case of simple correlation), it is called zero order coefficients. In case the variables are three, with one variable held constant it is called first order coefficients. Similarly partial correlation for four variables, with two held as constant, is called as second order coefficient, and so on. If the formula of partial correlation coefficient is examined closely, it can be seen that the first order correlation coefficient is based on zero order correlation coefficients. Similarly the second order partial correlation coefficient (having four variables, with two held constant) will be expressed through first order coefficients, as shown below:

$$r_{yx_1.x_2x_3} = \frac{r_{yx_1.x_3} - r_{yx_2.x_3}r_{x_1x_2.x_3}}{\sqrt{(1-r^2_{yx_2.x_3})(1-r^2_{x_1x_2.x_3})}}$$

MULTIPLE CORRELATION

In partial correlation, the interest was in finding from the multivariate distribution, the correlation between two variables, while keeping all the other variables constant. Contrary to this, if the interest is to find the correlation between a dependent variable and all other independent variables collectively, then it becomes a case of multiple correlation. The choice of dependent variable gets defined from the subject of research or the researcher.

If y, x_1 and x_2 are three variable, with variable y dependent on variables x_1 and x_2, then the coefficient of multiple correlation is given as:

$$r_{y.x_1x_2} = \sqrt{\frac{r_{yx_1}^2 + r_{yx_2}^2 - 2r_{yx_1}r_{yx_2}r_{x_1x_2}}{1 - r_{x_1x_2}^2}}$$

In case variable x_1 is considered dependent on variables y and x_2, then the coefficient of multiple correlation formula is modified as given as:

$$r_{x_1.yx_2} = \sqrt{\frac{r_{x_1y}^2 + r_{x_1x_2}^2 - 2r_{yx_1}r_{yx_2}r_{x_1x_2}}{1 - r_{yx_2}^2}}$$

In coefficient of multiple correlation, the subscript before the dot signifies the variable which has been considered as dependent and all the variables after the dot are independent variables whose combined impact is being determined. Further, in case of multiple correlation the value of $r_{x_1.yx_2} = r_{x_1.x_2y}$, i.e. the sequence in the subscript after the dot is of no significance.

Unlike simple correlation where the value of correlation can vary from −1 to +1, the value of multiple correlation always lies between 0 and 1, with 0 signifying no correlation, whereas 1 signifying perfect correlation. No correlation implies *no linear correlation* exists between the variables, but it does not rule out the possibility of nonlinear relationship.

Mathematically, multiple correlation coefficient $r_{y.x_1x_2}$ is the simple correlation between y and the combined effect of x_1 and x_2 on y. Stated differently it is the correlation coefficient between y and its estimated value through regression equation of y on x_1 and x_2. This implies that the multiple correlation coefficients indicates the closeness of relationship between the observed values of y (dependent variable) and the expected values of y calculated by the regression equation of y on x_1 and x_2, the independent variables.

From this it follows that if $r_{y.x_1x_2} = 1$, it is a perfect correlation case which has no residuals in the regression equation and such a regression equation is also called a perfect prediction equation. If $r_{y.x_1x_2} = 0$, a case of complete uncorrelated variables, it implies that all total and partial correlations among y, x_1 and x_2, are also zero. Another important property is that the multiple correlation coefficient is always greater than or equal to any of the underlying total correlations of the lower order, i.e.

$$r_{y.x_1x_2} \geq r_{yx_1} \text{ or } r_{yx_2} \text{ or } r_{x_1x_2}$$

Like partial correlation, in multiple correlation, the interpretation is based on the coefficient of determination, which is nothing but the square of the multiple correlation coefficient.

PARTIAL AND MULTIPLE REGRESSION

In simple regression analysis between two variables, one dependent and other independent, the regression equation was given as '$y = a + bx$' where variable y is dependent on independent variable x. By interchanging 'x' and 'y' there were two regression lines generated—one each for 'x on y' and 'y on x'.

Chapter 17: Partial and Multiple Correlation and Regression

In multiple regression, the number of variables under consideration are, more than two. Among these multiple variables, one variable is considered as the dependent variable and rest all variables are considered as independent. For explanation, consider a case of three variable y, x_1 and x_2. Considering y as the dependent variable and x_1, x_2 as independent, the regression equation (y on x_1 and x_2) for three variables is given as:

$$y = a + b_{yx_1.x_2} x_1 + b_{yx_2.x_1} x_2$$

Where,
y = Dependent variable; x_1 and x_2 = independent variables.
a = Intercept of the regression plane when all independent variables are 0.
$b_{yx_1.x_2}$ = Partial regression coefficient, indicates change in y due to x_1 when x_2 is held constant.
$b_{yx_2.x_1}$ = Partial regression coefficient, indicates change in y due to x_2 when x_1 is held constant.

The above regression equation can be solved by using the following equations:

$$\Sigma y = Na + b_{yx_1.x_2} \Sigma x_1 + b_{yx_2.x_1} \Sigma x_2$$

$$\Sigma yx_1 = a\Sigma x_1 + b_{yx_1.x_2} \Sigma x_1^2 + b_{yx_2.x_1} \Sigma x_1 x_2$$

$$\Sigma yx_2 = a\Sigma x_2 + b_{yx_1.x_2} \Sigma x_1 x_2 + b_{yx_2.x_1} \Sigma x_2^2$$

By solving these algebraic equations, the values of constants a, $b_{yx1.x2}$ and $b_{yx2.x1}$ can be calculated. For directly calculating these values, without the use of complex algebraic equation, the same can be determined by using the following formulas:

$$b_{yx_1.x_2} = \frac{(\Sigma yx_1 - n\bar{y}\bar{x}_1)(\Sigma x_2^2 - n\bar{x}_2^2) - (\Sigma yx_2 - n\bar{y}\bar{x}_2)(\Sigma x_1 x_2 - n\bar{x}_1\bar{x}_2)}{(\Sigma x_1^2 - n\bar{x}_1^2)(\Sigma x_2^2 - n\bar{x}_2^2) - (\Sigma x_1 x_2 - n\bar{x}_1\bar{x}_2)^2}$$

$$b_{yx_2.x_1} = \frac{(\Sigma yx_2 - n\bar{y}\bar{x}_2)(\Sigma x_1^2 - n\bar{x}_1^2) - (\Sigma yx_1 - n\bar{y}\bar{x}_1)(\Sigma x_1 x_2 - n\bar{x}_1\bar{x}_2)}{(\Sigma x_1^2 - n\bar{x}_1^2)(\Sigma x_2^2 - n\bar{x}_2^2) - (\Sigma x_1 x_2 - n\bar{x}_1\bar{x}_2)^2}$$

$$a = \bar{y} - b_{yx_1.x_2} \bar{x}_1 - b_{yx_2.x_1} \bar{x}_2$$

(The use of these formula's will be shown as an example in following pages).

After establishing the multiple regression equation, the next step is to determine its standard error. As stated earlier the standard error provides the estimate of error involved when the regression equation is used to forecast values of the dependent variable and is given as:

Standard error of y on x_1 and x_2 $(S_{y.x_1 x_2}) = \sqrt{\dfrac{\Sigma(y - \hat{y})^2}{n-3}}$

Where y is the actual observed values, n is the number of observations and $\hat{y}$ is the estimated values of y through regression equation y on x_1 and x_2.

Coefficient of Multiple Determination (Multiple Coefficient of Determination (R^2))

Coefficient of multiple determination is defined as the ratio of the explained variation by the regression equation to the total variation of the dependent variable, i.e.

the proportion of variance in the dependent variable which is explained by the independent variables.

Mathematically, if y is the dependent variable and x_1, x_2 are independent variables, if then, the total, explained and unexplained variations can be denoted as:

$$\Sigma(y-\bar{y})^2 = \Sigma\left(\hat{y}-\bar{y}\right)^2 + \Sigma(y-\hat{y})^2$$

i.e. Total variation = Explained variation + Unexplained variation

Where y = Observed actual values
$\bar{y}$ = Mean of y
$\hat{y}$ = value of y, as estimated from the regression equation

Using these, the coefficient of multiple determination can be mathematically stated as:

$$R^2 = \frac{\Sigma\left(\hat{y}-\bar{y}\right)^2}{\Sigma(y-\bar{y})^2} = \frac{\text{Explained variation}}{\text{Total variation}}$$

Alternatively,

$$R^2 = 1 - \frac{\text{Unexplained variation}}{\text{Total variation}} = 1 - \frac{\Sigma(y-\hat{y})^2}{\Sigma(y-\bar{y})^2}$$

The value of coefficient of multiple determinations always lies between 0 and 1. If $R^2 = 1$, it implies that there is no unexplained variation and total variation in the dependent variable have been accounted by its explained variation.

Adjusted Coefficient of Multiple Determination ($\bar{R}^2$)

Multiple regression is impacted by the number of independent variables the data has, and also the sample size of each data set. Therefore, two regression equations may not be truly comparable on the basis of their R^2 values. To make them comparable, the adjusted coefficient of multiple determination is utilized and for a distribution it is given by:

$$\bar{R}^2 = 1 - \left(\frac{n-1}{n-k-1}\right)(1-R^2)$$

Where, $\bar{R}^2$ = Adjusted coefficient of multiple determination
n = Sample size and k = Number of independent variables.

For a given distribution, if the number of independent variables is kept constant, the value of $\bar{R}^2$ will tend to get equal to the value of R^2 as the sample size increases.

Dummy Variable

In the discussion so far, only quantitative variables have been considered, i.e. variable which have a specific quantity attached to them or can be measured. In certain research subjects, at times a qualitative variable which cannot be quantified, i.e. an attribute which is either present or absent, may be an important variable which needs to be studied too. *For example*: In a research subject on car accidents, a researcher may try to study, apart from the age of driver, the gender or marital status of the driver. Assume in this situation, the dependent variable is the *number of car accidents*, whereas the independent variables are *age* and *gender-male or female*'. Is it possible to establish a multiple regression equation between these three variables, where one of the variables cannot be quantified? The answer is *Yes*; it is possible by considering the qualitative variable as a *dummy variable*.

Dummy variable is the variable which represents the qualitative characteristics or attributes present or absent. It modifies the form of the variable from *qualitative* to *binary numeric*, with one characteristic assigned a value of '0' and other characteristic given a value as '1'. In the example, the gender is the dummy variable, where being a male can be assigned a value 1 and female as 0 (even vice versa can be done). By taking binary numbers, the dummy variable acts like a switch which turns on or off, and accordingly influences the calculations.

Subsequent to this binary coding of one of the independent variables, while the other independent variable and dependent variable are expressed in their quantified form, the multiple regression calculations are performed, as done routinely, to establish the multiple regression equation. The interpretation of this equation can be elaborated as below:

Hypothetically, assume that the multiple regression equation so established is given as:

$$y = 15 - 0.065x_1 + 12.5x_2$$

Where hypothetically, y = number of accident, x_1 = Age of the Driver and x_2 = Gender of the driver (Male = 1 and Female = 0).

This equation can be read as (a) As the age of the driver increases, the number of accident caused gets reduced—increase in unit age, causes *number of accident* to be reduced by a factor of 0.065 and (b) For same age of the driver, a male driver will cause 12.5 times more accidents than a female driver.

Such studies are routinely explored in insurance industry to analyze the risk behavior and fine tune their product-price offerings.

MULTICOLLINEARITY

Multicollinearity refers to the presence of high correlation between the independent variables. Regression coefficients where the independent variables are highly correlated tend to be unreliable. A very high value of R^2 indicates possible presence of multicollinearity. In such situations, the regression equation can be used to estimate the value of dependent variable, but isolated use of partial regression values, contained in the regression equation should be avoided. To overcome this, one possible option is to discard one of the independent variables or use principal component analysis. (Detailed note on principal component analysis is outside the scope of the present discussion). In brief, in principal component analysis (PCA), new sets of independent variables are formed, which have relationship with the original variables in such a manner, that the multicollinearity is eliminated.

In conclusion, regression analysis can help to achieve the following:
- Establish the equation of relationship between the dependent and independent variables (regression equation).

Section 5: Correlation, Regression, Partial and Multiple Correlation ...

- Measure the error between the actual values of the dependent variable and the value of dependent variable calculated through the regression equation (calculated through the standard error)
- To understand the explained and unexplained components of the total variation in the dependent variable, thereby be able to calculate the multiple coefficient of determination.
- Can be used as a forecasting tool too.

Example: Consider the following hypothetical data on three variables. For the given data (a) Establish the regression equation, considering variable y (Height of male child) as dependent on variable x_1 (Father's height) and x_2 Mother's height), i.e. regression y on x_1 and x_2. (b) Also calculate the multiple coefficient of determination. (The heights mentioned below are in inches).

y	61	67	71	68	67	63	69	68	67	69
x_1	63	66	70	69	65	68	70	70	72	67
x_2	63	63	66	64	61	63	66	64	67	63

Solution:

a. Establishing the proper table for deriving the desired regression equation:

S. No.	y	x_1	x_2	x_1^2	x_2^2	y^2	yx_1	yx_2	x_1x_2
1	61	63	63	3969	3969	3721	3843	3843	3969
2	67	66	63	4356	3969	4489	4422	4221	4158
3	71	70	66	4900	4356	5041	4970	4686	4620
4	68	69	64	4761	4096	4624	4692	4352	4416
5	67	65	61	4225	3721	4489	4355	4087	3965
6	63	68	63	4624	3969	3969	4284	3969	4284
7	69	70	66	4900	4356	4761	4830	4554	4620
8	68	70	64	4900	4096	4624	4760	4352	4480
9	67	72	67	5184	4489	4489	4824	4489	4824
10	69	67	63	4489	3969	4761	4623	4347	4221
Σ	670	680	640	46308	40990	44968	45603	42900	43557
	$\bar{y}$=67	$\bar{x}_1$=68	$\bar{x}_2$=64						

The multiple regression equation is given as:

$$y = a + b_{yx_1.x_2} x_1 + b_{yx_2.x_1} x_2$$

The constants can be derived as follows:

$$b_{yx_1.x_2} = \frac{(\Sigma yx_1 - n\bar{y}\bar{x}_1)(\Sigma x_2^2 - n\bar{x}_2^2) - (\Sigma yx_2 - n\bar{y}\bar{x}_2)(\Sigma x_1x_2 - n\bar{x}_1\bar{x}_2)}{(\Sigma x_1^2 - n\bar{x}_1^2)(\Sigma x_2^2 - n\bar{x}_2^2) - (\Sigma x_1x_2 - n\bar{x}_1\bar{x}_2)^2}$$

$$b_{yx_1.x_2} = \frac{(45603 - 10\times67\times68)(40990 - 10\times64^2) - (42900 - 10\times67\times64)(43557 - 10\times68\times64)}{(46308 - 10\times68^2)(40990 - 10\times64^2) - (43557 - 10\times68\times64)^2}$$

Chapter 17: Partial and Multiple Correlation and Regression

$$b_{yx_1.x_2} = \frac{550}{671} = 0.8197$$

Using the other equation given below, value of the second constant can be calculated.

$$b_{yx_2.x_1} = \frac{(\sum yx_2 - n\bar{y}\,\bar{x}_2)(\sum x_1^2 - n\bar{x}_1^2) - (\sum yx_1 - n\bar{y}\,\bar{x}_1)(\sum x_1x_2 - n\bar{x}_1\bar{x}_2)}{(\sum x_1^2 - n\bar{x}_1^2)(\sum x_2^2 - n\bar{x}_2^2) - (\sum x_1x_2 - n\bar{x}_1\bar{x}_2)^2}$$

$$b_{yx_2.x_1} = \frac{-231}{671} = -0.3443$$

Using the partial regression coefficients, the value of intercept 'a' can be calculated as:

$$a = \bar{y} - b_{yx_1.x_2}\bar{x}_1 - b_{yx_2.x_1}\bar{x}_2$$

i.e. $a = 67 - 0.8197 \times 68 - (-0.3443) \times 64$
Solving, $a = 33.2951$
Hence the multiple regression equation is

$$y = 33.2951 + 0.8197x_1 - 0.3443x_2$$

b. For calculating the multiple coefficient of determination, creating the necessary table:

S. No.	y	ŷ	(y−ŷ)	(y−ŷ)²	(y−ȳ)	(y−ȳ)²
1	61	63.25	−2.25	5.04	−6	36
2	67	65.71	1.29	1.68	0	0
3	71	67.95	3.05	9.30	4	16
4	68	67.82	0.18	0.03	1	1
5	67	65.57	1.43	2.03	0	0
6	63	67.34	−4.34	18.87	−4	16
7	69	67.95	1.05	1.10	2	4
8	68	68.64	−0.64	0.41	1	1
9	67	69.25	−2.25	5.04	0	0
10	69	66.52	2.48	6.13	2	4
	Σ = 670		Σ = 0	Σ = 49.64	Σ = 0	Σ = 78
	Mean ȳ = 67			Unexplained variation		Total variation

The values in column ŷ (y estimated) have been calculated on the basis of the regression equation established in the previous part of this solution by replacing the given values of x_1 and x_2. For example, for the first entry: $33.2951 + 0.8197 \times 63 + (-0.3443) \times 63$ gives the y estimated value as 63.25.

From the table, calculating the multiple coefficient of determination:

$$R^2 = 1 - \frac{\text{Unexplained variation}}{\text{Total variation}}$$

i.e.

$$R^2 = 1 - \frac{49.64}{78} = 0.3636$$

Section 5: Correlation, Regression, Partial and Multiple Correlation ...

c. Standard error

Standard error of y on x_1 and x_2 $(S_{y.x_1x_2}) = \sqrt{\dfrac{\Sigma(y-\hat{y})^2}{n-3}}$

$= \sqrt{\dfrac{49.64}{10-3}} = \sqrt{7.09}$

$S_{y.x_1x_2} = 2.663$

Interpretation of results:
a. The multiple regression established is given as:
$$y = 33.2951 + 0.8197x_1 - 0.3443x_2$$
Where y = height of the male child, x_1 = height of father and x_2 = the height of mother, (in inches). The value of 0.8197 implies that if height of mother is held constant, then for every unit change (1 inch) in the height of the father, causes a change of 0.8197 inches in the height of the male child. Similarly the value of –0.3443 implies that if the height of father is held constant, then every change of 1 inch in the height of mother causes a change of –0.3443 inches in the height of the male child.
b. The value of R^2 = 0.3636 implies that 36.36% of the variation in the height of the male child is explained jointly by the variation in the heights of father and mother.
c. The error of estimate is 2.663 inches, when the regression equation is used to estimate the value of the dependent variable, height of the child on the basis of the heights of both his parents.

The importance of multiple regression can be understood by calculating the simple regression equation for the same data values. For the above given data, the simple regression equations along with R^2 values are summarized as below.

Regression	Equation	R^2 value
Simple – Y on X_1	$Y = 31.06 + 0.551 x_1$	0.349
Simple – Y on X_2	$Y = 46.82 + 0.256 x_2$	0.171
Multiple Y on X_1 and X_2	$Y = 33.29 + 0.82x_1 - 0.34x_2$	0.364

(Values rounded off)

Through simple regression where only y and x_1 are considered, the unit change in father height tends to change 0.551 inch in child height, but when the second independent variable is added, i.e. mother height, the unit change in father's height tends to change 0.82 inches in the child's height. In the simple regression equation 'y on x_2' a unit change in mother tends to change 0.256 inches in child's height, but the moment the second variable, i.e. father height is also considered, the unit change in mother's height causes a negative change of 0.34 inches in the child's height, which is completely different when these two variable were interacting in isolation. This highlights the importance of multiple regression as compared to simple regression.

CHAPTER 18

Theory of Attributes

INTRODUCTION

Independent to the study of quantitative measurement of variables, there are at times variables whose qualitative characteristic is the subject of study. Such qualitative data is considered under theory of attributes. An attribute refers to a characteristic or quality like blindness, honesty, specific habit, etc. These cannot be measured, but can be identified as present or absent in the population.

To understand the notation and terminology of theory of attributes, consider the following example:

Example: A population is surveyed for the employment status and literacy. The results are of 100 respondents are as follows:

Literate and employed	:	45
Literate and unemployed	:	10
Illiterate and employed	:	15
Illiterate and unemployed	:	30

Representing the above data in matrix format:

	Literate A	Illiterate α	Total
Employed B	$(AB) = 45$	$(\alpha B) = 15$	$\Sigma B = 60$
Unemployed β	$(A\beta) = 10$	$(\alpha\beta) = 30$	$\Sigma \beta = 40$
Total	$\Sigma A = 55$	$\Sigma \alpha = 45$	$\Sigma(A+\alpha) = \Sigma(B+\beta) = 100$

TERMINOLOGY OF ATTRIBUTE THEORY

Studying the terminology of attribute theory, by using the above provided details:
- *Attributes:* Attributes are the characteristics under study. Literacy and employment are the two attributes or characteristics which are identified as present or absent in the example being considered.
- *Symbols:* The attributes are denoted as A, B, α, etc. The English capital letters A, B, C, D…. denote presence of the attribute, whereas the Greek small letter α, β, γ, δ…denote the absence of that attribute. Therefore AB would stand for presence of both A and B attributes, $\alpha\beta$ would indicate absence of both the attributes,

while αB would signify absence of A and presence of B attribute, and similarly for other values.
- *Class:* The groups obtained by demarcation of population on basis of presence or absence of an attribute are called the *Classes*. For example, A, B, AB, αβ, etc. are the classes.
- *Class symbol:* The letters/symbols explained above are called *Class symbol*.
- *Positive class:* Classes which show only presence of an attribute are called Positive classes, such as A, AB, ABC, B, BC, etc.
- *Negative class:* Classes which show only absence of any attribute are called Negative classes such as α, β, αβγ, αβ, etc.
- *Complementary class:* The two classes 'A', indicating presence of an attribute and 'α' the absence of the same attribute, are complementary classes to each other. Similarly B and β, C and γ are further examples of complementary classes.
- *Class frequency:* Number of observations pertaining to the class is called Class frequency and is denoted by placing brackets around the class symbol. For example, for class symbol AB, the class frequency is $(AB) = 45$. Accordingly frequencies of Positive Classes are called Positive Frequencies and for Negative classes they are called Negative Frequencies. The remaining frequencies are called Contrary Frequencies.
- *Order of classes:* It denotes the number of attributes specified. Class having one attribute like A or B or C, etc. is known as Class of First Order. Similarly classes having two attributes such as AB, AC, αβ, etc. are known as Classes of Second Order. Classes having three attributes such as ABC, αBC, Aβγ, are called Classes of Third Order, and so on. The 'Total population (=N)' without having any demarcation of population on basis of any attribute is known as *Class of Zero Order*.

Total Frequencies and Class Order

i. For a given number of attributes, the total number of frequencies are:

   Total frequencies = 3^n, where 'n' is the number of attributes

   This can be seen from below depiction:
   For zero attributes, i.e. '$n = 0$', the total number of frequency = 1 i.e. N (total population)
   For one attribute, i.e. '$n = 1$', the total number of frequency = $3^1 = 3$ (i.e. N, A and α).
   For two attributes, i.e. '$n = 2$', the total number of frequencies = $3^2 = 9$
   (These are frequencies for classes N, A, B, α, β, AB, αβ, Aβ, αB)
   For '$n = 3$', the total number of frequencies are = $3^3 = 27$. These 27 frequencies are:
   zero order frequencies-N,
   1st order frequency: A, B, C, α, β, γ (6 no's);
   2nd order frequencies: AB, αβ, Aβ, αB, AC, Aγ, αγ, αC, BC, Bγ, βC, βγ (12 no's), and
   3rd order frequencies: ABC, αβγ, ABγ, AβC, Aβγ, αBC, αBγ, αβC (8 no's)

ii. Mathematically, the class frequencies can be expressed through the class frequencies of higher order. Some of these are discussed here for understanding:
   $N = (A) + (α)$ or $(B) + (β)$ (Zero order frequency being represented as first order frequencies)

$(A) = (AB) + (A\beta)$ } (1st order frequency being represented by 2nd
$(B) = (\alpha B) + (A\beta)$ } order frequencies)

$(AB) = (ABC) + (AB\gamma)$ } (2nd order frequencies being represented by
$(\alpha\beta) = (\alpha\beta\gamma) + (\alpha\beta C)$ } 3rd order frequencies)

The above can be extended, for example:
$(A) = (AB) + (A\beta) = (ABC) + (AB\gamma) + (A\beta C) + (A\beta\gamma)$

iii. The classes of higher order for the specified number of attributes are called the ultimate frequencies. Like, for $n=2$, i.e. two attributes, the highest order classes would be the one which comprises of both the attributes $(AB, \alpha\beta, A\beta, \alpha B)$, hence for two attributes the ultimate frequencies are $AB, \alpha\beta, A\beta, \alpha B = 4$ in number. Similarly for three attributes, $n = 3$, the highest order classes will be $(ABC, \alpha\beta\gamma, AB\gamma, A\beta C, A\beta\gamma, \alpha BC, \alpha B\gamma, \alpha\beta C) = 8$ in number. It is generalized that the total number of ultimate classes and frequencies for a specified number of attributes is:

Total number of ultimate classes/frequencies
$= 2^n$, where 'n' is the number of attributes

Consistency of Data

Before studying the association or independence of attributes it is necessary to validate the consistency of the data. The data is called consistent if none of the cell frequencies is a negative number. As all frequencies can be represented by higher order frequencies, the check narrows down to checking of all ultimate frequencies, therefore for data to be consistent, all ultimate frequencies need to non-negative.

Example: Verify if the following data is consistent in the two given cases.
(i) $N = 500, (A) = 310, (B) = 200$ and $(AB) = 30$ and
(ii) $N = 500, (A) = 300, (B) = 250, (AB) = 25$
i. Calculating $(\alpha\beta) = N-(A)-(B)+(AB) = 500-310-200+30 = 20$
$(\alpha B) = N-(A)-(\alpha\beta) = 500-310-20 = 170$
$(A\beta) = N-(B)-(\alpha\beta) = 500-200-20 = 280$
As ultimate frequencies are non-negative, hence the data is consistent.

ii. Calculating $(\alpha\beta) = N - (A)-(B) + (AB) = 500-300-250+25 = -35$
As the ultimate frequency in negative, this data set is 'inconsistent.'
Consistency can also be verified by drawing a matrix table (on the lines of matrix given in the initial part of this chapter) by using the given data and filling the rest by addition/subtraction of given values.

Independence of Attributes

Independent attributes signify no relationship between the attributes. The term 'No relationship' does not mean that attribute A cannot be present or is not present among the population possessing attribute B, but it indicates that proportion of 'Attribute A possessing population' in 'Attribute B possessing population' is similar to proportion of 'Attribute α possessing population' in 'Attribute β possessing population'; and vice versa.

Mathematically,

$$\frac{(AB)}{(B)} = \frac{(A\beta)}{(\beta)} \text{ or } \frac{(AB)}{(A)} \frac{(\alpha B)}{(\alpha)}$$

Alternatively if the two attributes satisfy the following equation, then they are said to be independent:

$$(AB) = \frac{(A)(B)}{N}$$

ASSOCIATION OF ATTRIBUTES

Contrary to independence of attributes, two attributes are called 'associated' if

$$(AB) \neq \frac{(A)(B)}{N}$$

if $(AB) > \frac{(A)(B)}{N}$, attributes are positively associated

if $(AB) < \frac{(A)(B)}{N}$, attributes are negatively associated.

The association between two attributes can also be calculated using the below formula:

$$\delta = \frac{1}{N}\left[(AB)(\alpha\beta) - (A\beta)(\alpha B)\right]$$

if $\delta = 0$ it implies attributes are independent,
if $\delta > 0$ the attributes are positively associated
if $\delta < 0$ the attributes are negatively associated
if $(AB) = (A)$ or $(AB) = (B)$, it reflects complete association of attributes and if $(AB) = 0$ or $(\alpha\beta)=0$, it reflects complete disassociation of attributes.

The theory of association of attributes permits initial inference to be drawn about the attributes, but, unlike correlation, it does not provide any details about the degree of association. The association of attributes and significance of association of attributes is studied through test of significance of variables (Large samples) or through Chi-Square test.

Determining Association

Association between two attributes can be determined by any of these methods:

Expectation Method (Or Probability Method)

If $(AB) = \frac{(A)(B)}{N}$, attributes are independent

If $(AB) > \frac{(A)(B)}{N}$, attributes have positive association

If $(AB) < \frac{(A)(B)}{N}$, attributes are negative association

This method is also called probability method, because if probability of $(A) = (A)/N$ and probability of $(B) = (B)/N$, then their combined probability $= (A)/N \times (B)/N$.

As expectation of (AB) is given as:

$$(AB) = N \times \frac{(A)}{N} \times \frac{(B)}{N} = \frac{(A)(B)}{N}$$

which is same as the above formula. Hence, it is also termed as probability method. Lacking in this is that this method only provides direction of association and does not state anything about the degree of the association.

Proportion Method

This method is as explained in the independence of attributes and considers the proportions. Mathematically,

If $\dfrac{(AB)}{(B)} = \dfrac{(A\beta)}{(\beta)}$, the attributes have no association

If $\dfrac{(AB)}{(B)} > \dfrac{(A\beta)}{(\beta)}$, the attributes have positive association

If $\dfrac{(AB)}{(B)} < \dfrac{(A\beta)}{(\beta)}$, the attributes have negative association

Like expectation method, this method also provides the direction and not degree of association.

Yule's Coefficient of Association

Yule's coefficient of association allows calculation of degree of association apart from the direction. The formula of Yule's co-efficient of correlation is:

$$\text{Yule's coefficient of association } (Q) = \frac{(AB)(\alpha\beta) - (A\beta)(\alpha B)}{(AB)(\alpha\beta) + (A\beta)(\alpha B)}$$

where $-1 < Q < +1$ and
If $Q = 0$, it signifies no association,
If $Q > 0 \Rightarrow$ positively associated
If $Q < 0 \Rightarrow$ negatively associated
If $Q \pm 1 \Rightarrow$ completely positively/negatively associated

Being a ratio, its value remains unchanged if all the cell frequencies are multiplied or divided by a constant.

Coefficient of Colligation

Y, Coefficient of colligations is given as:

$$\gamma = \frac{1 - \sqrt{\dfrac{(A\beta)(\alpha B)}{(AB)(\alpha\beta)}}}{1 + \sqrt{\dfrac{(A\beta)(\alpha B)}{(AB)(\alpha\beta)}}}$$

where $-1 < \gamma < +1$ and
If $\gamma = 0$, it signifies no association,
If $\gamma > 0 \Rightarrow$ positively associated
If $\gamma < 0 \Rightarrow$ negatively associated
If $\gamma \pm 1 \Rightarrow$ completely positively/negatively associated

Relationship between Coefficient of Colligation and Yule's Coefficient of Association

Mathematically, it can be derived that:

$$Q = \frac{2\gamma}{1+\gamma^2}$$

Coefficient of Contingency

In the methods discussed so far, all of these methods pertain to dichotomous classification, i.e. population being divided over an attribute which was either present or absent. There are possibilities that an attribute could have more subdivisions instead of just being present or absent. For example, if the study is about employment, then this can be further subdivided into a part time employed, full time employed, unemployed, etc. To study such classes, a contingency Table is prepared, which is an extended form of the 2 × 2 matrix prepared earlier.

For such data, the overall degree of association between the two attributes is given by Pearson's coefficient of mean square contingency, under null hypothesis H_0 that the two attributes are independent and have no association.

Pearson's coefficient of mean square contingency

$$C = \sqrt{\frac{\chi^2}{N+\chi^2}}$$

where N = Total number of observations, and
χ^2 = Chi-square value

This coefficient is seldom used for 2 × 2 tables, because for such a table it maximum value is approximately 0.71, i.e. a value much lower than the limits. (To calculate Chi square value, refer to the chapter on Chi-square).

Example: A survey of 250 families on heights of parents and children showed the following results:
Tall parents and tall children : 90
Tall parents : 150
Tall children : 110

Complete the data by filling in the missing figures, validate consistency of data and verify if any association exists between the heights of the parents and their children heights.

Solution: Let attribute A = Tall parents, and Attribute B = Tall children, then α = Not tall parents and β = Not tall children (Not 'short' because then it could imply that another class 'medium' could have been there—by writing 'Not Tall' it eliminates all other possibilities)

Therefore as per given data $N = 250$, $(AB) = 90$, $(A) = 150$ and $(B) = 110$
Calculating $(\alpha\beta) = N - (A) - (B) + (AB)$
$= 250 - 150 - 110 + 90 = 80$
Representing the available data in 2×2 table:

	Tall parents (A)	Not tall parents (α)	Total
Tall children (B)	(AB) = 90	(αB) = 20	(B) = 110
Not tall children (β)	(Aβ) = 60	(αβ) = 80	(β) = 140
Total	(A) = 150	(α) = 100	N = 250

Based on the given values, the rest of the values can be calculated as follows:
$(A\beta) = (A) - (AB)$ and $(\alpha B) = (B) - (AB)$.
As all the values have been found to be non-negative, the data is consistent.

Calculating Association

i. Expectation method
$(AB) = 90$
$(A)(B)/N = (150 \times 110)/250 = 66$
i.e. $(AB) > (A)(B)/N$ therefore there exists positive association between the two attributes of tall height of parents and tall heights of their children.

ii. Proportion method:
$(AB)/(B) = (A\beta)/(\beta)$
$(AB)/(B) = 90/110 = 0.818$
$(A\beta)/(\beta) = 60/140 = 0.429$
As, $(AB)/(B) > (A\beta)/(\beta)$. The association between the two attributes is positive.

iii. Yule's coefficient of association:

$$\text{Yule's coefficient of association (Q)} = \frac{(AB)(\alpha\beta) - (A\beta)(\alpha B)}{(AB)(\alpha\beta) + (A\beta)(\alpha B)}$$

Therefore,

$$Q = \frac{(90)(80) - (60)(20)}{(90)(80) + (60)(20)} = \frac{6000}{8400} = 0.714$$

Hence, the two attributes are positively associated and the degree of association, as per Yule's coefficient of association is 0.714.

iv. Coefficient of colligation:

$$\gamma = \frac{1 - \sqrt{\frac{(A\beta)(\alpha B)}{(AB)(\alpha\beta)}}}{1 + \sqrt{\frac{(A\beta)(\alpha B)}{(AB)(\alpha\beta)}}}$$

$$\Rightarrow \gamma = \frac{1-\sqrt{\frac{(60)(20)}{(90)(80)}}}{1+\sqrt{\frac{(60)(20)}{(90)(80)}}} = \frac{1-0.408}{1-0.408} = \frac{0.592}{1.408} = 0.42$$

Hence, the two attributes are positively associated and the degree of association, as per coefficient of colligation method is 0.42.

Check:

$$Q = \frac{2\gamma}{1+\gamma^2}$$

$$Q = \frac{2\gamma}{1+\gamma^2} = \frac{2 \times 0.42}{1+0.42^2} = \frac{0.84}{1.1764}$$

$= 0.714$ (Which is same as calculated directly)

Correlation and Association

Correlation provides direction and degree of linear relationship between two variables, but is not able to assess the nonlinear relationships that data might have. For nonlinear relationships, correlation will not be able to identify or indicate the same. In case of association, it provides a general relationship between two variables irrespective of it being linear or nonlinear. Which of these two to be used, depends on the subject and observations, but generally for scientific research, statisticians prefer correlation, whereas for very large sets of data and where is interest is more on presence or absence of association, like psychological studies, marketing research, association of attributes are used.

It is worth a remark that before analyzing the data for association, consistency has to be ensured and also the association parameters should be clearly defined so that no ambiguity exists in the classification of data.

This chapter must have dispelled some commonly held misbeliefs by practitioners. In medical studies different methods of studying association between variables are used but surprisingly many researchers do not look at the type of variable before applying the appropriate method and hence miss the essence. Many nuggets of information have been picked up (and sometimes found to be nonreportable as science had already discovered them from other reports) by other data scientists doing secondary research or practicing techniques on data-sets.

Section 6

Statistical Testing

- Hypothesis Testing
- Tests of Significance (Z and Student's 't'-Test)
- F-Test and Analysis of Variance and Covariance
- Chi-square Test
- Non-Parametric Tests
- Which Statistical Test to Use?
- Special Considerations about Inferential Issues in Biostatistics
- Measurement and Error Analysis

CHAPTER 19

Hypothesis Testing

Hypothesis testing is a critical part of research analysis. Many a times, due to faulty setting up of hypothesis, the statistical analysis of the data does not produce any significant or sometimes even contrary results. It is important that the hypothesis and its related concepts are well understood, so that appropriate 'test of significance' can be applied and proper statistical inferences can be drawn. This chapter deals with the concept of hypothesis and its significance.

A hypothesis is defined by the Merriam-Webster dictionary as "a tentative assumption made in order to draw out and test its logical or empirical consequences". It is a tentative statement about a testable relationship between two or more variables which has not been proven to be correct beyond reasonable doubt.

INTRODUCTION

Sampling, as discussed in earlier units, has application all across in practically every field as well as our daily lives. It is primarily done to understand the population characteristics better. To recapitulate the concept, in statistics, 'Population' refers to the entire group under study. This entire group may consist of people, things, or anything being studied, and at times it is also referred as 'Universe'. Due to various reasons like size of the universe, cost involved, time, destruction of unit during testing (bursting of cracker to test its quality), etc., it is not possible to study every individual unit of the population. Hence, sampling is resorted to, where sample is drawn from the population and studied for its characteristics. Thereafter, the sample characteristics are used to approximately define, estimate and determine the population characteristics. It is very much possible that, based on a sample study, rejection of an acceptable option or acceptance of a rejectable option can happen. Such an error involved in these approximations are called sampling errors.

In spite of inherent sampling errors in sampling methodology, it still remains the best tool available to study the population characteristic. The process of drawing inference on the basis of sampling and using various tests to reach conclusion about the population are collectively referred as 'Statistical Inferences.'

Statistical Inferences

Statistical inferences, where hypothesis testing is a core and an integral part, are inferences and not decisions; these are tools to aid decision making. Statistical inferences are applied in two broad areas:

1. Estimation

In estimation, the objective is to estimate the characteristics (parameters) of population, from which the sample has been drawn, by use of statistics obtained from the sample.

2. Hypothesis Testing

The objective in Hypothesis Testing is to verify any statement or assertion about the population on the basis of information derived from a sample of that population.

Before the concept of Estimation and Hypothesis Testing is discussed, some basic concepts and terminology needs to be dealt with, so that it becomes easier to understand these, as well as the advanced statistical concepts including Significance tests.

BASIC CONCEPTS

Population versus Statistic

To differentiate between the constants of a population and those of a sample, the population constants are referred to as "Parameters" while those of a sample are called "Statistic". Accordingly for population the parameters, mean is denoted as 'μ' and variance = σ^2. For sample, the statistic mean is denoted as $\bar{x}$ and variance as s^2. Thus, a sampling distribution, consisting of various sets of samples can have its statistic denoted as $\bar{x}_1, \bar{x}_2, \bar{x}_3, \bar{x}_4$ ... and $s_1^2, s_2^2, s_3^2, s_4^2$ ...

Sample Size

The number of items in a sample is called the sample size.

Unbiased Estimate

If the estimate from sample statistic is equal to the population parameter, then such statistic is called an unbiased estimate of parameter.

Sampling Distribution

A given population can generate infinite number of samples of same or different sample sizes. From a finite population of size N, if samples of size n are drawn, the number of such samples which can be drawn are equal to ^NC_n (also written as $\binom{N}{n}$) –Combination). For each such sample, statistic like mean $\bar{x}_1, \bar{x}_2, \bar{x}_3$, and variance s_1^2, s_2^2, s_3^2, can be calculated. The set of values of such 'statistic' collectively are called as "Sampling Distribution of the Statistic."

Standard Error

It is defined as the 'standard deviation of the sampling distribution of the statistic' and denoted as SE.

Consider that samples are drawn from a population in order to investigate it. The sample statistic, so calculated from the single sample or from the sampling distribution, is supposed to reflect the true parameters of the population. In reality, this cannot be always true, due to inherent limitations of sampling. Standard error of a sample, as a tool, helps to estimate the interval in which population parameter may be found and indicates the degree of precision with which the sample statistic represents the population parameter.

Though Standard error of the mean is most commonly used, Standard Error (for large samples) exists for other statistics, as given below:

Statistic	Standard Errors Formula
Mean ($\bar{x}$)	$\dfrac{\sigma}{\sqrt{n}}$
Standard deviation (s)	$\sqrt{\dfrac{\sigma^2}{2n}}$
Variance s^2	$\dfrac{\sigma^2}{\sqrt{2n}}$
Median	$\dfrac{1.25331\sigma}{\sqrt{n}}$
Observed sample proportion	$\sqrt{\dfrac{PQ}{n}}$

Where n = Sample size, σ^2 = Population variances and
P = Population proportion and Q = (1 – P)

Standard error should not be confused with the Standard deviation of a distribution. Whereas standard deviation is a measure of dispersion of values in the sample or of a population, the standard error relates to the 'statistic' ability of the sampling distribution to convey information about 'population' parameters. Naturally, as the sample size increases, the standard error decreases.

The reciprocal of standard error, i.e. $\dfrac{1}{\text{Standard Error}}$ is the measure of reliability of the sampling distribution. Conversely, the value of standard error gives the unreliability of the sampling distribution, implying that higher the standard error, higher is the difference between actual and expected values. Consider a hypothetical situation, where the Standard deviation given is 20 and for a sample size as 100.

As per standard error of mean formula,

$$\text{Standard Error} = \dfrac{\sigma}{\sqrt{n}} = \dfrac{20}{\sqrt{100}} = \dfrac{20}{10} = 2.0$$

Assume that there is a need to reduce the standard error to 1.0, while ensuring the standard deviation of the sample does not change. As stated earlier, with increase in sample size, the standard error decreases. Hence, to reduce the standard error, the sample size needs to be increased. Calculating the new sample size by using the same equation:

$$\sqrt{n} = \frac{\sigma}{\text{Standard Error}} = \frac{20}{1} = 20$$

Squaring both sides, we get $n = 20^2 = 400$, i.e. 4 times of earlier sample size.

From the above example and the equation form, it can be generalized that to improve the Standard error of mean by half (or double its precision), the sample size needs to be increased 4 times.

Standard error assists in identifying the limits within which the population parameter values are expected to lie. As large samples and sampling distributions are assumed to be distributed normally, the values of the population parameter are expected to be within 68.27%, 95.45% and 99.73% within ± 1SE, ± 2SE and ± 3SE, respectively.

Standard error is very important and critical instrument in testing of hypothesis of large samples and shall be discussed in the next sections.

Estimation (Theory of Estimates)

Estimation deals with the understanding of population parameter by studying the sample statistic. Estimation tools help in obtaining characteristics of population with desired degrees of precision. Population parameter denoted as θ, can be estimated by the following ways.

Point Estimates

As name suggests, it is a single point estimate of the population parameter. A preconceived method is used to estimate the value which could be a true representative of the specific population parameter. This value could be the mean, median, mode, measure of variability, etc. The point estimate depends entirely on the sample values.

Interval Estimates

If the estimated population parameters are based on two statements or values range, they are called Interval estimates. Naturally being a range, the interval estimates will have two specific values (the limits), within which the population parameter is expected to lie. The two limits, upper and lower are called as 'Confidence limits' and the interval between them – 'Confidence interval'.

For example, if the statement is, 'based on multiple factors it is estimated that tomorrow, the rainfall would be 20 mm'—it is a case of Point estimate. In case the statement is 'the estimated rainfall tomorrow is about 18–24 mm', then it is a case of Interval estimate. Which of the two estimates is to be used, depends on the objective and criticality of the result of the subject of study.

Characteristics of Estimator

An estimator signifies the statistical measure used for estimation and naturally is different from an estimate. The value or interval values derived from 'estimation' by use of an 'estimator' are called the 'estimates'. For example, if the estimation has been done by using mode, then mode is the estimator and the value of mode is the estimate.

For determining the population parameter, an ideal estimator should be:

Consistent: The estimator should be consistent in the manner that as the sample size $n \to \infty$, or sample size $n \to N$, the population size; the probability should approach unity. From this perspective Mean is a consistent estimator because as $n \to N$, $\bar{x} \to \mu$. From this perspective, Median is a consistent estimator only when the population is symmetrical.

Unbiased: If the estimator generated expected values are identical to the population parameter being estimated, then such an estimator signifies an unbiased estimator. Mathematically, $E(\theta) = \theta$. Being unbiased is more properly associated with finiteness of sample size, as against consistency, which was more associated with the infiniteness of sample size where $n \to \infty$.

Efficient: Efficiency of an estimator is based on the variances of the sampling distribution of the estimator. The estimator with lower variance will be more efficient measure of the estimator as compared to the estimator giving a higher variance. Lower variance will signify higher concentration around the population parameter being estimated, hence, better. For example, for a normal population, both mean and median as estimators are unbiased and consistent, and their variances are given as:

$$\text{Variance }(\bar{x}) = \frac{\sigma^2}{n} \text{ for all '}n\text{' and}$$

$$\text{Variance (Median)} = \frac{1.57\sigma^2}{n} \text{ for large '}n\text{'}$$

(calculated from the standard error of the estimators given earlier in this chapter).

Comparing the two variances, $V(\bar{x}) < V$ (median), it can be inferred that, for a normal distribution, 'mean' is a more efficient estimator than 'median'. It follows that the 'Most efficient estimator' among all the class of estimators, will have least sampling variance.

Sufficient: An estimator, which conveys all the information contained in the sample, about the parameter, is stated to be a sufficient estimator of population. If such an estimator is available, then use of any other estimator is not necessary.

Methods of Estimation

There are many methods for estimation. Important among them are the method of maximum likelihood estimator, Method of minimum variance, Method of least squares, Method of minimum Chi-square and Method of inverse probability. The important among these concepts have been covered under different sections of the book and hence are not detailed. Further in medicine and indeed in biosciences, use of hypothesis testing is more prominent.

Hypothesis Testing

As stated earlier, the objective in Hypothesis testing is to verify any statement or assertion about the population on the basis of information derived from a sample drawn from that population. Being an established procedure, the steps in testing of hypothesis are as follows:

Step 1: Set up a Null hypothesis H_0

Step 2: Set up an alternate hypothesis H_1. Alternate hypothesis, helps to determine, whether to use single tailed (right or left) or the two-tailed tests.

Step 3: Choose appropriate 'Level of significance (α)' on the basis of reliability and permissible risk. α, the level of significance should be fixed before the sample is drawn.

Step 4: Compute the test statistics $z = \dfrac{t - E(t)}{SE}$ under null hypothesis.

Step 5: Conclude by comparing. If the computed values of 'z' is less than the tabulated significant value $Z_{tabulated}$, i.e. if "$|Z| \leq Z_{tabulated}$" at a given level of significance, it implies that the difference is non-significant. This signifies that the difference between the observed value 't' and the expected value $E(t)$ in step 4, is due to sampling fluctuations of the sample drawn and the sample data does not provide statistically sufficient evidence against the Null hypothesis, hence, the Null hypothesis needs to be statistically accepted. Contrary to this if "$|Z| > Z_{tabulated}$" at a given level of significance, implies that the difference is significant and cannot just occur due to sampling fluctuations. Therefore, null hypothesis needs to be rejected and alternate hypothesis needs to be accepted.

Each of the above steps is crucial and need to be followed properly to arrive at appropriate conclusions. The following pages detail the underlying concepts like what is a null hypothesis, how to set it up, etc., so that steps are also understood (Step 4: Test of Significance being a complete topic in itself, is being discussed in separate chapters).

Null Hypothesis (denoted as H_0) and Setting it Up

The first step outlined is 'to set up a Null hypothesis'. Null hypothesis is a statement or assertion about the population parameter, that there exists 'no significant difference' between sample statistic and population parameter, i.e. Null, Void or Invalid difference. It should be noted that 'no significant difference' is not same as 'no difference.'

Generally, there always exist differences between the sample and population values, but whether this difference is significant or not, is the consideration. Can the values observed be explained by chance alone? Null hypothesis statement implies that whatever difference exists between the sample statistic and population parameter, it is due to sampling fluctuations.

Null hypothesis is the hypothesis which tests for possible rejection of new theory/fact, etc., i.e. protects the existing norm. This is due to the fact that the objectives of most of the statistical analysis, is to establish new theories from research or in other words-to establish significant differences of new, from old and existing. Only when, an existing fact is rejected, a new fact can be established. The new fact has to be 'statistically significant' from the existing, i.e. it should not arise due to the sampling fluctuations. Therefore, the Null hypothesis states that 'no significant difference exists' and for new fact or theory to get established, Null hypothesis has to get 'rejected' by proving that the results are 'statistically significant'.

The key critical word is the term "Statistically significant." This term, if understood in right perspective, can allow complete understanding of the concept of hypothesis, test of significance, etc. A new concept or theory has to be 'statistically significant' for it to be accepted and be able to establish itself in place of existing one. *It can be proved to be 'statistically significant' if it can be proved that the results are beyond the limits of the sampling fluctuations and hence cannot be attributed to sampling alone.* These limits are defined through the level of significance, which is discussed in detail in the following pages.

For example: To test whether the new therapy is effective or not in treatment of a disease, the H_0 can be a statement like 'that the new therapy is not effective for the treatment of the disease.' Such a statement would not be an appropriate H_0, because it reflects bias of the researcher as it may be a preconceived notion of being 'not effective.' The ideal H_0 statement would be that "there is no difference between the old and the new therapy in the treatment of the disease." This neutral stand in framing the H_0, before collection of sample observation is the mainstay of H_0 and hence should be given due importance.

Alternate Hypothesis

Alternate hypothesis, denoted as H_1, is as important as Null hypothesis, as it defines the purpose of acceptance or rejection of the Null hypothesis. Without alternate hypothesis, Null hypothesis is insufficient. Alternate hypothesis should be explicitly stated and the two hypotheses H_0 and H_1 should be mutually exclusive. In most statistical analysis, the H_1 is a statement which the researcher hopes to be proven true.

For example, if H_0 is set up as the mean of the population (μ) is (say) 10, then the H_1 could be that mean is not equal to 10 or it is either lower or higher than 10. This creates three different hypothesis as stated below:

$H_0 : \mu = 10$ and $H_1 : \mu \neq 10$ or
$H_0 : \mu = 10$ and $H_1 : \mu > 10$ or
$H_0 : \mu = 10$ and $H_1 : \mu < 10$

For testing these three hypothesis's, three different tests would be needed, i.e. for (i) it would be Two Tailed, for (ii) it would be Right Tailed Test and for (iii) a Left Tailed Test. Thus, the concept of 'non-inferiority trial', 'equivalence trial', etc. come into play with which the practitioners are more conversant. The difference lies in the setting up of the hypothesis.

This signifies the importance of setting up the correct H_0 and H_1, as the test of significance to be conducted depends on them. If not properly established, it may lead to faulty sampling or bias, improper tests could get applied, and ultimately this can result in discarding of the inferences.

Level of Significance

Denoted as 'α' (alpha), it is the confidence with which the Null hypothesis is accepted or rejected and is expressed as a percentage. The term 'level of significance' is often misinterpreted and hence needs thorough understanding.

'Statistically significant' refers to a situation, that results obtained are due to certain effects or relationship or some other factor (known or unknown) and not just due to chance variation or random flukes in the data. Level of significance is the level at which the results are termed as 'statistically significant' or 'statistically insignificant'. More so, it can be conveyed as 'reliability' of the results.

Assume that the level of significance has been decided at 5%, i.e. 0.05. This figure indicates that a level of confidence is 95% and there is a 5% probability of making an error. The level of significance at 1% would imply a level of confidence of 99% and probability of error at 1%, i.e. a more stringent limit. The probability of error stated here reflects the 'probability of rejecting null hypothesis when it is true.'

Consider the previous example, where H_0 was stated as 'there is no difference between old and new therapy in treatment of the disease.' Let us assume that this statement is true. But on basis of the sample results, where there were significant differences found, the null hypothesis has been rejected, i.e. conclusion made is that the new therapy is more effective than old therapy. This results in an error. At 1% level of significance, it indicates that there is 1% probability that inspite of null hypothesis being true, it gets rejected.

Based on sample statistic, 'at what level of differences, would H_0 be rejected' is inferred from the level of significance. **Figure 19.1** indicating 5% level of significance will explain the concept further.

The 5% level of significance can be present in both directions, hence, 2.5% area on either side, i.e. 0.025 area on both sides of the mean. At 1% level of significance, the area under the curve gets further reduced to 0.005 on each side of the tail. Hence, it is possible that Null hypothesis may get rejected at 5% level of significance, but remains accepted at 1% level of significance, i.e. relatively stronger evidence is required at 1% significance level than 5% significance level in order to reject the Null hypothesis or the difference to be counted as "Statistically significant".

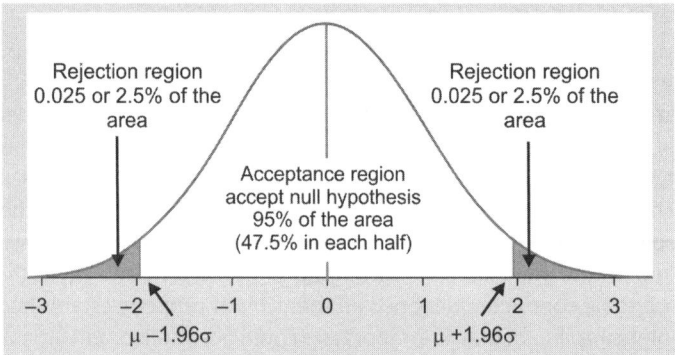

Fig. 19.1: Acceptance and rejection area at 5% level of significance two-tailed (2.5% in each tail)

The level of significance should be ideally, established before the collection of sample. Though it is possible to establish any level of significance, but it is generally considered at 5% or 1%, depending upon the subject of study and criticality of the results. More often than not, it is considered at 5%, which is universally accepted level for most of the subjects. The rule of the thumb in biomedical studies is 5% for most studies like those involving drugs, etc. However, a more stringent level of 1% is advocated by some authors for new procedures and devices. The level should be specified and reported. The practice of reporting the *p* value directly in most biomedical and for longer by other bioscience journals actually makes it easier for editors, reviewers, readers and administrators now to understand the import of the inferences.

Level of significance is also called the 'size of critical region' or 'region of rejection' as H_0 is rejected if the sample statistic falls in this region.

Errors in Testing of Hypothesis

Sample tests are conducted to ascertain population parameters. Hypotheses are established and tested at certain level of significance or confidence. The results from these may culminate in accepting or rejecting a null hypothesis, which may not be necessarily true or correct. For example, assume a new drug was proved to be effective during field trials and allowed to be marketed. Later, when patients were randomly selected and studied for the effectiveness of this drug, it indicated no significant changes from the earlier drug. Multiple tests were carried which found all the samples to be in order. This has happened with a number of drugs and the example is stated only to indicate that the conclusion drawn on the basis of the samples can sometimes indicate false information about the population, causing errors in the inferences drawn.

Hypothesis testing results in four possibilities, in which two are correct possibilities and balance two are erroneous and are commonly called as types of errors. These are:

Possibility	Remark
Reject null hypothesis H_0 when it is true.	(Error- H_0 should have been accepted)
Accept null hypothesis H_0 when it is true.	(Correct result)
Reject null hypothesis H_0 when it is false.	(Correct result)
Accept null hypothesis H_0 when it is false	(Error- H_0 should have been rejected)

Diagrammatically, these can be represented as given below.

		Result deduced from sample	
		Reject H_0	Accept H_0
Actual status of population	H_0 true	Incorrect assessment Type I Error (α)	Correct assessment
	H_0 false	Correct assessment	Incorrect assessment Type II Error (β)

Type I Error: Situation where H_0 gets rejected even when it is true is called as Type I error. It is denoted as α, and its probability limit is also known as 'Level of Significance' of the test. In other words, though there are no significant differences, but the tests indicate significant differences.

Type II Error: Denoted as β, it arises when H_0 is accepted, though it is false. In other words, it implies that though there are significant differences, the tests indicate no significant differences.

Mathematically, these conditions are stated as follows:
Type I error = P (Reject H_0 when it is true) = P (Reject H_0/H_0) = α
Type II error = P (Accept H_0 when it is False) = P (Accept H_0/H_1) = β

In commercial and business sense, α and β are also referred as Producers' and Consumers' risk, respectively.

Producers' risk = Type I error = Rejecting a lot when it is good.
Consumers' risk = Type II error = Accepting a lot when it is bad.

Power of the Test

Ideally, both the errors in hypothesis testing should be small. Type I error, where H_0, inspite of being true gets rejected while H_1 gets accepted is considered a larger consequential risk than Type II error, where H_0 is accepted inspite of being false. Type II error is more of an opportunity loss. This can be explained through a generalized example, as given below:

A medicine is being tested for being better in treatment of a particular ailment. The H_0 would be established as:

H_0: There is no difference between the existing and the new medicine in treating the ailment.

H_1: Alternate hypothesis—The new medicine is better.

Assuming that in reality—the new medicine is no better than existing medicine. In this situation Type I error—(Reject H_0 when it is true) will translate into a statement that 'Reject the H_0 statement— that new medicine is no different from existing medicine', i.e. accept alternate hypothesis H_1 – which states that 'new medicine is better than existing. On basis of this, the new medicine could be introduced as a part of treatment.

Compared to this, assume that in reality the medicine is actually better than existing medicine. Keeping H_0 and H_1 same as stated in above paragraph, the statements pertaining to Type II error – 'accept H_0 when it is false' will imply that 'accept H_0 which states that the new medicine is no better than existing.' On basis of the new medicine would not be introduced as it is no better.

Thus, it can be seen that Type I is of more consequential loss, whereas in Type II there is an opportunity loss that a new medicine was rejected inspite of being beneficial due to sample test. This is an oversimplified explanation to illustrate the topic. Actually, this would depend on the hypothesis set up. However, for the present discussion this example suffices and we leave it there.

Ideally, any test applied to the sample should control both types of errors toward minimum, i.e. probability of Type I error (α) and probability of Type II error (β) should be minimized. Ironically, for a fixed sample size 'n', α and β behave in opposition—reduction in one results in the increase of another. Thus, it is not practically possible. Therefore, the next best option is to minimize α, as Type I error is more serious. Accordingly, the aim is to keep α to a predetermined low level (level of significance 5% or 1% even better) and then choose a test which minimizes β or maximizes (1 − β).

(1 − β) is called the power function of the test hypothesis and the value of the power function is called the power of the test at that point.

One and Two-tailed Tests

A two-tailed test of hypothesis will reject the null hypothesis, if the sample statistic is significantly higher or lower than the population parameter. As shown in the previous figure, in a two-tailed test the critical or rejection region is situated in both the tails. If the test of hypothesis is being conducted at 5% level of significance, then half of this, i.e. 2.5% or 0.025 of the area under the curve, would lie in each tail. As known, the total area under the curve is unity. The distribution being symmetrical, half of this, i.e. 0.50 lie in each side of the center. At 5% level of significance, 0.025 of the area lies in each tail. This implies that the 0.0475 area is the acceptance region. From the normal curve tables, the area of 0.0475 on both sides corresponds to (± 1.96 × Standard Error) on both sides of the curve. Similarly for tests at 1% significance level, the critical region is 0.005 on both sides, i.e. (± 2.58 × Standard Error). Mathematically, the two tailed test would be stated as follows:

H_0: $\mu = \mu_0$
H_1: $\mu \neq \mu_0$ (i.e. $\mu > \mu_0$ or $\mu < \mu_0$, hence two-tailed test)

Descriptively, these notations can be stated as:

Null hypothesis: Mean of the sample is equal to the population parameter.

Alternate hypothesis: Mean of the sample is not equal to population parameter (i.e. it can be either lower or higher than the population parameter).

Unlike two-tailed test, in the case of one-tailed test, the critical region is located only in One tail-either left tail or right tail. In the above example, it would be one tailed test, if alternate hypothesis is:

H_1: $\mu < \mu_0$ — Left tail test
H_1: $\mu > \mu_0$ — Right tail test

Graphically, these are as shown in **Figures 19.2A and B**:

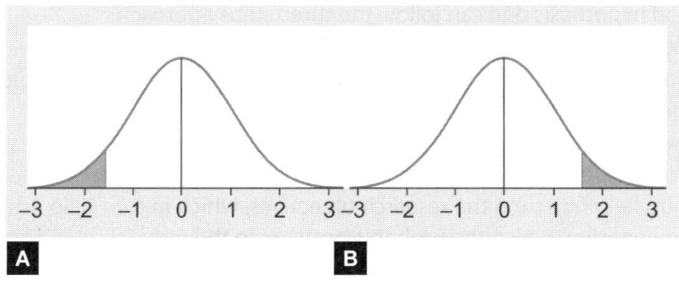

Figs 19.2A and B: Left tail test, right tail test and critical region (shaded portion)

In one-tailed test, the entire critical region lies on one side. Therefore, a 5% level of significance would imply that 5% area would be located entirely in one tail only, as shown above. Thus, in a One tailed test, the critical region is '2α' (double) than that of two-tailed test, in the tail under consideration.

The use of one-tailed test or two-tailed test entirely depends on the alternate hypothesis H_1.

Critical Values

The values that separate the critical region (Rejection region) from the acceptance region are called the critical values. These values depend on the level of significance used in the test and the alternate hypothesis. Given below are critical values for commonly used levels of significance for both, One tail and Two tail tests:

Critical values (z)	α Level of significance		
	1%	5%	10%
Two-tailed test	2.58	1.96	1.645
One-tailed (Right)	2.33	1.645	1.28
One-tailed (Left)	−2.33	−1.645	−1.28

These significance values are for normal population, where 'n' is large and cannot be used in cases where 'n' is small. For cases with small 'n' ($n < 30$), significant values derived from exact sampling distributions, i.e. t, F or x^2, which are covered separately, should be used.

PRACTICAL CONSIDERATIONS IN FRAMING AND TESTING HYPOTHESIS

A hypothesis is the statement of expectation of the researchers; the goal of research effort is to find whether the expectation is correct or are the differences observed merely due to element of chance. A hypothesis may be drawn from previous research or a specific theory or it may be purely exploratory in nature. When too little is known about a subject, the first step is to understand the contours of the problem by doing descriptive studies or surveys. Armed with the background from previous literature or the pilot study or the descriptive research, the research problem can be formulated and then defined in the form of a testable statement called hypothesis.

For a good hypothesis one can follow the three steps approach:
1. It should be based on verifiable research of a topic.
2. It should be capable of being measured and thus have what are called independent and dependent variables.
3. It should be testable.

Fixation of research topic, scanning of secondary information, and discussions with peer group help fine tune the research objectives, which in turn help establishing of proper hypothesis. Also the probable answers to the research questions help in deciding whether it requires a One-tailed or Two-tailed test and the desired sample size. Thus, clues from the existing findings and/or theories about the topic form the basis of hypothesis.

As discussed above, hypotheses can either be directional (more than or less than type relationship, one-tailed) or non-directional. A non-directional hypothesis is one in which the independent variable affects the dependent one in some way but the hypothesis does not specify the direction in which the difference will occur (can be more or less than-two tailed). A directional hypothesis though providing more information about the nature (or "direction") of the relationship, is usually dependent on robust assumption of direction of association. In the absence of such an assumption the non-directional one does yeoman's service. Ideally, the hypothesis should only suggest one relationship and more specifically should only have one independent variable. If there are more than one independent variable the source of any effects observed require highly advanced methods for deriving the inference and even then the source of the difference may not be exactly pinned to the real independent variable.

Generating the general hypothesis on paper is a good first step. Thereafter, the process of refining it starts. Hypothesis should be made as specific as possible so that the ideas being tested, the population in which the research is conducted and evidence of a relationship between the variables is laid out.

Research activity does not prove that a hypothesis is right or wrong. It merely looks for evidence that the opposite of the stated hypotheses is probably not true. If the opposite is probably not true, the hypothesis probably is true.

While it is important to be specific in framing hypotheses this should not be taken so far that it limits the generalizability of the results to anything outside the specific current study. Hypothesis should never use the terms like "we feel," "we believe" or "we think". Any bias or subjective opinion, belief or feeling is best kept out of research.

Keeping the fundamentals in mind about null hypothesis, Alternate hypothesis, Level of significance, etc., next steps are about applying the test statistic on the observation. These tests are called the 'Test of significance' and being various, they are considered in detail in the chapter on 'Test of significance'.

CHAPTER 20

Tests of Significance (Z and Student's 't'-Test)

After the successful establishment of the Null hypothesis, alternate hypothesis and level of significance the next step in testing of the hypothesis is of computing the test statistic or test criteria. As outlined previously, these tests 'statistically validate' the hypothesis and help researcher to draw inferences from the observational data collected and collated. Concepts like acceptance region, rejection region and level of confidence or probability, etc. have been detailed out in the previous chapter, hence are not being discussed here. But before discussing the various 'Tests of significance' the following points need to be mentioned again:

- The 'test statistic' so calculated through these tests, when compared with theoretical tabulated values, helps to understand whether the differences or similarities in observational data are 'Statistically significant' or not.
- The objective of these tests is not to prove or disprove any hypothesis but to evaluate the significance of the differences, similarities or relationships.
- Proper set up of Alternate and Null hypothesis is critical for the use of these tests, as comparison of tabulated and calculated values (One or Two tailed), as well as drawing of proper inferences depend directly on them.

The tests of significance are broadly classified as 'parametric' or 'nonparametric'. The test methods applicable to data, where the normality (or some other form of probability distribution) of the population from which the sample is drawn is the underlying assumption, are called as the Parametric tests. Contrary to this, Non-parametric test does not make any assumption about the form and distribution of the population.

The importance of statistical tests in drawing 'inferences', is already well known. But it should always be kept in mind, that these tests convey the statistical significance and are not decisions about the hypothesis in themselves. As the tests are evaluated against a specific significance level or level of confidence, i.e. probabilities, the results therefore should be read in probabilistic terms and not in certain terms.

In this chapter, only Parametric tests are discussed, hence before applying any of the tests mentioned in this chapter, normal distribution in the population should be ensured.

The numbers of tests are many and not all of them are regularly used. Each of these broad nomenclatures contains lot of tests specific to situations like:

- What is being tested (mean, proportion, correlation, etc.)
- What is number of samples under consideration (1, 2 or k number of samples, etc.)
- What is the sample size (Large or small)
- What does the data pertain to [Qualitative (Attribute) or Quantitative subject]

Given such a complex scenario, it is natural that errors can occur while selecting the most appropriate test to be applied and hence necessary caution needs to be exercised on account of this.

For ease of understanding, the various tests of significance have been broadly classified as below:

- Large Samples Tests
  - Test of significance of attributes (Sampling of attributes)
    - Test for number of success or presence of an attribute
    - Test for single proportion or proportion of success
    - Test of significance of difference in proportion
  - Test of significance of variables (sampling of variables)
    - Z- test for population mean (known variance)
    - Z-test of significance for difference of means
    - Z-test for difference of standard deviation.
- Small sample tests
  - t-test for population mean
  - t-test for difference of means (two population means)
  - Paired t-test for difference of means.
  - t-test for testing significance of sample correlation coefficient
  - t-test for testing significance of the regression coefficient
  - Fisher's z-transformation for tests of correlation coefficient.

LARGE SAMPLE TESTS

Large samples tests exist for both attributes and variables. In the chapter on Probability distribution, it was stated that most of the distributions like Binomial, Poisson, etc. tend to normal distribution as $n \to \infty$. Generally any sample size greater than 30 is considered as a large sample. Though some literature state that use of large sample tests should be applied only if sample size is >100. Therefore, though these tests can be applied for sample sizes >30, but care should be taken to ensure that the sample size is adequate. If the population follows normal distribution, then a sample size of 30+ may suffice, but if the underlying distribution is of Binomial distribution or Poisson distribution, where these tend to follow Normal distribution for large sample sizes, then large sample sizes (100+) would be better.

As these follow Normal distribution, the test statistic is broadly based on the area property of the normal curve, i.e.

$$\text{if } X \sim N(\mu, \sigma^2), \text{ then } Z = \frac{X - \mu}{\sigma} \text{ or } X = \frac{X - E(X)}{\sqrt{V(X)}} \sim N(0, 1)$$

The critical values of Standard Normal Curve hold true for drawing statistical inference.

With this basic understanding, the actual tests can now be taken up.

Test of Significance for Attributes

Broadly speaking, an attribute is a characteristic being studied for its presence or absence, i.e. sampling from population which has mutually exclusive and collectively exhaustive characteristic, wherein one portion of the population possess the characteristic and the other does not. In other words, the study pertains to presence or absence of an attribute. Naturally, there would be a probability attached to presence (say, success) and absence (say, failure) for each draw of the sample (event). This translates into a Bernoullian trial, as explained in topic on Binomial distribution, where the probability of success is given as $P(r) = {}^{N}C_r P^r Q^{(n-r)}$, where $r = 0, 1, 2, 3, \ldots$

The different types of Test of significance for attributes are as follows:

Test for Number of Success (Presence of an Attribute)

The Test statistic is given as:

$$Z = \frac{t - E(t)}{SE}$$

Where t = Observed number of success or attributes presence
$E(t)$ = Expected number of success/attributes presence and SE = Standard error

Due to Binomial distribution behavior of such data,
The Standard error of the data is: Standard error = $\sqrt{npq}$;
and the expected value: $E(t) = np$
Where n = sample size, p = probability of success (attribute's presence) and q = Probability of failure (attribute's absence) and $(q = 1 - p)$

Therefore, the test statistic can be calculated as:

$$Z = \frac{t - E(t)}{SE} = \frac{t - E(t)}{\sqrt{npq}} = \frac{t - np}{\sqrt{npq}}$$

Usage of the formula can be further understood from the below example.

Example: A coin is tossed 500 times and it turns heads 320 times and tails 180 times. Is the coin biased?

Solution: H_0: The coin is unbiased
H_1: The coin is biased
Level of significance considered at 5%
$n = 500$
p = Probability of success (Head turning up) = ½ = 0.5
$q = (1 - p) = 0.5$

The expected frequency for an unbiased coin should be $N \times p = 500 \times 0.5 = 250$ each for heads or tails.

Difference between observed and expected number of heads = 320 − 250 = 70

Standard error = $\sqrt{npq} = \sqrt{500 \times 0.5 \times 0.5} = \sqrt{125} = 11.18$

Computing test of statistic = $Z = \dfrac{t - E(t)}{SE} = \dfrac{320 - 250}{11.18} = 6.26$

Chapter 20: Tests of Significance (Z and Student's 't'-Test)

The value at 5% level of significance is 1.96 (Refer normal curve probability tables given in appendix). As test statistic value of 6.26 is higher than 1.96, the H_0 is rejected, and therefore H_1 is accepted. For the data given we can conclude that the coin is biased at 5% level of significance.

Test for Single Proportion

In case, instead of number of success (presence or absence of an attribute), the interest lies in proportion of success, the test is called as 'Test for Single Proportion'.

As the proportion of success $= p/n$, and the proportion of failure $= q/n$,

Therefore Standard error $= \sqrt{npq}$ would get modified by replacing p and q with p/n and q/n, respectively.

$$\text{Standard error} = \sqrt{n\frac{p}{n}\frac{q}{n}} = \sqrt{\frac{pq}{n}}$$

Hence, the Test Statistic for Proportion becomes:

$$Z = \frac{t - E(t)}{SE} = \frac{t - E(t)}{\sqrt{\frac{pq}{n}}}$$

If sampling is from a finite population of size N, then the

$$\text{Standard error (SE)} = \sqrt{\frac{(N-n)pq}{(N-1)n}}$$

Being a normal variate the Probable limits for the observed proportion of success are $E(t) \pm 3\ SE$.

Example: In a sample of 100 persons, presence of an ailment was recorded in 7 cases. (i) Calculate the proportion of the population which might be affected by this ailment, (ii) Can it be said that more than 10% of the population has the ailment at 5% level of significance.

Solution: In this case neither is the total population size available nor its mean value, hence test for single proportion is to be used.

i. From sample, the proportion of persons having the ailment 'p' = 7/100 = 0.07, hence $q = 1 - p = 0.93$
   With these values calculating the Standard error (SE)

$$= \sqrt{\frac{pq}{n}} = \sqrt{\frac{(0.07)(0.93)}{100}} = 0.0255$$

Proportion of population that might be affected = Probable limits = $E(x) \pm 3\ SE$

As Expected value = 'np', replacing p with proportion, i.e. $\frac{p}{n}$,

The Expected value = $n \times \frac{p}{n} = p$, Proportion, i.e. $E(t) = 7$ for this case.

Therefore, Probable limit = 7 ± 3 Standard error
= 7 ± 3 (0.0255) = 7 ± 0.0765, i.e. 7.0765 and 6.9235

Hence, the percentage of population that could be affected could be between 7.08% and 6.92%.

ii. Let H_0: Not more than 10% population is affected by the ailment
H_1: More than 10% population is affected by the ailment

From Alternate hypothesis, it can be inferred it to be a Right Tailed test. Significance values at 5% level for right tail is +1.645.

Also the observed proportion is 7% and expected (being tested for) is 10%, therefore computing the value for:

$$Z = \frac{t - E(t)}{\sqrt{\frac{pq}{n}}} = \frac{0.07 - 0.10}{\sqrt{\frac{(0.07)(0.93)}{100}}} = \frac{-0.03}{0.0255} = -1.176$$

Since $Z_{Calculated} < Z_{0.05\,Tab}$, Null hypothesis is accepted, i.e. not more than 10% population is affected by the ailment.

Test of Significance of Difference in Proportion

This test is utilized when comparison of two distinct populations is being done, in terms of presence or absence of an attribute. If,

n_1 and n_2 = Sample sizes of the two populations, and
p_1 and p_2 = Proportion of success (attribute's presence) in each of them, with $p_1 = \frac{x_1}{n_1}$ and $p_2 = \frac{x_2}{n_2}$

(x_1 and x_2 are the number of occurences of the attribute)
Hence, the test statistic for difference of proportion is given as:

$$Z = \frac{p_1 - p_2}{\sqrt{PQ\left(\frac{1}{n_1} + \frac{1}{n_2}\right)}} = \frac{p_1 - p_2}{SE}$$

where, $P = \frac{n_1 p_1 + n_2 p_2}{n_1 + n_2} = \frac{x_1 + x_2}{n_1 + n_2}$, and $Q = 1 - P$

Example: A survey was done in two cities A and B about drinkers and nondrinkers. In city A from a sample of 1,000 people, 400 were drinkers; whereas in city B, from a sample of 1,500, 490 were drinkers. Does the two cities have significant difference in the number of drinkers?

Solution: Let H_0: $p_1 = p_2$, there is no significant difference between the number of drinkers in 2 cities.

H_1: Significant difference exist between the number of drinkers in the two cities.
Data provided : $n_1 = 1,000$; $x_1 = 400$; $p_1 = 400/1,000 = 0.4$
$n_2 = 1,500$ $x_2 = 490$; $p_2 = 490/1,500 = 0.33$

∴ Pooled estimate $P = \dfrac{400+490}{1{,}000+1{,}500} = \dfrac{890}{2{,}500} = 0.356,$

and $Q = 1 - P = 1 - 0.356 = 0.644$
Calculating Standard Error =

$$\sqrt{PQ\left(\dfrac{1}{n_1}+\dfrac{1}{n_2}\right)} = \sqrt{(0.356)(0.644)\left(\dfrac{1}{1{,}000}+\dfrac{1}{1{,}500}\right)} = 0.0195$$

Thus,

$$Z = \dfrac{p_1 - p_2}{SE} = \dfrac{|0.4 - 0.49|}{0.0195} = \dfrac{0.09}{0.0195} = 4.615$$

At 5% level of significance the tabulated value is 1.96, whereas calculated value being 4.615, which is higher than the tabulated. Hence, H_0 is rejected, i.e. significant difference exists between the number of drinkers in the two cities.

Test of Significance for Variables (Large Samples)

As against the previous tests which dealt with presence or absence of an attribute (relied primarily on the binomial distribution), the tests for variable deal with sampling of variables, which may take any value, such as weight, height, income, etc. Unlike sampling of attributes, the classification of occurrence into presence or absence is not possible in this case.

Generally, these tests are applied in situations like, checking reliability of estimates, estimating population parameter from the samples or comparing expected and observed values and deducing whether this deviation is due to sampling fluctuations or not. These tests also help to understand the reliability of the estimates.

Similar to previous section, the test statistic is calculated on the basis of normal variate. The main tests of significance for large samples are provided below:

Z-Test for Population Mean (Known Variance)

This test is used to check if significant differences exist between an assumed population mean and the sample mean.
The test statistic is given as:

$$Z = \dfrac{\bar{x} - \mu}{\sigma/\sqrt{n}} = \dfrac{\text{Sample mean} - \text{Population mean (assumed)}}{\text{Standard error of mean}}$$

Where, σ = Standard deviation of the population; and n is the sample size.
This test requires either the variance (σ^2) or Standard deviation (σ) of the population to be known. In case population variance is unknown, then the sample variance can be used. The inferences are drawn on the basis of Normal Probability Curve. Consider the below example:

Example: A sample of 400 patients, treated by a hospital was surveyed for their expenses for treatment at the said hospital. The mean expense was found to be ₹ 20,000 with a standard deviation of ₹ 1,600. Will it be a correct statement that the average expense of all patients is ₹ 21,000. Also find the limits at 99% confidence level for the charges of the hospital.

Solution: Let H_0: No significant difference exists between the sample mean and assumed population mean.

H_1: The population mean is not equal to ₹ 21,000

From data provided : $n = 400, \bar{x} = 20,000; \sigma = 1,600$

Standard error : $SE_{Mean} = \dfrac{\sigma}{\sqrt{n}} = \dfrac{1,600}{\sqrt{400}} = \dfrac{1,600}{20} = 80$

Test statistics $Z = \dfrac{\bar{x} - \mu}{SE} = \dfrac{|20,000 - 21,000|}{80} = 12.5$

As Tabulated $Z_{0.05} = 1.96$, is less than calculated Z, the Null hypothesis is rejected, i.e. ₹ 21,000 cannot be taken as the population mean.

99% of values are expected to lie in $\bar{x} - 2.58\, SE \leq \mu \leq \bar{x} + 2.58\, SE$
$= 20,000 \pm 2.58 \times 80 = ₹\ 19,793.6$ to ₹ 20,206

Z-Test of Significance for Difference of Means

This test statistic is utilized to determine whether the difference between the means of two populations is significant or not.

The test statistic under Null hypothesis $H_0: \mu_1 = \mu_2$ is given as:

$$Z = \dfrac{(\bar{x}_1 - \bar{x}_2)}{\sqrt{\dfrac{\sigma_1^2}{n_1} + \dfrac{\sigma_2^2}{n_2}}}$$

Where, $\bar{x}_1$ is the mean of the sample having sample size of n_1 and drawn from a population whose mean is μ_1 and variance = σ_1^2. Similarly, $\bar{x}_2$ is the mean of the sample having sample size of n_2 and drawn from a population whose mean is μ_2 and variance = σ_2^2. The test statistic is compared with the standard normal variate values for drawing inference.

At times, the data may present upon situation where the variances are known or not known and may be equal or not, accordingly, the calculation formula is suitably modified, as shown below:

- If standard deviation of the two populations are known and equal, i.e. $\sigma_1^2 = \sigma_2^2 = \sigma^2$, the Test statistic is given as:

$$Z = \dfrac{(\bar{x}_1 - \bar{x}_2)}{\sigma\sqrt{\dfrac{1}{n_1} + \dfrac{1}{n_2}}}$$

If in this case, it is known that the variance σ^2 of populations are equal but the exact value of variance is not known, then its estimate-based on sample variance can be used and is given as:

$$\sigma^2 = \dfrac{n_1 s_1^2 + n_2 s_2^2}{n_1 + n_2}$$

Where s_1^2 and s_2^2 are the standard deviations of the two samples.

- If the variances are neither equal ($\sigma_1^2 \neq \sigma_2^2$), nor values of σ_1 and σ_2 are known. In such cases, the variance is estimated from the sample variances. Since samples are large, the results provide close approximation and can be used, as per below test statistic:

$$Z = \frac{(\bar{x}_1 - \bar{x}_2)}{\sqrt{\frac{s_1^2}{n_1} + \frac{s_2^2}{n_2}}}$$

Example: Two large samples are drawn to study the weights of population. The sample results are as below:

	Sample A	Sample B
Sample size	250	300
Mean weight (Kg)	72	69
Standard deviation	4	5

Is there a difference in the two populations mean weights.

Solution: Let Null hypothesis $H_0: \mu_1 = \mu_2$ the two population means have no difference. Therefore, $H_1: \mu_1 \neq \mu_2$

As the variances of population are unknown, the sample standard deviations are used to calculate the test statistic, i.e.

$$Z = \frac{(\bar{x}_1 - \bar{x}_2)}{\sqrt{\frac{s_1^2}{n_1} + \frac{s_2^2}{n_2}}} = \frac{72 - 69}{\sqrt{\frac{16}{250} + \frac{25}{300}}} = \frac{3}{0.384} = 7.81$$

As the Calculated Test Statistic $|Z| = 7.81 > 1.96$ (Tabulated Test Statistic at 5% level of significance), the Null hypothesis is rejected and alternate hypothesis is accepted. This implies that there is statistically significant difference between the mean weights of the two populations.

Z-Test for Difference of Standard Deviation

This test is used to determine if the standard deviation of the two populations differ significantly or not. The Test Statistic, under $H_0: \sigma_1 = \sigma_2$ is given as:

$$Z = \frac{(s_1 - s_2)}{\sqrt{\frac{\sigma_1^2}{2n_1} + \frac{\sigma_2^2}{2n_2}}}$$

Where, s_1 and s_2 are the standard deviations of the two samples and each having sample sizes of n_1 and n_2, respectively. σ_1 and σ_2 is the population standard deviations.

Considering in most of the practical research data, the population standard deviation is not known, therefore for large samples, the sample standard deviations can be used as the estimates of the population parameter. In such event, the test statistics can be written as:

$$Z = \frac{(s_1 - s_2)}{\sqrt{\dfrac{s_1^2}{2n_1} + \dfrac{s_2^2}{2n_2}}}$$

Example: With the data values provided in the previous example, test if the difference between the standard deviations significant.
Solution: Let the Null hypothesis be $H_0: \sigma_1 = \sigma_2$ and $H_1: \sigma_1 \neq \sigma_2$

As the data provided contains the standard deviations of the two samples, using the test statistic and calculating the Z value:

$$Z = \frac{(s_1 + s_2)}{\sqrt{\dfrac{s_1^2}{2n_1} + \dfrac{s_2^2}{2n_2}}} = \frac{4-5}{\sqrt{\dfrac{16}{2 \times 250} + \dfrac{25}{2 \times 300}}} = -3.68$$

As the Calculated Test Statistic $|Z| = 3.68 > 1.96$ (Tabulated Test Statistic at 5% level of significance), the Null hypothesis is rejected and alternate hypothesis is accepted. This implies that statistically there is significant difference between the standard deviations of the two populations.

Note: The reader should remember, being a standard normal curve that the tabulated value of Z at a given level of significance ranges from negative to positive region around the mean. At 5% level of significance, though in the examples of the value has been shown as 1.96, but in actuality it ranges from –1.96 to +1.96. Therefore, to simplify the comparison, the modulus of the calculated value is compared with the positive limit of the tabulated value. The Sign of the Critical Limit, assumes importance when One-Tailed Tests—right or left, are being conducted. In such cases, it is necessary to compare with exact values and modulus should not be used.

SMALL SAMPLE TESTS (STUDENT'S T-TESTS)

Tests for small samples differ from large sample tests, primarily because of the underlying assumptions of large samples may not hold. The challenge with small samples is that they cannot be assumed to have close approximation to normality and might not possess sufficiently close approximation of the population parameters. Inspite of these restrictions, at times and in certain cases, large samples may not be feasible; therefore, small sample tests are required. Researcher should take a conscious decision before resorting to small samples as these are more prone to the sampling fluctuations.

Even though the sampling distribution might not be distributed normally, but in small samples tests, it is assumed that the population from which the sample is drawn does not markedly deviate from normality. Therefore, the small sample tests should be exercised with caution in extremely skewed distribution like J-shaped, as the inferential results may not hold true. More often than not, the interest in studying small samples lies in testing the hypothesis instead of estimation of population parameters.

As noted in the previous section large sample tests are based on the normal distribution, while small sample tests are based on the Students *t*-distribution and are also referred to as the *t*-tests or Student's *t*-tests.

Student's t-distribution

The theoretical *t*-distribution was developed by WS Gosset, who published it in *Biometrika* in 1908, under his pseudonym 'Student', and hence the name 'Student's *t*-test'.

Student's *t*-distribution is used when the sample size is small (<30) and/or the population standard deviation is unknown. The probability density function of *t*-distribution is given as:

$$f(t) = \frac{\Gamma\left(\frac{v+1}{2}\right)}{\sqrt{v\pi}\,\Gamma\left(\frac{v}{2}\right)} \left(1 + \frac{t^2}{v}\right)^{-\left(\frac{v+1}{2}\right)}$$

where, v = degree of freedom ($n - 1$); and is the gamma function.

It can be mathematically derived and proved that as the sample size 'n' tends towards ∞, i.e. for large degrees of freedom ($n - 1$), the *t*-distribution tends to standard normal distribution. Other properties of Student's distribution include its range being from $-\infty$ to $+\infty$ and its function $f(t)$ is symmetrical about the $t = 0$. The constants of the *t*-distribution are given as:

- Mean = 0 for degrees of freedom $v > 1$, otherwise undefined.
- The median and mode are also 0. The skewness is 0 for $v > 3$, otherwise undefined.
- Variance = $\dfrac{v}{v-2}$ for $v > 2$, and

- $\beta_2 = \dfrac{3(v-2)}{v-4}$ for $v > 4$ (Measure of kurtosis)

From these constants it can be easily inferred that for $v > 4$, $\beta_2 > 3$ and the variance for $v > 2$, would be > 1, and as such the Student's *t*-distribution curve **(Fig. 20.1)** would be relatively flatter at top, but higher at tails than the normal distribution curve. The graph of Student's *t*-distribution in relation to normal curve is as shown below. From the graph it can be observed that as 'n' increases from 3 to 10, the curve gets closer to normal curve. With a further increase in 'n' it will tend to get even closer.

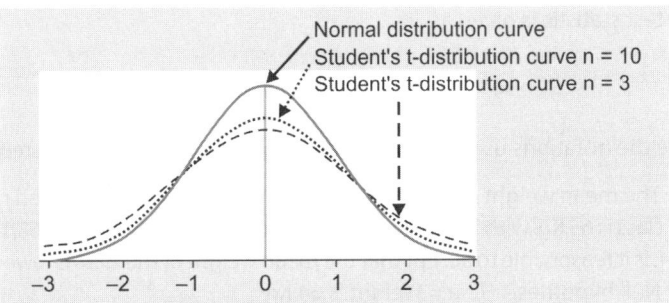

Fig. 20.1: Student's *t*-distribution curve

Student t-test Statistic

The 't-statistic' is defined as:

$$t = \frac{\bar{x} - \mu}{S/\sqrt{n}} = \frac{\bar{x} - \mu}{S} \times \sqrt{n}$$

Where
$\bar{x}$ = Mean of the sample; n = Sample size
μ = Mean of the population (assumed or actual)

Sample standard deviation $S = \sqrt{\dfrac{\Sigma(x - \bar{x})^2}{n-1}}$

This t-statistic is used in the various t-test of significance. The calculated value of t-statistic is compared with the tabulated value (t-distribution tables given in appendix) at the desired level of significance. If the calculated value of 't' < tabulated value of 't', at a specific level of significance, then the differences are not statistically significant and therefore the sample might have been drawn from the population with mean μ. From the above, the fiducial limits of the population mean can be derived as:

Fiducial limits of population mean as per t–statistic = $\bar{x} \pm \dfrac{S}{\sqrt{n}} t_\alpha$

where t_α = tabulated value of t for $df = (n - 1)$ at α significance level

The significant values of t-statistic at level of significance α for a single tailed test can be obtained from the significant values of the two tailed test by considering the significant value at 2α level of significance in two tailed test. For example, the significant value of t-statistic at 5% level of significance in One tailed test = significant value at 10% level of significance in a Two tailed test. With these basics, the t-tests can now be discussed:

t-Test for Population Mean

This test is used to determine the significance of the difference between the sample mean $\bar{x}$ and the assumed or actual population mean μ, where the variance of the population is unknown (In case variance is known, Z-test should be used).

The t-test statistic is given as

$$t = \frac{\bar{x} - \mu}{S} \times \sqrt{n}$$

Where the notations used and inferential process are same as depicted above.

Example: The mean weight of a sample consisting of 10 students selected randomly from the class is 65 Kg, with individual sample value being 67, 64, 62, 66, 68, 69, 62, 63, 65 and 64. Is it reasonable to accept that the mean weight of the class is lower than 64.
Solution: Null hypothesis H_0: μ = Weight is 64 Kg

Alternate hypothesis H_1: $\mu < 64$ Kg (One tailed test)
Preparing the necessary table for calculating the sample standard deviation:

Weight x	$x - \bar{x}$	$(x - \bar{x})^2$
67	2	4
64	-1	1
62	-3	9
66	1	1
68	3	9
69	4	16
62	-3	9
63	-2	4
65	0	0
64	-1	1
$\Sigma = 650$		$\Sigma = 54$

$$\bar{x} = \frac{650}{10} = 65$$

As $S = \sqrt{\dfrac{\Sigma(x - \bar{x})^2}{n-1}} = \sqrt{\dfrac{54}{10-1}} = 2.45$

Using t-test statistic:

$$t = \frac{\bar{x} - \mu}{S} \times \sqrt{n}$$

$$\therefore t = \frac{65 - 64}{2.45} \times \sqrt{10} = 1.29$$

The tabulated value of 't' at 5% level of significance for one tailed test at (10 – 1) = 9 degrees of freedom is 1.83. Since the calculated value is lower than the tabulated values, the Null hypothesis is accepted at 5% level of significance and alternate hypothesis is rejected. Therefore, it is not reasonable to state the population mean is lower than 64 Kg.

t-Test for Difference of Means (Two Population Means)

This test is used to determine the significance of difference between the means of two populations, with the assumption that the variances of the two populations though unknown, but are equal.

The test statistic is given as:

$$t = \frac{(\bar{x} - \bar{y}) - (\mu_x - \mu_y)}{S\sqrt{\dfrac{1}{n_1} + \dfrac{1}{n_2}}}$$

Where,
$\bar{x}$ and $\bar{y}$ = Means of two independent samples;
n_1 and n_2 = their respective sample sizes;
μ_x and μ_y = their respective population means; and

$$S = \sqrt{\frac{\Sigma(x-\bar{x})^2 + \Sigma(y-\bar{y})^2}{n_1 + n_2 - 2}}$$

Degrees of freedom in this test are $(n_1 + n_2 - 2)$

The test statistic 't' is applied under the Null hypothesis that the samples have been drawn from the normal populations having means μ_x and μ_y, with the assumption that the population variances are unknown but equal, i.e. $\sigma_x^2 = \sigma_y^2$

The practical utility of this test statistic lies in testing the following, whether:
i. The two samples have been drawn from populations having same mean $\mu_x = \mu_y$
or
ii. Two sample means $\bar{x}$ and $\bar{y}$ do not differ significantly.

Therefore, when $\mu_x = \mu_y$ is assumed under Null hypothesis, the above given test statistic can be simplified as below and used accordingly. The inferences from calculated and tabulated values are same as previous test.

$$t = \frac{(\bar{x} - \bar{y})}{S\sqrt{\frac{1}{n_1} + \frac{1}{n_2}}}$$

Where the notations are same as mentioned earlier.

Example: Hypothetically, two sample values are given as below:

Sample A: 10 9 8 11 12 14 16 15 12 13
Sample B: 8 7 11 14 11 12 13 14 15 10

Test whether the two population means are equal or not.

Solution: Null hypothesis H_0: The two population means are equal $\mu_x = \mu_y$
Alternate Hypothesis H_1: The two population mean are unequal $\mu_x \neq \mu_y$
Preparing the necessary table for carrying out the required calculations:

x	x−$\bar{x}$	(x−$\bar{x}$)²	y	y−$\bar{y}$	(y−$\bar{y}$)²
10	−2	4	8	−3.5	12.25
9	−3	9	7	−4.5	20.25
8	−4	16	11	−0.5	0.25
11	−1	1	14	2.5	6.25
12	0	0	11	−0.5	0.25
14	2	4	12	0.5	0.25
16	4	16	13	1.5	2.25
15	3	9	14	2.5	6.25
12	0	0	15	3.5	12.25
13	1	1	10	−1.5	2.25
Σ = 120		Σ = 60	Σ = 115		Σ = 62.5

Calculating sample means $\bar{x} = \dfrac{120}{10} = 12$; and $\bar{y} = \dfrac{115}{10} = 11.5$

Calculating

$$S = \sqrt{\dfrac{\Sigma(x-\bar{x})^2 + \Sigma(y-\bar{y})^2}{n_1 + n_2 - 2}} = \sqrt{\dfrac{60 + 62.5}{10 + 10 - 2}} = \sqrt{\dfrac{122.5}{18}} = 2.61$$

Calculating *t*-test statistic:

$$t = \dfrac{(\bar{x} - \bar{y})}{S\sqrt{\dfrac{1}{n_1} + \dfrac{1}{n_2}}} = \dfrac{12 - 11.5}{2.61\sqrt{\dfrac{1}{10} + \dfrac{1}{10}}} = \dfrac{0.5}{1.17}$$

by solving $t = 0.428$

Tabulated *t* value at 5% level of significance for two tailed test, at $(n_1 + n_2 - 2) = 18$ (degrees of freedom is given as 2.1009).

As the Calculated value of *t* < Tabulated value, the Null hypothesis is accepted, i.e. the two populations have equal mean and there is no statistically significant differences between them.

Paired t-Test for Difference of Means

This test is used to determine the significance of difference between two population means where the sample observations are in pairs.

In contrast to the previous test, here the two samples are not independent and there are no assumptions being made for the population variances. As the observations are taken in pairs, the sample sizes are and should be equal to each other.

This test can be applied in instances like efficiency of skill improvement program (scores before and after specific training), drug efficiency—pain management before and after, medical condition before and after treatment, etc. The advantage of such tests which have one respondent is that influence of external factors mostly remains same, i.e. are automatically controlled.

The *t*-test for paired observations is given as:

$$t = \dfrac{\bar{d}}{S/\sqrt{n}}$$

Where $\bar{d}$ is the mean of the differences, i.e. $\bar{d} = \dfrac{1}{n}\Sigma d_i$

d_i = deviation between the observed values of matched pairs

$$S = \sqrt{\dfrac{\Sigma(d - \bar{d})^2}{n - 1}} \text{ or } \sqrt{\dfrac{\Sigma d^2 - n(\bar{d})^2}{n - 1}} \text{ ; and}$$

n = Number of observations (Sample size)

This test statistic is based on $(n-1)$ Degrees of Freedom.

Example: A nicotine patch is reported to cause change in the pulse rate in men. To evaluate a sample of ten men belonging to same age group is considered and their pulse rates before (sample A) and after putting the patch (sample B) is recorded. Evaluate if the nicotine patch causes change in pulse rate or not.

Sample A: 72 75 71 78 69 61 66 69 71 74
Sample B: 70 79 73 76 73 64 67 69 73 75

Solution: The Null hypothesis H_0: Nicotine patch does not cause any change in the pulse rate; and Alternate Hypothesis H_1: Nicotine patch causes change in the pulse rate (Two Tailed). Representing the observational data in table format for ease of calculations:

Sample A (Before)	Sample B (After)	Deviation d_i	d_i^2
72	70	−2	4
75	79	+4	16
71	73	+2	4
78	76	−2	4
69	73	+4	16
61	64	+3	9
66	67	+1	1
69	69	0	0
71	73	+2	4
74	75	+1	1
		$\Sigma d_i = 13$	$\Sigma d_i^2 = 59$

$$\bar{d} = \frac{13}{10} = 1.3$$

$$S = \sqrt{\frac{\Sigma d^2 - n(\bar{d})^2}{n-1}} = \sqrt{\frac{59 - 10(1.3)^2}{10-1}}$$

by solving for, $s = 2.163$

Based on these the *t*-test statistic can be calculated as:

$$t = \frac{\bar{d}}{S/\sqrt{n}}$$

$$= \frac{1.3}{2.163/\sqrt{10}} = \frac{1.3}{2.163} \times \sqrt{10} = 1.901$$

The tabulated value of *t* at 5% level of significance for $(n-1)$ *df*, i.e. $10-1 = 9$ degrees of freedom is given as 2.262 > calculated *t*-value of 1.901. Therefore, Null hypothesis is accepted. Hence, statistically the nicotine patch is not causing any significant change in the pulse rate.

t-Test for Testing Significance of Sample Correlation Coefficient

This test is used to determine whether the difference between the sample correlation coefficient and zero (uncorrelated) is statistically significant or not.

This test assumes that the sample observations for which the correlation coefficient has been calculated, are from bivariate normal population and have a linear relationship between them.

Under Null hypothesis $H_0 : \rho = 0$, the t-test statistic for testing significance of sample correlation coefficient, at $(n-2)$ degrees of freedom (df) is given as:

$$t = \frac{r}{\sqrt{(1-r^2)}} \times \sqrt{(n-2)}$$

Where r = Observed correlation coefficient between the two samples and n = number of pairs of observation.

The calculated and tabulated values of 't' are compared at desired level of significance to arrive at inferences. In case, the difference is significant (Calculated t > tabulated t), the Null hypothesis is rejected, i.e. the observed correlation coefficient is significant in the population.

In case the difference is insignificant (Calculated t < Tabulated t), the Null hypothesis $H_0 : \rho = 0$ is accepted and accordingly it is inferred that the correlation coefficient is zero in the two variables, i.e. the two variables are uncorrelated.

Example: The correlation coefficient between 20 pairs of observation of two variables is given as 0.555. Test the significance of this correlation coefficient in the two populations.

Solution: The following details have been provided $n = 20, r = 0.555$
Setting up the Null hypothesis $H_0: \rho = 0$, the two populations are uncorrelated and
H_1 : The correlation coefficient is significant
Calculating the t-value:

$$t = \frac{r}{\sqrt{(1-r^2)}} \times \sqrt{(n-2)} = \frac{0.555}{\sqrt{1-0.555^2}} \times \sqrt{20-2} = 2.831$$

The tabulated value of t at 5% level of significance and at 18 df = 2.1009.

As t-calculated > t-tabulated, the Null hypothesis is rejected. Therefore, it can be inferred that the variable are correlated in the population.

t-Test for Testing Significance of the Regression Coefficient

This test is used to determine the significance of the regression coefficient.

Regression equation y on x is given as, $y = a + bx$ where b is the slope of the line or the regression coefficient. Even if the slope is 0, i.e. $b = 0$, the variable y may still have a value based on the y-intercept or 'a'. This test determines if the slope of the line, i.e. regression coefficient is significantly different from 0 or not. If in the sample, it is close to zero, then it is possible that in the population could be around zero (statistically), but if the regression coefficient is statistically different than 0, then its value in the population would be significant.

Section 6: Statistical Testing

Under Null hypothesis $H_0: b = 0$ and $H_1: b \neq 0$, the t-test for regression coefficient's significance with $(n-2)$ degree of freedom is given as:

$$t = \frac{b}{s_b} = \frac{\text{Regression coefficint } b}{\text{Standard error of } b}$$

Where, value of regression coefficient is calculated by methods discussed in the chapter on regression, while

$$s_b = \sqrt{\frac{\sum(y_i - \hat{y}_i)^2}{(n-2)\sum(x_i - \bar{x})^2}}$$

Remember $\hat{y}_i$ is the estimated values of variable y, as derived from the regression equation $y = a + bx$.

The calculated value and tabulated value of t- at specified level of significance and $(n-2)$ degree of freedom are compared. If t-calculated < t-tabulated, Null hypothesis is accepted and vice versa. The above test statistic pertains to equation 'y on x'. For t-test of 'x on y' equation, same steps can be utilized after appropriate interchanges, but caution should be exercised to ensure that the slope of equation used is too of equation 'x on y'.

Example: Two hypothetical variable values are provided as below. Test if the regression coefficient for regression equation y on x is significant or not.

    Variable X: 4   3   4   7   5   2
    Variable Y: 6   9   9   9   9   4

Solution: Let Null hypothesis be H_0: Regression coefficient $b = 0$, and $H_1: b \neq 0$.

Preparing the table for conducting further calculations.

Variable X	Variable Y	X^2	Y^2	XY
4	6	16	36	24
3	9	9	81	27
4	9	16	81	36
7	9	49	81	63
5	9	25	81	45
2	4	4	16	8
$\sum X = 25$	$\sum Y = 46$	$\sum X^2 = 119$	$\sum Y^2 = 376$	$\sum XY = 203$

- Calculating the regression equation 'y on x' for finding the value of regression coefficient b or slope.

$\sum y = Na + b\sum x$ and $\sum xy = a\sum x + b\sum x^2$

i.e. $46 = 6a + 25b$ and $203 = 25a + 119b$
Solving these equations for a and b, we get the values as:
$a = 4.48$ and $b = 0.764$
and the Regression equation for Y on X is $y = 4.48 + 0.764x$

- Calculating s_b standard error through the below table.

Variable x	Variable y	$\hat{y} = 4.48 + 0.764x$	$(y - \hat{y})$	$(y - \hat{y})^2$	$(x - \bar{x})$	$(x - \bar{x})^2$
4	6	7.54	−1.54	2.36	−0.17	0.03
3	9	6.77	2.23	4.96	−1.17	1.36
4	9	7.54	1.46	2.14	−0.17	0.03
7	9	9.83	−0.83	0.69	2.83	8.03
5	9	8.30	0.70	0.49	0.83	0.69
2	4	6.01	−2.01	4.03	−2.17	4.70
$\Sigma = 25$				$\Sigma = 14.67$		$\Sigma = 14.83$

Mean of $x = 25/6 = 4.167$

$$S_b = \sqrt{\frac{\Sigma(y_i - \hat{y}_i)^2}{(n-2)\Sigma(x_i - \bar{x})^2}} = \sqrt{\frac{14.67}{(6-2)14.83}} = 0.497$$

Hence, the t-calculated $= t = \dfrac{b}{S_b} = \dfrac{0.764}{0.497} = 1.537$

Tabulated t-value for d.f. (6−2) = 4 at 5% level of significance is 2.776.

As tabulated value of t > calculated value of t, Null hypothesis is accepted, which implies that the regression coefficient is insignificant at 5% level of significance.

Fisher's z-Transformation for Testing Correlation Coefficients

In the t-test for observed correlation coefficient, the test was conducted to determine the significance of r from zero, i.e. $\rho = 0$. As proved by Fisher, when $\rho \neq 0$, the distribution of r is not normal but skewed. Therefore, for $\rho \neq 0$, it was suggested to use the following transformation, commonly known as Fisher's z-transformation (This should not be confused with the Z-tests).

$$Z = \frac{1}{2}\log_e\left(\frac{1+r}{1-r}\right) = 1.1513 \log_{10}\left(\frac{1+r}{1-r}\right) = \tanh^{-1} r$$

The distribution of Z is approximately normal for even small samples and its mean is given as:

$$\xi = \frac{1}{2}\log_e\left(\frac{1+\rho}{1-\rho}\right) = 1.1513 \log_{10}\left(\frac{1+\rho}{1-\rho}\right) = \tanh^{-1} \rho$$

And variance $= \dfrac{1}{(n-3)}$

Fisher's z-transformation can be applied in following cases:
a. To test whether significant difference exists between the observed correlation coefficient r and an assumed value of population correlation coefficient ρ; under

the Null hypothesis that there exist no significant difference between these. With the above given equations of Z and ξ, the test statistic is given as:

$$\text{Fisher's z test for correlation coefficient} = \frac{Z - \xi}{\sqrt{\dfrac{1}{(n-3)}}}$$

b. To test the difference between two independent sample correlation coefficients r_1 and r_2; under the Null hypothesis that the two sample correlation coefficients do not differ significantly or belong to same population or belong to two different populations with same correlation coefficients.

With above given notations and equation of Z, the test statistic is given as:

$$\text{Fisher's z transformation two samples} = \frac{Z_1 - Z_2}{\sqrt{\dfrac{1}{(n_1 - 3)} + \dfrac{1}{(n_2 - 3)}}}$$

As stated above, Z closely approximates normal distribution, therefore inferences are drawn by using the normal curve table/critical values, i.e. 1.96 at 5% or 2.58 at 1% for two tailed test, and so on for other values. The below examples would make the calculation clear.

Example: The correlation coefficient $r = 0.45$ is from a sample size of 15. Test if the population has a correlation coefficient $\rho = 0.65$.

Solution: As the test is not for $\rho = 0$, t-test cannot be applied. Applying z-transformation under the Null hypothesis that there is no significant difference between $r = 0.45$ and $\rho = 0.65$; and the population correlation coefficient can be regarded as $\rho = 0.65$.

Calculating,

$$Z = 1.1513 \log_{10}\left(\frac{1+r}{1-r}\right) = 1.1513 \log_{10} \frac{1+0.45}{1-0.45} = 0.485$$

$$\xi = 1.1513 \log_{10}\left(\frac{1+\rho}{1-\rho}\right) = 1.1513 \log_{10}\left(\frac{1+0.65}{1-0.65}\right) = 0.775$$

Applying Z Transformation =

$$\frac{Z - \xi}{\sqrt{\dfrac{1}{(n-3)}}} = \frac{0.485 - 0.775}{\sqrt{\dfrac{1}{15-3}}} = -1.004$$

At 5% level of significance the tabulated value is 1.96. Modulus of the calculated value is less than the tabulated value, the Null hypothesis is accepted, i.e. there is no significant difference and the population correlation coefficient may be taken as 0.65.

Example: Two samples have correlation coefficients 0.25 and 0.60 and sample sizes being 12 and 20 respectively. Determine if the two samples correlation coefficients differ.

Solution: Under the Null hypothesis that the sample correlation coefficients do not differ significantly (same), the calculations are done as follows:

$$Z_1 = 1.1513 \log_{10}\left(\frac{1+r_1}{1-r_1}\right) = 1.1513 \log_{10}\left(\frac{1+0.25}{1-0.25}\right) = 0.255$$

$$Z_2 = 1.1513 \log_{10}\left(\frac{1+r_2}{1-r_2}\right) = 1.1513 \log_{10}\left(\frac{1+0.60}{1-0.60}\right) = 0.693$$

Using the test statistic:

$$= \frac{Z_1 - Z_2}{\sqrt{\frac{1}{(n_1-3)} + \frac{1}{(n_2-3)}}} = \frac{0.255 - 0.693}{\sqrt{\frac{1}{(12-3)} + \frac{1}{(20-3)}}} = -1.06$$

As the modulus of the calculated value < tabulated value of 1.96 at 5% level of significance, the Null hypothesis is accepted. The data does not provide enough statistical evidence to conclude that the two samples differ.

WHAT IS *P*-VALUE?

The concept of *P*-value as a method of accepting or rejecting of the hypothesis is what biomedical scientists are more conversant with. We assiduously bring up the topic now because practitioners usually complain of confusion created by use of *t* or *z* values and *P*-values in hypothesis testing.

Classically and as discussed through this entire book, the significance testing is done against a pre-specified level of significance, say 5%. The tabulated critical values at this significance level and the calculated value from the test statistics are compared, and inferences drawn. In classical and historical approach, establishing critical value tables, without the use of computers was a tedious and complicated task. Therefore, critical values tables were available only for limited levels of significance like 1%, 3%, 5%, 10%, etc. Books, only having statistical tables were also common and included tables like square root, cube root, reciprocal, power, etc. apart from critical values. The introduction of computers helped to overcome these challenges. All statistical calculations were now possible with accuracy and speed. In terms of statistical tests, the critical value for any intermediate level of significance was also made possible.

Consider the following case. A research data is evaluated and is being tested at 5% level of significance through the *z*-test (assume any *z*-test). The calculated value of the test statistic turns out to be 2.33, whereas the tabulated critical level at 5% is 1.96. On basis of this, the Null hypothesis is rejected. A question arises, that even though the tests have rejected the Null hypothesis at 5% level of significance, but with same data, what should have been the level of significance at which the Null hypothesis would have been accepted. This gets answered by the *p*-value generated by computers conducting the inferential tests.

When the statistical tests are run through computers, the output is not as per the classical approach, but a *p*-value gets generated. This *p*-value signifies the maximum level of significance at which the Null hypothesis would get accepted. This p-value so generated is compared with the level of significance pre-specified. If the p-value is less than the pre-specified level of significance, the Null hypothesis is rejected. On the contrary, if the p-value is more than the level of significance against which the test has been conducted, it results in acceptance of the Null hypothesis. Due to direct inference about the hypothesis, from the p-value and level of significance, the need to refer to statistical tables is obviated. This also explains the confusion that doctors or biological scientists have when z or t values used by mathematicians in discussing hypothesis testing instead of the p-values the former are more comfortable with. In practice, what is computed is the t or the z value and the p-values are derived as by products. Since both groups are right in their assertions there arises a need for this chapter and this book.

CONCLUSION

The tests of significance are strong tools in inferential statistics. Proper care is necessary for selecting the appropriate test and ensuring that the underlying assumptions are not violated, in order to have proper results. Many a times, in certain research subjects, the required assumptions, especially normality cannot be assumed. To handle such cases and many other situations like availability of data in ranks, etc. other tests are necessitated. The following chapters deal with such other tests, like F-test, Chi-square, etc.

CHAPTER 21

F-Test and Analysis of Variance and Covariance

Named after RA Fisher, F-test is used to determine whether the difference between the two independent estimates of the population variances is significant or not. Conversely, it is also used to determine whether the two samples can be regarded to have been drawn from normal populations having the same variance. These objectives to evaluate the significances, are achieved by calculating the F-ratio.

F-Statistic is defined as:

$$F = \frac{\chi_1^2/\upsilon_1}{\chi_2^2/\upsilon_2}$$

where χ_1^2 and χ_2^2 are two independent Chi-square variates and υ_1 and υ_2 being their respective degrees of freedom.

Hence, F-distribution depends only on degrees of freedom and does not involve any other parameter.

CONSTANTS AND PROPERTIES OF F-DISTRIBUTION

$$\mu_1 = \frac{\upsilon_2}{\upsilon_2 - 2}; \; \mu_2 = \frac{2\upsilon_2^2(\upsilon_1 + \upsilon_2 - 2)}{\upsilon_1(\upsilon_2 - 2)^2(\upsilon_2 - 4)}$$

Mode of F-distribution:

$$\text{Mode of F-distribution} = \frac{\upsilon_2(\upsilon_1 - 2)}{\upsilon_1(\upsilon_2 + 2)}$$

(Since $F > 0$, from the above equation, mode only exists if $\upsilon_1 > 2$)

Restating the above equation as $\dfrac{\upsilon_2}{(\upsilon_2 + 2)} \times \dfrac{(\upsilon_1 - 2)}{\upsilon_1}$; further implies that Mode of F-distribution which exists only when $\upsilon_1 > 2$, will always be < 1, because the numerator is equal to (denominator minus 2) for the second part and denominator is (numerator + 2) for the first part. This will result in fractions <1 in both parts, which when multiplied will always result in a value lower than 1. Other properties of F-distribution are:

- Point of Inflexion of $F(v_1, v_2) = 2$ (Mode), provided $v_1 > 4$
- F-distribution a highly positively, skewed distribution
- The probability of F-curve of F-distribution increases steadily and reaches maximum at Modal Value (<1) and then slowly decreases to become asymptote to x-axis on its right tail.
- At same significance level, $F_\alpha(v_1, v_2) \neq F_\alpha(v_2, v_1)$, but at reciprocal level of significance they are equal, i.e. $F_\alpha(v_1, v_2) = F_{1-\alpha}(v_2, v_1)$, for example:

$$F_{0.05}(5, 10) \neq F_{0.05}(10, 5);$$
$$\text{but } F_{0.05}(5, 10) = F_{0.95}(10, 5)$$

- Relationship between F-distribution and Students t-distribution: $F_{(1, n)} = t_n^2$.

APPLICATIONS

F-test can be applied in many situations. The F-test for equality of Population Variance is dealt with in detail. The most important application of F-test is in "F-test for equality of several means" commonly known as ANOVA, which is also considered in detail in this chapter.

F-Test for Equality of Population Variances

This test is used for answering the following questions:
a. Whether the two independent samples have been drawn from normal populations with same variance or not?
b. Whether the two independent estimates of population variance are homogeneous or not?

To test the significance, the F-ratio is calculated as follows:

$$F = \frac{s_1^2}{s_2^2}$$

where $s_1^2 = \frac{1}{n_1 - 1} \sum_{i=1}^{n_1} (x_{1_i} - \bar{x}_1)^2$ and $s_2^2 = \frac{1}{n_2 - 1} \sum_{i=1}^{n} (x_{2_i} - \bar{x}_2)^2$

Being a ratio, it is also referred as *Variance Ratio Test*. In the ratio, though the formula states s_1 as the numerator and s_2 as the denominator, but in practice the greater of the variances between s_1 and s_2 is taken as the numerator and the other as the denominator. F-test follows $(n_1-1)(n_2-1)$ degrees of freedom. $F_{calculated}$ is compared with $F_{tabulated}$ at the desired level of significance for the specific degrees of freedom and if $F_{calculated} > F_{tabulated}$, H_0 is rejected, and accepted if vice versa.

Example: Given below are observations from two samples drawn from two normal populations. Test whether the two samples drawn have same variance at 5% level of significance.

Sample A:	40	44	45	40	41	42	
Sample B:	65	60	67	65	63	68	67

Solution: Consider H_0—The two samples have no difference and have same variance. Representing the values in tabular form for ease of calculations:

	Sample A			Sample B		
x_1	$(x_1 - \bar{x}_1)$	$(x_1 - \bar{x}_1)^2$	x_2	$(x_2 - \bar{x}_2)$	$(x_2 - \bar{x}_2)^2$	
40	-2	4	65	0	0	
44	2	4	60	-5	25	
45	3	9	67	2	4	
40	-2	4	65	0	0	
41	-1	1	63	-2	4	
42	0	0	68	3	9	
			67	2	4	
$\Sigma = 252$ and $n = 6$		$\Sigma = 22$	$\Sigma = 455$ and $n = 7$		$\Sigma = 46$	

$$\bar{x}_1 = \frac{252}{6} = 42; \quad \bar{x}_2 = \frac{455}{7} = 65$$

Therefore $s_1^2 = \frac{1}{(6-1)} \times 22 = 4.4$ and $s_2^2 = \frac{1}{(7-1)} \times 46 = 7.67$

Therefore $F_{Calculated} = 7.67/4.4 = 1.743$

$F_{Tabulated}$ for $(7-1)(6-1)$ degrees of freedom, i.e. $F_{Tab.}$ $(6,5)_{0.05} = 4.9503$ (between s_1^2 and s_2^2, the higher of these two is taken as numerator while calculating F, and accordingly the degrees of freedom are written first for the numerator).

As $F_{Calculated} < F_{Tabulated}$, therefore, the H_0 is accepted, i.e. there is no significant difference between the two samples in terms of their variances.

F-Test to Test if Two Samples are from the Same Population

For testing whether two samples are possibly from the same population, test of significance for both the population parameters, i.e. mean and variance, are required. t-test is conducted for testing the difference of mean being significant or not, while F-test is done to check the significance of difference of variances. As t-test assumes that the population and/or sample variances are equal (no significant difference exists), hence F-test is conducted first to satisfy this condition of t-test. If there is significant difference existing in variances, t-test will not be required.

Example: Two samples have the following observations:
Sample A: 165 168 172 173 177 169 171 165
Sample B: 169 177 176 180 181 178 176 175 174 184
Evaluate to find if these two samples belong to the same normal population.

Solution: To test whether the samples belong to the same normal population, both significance of difference of means and variances, need to be tested. Therefore, first applying F-test for significance of difference of variance, and thereafter t-test for significance of difference of means, by tabulating and processing of data, as shown below.

	Sample A			Sample B	
x_1	$(x_1-\bar{x}_1)$	$(x_1-\bar{x}_1)^2$	x_2	$(x_2-\bar{x}_2)$	$(x_2-\bar{x}_2)^2$
165	−5	25	169	−8	64
168	−2	4	177	0	0
172	2	4	176	−1	1
173	3	9	180	3	9
177	7	49	181	4	16
169	−1	1	178	1	1
171	1	1	176	−1	1
165	−5	25	175	−2	4
			174	−3	9
			184	7	49
$\Sigma = 1360$ and $n = 8$		$\Sigma = 118$	$\Sigma = 1770$ and $n = 10$		$\Sigma = 154$

$$\bar{x}_1 = \frac{1360}{8} = 170; \quad \bar{x}_2 = \frac{1770}{10} = 177$$

Therefore $s_1^2 = \dfrac{1}{(8-1)} \times 118 = 16.86$ and $s_2^2 = \dfrac{1}{(10-1)} \times 154 = 17.11$

Therefore $F_{Calculated} = 17.11/16.86 = 1.0148$

$F_{Tabulated}$ for $(10 - 1)(8 - 1)$, i.e. $F_{Tab.}(9,7)_{0.05} = 3.6767$

As $F_{Calculated} < F_{Tabulated}$, therefore, the H_0 is accepted, i.e. there is no significant difference between the two samples in terms of their variances.

Now applying t-test for difference of means

Establishing H_0: Mean of two populations has no significant difference ($\mu_1 = \mu_2$) and $H_1: \mu_1 \neq \mu_2$

$$t = \frac{(\bar{x}_1 - \bar{x}_2)}{\sqrt{s^2\left(\frac{1}{n_1} + \frac{1}{n_2}\right)}}; \text{ where } s^2 = \frac{\Sigma(x_1-\bar{x}_1)^2 + \Sigma(x_2-\bar{x}_2)^2}{n_1+n_2-2}$$

Calculating $s^2 = \dfrac{118+154}{8+10-2} = 17$

Hence, $t = \dfrac{170-177}{\sqrt{17\left(\frac{1}{8} + \frac{1}{10}\right)}} = -\dfrac{7}{\sqrt{3.825}} = -3.579$

Modulus of $t = 3.579$

Calculated 't' value = 3.579

Tabulated t-value for $(8 + 10 - 2) = 16$ degrees of freedom at 5% level of significance = 2.12

Since calculated t-value > tabulated t-value, the null hypothesis is rejected and therefore significant difference exists between the means of the two sample, and statistically the two samples do not belong to the same normal population.

F-Test for Equality of Several Means

This test is conducted by the analysis of variance (ANOVA) technique.

Analysis of Variance (ANOVA)

Analysis of Variance (ANOVA) is a tool to study whether the means of more than two populations are equal or not. Though t-test of significance is also designed to test the same, but the limitation of t-test is that it relates to only two samples. If the number of samples is greater than two, the t-test formula is not equipped to test the significance. For such situations, ANOVA has been designed. Originally developed for use in agricultural sciences, ANOVA has found widespread application in almost every field. Analysis of variance helps in understanding the sources of total variation, through studies known as 'treatment' effects on sample. Treatment refers to the factor that the sample is being subjected to and the factor is controlled, i.e. researcher knows it and is controlling it.

One-way classification: This method is used if the data is classified according to only one criterion. In these, for testing of the equality of means, the hypotheses are established as:

Null hypothesis $H_0: \mu_1 = \mu_2 = \mu_3 = \mu_4 \ldots = \mu_n$ (all sample means are equal)
Alternate hypothesis $H_1: \mu_1 \neq \mu_2 \neq \mu_3 \neq \mu_4 \ldots \neq \mu_n$

Calculations of ANOVA are done and summarized in ANOVA-One-way classification table, as shown below:

Variation source	SS (Sum of squares)	Degree of freedom (v)	Mean sum of square	F-Variance ratio
Between samples	SS B	c–1	MSB = SSB/(c–1)	MSB/MSW
Within samples	SS W	n–c	MSW = SSW/(n–c)	

Where c = total number of samples and n = total number of observations

The F variance ratio calculated from above is compared with the tabulated F values at the degrees of freedom (considering correct numerator and denominator). If F calculated > F tabulated, the null hypothesis is rejected and vice versa.

The steps and calculations for ANOVA would become clearer by considering the following example:

Example: Incremental increase in weight in grams due to four different diets over a six months period is given below. Perform Analysis of variance to test if there is significant difference in weight increase made by them.

Diets	Increase in weights (grams)					
A	250	300	274	290	320	240
B	275	375	400	408	340	320
C	222	240	270	290	300	310
D	400	370	210	375	350	360

Solution: H_0: There is no significant difference in weight increase due to the different diets

H_1: There is significant difference in the weight increase due to different diets
> Tabulating the Data:

	Diets			
	A	B	C	D
Increase in weight in six months period	250	275	222	400
	300	375	240	370
	274	400	270	210
	290	408	290	374
	320	340	300	350
	240	320	310	360
$\Sigma =$	1674	2118	1632	2064
Sample Mean ($\bar{x}$)	279	353	272	344
Overall Mean ($\bar{\bar{x}}$)	(279 + 353 + 272 + 344)/4 = 312			

Calculating sum of squares of deviation between Samples (SSB) (Between diets):

It involves calculating deviation between sample mean and overall mean, i.e. for Diet A: (312–279) = 33. Square of deviation would be equal to 33^2 = 1089. As the sample mean is based on 6 values, hence the total deviation would be 6 × 1089 = 6534.

Similarly for Diet B: (312–353) = –41, Square = $–41^2$ = 1681; 1681 × 6 = 10086
Diet C: (312–272) = 40, Square = 40^2 = 1600; 1600 × 6 = 9600
Diet D: (312–344) = –32. Square = $–32^2$ = 1024 × 6 = 6144
Sum of squares of deviation between samples = 6534 + 10086 + 9600 + 6144 = 32364
Degrees of Freedom = (c–1)
where c = total no. of samples, i.e. (4–1) = 3
Calculating sum of squares of deviation within the sample (SSW)

It involves calculating the deviation of the values within the sample from its mean, thereafter calculating the squares of these deviations, which are then summed up. For ease of calculation tabulation is preferred at this step:

1st Entry in 1st Column = value of x from given data = 250, sample mean of the sample of which this entry is part of = 279; thus difference = (250–279) = –29.

Diet A		Diet B		Diet C		Diet D	
$x - \bar{x}$	$(x - \bar{x})^2$	$x - \bar{x}$	$(x - \bar{x})^2$	$x - \bar{x}$	$(x - \bar{x})^2$	$x - \bar{x}$	$(x - \bar{x})^2$
–29	841	–78	6084	–50	2500	56	3136
21	441	22	484	–32	1024	26	676
–5	25	47	2209	–2	4	–134	17956
11	121	55	3025	18	324	30	900
41	1681	–13	169	28	784	6	36
–39	1521	–33	1089	38	1444	16	256
	$\Sigma = 4630$		$\Sigma = 13060$		$\Sigma = 6080$		$\Sigma = 22960$

Sum of squares of difference within samples = 4630 + 13060 + 6080 + 22960 = 46730
Degrees of freedom = (n – c), where n= total no. of observations, i.e. 24 and c = no. of samples, i.e. 4, thus degrees of freedom are (24–4) = 20

Chapter 21: F-Test and Analysis of Variance and Covariance

Preparing ANOVA Table on the basis of the calculations, and calculating F Ratio:

Variation source	SS (Sum of squares)	Degree of freedom (v)	Mean square	F-Variance ratio
Between samples	32364	3	= 32364 ÷ 3 = 10788	10788/2336.5 = 4.617
Within samples	46730	20	2336.5	
Total	79094			

Conclusion: The calculated value of $F(3,20) = 4.617$; the tabulated value of $F_{0.05}(3,20) = 3.0984$; i.e. Tabulated value < Calculated value, thereby H_0 is rejected. This implies that there is significant difference in weight increase due to different diets.

The calculations of ANOVA look cumbersome as encountered above. To reduce the amount of calculations, and thereby chances of error, a 'Correction Factor' method exists which simplifies the process up to calculating ANOVA table. For the previous example, the same is solved again through the 'Correction Factor' method.

Diet A		Diet B		Diet C		Diet D	
x_1	x_1^2	x_2	x_2^2	x_3	x_3^2	x_4	x_4^2
250	62500	275	75625	222	49284	400	160000
300	90000	375	140625	240	57600	370	136900
274	75076	400	160000	270	72900	210	44100
290	84100	408	166464	290	84100	374	139876
320	102400	340	115600	300	90000	350	122500
240	57600	320	102400	310	96100	360	129600
$\Sigma =$ 1674	$\Sigma =$ 471676	$\Sigma =$ 2118	$\Sigma =$ 760714	$\Sigma =$ 1632	$\Sigma =$ 449984	$\Sigma =$ 2064	$\Sigma =$ 732976

Correction Factor (CF)

$$= \frac{(\text{Sum of all observations})^2}{\text{Number of observations}} = \frac{(1674+2118+1632+2064)^2}{24}$$

$$= 2336256$$

Other calculations based on correction Factor is as given:

Total sum of squares

$$= \Sigma x_1^2 + \Sigma x_2^2 + \Sigma x_3^2 + \Sigma x_4^2 - CF$$

$$= 471676 + 760714 + 449984 + 732976 - 2336256$$

$$= 79094$$

Sum of squares between samples

$$= \frac{(\Sigma x_1)^2}{N} + \frac{(\Sigma x_2)^2}{N} + \frac{(\Sigma x_3)^2}{N} + \frac{(\Sigma x_4)^2}{N} - CF$$

$$= \frac{1674^2}{6} + \frac{2118^2}{6} + \frac{1632^2}{6} + \frac{2064^2}{6} - 2336256$$

$$= 32364$$

Sum of squares within samples = (Total sum of squares) – (Sum of squares between samples)

$$= 79094 - 32364 = 46730$$

These values are exactly the same, as calculated in the previous method. The degrees of freedom remain same, and the next step of creating the ANOVA table and conclusion are exactly same as depicted in the previous method.

As F-test is a ratio, the values of the samples can be divided/ multiplied by a constant without affecting the outcome or the inference.

Two-way classification: Against a single classification ANOVA, where the response evaluation is for one variable only (like in previous example it was increase in weight due to different diets), the Two-way classification deals with two treatments or two factors. This is more useful as most of the times, there is not only one but more than one factor which affect the values. Like in previous example, the weight increase could also be affected by the activity levels which may vary from month to month. In some months, the activity level (calorie requirement) is higher than the others. The ANOVA two-way classification table is as given here:

Variation source	SS (Sum of squares)	Degree of freedom (v)	Mean square	F-variance ratio
Between samples	SSB	$(c-1)$	$MSSB = SSB/(c-1)$	i. MSSB/MSSR
Within samples	SSW	$(r-1)$	$MSSW = SSW/(r-1)$	ii. MSSW/MSSR
Residual/Error	SSR	$(c-1)(r-1)$	$MSSR = SSR/(c-1)(r-1)$	
Total	SST	$(n-1)$		

Where   SSB = Sum of squares of difference between sample
SSW= Sum of squares of difference within sample
SSR = Sum of square of residuals
c = Number of samples
r = Number of observations in a sample
n = Total number of observations across all samples

Sum of squares of residual reflects the error due to chance and is calculated by (SST − SSB − SSW). Two values of F-ratio is created and both have their respective degree of freedom and form the basis of the inference for significance of the two treatments.

Example: Consider the previous example of weight increase due to Diets A, B, C and D. The six months' values were also provided. Assume that these months are from Oct-March (half-year), perform a two-way ANOVA at 5% level of significance to assess the significance of diet and month on increase in the weights, i.e. null hypothesis that no significant difference exists between weight increase vis-à-vis diets and month.

Solution: Reproducing the earlier data table and modified for additional inputs:

Month	Diets				Month total
	A	B	C	D	
October	250	275	222	400	1147
November	300	375	240	370	1285
December	274	400	270	210	1154
January	290	408	290	374	1362
February	320	340	300	350	1310
March	240	320	310	360	1230
Total	1674	2118	1632	2064	7488

Chapter 21: F-Test and Analysis of Variance and Covariance

Using the Correction Factor method to carry out the calculations:

$$\text{Correction Factor (CF)} = \frac{(\text{Sum of all observations})^2}{\text{Number of observations}}$$

$$= \frac{(7488)^2}{24} = 2336256$$

Other calculations based on Correction Factor are as given:
Total sum of squares = Sum of square of every observation − CF (as calculated previously)

$$= 2415350 - 2336256$$
$$= 79094$$

Sum of squares between samples (between diets):
Sum of squares between samples

$$= \frac{(\Sigma x_1)^2}{N} + \frac{(\Sigma x_2)^2}{N} + \frac{(\Sigma x_3)^2}{N} + \frac{(\Sigma x_4)^2}{N} - CF$$

$$= \frac{1674^2}{6} + \frac{2118^2}{6} + \frac{1632^2}{6} + \frac{2064^2}{6} - 2336256$$

$$= 32364$$

Sum of squares within samples (within month, denoted as y):

$$= \frac{(\Sigma y_1)^2}{N} + \frac{(\Sigma y_2)^2}{N} + \frac{(\Sigma y_3)^2}{N} + \frac{(\Sigma y_4)^2}{N} + \frac{(\Sigma y_5)^2}{N} + \frac{(\Sigma y_6)^2}{N} - CF$$

$$= \frac{(1147)^2}{4} + \frac{(1285)^2}{4} + \frac{(1154)^2}{4} + \frac{(1362)^2}{4} + \frac{(1310)^2}{4}$$

$$+ \frac{(\Sigma 1230)^2}{4} - 2336256$$

$$= 2345649 - 2336256 = 9393$$

Sum of squares of residual = SST−SSB−SSW
$$= 79094 - 32362 - 9393 = 37339$$

Preparing the ANOVA two-way table for the details calculated:

Variation source	SS (Sum of squares)	Degree of freedom (v)	Mean square	F-variance ratio
Between samples	32364	(4−1) = 3	MSSB = 32364/3 = 10788	i. MSSB/MSSR = 10788/2489.3 = 4.33
Within samples	9393	(6−1) = 5	MSSW = 9393/5 = 1878.6	ii. MSSW/MSSR = 1878.6/2489.3 = 0.75
Residual/error	37339	(4−1)(6−1) = 15	MSSR = 37339/15 = 2489.3	
Total	79094	(24−1) = 23		

Conclusion
- Between samples (weight increase due to diets) $F_{\text{Calculated}} = 4.33$, while tabulated $F_{0.05}(3,15) = 3.2874$; as $F_{\text{Calculated}} > F_{\text{Tabulated}}$, H_0 is rejected, i.e. significant difference exists between the increase in weights due to the different diets.

- Within samples (Months) $F_{Calculated} = 0.75$, while Tabulated $F_{0.05}(5,15) = 2.9013$, which is higher than calculated value of F, hence H_0 is accepted, i.e. no significant difference exists in weight increase due to month.

Post Hoc Tests in ANOVA—Pairwise Comparison

Based on the ANOVA test, the Null hypothesis, that there exists no significant difference and all means are equal, is either accepted or rejected. Consider a case where the ANOVA test is applied, and based on the results, the Null Hypothesis is rejected. Therefore, its alternate hypothesis is accepted, i.e. significant difference exists in the means.

Though when H_1 is accepted, it implies that the null hypothesis $H_0 : \mu_1 = \mu_2 = \mu_3 = \mu_4 \ldots = \mu_n$ (sample means are equal) does not hold good. On closer inspection, it can be seen that even a presence of one inequality in the equation will render it unequal. ANOVA tests do not reveal which of the pair of means are significantly different and which ones are equal. For example, it is possible that $\mu_1 = \mu_2$ but $\mu_2 \neq \mu_3$ or $\mu_2 = \mu_3$ but $\mu_3 \neq \mu_1$. Due to any of these situations, the null hypothesis ($H_0 : \mu_1 = \mu_2 = \mu_3$) gets rejected. 'PostHoc tests' help to determine the pairs which have significantly different means and which ones not.

Consider the example detailed during the discussion on one-way ANOVA test, about four diets A, B, C and D. As per the ANOVA test results there, the null hypothesis was rejected and alternate hypothesis accepted, signifying that the variation in the weight increase due to the diets are significantly different. PostHoc Tests enable the researcher to find which diets caused similar increases and which ones were different, or was it a situation that all four diets caused different increases. It is possible to have other tests of mean for pairs, but using them is difficult. Consider if there were 7 treatments, which give rise to 21 different pairs of treatments and at 5% confidence level, 1 (5% of 21 = 1.xx) of the results could have been wrongly accepted. Post Hoc Test allows obviating the need to conduct so many tests.

Post Hoc test is to be performed only after ANOVA Test has been conducted and on basis of ANOVA Test, the null hypothesis has been rejected.

The main Post Hoc Methods to tests the means after experimentation are:
- Tukey's Honestly Significant Difference (HSD) Test
- Fisher's Least Significant Difference (LSD) Test
- Bonferroni Method

Tukey's Method is most widely accepted; and accepted in peer review as a standard method, hence only this method is being discussed here.

Tukey's HSD method: This test requires that number of observations 'n' for each treatment should be equal or equal sample sizes per group. The formula to calculate Tukey's HSD is given as:

$$\text{Tukey's HSD} = q\sqrt{\frac{MSE}{n}}$$

Where, q = Studentised range from table at given significance level and degrees of freedom = {k, k(n – 1)}
MSE = Mean sum of squares within samples (or MSSW)
n = number of observations, and
k = number of groups

Chapter 21: F-Test and Analysis of Variance and Covariance

The value of Tukey's HSD is then compared with the difference between the sample means for all the pairs. If Tukey's HSD value is lower, it implies significant difference exists between the two group means, and if Tukey's HSD value is higher, it implies that significant differences do not exist at the given level of confidence.

Example: Reproducing the last table from the earlier example of weight increase due to Diets A, B, C and D:

Variation source	SS (Sum of squares)	Degree of freedom (v)	Mean sum of square	F-variance ratio
Between samples	32364	3	10788	10788/2336.5 = 4.617
Within samples	46730	20	2336.5	
Total	79094			

From the above example, extracting the relevant details required for performing Tukey's HSD Test:

Mean sum of squares within samples = 2336.5
'q' from Studentised range table (in appendix) for 'k' groups and 'k(n-1)' d.f. = {4, 4(6–1)}
= for 4 groups and 23 d.f. = 3.914 at 5% level of significance
'n' = number of observations = 6

Therefore, Tukey's HSD $= q\sqrt{\dfrac{MSE}{n}} = 3.914\sqrt{\dfrac{2336.5}{6}}$

$= 3.914 \times 19.73 = 77.24$

For comparison of the results, reproducing the table used earlier in the example:

	Diets			
	A	B	C	D
Increase in weights over six months period	250	275	222	400
	300	375	240	370
	274	400	270	210
	290	408	290	374
	320	340	300	350
	240	320	310	360
$\Sigma =$	1674	2118	1632	2064
Sample mean ($\bar{x}$)	279	353	272	344
Overall mean ($\bar{\bar{x}}$)	(279 + 353 + 272 + 344)/4 = 312			

From above table:
- Difference in means of Diets A and B = 353–279 = 74
- Difference in means of Diets A and C = 279–272 = 7
- Difference in means of Diets A and D = 344–279 = 65
- Difference in means of Diets B and C = 353–272 = 81 (> 77.24 the Tukey's HSD Value)
- Difference in means of Diets B and D = 353–344 = 9
- Difference in means of Diets C and D = 344–272 = 72

The difference between the Diets B and C, where the difference of mean is higher than Tukey's HSD is significantly different, and for all other cases the difference was not significant. Therefore it can be inferred that Diets B and C are causing the inequality or significant differences.

Analysis of Covariance

Analysis of Covariance (ANCOVA) is an extension of ANOVA. Most of the times, any research involves an independent variable and its influence on the dependent variable. For example, effect of a treatment, effect of education, effect of fertilizer and so on. In situations where the dependent variable is influenced only by the independent variable, ANOVA calculations, discussed earlier, are sufficient to derive whether the differences are significant or not. But in practical situations, it is very difficult to assume that the dependent variable is not getting influenced by any other parameter. In such a scenario, when the dependent variable is impacted by the independent variable and also some other variable, equally significant, then the simple ANOVA test would not be able to truly identify the significance of the main independent variable's influence on the dependent variable.

Consider an independent variable, X, and a dependent variable Y, which is dependent on X. Along with X, there is another independent variable Z, which also influences variable Y. If an ANOVA test is performed, to evaluate if X's influences on Y is significant or not, the results could be that there is (say) significant influence. But it could be possible that the significant variation (influence) has been caused by variable Z and not necessarily by X. The other independent variable is called as the 'Covariate.' For such cases 'Analysis of Covariance ANCOVA' is performed which allows for controlling the impact of Z on Y, while the effect of treatment X on Y is being studied. For example, in the earlier example of diets and the increase in weight due to them, it could be that the subjects lifestyle which could be the cause of significant difference in the increase in weights.

During the study of treatments or influencing factors (Independent variable) effect on dependent variable, the methodology is to try and make the independent variable a constant and thereafter evaluate the effect on the dependent variable. Similarly, through ANCOVA, the objective is to 'statistically' remove the covariates influence (variance attributable to the covariate) on the dependent variable, so that the variances caused by the 'independent variable under study', are known. Regression, as a statistical tool, is used in ANCOVA to even out the effect of covariate, and it is assumed that the covariate is linearly related to the dependent variable. It is also assumed that the covariate is not affected by the independent variable under study and the regression line of the groups under study, are assumed to be parallel ('homogeneity of regression slopes'). Data on covariates ideally should be gathered before the experiment.

Variability partitioning: In ANOVA, the variability is divided in two parts—Variance due to experimental and individual difference. Against this, in ANCOVA the variability seen is divided in three parts- Variance due to experiment, Error and Variance due to covariate.

Chapter 21: F-Test and Analysis of Variance and Covariance

The following example would bring clarity to the calculations and procedure to perform the Analysis of Covariance, ANCOVA.

Example: A research is being conducted to evaluate the Weight Gain under four different Diet Programs A, B, C, D. It is also understood that Weight Gain can be impacted by the lifestyle of the subject. Three type of lifestyles are considered for study, viz. Fitness-Oriented, Routine Active and Sedentary. The results of the diets were observed for weight gain after 12 weeks, are given below. Conduct ANCOVA to evaluate the significance of variation due to different diets in weight gain.

Weight Gain Observed (Kg) after 12 weeks

	Diet A	Diet B	Diet C	Diet D
Fitness oriented	4	6	2	4
	4	4	8	8
	2	6	6	6
Routine active	4	8	4	4
	6	4	6	8
	4	6	8	8
Sedentary	8	6	6	8
	6	6	10	10
	6	6	10	8

Solution: The test involves validity of Diet program (Independent variable) on the weight gain (Dependent variable), while the life-style of the person (Covariate) may also be affecting the weight gain. Using ANCOVA analysis for testing the variations:

Null hypothesis
For diets H_0: There is no difference in weight gained due to different diets, and
H_1: The weight gained due to different diets are different/not equal

For lifestyle H_0: The weight gained under the different lifestyles is equal
H_1: The weight gained is not equal under the different lifestyles

For interaction H_0: There is no difference in weight gained in any of the interactions
H_1: The weight gained in different interactions in not equal

Reproducing the table, and adding relevant rows and columns, from the above table, calculating:

Observed values (Given)

	Diet A	Diet B	Diet C	Diet D	Marginal Total-lifestyle
Fitness oriented	4	6	2	4	60
	4	4	8	8	
	2	6	6	6	
Routine active	4	8	4	4	70
	6	4	6	8	
	4	6	8	8	
Sedentary	8	6	6	8	90
	6	6	10	10	
	6	6	10	8	
Marginal Total-diet	44	52	60	64	$\Sigma = 220$

Total observations: 36

Sum of squares

Diet A	Diet B	Diet C	Diet D	Marginal Total-lifestyle
16	36	4	16	344
16	16	64	64	
4	36	36	36	
16	64	16	16	444
36	16	36	64	
16	36	64	64	
64	36	36	64	708
36	36	100	100	
36	36	100	64	
240	312	456	488	1496

Correction Factor (CF) = (Sum of all observations)² ÷ Total number of observations
 = (220)² ÷ 36
 = 1344.44

Total sum of squares = (Sum of squares of all observations) − Correction factor
 = 1496 − 1344.44
 = 151.56 (TSS) with degrees of freedom (36−1) = 35

Sum of squares between lifestyles
 = (Σ (row total squares) ÷ No. of observations included) − CF
 = {(60² + 70² + 90²) ÷ 12} − 1344.44
 = 38.89 (SSB) with degrees of freedom (3−1) = 2

Sum of squares within diets
 = (Σ (Column total squares) ÷ No. of observations included) − CF
 = {(44² + 52² + 60² + 64²) ÷ 9} − 1344.44
 = 26.22 (SSW) with degrees of freedom (4−1) = 3

Sum of squares due to interaction (SSI): For calculation of this, necessary tables are prepared:

$$SSI = n \sum (\bar{x}_{ij} - \bar{x}_i - \bar{x}_j + \bar{\bar{x}})^2$$

	Actual observations				Mean of observations in interaction ($\bar{x}_{ij}$)					$(\bar{x}_{ij} - \bar{x}_i - \bar{x}_j + \bar{\bar{x}})^2$			
	Diet A	Diet B	Diet C	Diet D	Diet A	Diet B	Diet C	Diet D	Row Mean	Diet A	Diet B	Diet C	Diet D
Fitness-oriented	4	6	2	4	3.33	5.33	5.33	6.00	5.00	0.20	0.44	0.05	0.00
	4	4	8	8									
	2	6	6	6									
Routine active	4	8	4	4	4.67	6.00	6.00	6.67	5.83	0.00	0.25	0.15	0.03
	6	4	6	8									
	4	6	8	8									
Sedentary	8	6	6	8	6.67	6.00	8.67	8.67	7.50	0.15	1.36	0.37	0.03
	6	6	10	10									
	6	6	10	8									
Column mean					4.89	5.78	6.67	7.11	$\bar{\bar{x}} = 6.11$	$\sum(\bar{x}_{ij} - \bar{x}_i - \bar{x}_j + \bar{\bar{x}})^2 = 3.04$			

Sample calculation: Mean of observation in interaction = (4 + 4 + 2) ÷ 3 = 3.33
Sample calculation: $(\bar{x}_{ij} - \bar{x}_i - \bar{x}_j + \bar{\bar{x}})^2$
For Diet A, Fitness oriented = $(3.33 - 5.00 - 4.89 + 6.11)^2 = 0.20$
    For Diet A, Routine active = $(4.67 - 5.83 - 4.89 + 6.11)^2 = 0.00$ (rounded)
    From the above, table 'n' (Number of observations for each interaction) = 3 and
$\Sigma = 3.04$
    Therefore, SSI = 3 × 3.04 = 9.12 with degrees of freedom (3−1)(4−1) = 6

Sum of squares due to error (SSE) = TSS − SSB − SSW − SSI
= 151.56 − 38.89 − 26.22 − 9.12
= 77.33

Based on all the calculations done so far, ANCOVA table (Similar to ANOVA) is prepared as below:

Variation source	SS (Sum of squares)	Degree of freedom (v)	Mean sum of square = SS ÷ v	F-variance ratio calculated	F- Tabulated α = 0.05	Decision
Between- lifestyle	38.89	2	19.45	= 19.45÷3.22 = 5.95	3.40	Reject H_0
Within- diets	26.22	3	8.74	= 8.74÷3.22 = 2.71	3.01	Accept H_0
Due to interaction	9.12	6	1.52	= 1.52÷ 3.22 = 0.47	2.51	Accept H_0
Due to error	77.33	24	3.22			
Total	151.56	35				

From the above data, it can be concluded that:
For Diets:
- H_0: There is no difference in weight gained due to different diets–accepted
- H_1: The weight gained due to diets are not equal

For Lifestyle
- H_0: The weight gained under different lifestyles is equal–rejected
- H_1: The weight gained is not equal under different lifestyles– accepted

For Interaction
- H_0: There is no difference in weight gained in any of the interactions–accepted
- H_1: The weight gained in different interactions in not equal.

    Statistically, it can be concluded that the there is no significant difference among the weights gained through different diets and interaction of variable, but there is significant differences in weight gained due to lifestyles.

CHAPTER 22

Chi-square Test

'Chi'-pronounced as 'ki' as in 'kindly'. It is the most widely used non-parametric test of significance.

While test like t, Z and F, were based on population parameters, hence considered parametric test, Chi-square (χ^2) test does not involve population parameters and therefore, considered as nonparametric test.

This test is based on observed (actual), expected (theoretical) frequencies and the degrees of freedom.

Mathematically

$$\text{Chi-square } (\chi)^2 = \sum_{i=1}^{k}\left(\frac{O-E}{E}\right)^2$$

Where O = Observed frequency and E = Expected frequency for $v = n - 1$ (degrees of freedom $= n - 1$), and n = Number of observations.

DEGREES OF FREEDOM

χ^2 test always takes into consideration the degrees of freedom in the values. Degree of freedom is defined as the number of independent observations in the sample, reduced by the number of constraints. Simplifying, it is the number of entries in the data which can be assigned independently without any restriction. For example, if sum of 10 positive numbers is 50, then the first nine numbers can be decided independently, but for the tenth number there will be no discretion, but will get decided automatically as the sum of these ten numbers has to 50. Thus, it is stated that this situation has $10 - 1 = 9$ or generalized as $(n - 1)$ degrees of freedom. In cases of data being in matrix form, which is generally encountered in research, the degrees of freedom is calculated $(r - 1)(c - 1)$, where r denotes the number of rows and c - the number of columns in the matrix. Same is explained below:

For example, a 2×2 matrix, where the final total and marginal totals are known, is given below:

2	3	5
6	4	10
8	7	15

Once any cell out of the four cells is filled in, the rest of the values of the 2 × 2 matrix are automatically filled. This means that the degrees of freedom for a 2 × 2 matrix is 1. Consider a 3 × 3 matrix. In this case there is more freedom to choose the values in many more cells, as shown below:

7	4	1	12
5	3	6	14
5	4	15	24
17	11	22	50

In both the matrix above, the highlighted entries have been independently decided, whereas the nonhighlighted got automatically calculated due to marginal totals which were fixed. In 3 × 3 there are 4 values (appearing in bold) which could be filled in by choice and rest 5 entries get calculated once the 4 values are chosen. These 4 values can belong to any cells and not necessarily to the ones shown in this example. Generalizing, the result of 2 × 2 matrix and 3 × 3 matrix, the degrees of freedom for matrix data is product of $(r-1)(c-1)$, as stated above.

SALIENT POINTS OF CHI-SQUARE

- The value of χ^2 ranges from $0 \to \infty$, with zero indicating complete coincidence of observed and expected frequencies.
- Larger the value of χ^2, larger is the difference between the observed and expected frequencies.
- χ^2 calculated is compared with χ^2 tabulated, at the required level of significance.
  - If χ^2 calculated > χ^2 tabulated, null hypothesis H_0 is rejected as the difference between observed and expected values is 'statistically significant'.
  - If χ^2 calculated < χ^2 tabulated, null hypothesis H_0 is accepted as the difference between observed and expected values is not statistically significant and could be attributed to sampling fluctuations.
- χ^2 is not distribution dependent, i.e. it can used for any type of population distribution. Due to its distribution-free property, no assumptions are made about the population parameters.
- Inspite of these benefits, χ^2 should not be considered better than parametric tests, but should be used where parametric tests cannot be applied. If both parametric and nonparametric tests are available, first preference should be for parametric tests.
- $\Sigma (O - E)$ is always zero.
- χ^2 is a limiting approximation of multimonial distribution. If χ^2 is a Chi-square variate with 'n' degrees of freedom, then $\frac{x^2}{2}$ is a gamma variate with parameter $\frac{n}{2}$. (Proofs of these statements are mathematical and beyond the scope of this book).
- Chi-square can be applied to both discrete as well as continuous variables.
- The sum of independent Chi-square distribution variates is also a χ^2 variate. The converse is also true.

CHI-SQUARE DISTRIBUTION

The probability function of χ^2 is given as:

$$f(\chi^2) = C(\chi^2)^{\frac{v}{2}-1} \cdot e^{-x^2/2}$$

Where, $e = 2.7183$
v = Degree of freedom
C = Constant dependent on v

Constants of Chi-square Distribution

Mean of $\chi^2 = v$ Variance = $2v$
$\mu_1 = 0$ $\mu_2 = 2v$
$\mu_3 = 8v$ $\mu_4 = 48 + 12v^2$
$\beta_1 = 8/v$ $\beta_2 = 3 + 12/v$

(v = Degree of freedom)

The mode of Chi-square distribution, with 'n' degrees of freedom is ($n - 2$), and skewness = $\sqrt{\frac{2}{n}}$; which implies that for '$n \geq 1$' the Chi-square distribution is positively skewed.

Chi-Square Distribution Curve

The parameter on which Chi-square distribution relies is its degrees of freedom. The Chi-square distribution curve for various degrees of freedom is given below:

For $v = 1$ and $v = 2$, the curve is monotonically decreasing. The mode is at $v = 2$, where it achieves the highest value. Also the curve is asymptote of X-axis. Observing the various curves, it can be noticed that as degrees of freedom increase, the curve starts at 0, achieving maximum value at ($n - 2$), while becoming increasing symmetrical. For large degrees of freedom, Chi-square distribution curve tends to approximate normal distribution. In the curve graph **(Fig. 22.1)**, it can be seen that with increasing degrees of freedom, the curve skew is shifting towards right and appearing like a normal curve for 30° of freedom.

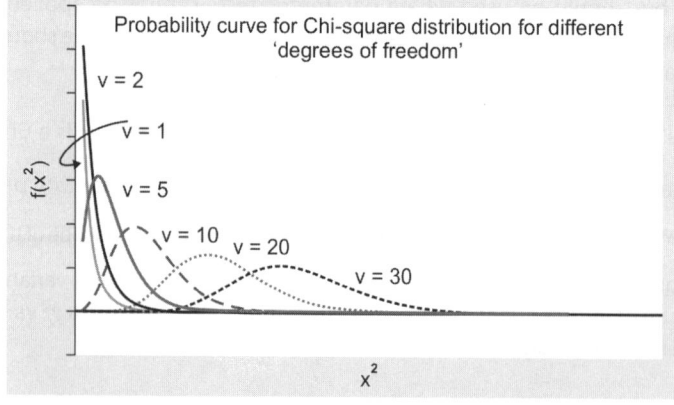

Fig. 22.1 Chi-square curves for different degrees of freedom

The total area under the χ^2 is considered unity. Like t-test and normal curves, instead of having multiple tables of χ^2 values, there is a single table of χ^2 for varying degrees of freedom and at different levels of significance. These tables are provided for up to 30 degrees of freedom, as beyond that the 'normal distribution' tables can be referred as for large samples, χ^2 tends to approximate normal distribution.

Conditions for Validity of Chi-square

- Sample observations need to be independent.
- Constraints on the frequencies should be linear.
- N- total population size should be large (> 50 on minimum basis).
- No expected frequency should be < 5, as at this level, the distribution loses its continuous character and results in overestimated χ^2. This is because, the Chi-square formula consist of dividing the 'square' of difference between observed and expected frequencies by expected frequency. Division of a squared number with a small number, results in overestimate especially when degrees of freedom are less than 3. Such cases are dealt by merging the cell value with its preceding or succeeding entry alongwith adjusting (reducing) the degree of freedom (refer Yate's correction in the later part of this chapter).

Critical Region

Figure 22.2 denotes the acceptance and the rejection region of the Chi-square test.

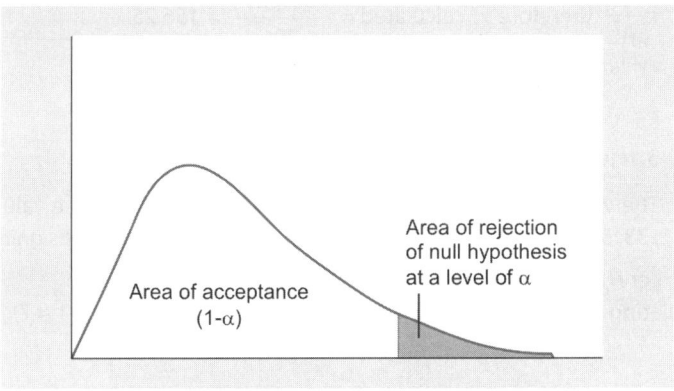

Fig. 22.2 Critical region in Chi-square test

APPLICATIONS OF χ^2 DISTRIBUTION

To Test if Hypothetical Value of Population Variance $\sigma^2 = \sigma^2_o$

To test if a random sample has been drawn from a normal population with known variance σ^2, i.e. to test if the hypothetical value of population variance is $\sigma^2 = \sigma^2_o$. The formula for this is as given below:

$$\chi^2 = \sum_{i=1}^{n}\left[\frac{(x_i - \bar{x})^2}{\sigma_0^2}\right] = \frac{1}{\sigma_0^2}\left[\sum_{i=1}^{n} x_i^2 - \frac{(\Sigma x_i)^2}{n}\right] = \frac{n}{\sigma^2}\left[\frac{\Sigma x_i^2 - (\Sigma x_i)^2}{n}\right] = \frac{ns^2}{\sigma^2}$$

i.e.
$$\chi^2 = \frac{ns^2}{\sigma^2}$$

Where, n = number of observations;

s = sample standard deviation;

σ^2 or σ_0^2 = population variance estimated at $(n-1)$ degrees of freedom.

It is to be kept in mind that the population parameters are not known in Chi-square, and therefore, this test is used to evaluate if a hypothetical value of variance, based on the sample parameters, could be the actual variance of the population or not.

Though sample size for application of Chi-square test, is more suitable for sample size of less than 30, for larger sample size, it can still be applied, by using Fisher's approximation and then applying normal test. Fisher's approximation is given as:

$$z = \sqrt{2\chi^2} - \sqrt{2n-1}$$

Example: The standard deviation = 25 for a random sample size of 100. If the sample is from a normal population, test if the population variance = 20 or not.

Solution: Null hypothesis $H_0 : \sigma = 20$

Given $s = 25, n = 100$,

As $\chi^2 = \frac{ns^2}{\sigma^2}$ therefore χ^2 calculated = $\frac{100 \times 25^2}{20^2} = 156.25$

Since n = large, applying Fisher's approximation:

$$z = \sqrt{2 \times 156.25} - \sqrt{2 \times 100 - 1} = 7.73$$

As $z > 3$, reject H_0; therefore the variance of the population $\neq 20$.

Example: The variance of a sample is considered as 2.8, with sample values being 30, 31, 32, 33, 37, 35, 36, 34. Is the given variance valid at 5% level of significance.

Solution: Let $H_0 : \sigma^2 = 2.8$ and $H_1 > 2.8$.

Calculating Chi-square value (degrees of freedom = $(n-1) = 8 - 1 = 7$).

x_i	$x_i - \bar{x}$	$(x_i - \bar{x})^2$
30	−3.5	12.25
31	−2.5	6.25
32	−1.5	2.25
33	−0.5	0.25
37	+3.5	12.25
35	+1.5	2.25
36	+2.5	6.25
34	+0.5	0.25
$\Sigma x_i = 268$		$\Sigma = 42$

$$\bar{x} = \frac{268}{8} = 33.5$$

Applying the formula of Chi-square:

$$\chi^2 = \sum_{i=1}^{n}\left[\frac{(x_i - \bar{x})^2}{\sigma_o^2}\right] = \frac{42}{2.8} = 15$$

Chi-square calculated at 7 d.f. = 15.
Chi-square tabulated for 7 d.f. at 5% level of significance ($\chi^2_{0.05\ tab}$) = 14.067.
As calculated value of χ^2 > tabulated value at given degree of freedom, H_0 is rejected and H_1 accepted. Hence, population variance is > 2.8.

Chi-Square Test of "Goodness of Fit"

This test is considered very powerful and widely used. It is utilized to test the difference between theory and actual values. It facilitates to find whether the difference between expected values and actual values is due to sampling fluctuations or is it due to in-adequateness of theory to fit the observed data. Stating it differently, this test helps to understand how 'appropriately' do the theoretical distributions like Normal, Poisson, Binomial, fit the distribution obtained from the observed sample, i.e. how good is the fit of the sample values with the theoretical distribution values.

$$\text{Chi-square }(\chi^2) = \sum_{i=1}^{k}\left(\frac{(O-E)^2}{E}\right)$$

Where O = Observed frequency
E = Expected frequency as per the theoretical distribution for v = n − 1 (degrees of freedom = n − 1), and
n = Number of observations.

Example: An insurance policy has 7 variants. The number of people opting for these variants, are as given. Test if all the variants are equally popular among the population.

Variant:	I	II	III	IV	V	VI	VII
No. of people:	790	850	900	950	870	920	950

Solution: H_0 : The variants are equally popular.
If H_0 is true then the expected frequency should be equal for all the variants, i.e. expected frequency = mean of the values = 6230/7 = 890 (equally popular implies each variant should have equal number of people opting for them).
Calculating Chi-square

Variant	Observed (O)	(O − E)	(O − E)²	(O − E)²/E
I	790	−100	10000	11.236
II	850	−40	1600	1.798
III	900	10	100	0.112
IV	950	60	3600	4.045
V	870	−20	400	0.449
VI	920	30	900	1.011
VII	950	60	3600	4.045
Totals	6230			Σ = 22.696

Expected value $(E) = 6230/7 = 890$
Degrees of freedom $= (n - 1) = (7 - 1) = 6$

χ^2 tabulated at 6 d.f. and 5% level of significance $= 12.592$
χ^2 calculated (discussed before in table) for 6 d.f. $= 22.696$

As χ^2 tabulated $< \chi^2$ calculated, hence H_0 is rejected and H_1 is accepted. Therefore, it is statistically concluded that all the variants are not equally popular.

Example: Four coins are tossed 500 times. The observations are as follows:

Number of heads:	0	1	2	3	4
Number of tails:	4	3	2	1	0
Number of observations:	90	88	110	105	107

Find if the coins are unbiased.

Solution: Let H_0: The coins are unbiased.

As probability of head or tail is $\frac{1}{2} = 0.5$, i.e. $p = q = 0.5$.

As this is a case of binomial distribution, the theoretical frequencies can be deduced from binomial distribution equation ($^nC_r p^r q^{(n-r)}$). Using this equation and multiplying by number of trials gives us the distribution frequencies, as follows:

Probability of 0 Heads $= P(0) = 1/16 \times 500 = 31.25$
Probability of 1 Heads $= P(1) = 4/16 \times 500 = 125.00$
Probability of 2 Heads $= P(2) = 6/16 \times 500 = 187.50$
Probability of 3 Heads $= P(3) = 4/16 \times 500 = 125.00$
Probability of 4 Heads $= P(4) = 1/16 \times 500 = 31.25$

Calculating Chi-square

Observed frequency (O)	Expected frequency (E)	(O – E)	(O – E)²	(O – E)²/E
90	31.25	58.75	3451.56	110.45
88	125.00	–37	1369.00	10.95
110	187.50	–77.5	6006.25	32.03
105	125.00	–20	400.00	3.20
107	31.25	75.75	5738.06	183.61
Total				Σ = 340.24

Calculated χ^2 at 4 d.f. $= 340.24$.
Tabulated χ^2 for 4 d.f. at 5% level of significance $= 11.070$.

As calculated $\chi^2 >$ tabulated $\chi^2_{0.05}$ hence null hypothesis is rejected and alternate hypothesis is accepted. The coins are statistically significantly biased and the distribution of observed values is not a good fit of binomial distribution.

Example: Fit a poisson distribution to the following observation recorded and test goodness of fit.

| Number of defects: | 0 | 1 | 2 | 3 | 4 |
| Frequency: | 214 | 92 | 20 | 3 | 1 |

Solution: The mean of poisson distribution is given as λ. Calculating the mean of the given values:

No. of defects (x)	Frequency (f)	fx
0	214	0
1	92	92
2	20	40
3	3	9
4	1	4
	$\Sigma f = 330 = N$	$\Sigma fx = 145$

Mean $\lambda = 145/330 = 0.439$

Fitting poisson distribution by using the equation $P(r) = (e^{-\lambda}\lambda^r)/r!$ (For detailed steps refer fitting poisson distribution covered under discrete theoretical distributions).

$P(0) = e^{-0.439} = 0.6447$ therefore $N \times P(0) = 330 \times 0.6447 = 212.75$
$N \times P(1) = N\,P(0) \times \lambda = 212.75 \times 0.439 = 93.4$
$N \times P(2) = N\,P(1) \times \lambda/2 = 93.4 \times 0.439/2 = 20.5$
$N \times P(3) = N\,P(2) \times \lambda/3 = 20.5 \times 0.439/3 = 3.0$
$N \times P(4) = N\,P(3) \times \lambda/4 = 3 \times 0.439/4 = 0.33$

Calculating Chi-square

Observed frequency (O)	Expected frequency (E)	(O – E)	(O – E)²	(O – E)²/E
214	212.75	2.75	7.563	0.036
92	93.40	–1.40	1.96	0.021
20	20.50	–0.50	0.25	0.012
3	3.00	0	0	0
1	0.33	0.67	0.449	1.360
				$\Sigma = 1.429$

Calculated $\chi^2 = 1.429$

Tabulated χ^2 for $(5 - 1 - 1) = 3$ df (See note at the end of example) at 5% level of significance = 7.815.

As calculated $\chi^2 <$ tabulated $\chi^2_{0.05}$ hence null hypothesis is accepted and alternate hypothesis is rejected. The observed values are statistically a good fit of poisson distribution.

Note: The degrees of freedom in this example is $(5 - 1 - 1) = 3$, instead of $(n - 1)$ as in previous example. The reason being that for estimating the population parameter (mean), one constraint is introduced, hence $(n - 1 - 1)$. In previous example, pertaining to Binomial Distribution, no constraint was introduced to estimate the population parameter. Had there been more constraints for determining the population parameter, like (say) standard deviation, it would have resulted in one more reduction of degree of freedom and so on. The formula of the distribution indicates the constraints being introduced. Further in this example, deliberately, grouping of frequencies was not done, in order to explain the concept of reduction of degree of freedom due to constraints. Grouping of frequencies reduces the 'n' and accordingly this needs to be considered appropriately by reducing the degree of freedom.

Tests for Independence of Attributes

Chi-square helps to assess whether two or more attributes are associated or not associated, or independent or not independent. For finding the association, the null hypothesis is taken as 'the attributes are independent.'

The manifold classification table is used to denote the various cell frequencies of the attributes, as shown below (for detailed understanding on attributes, refer to the chapter on 'Theory of Attributes'):

'Attribute' A → ↓ B	A_1	A_2	$A_{..}$	A_r	Marginal Total
B_1	$A_1 B_1$	$A_2 B_1$	$A_{.} B_1$	$A_r B_1$	ΣB_1
B_2	$A_1 B_2$	$A_2 B_2$	$A_{.} B_2$	$A_r B_2$	ΣB_2
$B_{..}$	$A_1 B_{..}$	$A_2 B_{..}$	$A_{.} B_{..}$	$A_r B_{..}$	$\Sigma B_{..}$
B_c	$A_1 B_c$	$A_2 B_c$	$A_{.} B_c$	$A_r B_c$	ΣB_c
Marginal total	ΣA_1	ΣA_2	$\Sigma A_{..}$	ΣA_r	= N

$A_1, A_2, \ldots A_r$ and $B_1, B_2 \ldots B_c$ are the various attributes and $\Sigma_{i=1}^{r} A_i = \Sigma_{i=1}^{c} B_i = N$. Such a table is called $(r \times c)$ Manifold Contingency Table, where $A_1 B_1$ is the number of people possessing both A_1 and B_1 attributes and so on.

From above, probability of a person possessing attribute:

$$A_i = P(A_i) = \frac{A_i}{N} \text{ and similarly for } B = P(B_i) = \frac{B_i}{N}$$

Probability of person possessing both $A_i B_i$ = compound probability = $P(A_i B_i) = \frac{A_i}{N} \times \frac{B_i}{N}$

Hence, expected number of persons having $A_i B_i = N \times P(A_i B_i)$

$$= N \times \frac{A_i}{N} \times \frac{B_i}{N} = \frac{A_i B_i}{N}$$

Using this formula all expected cell frequencies can be calculated.

For such data on attributes, Chi-square calculation formula is as below:

$$\chi^2 = \sum_{i=1}^{r}\sum_{i=1}^{c} \left[\frac{(A_i B_{i\,observed} - A_i B_{i\,expected})^2}{A_i B_{i\,expected}} \right] \text{ at } (r-1)(c-1) \text{ degree of freedom}$$

The below example would help to understand the formula:

Example: A pain treatment was tested on 400 people, with half of them given new treatment and rest, the existing (conventional) treatment. The results of the tests are as follows:

	Relieved pain	Did not relieve pain
New treatment	120	80
Old treatment	90	110

Test if there is difference between old and new treatment.

Chapter 22: Chi-square Test

Solution:
Let H_0 : There is no difference between old and new pain treatments.
H_1 : There is difference between the old and new treatments

Expressing the given data in 2 × 2 contingency table:

Observed data	Relieved pain	Did not relieve pain	Total
New treatment	120	80	200
Old treatment	90	110	200
Total	210	190	400

The expected frequency contingency table for above, would be (only $(r-1)(c-1)$, i.e. $(2-1)(2-1) = 1$ expected value needs to be calculated, as the rest would be automatically filled in as the marginal totals will remain same). Calculating expected frequency for $P(A_1B_1) = A_1B_1/N = (210 \times 200)/400 = 105$.

Expected data	Relieved pain	Did not relieve pain	Total
New treatment	105	95	200
Old treatment	105	95	200
Total	210	190	400

Chi-square calculation table:

Observed frequency (O)	Expected frequency (E)	(O – E)	(O – E)²	(O – E)²/E
120	105	15	225	2.143
90	105	–15	225	2.143
80	95	–15	225	2.368
110	95	15	225	2.368
				Σ = 9.022

Degree of freedom = $(2 – 1)(2 – 1) = 1$
χ^2 calculated = 9.022
χ^2 tabulated for 1 d.f at 5% level of significance = 3.841.
As χ^2 calculated > χ^2 tabulated, null hypothesis is rejected and H_1 is accepted, i.e. new treatment is effective.

Example: A survey was conducted to study the age group of incidence of cancer among men and women. The results obtained are as follows:

Age group	20–40	40–60	60–80	80–100
Men	14	17	20	5
Women	24	13	10	4

Test if there is any difference between incidence of cancer among men and women.

Solution: Let H_0 : There is no difference among age groups of men and women about incidence of cancer and H_1 : There is significant difference between incidence of cancer among different age groups of men and women.

Representing the above data in 4 × 2 contingency table:

Observed data	20–40	40–60	60–80	80–100	Total
Men	14	17	20	5	56
Women	24	13	10	4	51
Total	38	30	30	9	107

Expected values 4 × 2 contingency table:

Expected frequency $A_1B_1 = (38 \times 56)/107 = 19.89$
Expected frequency $A_2B_1 = (30 \times 56)/107 = 15.70$
Expected frequency $A_3B_1 = (30 \times 56)/107 = 15.70$
Rest of expected frequencies can be derived as marginal totals are available.

Expected data	20–40	40–60	60–80	80–100	Total
Men	19.89	15.70	15.70	4.71	56
Women	18.11	14.30	14.30	4.29	51
Total	38	30	30	9	107

Calculating the value of χ^2

Observed frequency (O)	Expected frequency (E)	(O – E)	(O – E)²	(O – E)²/E
14	19.89	–5.89	34.69	1.744
24	18.11	5.89	34.69	1.916
17	15.70	1.3	1.69	0.108
13	14.30	–1.3	1.69	0.118
20	15.70	4.3	18.49	1.178
10	14.3	–4.3	18.49	1.293
5, 4 (→9)	4.71, 4.29 (→9)	0	0	0
				$\Sigma = 6.356$

Degrees of Freedom $v = (4 – 1)(2 – 1) – 1 = 3 \times 1 – 1 = 2$ (because one degree of freedom is lost in grouping of frequency)

χ^2 calculated $= 6.356$
χ^2 tabulated for 2 d.f at 5% level of significance $= 5.991$.

As χ^2 calculated $> \chi^2$ tabulated, null hypothesis is rejected and H_1 is accepted, i.e. there is statistically significant difference in the incidence of cancer among different age groups of men and women.

Chi-square as a Test of Homogeneity

In test of homogeneity, χ^2 is used to determine whether the independent random variable samples have been drawn from the same population or not, i.e. homogeneous. Under consideration are two or more samples and the check is done whether they are from the same population or not. The calculations are similar to tests of independence.

Example: A study is conducted about the feedback of patients of a hospital after their discharge. The collected data from the survey is as below:

	Satisfied	Not satisfied	Average
Patients who were admitted in single AC Room	40	10	20
Patients who were admitted in single non-AC room	20	19	21
Patients who were admitted in shared AC room	15	28	22
Patients who were admitted in dormitory	12	37	41

Check if the feedback is homogeneous among the various patients about the quality of services provided.

Solution: Under null hypothesis that the feedback on quality of services provided was homogeneous, representing the given values in contingency table, along with calculated 'expected frequencies' as shown below:

Observed					Expected			
Satisfied	Not satisfied	Average	Total		Satisfied	Not satisfied	Average	Total
40	10	20	$\Sigma = 70$	AC Room	21.37	23.09	25.54	$\Sigma = 70$
20	19	21	$\Sigma = 60$	Non-AC	18.32	19.79	21.89	$\Sigma = 60$
15	28	22	$\Sigma = 65$	Shared	19.84	21.44	23.72	$\Sigma = 65$
12	37	41	$\Sigma = 90$	Dormitory	27.47	29.68	32.85	$\Sigma = 90$
$\Sigma = 87$	$\Sigma = 94$	$\Sigma = 104$	$\Sigma = 285$	Total	$\Sigma = 87$	$\Sigma = 94$	$\Sigma = 104$	$\Sigma = 285$

Expected frequency cell $A_1B_1 = (87 \times 70)/285 = 21.37$ and so on for other cells.
Degrees of freedom $= (4-1)(3-1) = 6$

Calculating Chi-square

Observed frequency (O)	Expected frequency (E)	(O – E)	(O – E)²/E
40	21.37	18.63	16.24
20	18.32	1.68	0.15
15	19.84	–4.84	1.18
12	27.47	–15.47	8.71
10	23.09	–13.09	7.42
19	19.79	–0.79	0.03
28	21.44	6.56	2.01
37	29.68	7.32	1.80
20	25.54	–5.54	1.20
21	21.89	–0.89	0.04
22	23.72	–1.72	0.12
41	32.85	8.15	2.02
			$\Sigma = 40.92$

χ^2 calculated = 40.92
χ^2 tabulated for 6 d.f. at 5% level of significance = 12.592

The calculated value is higher than tabulated value, therefore the null hypothesis is rejected, which implies that the feedback is not homogeneous and satisfaction levels vary among different class of patients (The reasons could then be explored for this difference, which could be due to difference in level of services and also the population strata using the different services).

CHI-SQUARE TEST OF HOMOGENEITY OF CORRELATION CO-EFFICIENT

This test is utilized when the objective is to test the hypothesis that correlation co-efficient of the sample is the estimate of the same normal population correlation coefficient, i.e. to test whether the given values of correlation coefficient are homogeneous or are from equally correlated population.

The formula for this test is as below:

$$\chi^2 = \sum ((n_i - 3)(z_i - \bar{z})^2), \text{ where } \bar{z} = \frac{\sum z_i (n_i - 3)}{\sum (n_i - 3)}$$

[z = correlation coefficient, n = sample sizes and d.f = $(n - 1)$]

The below example would help in understanding the above formula:

Example: Six samples of varying sizes were collected to assess the educational qualification and the income levels. The correlation co-efficient of the samples and their sample sizes are as given below:

Correlation coefficient: 0.751 0.619 0.810 0.512 0.491 0.787
Sample size: 17 21 40 35 20 27

Are the correlation coefficients homogeneous? If they are, calculate the common correlation coefficient.

Solution: Let H_0: The correlation coefficients are homogeneous.
H_1: The correlation coefficients are not homogeneous.

Using the tabulated form for arriving at desired summations:

Sample size 'n'	Correlation coefficient 'z'	(n – 3)	z (n – 3)	(z – $\bar{z}$)	(z – $\bar{z}$)^2	(n – 3)(z – $\bar{z}$)^2
17	0.751	14	10.514	0.080	0.0064	0.0896
21	0.619	18	11.142	–0.052	0.0027	0.0486
40	0.810	37	29.970	0.139	0.0193	0.7141
35	0.512	32	16.384	–0.159	0.0252	0.8064
20	0.491	17	8.347	–0.180	0.0324	0.5508
27	0.787	24	18.888	0.116	0.0135	0.3240
		Σ = 142	Σ = 95.245			Σ = 2.5335

Calculating $\bar{z}$ from above table:

$$\bar{z} = \frac{\sum z_i (n_i - 3)}{\sum (n_i - 3)} = \frac{95.245}{142} = 0.671$$

From the table, calculated $\chi^2 = \sum ((n_i - 3)(z_i - \bar{z})^2) = 2.5335$
Tabulated χ^2 for 5 d.f. at 5% level of significance = 11.070
As calculated χ^2 < tabulated χ^2; hence H_0 is accepted.

As it is accepted that the correlation coefficient are homogeneous, the combined or pooled correlation coefficient can be calculated as follows:

If population correlation coefficient = ρ, then let estimated population correlation be $\hat{p}$

$$\hat{p} = \frac{e^{2\bar{z}} - 1}{e^{2\bar{z}} + 1} = \tanh \bar{z}$$

For the values from the above example

$$\hat{p} = \frac{e^{2\bar{z}} - 1}{e^{2\bar{z}} + 1} = \frac{e^{2(0.671)} - 1}{e^{2(0.671)} + 1} = \frac{3.827 - 1}{3.827 + 1} = \frac{2.827}{4.827} = 0.586;$$

or directly

$$\hat{p} = \tanh \bar{z} = \tanh 0.671 = 0.586$$

The combined correlation coefficient of the data provided is 0.586.

Yates' Correction

Consider a 2 × 2 contingency table. The degree of freedom of such a table is (2 – 1)(2 – 1) = 1. In case one of the observed values is ≤ 5, then as per the grouping or pooling method, the frequencies have to be merged together with preceding or succeeding frequency. Result of this would be reduction of the d.f. by 1, making the d.f. equal to zero. For such cases 'Yates' correction' is applied, so that the continuity is maintained where the frequency is less than 5. This involves adding 0.5 to the expected frequency for the first cell and appropriate reduction by 0.5 so that the marginal totals remain valid.

Divided opinion exists on the use of Yates' correction. For large N, it makes little difference, but for small N, it tends to overstate the probability. As a balanced approach, it is advisable to be used when expected frequency is less than 5. In case the expected cell frequency is between 5 and 10 then both corrected and regular value of Chi-square should be calculated. In case the test conclusion from both the values is same, then it is acceptable to accept the results. In case these throw contrary conclusion, then either sample size should be increased or other nonparametric tests should be considered.

Example: Consider the following 2 × 2 table of observed values:

Observed			Expected			Expected (Corrected)		
18	17	$\Sigma = 35$	15.91	19.09	$\Sigma = 35$	16.41	18.59	$\Sigma = 35$
7	13	$\Sigma = 20$	9.09	10.91	$\Sigma = 20$	8.59	11.41	$\Sigma = 20$
$\Sigma = 25$	$\Sigma = 30$	$\Sigma = 55 = N$	$\Sigma = 25$	$\Sigma = 30$	$\Sigma = 55 = N$	$\Sigma = 25$	$\Sigma = 30$	$\Sigma = 55 = N$

Expected frequencies (Uncorrected) calculation: For cell $A_1 B_1 = (25 \times 35)/55 = 15.91$.

Expected frequency (Corrected) = 15.91 + 0.5 = 16.41. Rest of the corrected frequencies can be calculated as marginal totals remain same.

Calculating Chi-square—corrected and uncorrected

Observed	Uncorrected				Corrected			
	Expected	(O − E)	(O − E)²	(O − E)²/E	Expected	(O − E)	(O − E)²	(O − E)²/E
18	15.91	2.09	4.368	0.275	16.41	1.59	2.528	0.159
7	9.09	−2.09	4.368	0.624	8.59	−1.59	2.528	0.278
17	19.09	−2.09	4.368	0.257	18.59	−1.59	2.528	0.132
13	10.91	2.09	4.368	0.336	11.41	1.59	2.528	0.232
				Σ = 1.492				Σ = 0.801

Degree of freedom = (2 − 1)(2 − 1) = 1
χ^2 calculated = 1.492
χ^2 (corrected) calculated = 0.801
χ^2 tabulated for 1 d.f. at 5% level of significance = 3.841

As in both cases, corrected and uncorrected the calculated value is less than tabulated value, the null hypothesis would have been accepted.

REMARKS ON CHI-SQUARE

Chi-square statistic assumes variable to be independent, implying that the occurrence of an event does not affect the outcome of the other. Test of independence measures possibility of relationship between the variables. Test of attributes measures significance of qualitative values which cannot be quantified. Test of homogeneity is similar to test of independence, but the difference lies in the establishment of the null hypothesis.

Errors can come in the calculations of Chi-square due to small expected frequencies, not recording of nonoccurrence, failure to check that the sum of difference between observed and expected frequency to be equal to zero, incorrect categorization of data, incorrect calculation of degree of freedom, especially when constraints are present either from calculation from formula or due to grouping.

If these pitfalls are avoided, then Chi-square test is a very useful and strong nonparametric tool in the analysis of samples and population.

CHAPTER 23

Nonparametric Tests

INTRODUCTION

In the tests of significance discussed so far in earlier chapters, it was an underlying assumption that the population from which the sample has been drawn is a normally distributed one or having some other form of distribution. The tests were focused on testing statistical hypothesis about the parameters of the population or estimating them (except Chi-square test), especially in the case of small samples test such conditions are always assumed. These tests which deal with the parameters of the populations are called as 'parametric tests'.

In many cases, it is not possible or it is doubtful to assume that the population is distributed normally. In such situations nonparametric tests are appropriate and useful, as these do not depend on or make any assumption about the form and distribution characteristics of the population. Hence, these are also called 'distribution free tests'.

The nonparametric tests can be applied in situations like:
- When assumption of population being a normal distribution is not true or doubtful.
- When test is being done for randomness, ranks, association or independence, etc.

The nonparametric tests assume that the variable and its probability density function are continuous and all sample observations are independent.

ADVANTAGES

Advantages of nonparametric tests:
- The tests are simple in nature, easy to understand, calculate and infer. They do not require full understanding of sampling and probability theory.
- These are distribution free tests, which increases their applicability.
- These tests can be applied to most types of data, like nominal scaled qualitative data, ranked data (ordinary scaling) or ratio-scaled data. Parametric tests cannot handle such type of ranked data.

- These can be applied on small samples. At times researchers use these to conduct initial preresearch pilot study based on small samples.

DISADVANTAGES

Disadvantages of nonparametric tests:

In technical aspects, parametric tests are stronger than nonparametric tests. If all assumptions of parametric tests hold, it is best to perform the relevant parametric test instead of nonparametric test.

Nonparametric tests can only test hypothesis, estimating populations parameters is outside their scope and capability.

In short, the use of nonparametric tests is a trade-off between reliability, amount and kind of data available. Though, the accuracy and reliability tends to decrease, but the flexibility and capability to operate on less data increases. At the same time for qualitative data/ranked data, nonparametric tests are appropriate tools.

Nonparametric tests are based on the property of 'order statistic' wherein the descriptive statistics tools are applied on the ordered data, i.e. if the data values consist of $x_1, x_2, x_3, x_4, \ldots x_n$, then ordered sample values imply $x_1 < x_2 < x_3 < x_4 < \ldots < x_n$. Order Statistic Theory propounds that 'the distribution of area under the density function between any two ordered observations is independent of the form of the density function.'

Nonparametric test consists of many different tests, out of which only the relevant tests are being considered here.

RUN TEST: TEST FOR RANDOMNESS IN A SERIES OF OBSERVATIONS

This test is performed to test the randomness in the series of observations of a sample. It allows checking whether the observations reflect randomness or there is some pattern indicating nonrandomness of the observations. In this test, as a standard norm,

Null hypothesis $\quad H_0$: The observations are random and
Alternate hypothesis $\quad H_1$: The observations are not random.

Procedurally, as a first step the median of the series/observations is calculated. Thereafter, each value is compared with the median value and allotted a positive sign, if it is greater than median; or allotted a negative sign, if it is lower than the median value. In case, the total number of observations is odd, then the median value is ignored. The allotted signs are then counted for the number of 'Runs (R).' A run is defined as the sequence of same signs, bounded by the other sign. The number count of entries in each run is called the length of the run (l). The number of runs, indicate the randomness of the series. If the value of 'R' is low, i.e. too few runs, it indicates a pattern and a high value of 'R' implies high fluctuation which could be caused by unknown factor or may be, data manipulation. The value of 'R' is compared with the table for 'Acceptance Region for Values of R in Run Test for Randomness.' (Given in Appendix). If the value of R falls within the acceptance limits, the null hypothesis is accepted, and otherwise it is rejected.

Example: Consider the following series of observations. Test the randomness of this by run test.

Values: 55 76 45 67 75 76 68 65 69 59

Solution:
Null hypothesis H_0: The observations are random and
Alternate hypothesis H_1: The observations are not random

The median of this series is calculated as 67.5. The runs are identified as below:

Values	55	76	45	67	75	76	68	65	69	59
Runs	−	+	−	−	+	+	+	−	+	−
Run count	1	2	3		4			5	6	7

Sample size = 10 having 7 runs

As per the table, for a sample size of 10, the minimum and maximum value of run is given as 6 and 16. The number of runs in this example lies between the specified acceptance limit, hence, null hypothesis is accepted.

This method to check randomness can be applied to practically all series and offers an easy tool for checking the randomness.

THE SIGNED-RANK TEST FOR TESTING SPECIFIED MEAN OR MEDIAN OF A POPULATION

The signed-rank test is used to determine the significance of the difference between the specified values of mean or median, and the population's mean or median. It needs to be observed that 'specified' mean or median is not the 'actual' mean or median either of the sample or population. The specified value is the estimate of the mean or median of the population, which is tested for its significance by conducting the signed-rank test on the sample observation belonging to that population. If $\bar{x}$ or $\bar{x}_{Median}$ are the population mean or median, and $\bar{x}_o$ is the specified value, hypothesis for this test would be:

Null hypothesis H_o: Mean (Median) of the population is equal to the specified mean (median) of the population i.e. $\bar{x}$ or $\bar{x}_{Median} = \bar{x}_o$
Alternate hypothesis H_1: $\bar{x}$ or $\bar{x}_{Median} \neq \bar{x}_o$

The test statistic for signed-rank test is denoted by '*T*'. The calculated value of *T* is compared with the tabulated value (Table given in Appendix). The important aspect of this test is that the 'null hypothesis is accepted' if the $T_{Calculated} > T_{Tabulated}$, a procedure which is contrary to most of the other significance tests where if the test statistic is higher than tabulated value, the null hypothesis is rejected. Further while calculating the test statistic '*T*', any observation equal to the specified mean or median is not considered for calculations and is discarded. The following example shows the procedure for testing the specified mean and is applicable for median also.

Example: The mean weight of newborn male babies is specified as 5.5 Kgs. A sample of 10 observations is recorded. Test whether the specified mean weight is the true or not.
Weights in Kgs: 5.1 5.3 4.9 5.0 5.8 5.6 5.7 5.9 5.4 5.6

Solution: From the given details:
Null Hypothesis: $H_0: \bar{x} = 5.5\ (\bar{x}_0)$
Alternate Hypothesis: $H_1:: \bar{x} \neq 5.5$

x	Difference $x - \bar{x}_0$	Modulus of Difference	Rank of difference	Ranks with their signs + or −
5.1	−0.4	0.4	3.5*	−3.5
5.3	−0.2	0.2	6.5	−6.5
5.7	+0.2	0.2	6.5	+6.5
4.9	−0.6	0.6	1	−1
5.6	+0.1	0.1	9	+9
5.0	−0.5	0.5	2	−2
5.8	+0.3	0.3	5	+5
5.9	+0.4	0.4	3.5	+3.5
5.4	−0.1	0.1	9	−9
5.6	+0.1	0.1	9	+9

*In case of tie of ranks, the average of the ranks under consideration is allotted to all tied observations.

Sum of positive ranks = 33
Sum of negative ranks = 22

T statistic = Lower value between the sum of positive ranks and negative ranks, i.e. 22

Tabulated value of T at 5% level of significance, for $n = 10$ (*number of observations*) = 8

As calculated value of $T >$ tabulated value of T; the null hypothesis is not rejected and stands accepted. Therefore, from the sample observations, there is no sufficient evidence to reject the null hypothesis $H_0: \bar{x} = 5.5\ (\bar{x}_0)$.

WILCOXON SIGNED RANK TEST FOR PAIRED DATA

Wilcoxon signed rank is another nonparametric test and is relevant to matched or paired data. It is useful to compare before and after situations. It is similar to the sign test, but in addition to analyzing the difference signs, it also takes into account the magnitude of differences in the observations. Therefore, it does get affected by outliers.

Wilcoxon signed rank test assumes that the differences are approximately symmetrical, the observations are independent, the underlying variable is continuous and ranks can be applied to the observations (data is measured on ordinal, interval or ratio scale).

It tests for distribution being symmetric with 0 median, and the null hypothesis is taken as 'the median difference between pairs of observations is 0'. This is different from *t*-test of matched pairs where the null hypothesis was considered as mean difference between pairs is 0.

The Wilcoxon test statistic W is given as:

$$W = \sum_{i=1}^{n} R^+;\ \text{i.e. sum of the positive ranks}$$

The procedure involves calculating the difference between each paired observation, assigning ranks to the modulus value of differences, while omitting cases of zero difference, followed by reassigning the sign + or − to these ranks depending upon the difference being originally positive or negative. The sum of positive and negative ranks is computed. The calculated value of the W statistic is compared with the tabulated value of the Wilcoxon test critical values. The null hypothesis is rejected if the computed W test statistic is outside the upper and lower critical tabulated values.

For large samples, $n > 20$, the W is normally distributed and its test statistics can be stated as:

$$Z = \frac{W - \frac{n(n+1)}{4}}{\sqrt{\frac{n(n+1)(2n+1)}{24}}}$$

The normal curve critical values can be referred to in this case and inferences drawn accordingly.

In case of tied ranks (except 0, as 0 is excluded) the statistic is modified as below:

$$Z = \frac{W - \frac{n(n+1)}{4}}{\sqrt{\frac{n(n+1)(2n+1)}{24} - \frac{t_i^3 - t_i}{48}}}$$

Where, t_i is number of ties in a group and such reduction to be carried out for each group of tied ranks.

Example: Eight patients are being treated for lack of sleep. The observations pertaining to amount of sleep (in minutes) before and after the treatment is given. Test if there is significant difference in sleep due to the treatment.

Solution: Under null hypothesis that the median difference is zero, we perform the Wilcoxon signed rank test. Alternate hypothesis being the median is not zero and significant difference exists.

Patient	Sleep amount before treatment (minutes/day)	Sleep amount after treatment (minutes/day)	Difference	Modulus of difference	Ranks	Signed ranks
1	200	235	−35	35	4	−4
2	180	170	10	10	1	1
3	230	230	0	—	—	—
4	210	255	−45	45	6	−6
5	140	190	−50	50	7	−7
6	280	240	40	40	5	5
7	250	230	20	20	2	2
8	160	190	−30	30	3	−3
						$\sum R^+ = 8$

The tabulated critical value for n = 8 at 5% level of significance is 8 (Lower critical value = 8 and upper critical value = 47). As calculated value of W statistic = Tabulated value, null hypothesis is rejected.

KOLMOGOROV-SMIRNOV TEST FOR ONE SAMPLE

Kolmogorov- Smirnov test is used for testing the goodness of the fit of a distribution, i.e. to determine the significance of the difference between the observed and specified frequency distribution. This test validates whether the observed distribution fits well or may reasonably have come from the specified theoretical distribution. One may recall that Chi-square test is also available to test the goodness of fit, but in case of Chi-square test the underlying assumption is of observations coming from normal distribution. Such an assumption is not necessary for conducting Kolmogorov-Smirnov test.

The test statistic is denoted by D. The actual cumulative distribution is calculated for the sample observation, along the theoretical cumulative distribution of the assumed population. The maximum difference between the two distributions provides the test statistic D. If calculated D < tabulated D_α the null hypothsis is accepted, whereas if calculated D > tabulated D_α the null hypothesis is rejected and H_1 alternate hypothesis is accepted. The degree of freedom is equal to the number of observations. The null hypothesis is stated as H_0 : The sample comes from the specified theoretically distributed population.

Example: An area has 6 hospitals. Theoretically, these hospitals are expected to handle similar number of patients, as there is no difference in their infrastructure, accessibility, and charges, etc. A sample of 510 patients was surveyed about the hospital they got the treatment from. All other things being equal, evaluate if uniform number of patients are being treated by the various hospitals, based on the survey results as stated below:

Hospital:	1	2	3	4	5	6
No. of patients treated:	74	61	108	83	91	93

Solution: This is a case of uniform distribution, as each hospital is expected to treat equal/similar number of patients.

Null hypothesis: Uniform numbers of patients are being treated by each hospital (The sample comes from uniform distribution)

Section 6: Statistical Testing

Hospital	1	2	3	4	5	6
No. of patients treated (frequency distribution)	74	61	108	83	91	93
Observed cumulative distribution	74	135	243	326	417	510
Expected distribution (uniform distribution)	85	85	85	85	85	85
Expected cumulative distribution	85	170	255	340	425	510
Observed cumulative distribution function (cumulative probability) F_o	= 74 ÷ 510 = 0.145	0.265	0.476	0.639	0.818	1.000
Expected cumulative distribution function (cumulative probability) F_e	= 85 ÷ 510 = 0.167	0.333	0.500	0.667	0.833	1.000
Modulus of difference $|F_o - F_e|$	0.022	0.069	0.024	0.027	0.016	0.000
Test statistic D = Max value of $|F_o - F_e|$	0.069					

As degree of freedom = 510,
The tabulated value of D at 5% level of significance can be calculated as:

Tabulated $D_{0.05} = \dfrac{1.36}{\sqrt{n}} = \dfrac{1.36}{\sqrt{510}} = 0.0602$

As calculated D > tabulated D, null hypothesis is rejected, i.e. the numbers of patients treated by various hospitals are not uniform.

KOLMOGOROV-SMIRNOV TEST (FOR COMPARING TWO POPULATIONS)

Kolmogorov-Smirnov test for comparing two populations evaluates if the two sample distributions come from identical population distribution or not. Statistically, it probes the significance of the difference between two population distributions, based on two sample distributions. This test does not make any assumptions about the population, and is more reliable for sample sizes greater than 15. Similar to previous test, it involves determining cumulative distribution of both the samples, which are then compared. The null hypothesis is 'the two samples belong to identical population.' The maximum difference value becomes the test statistic D, which needs to be compared with the tabulated value at desired level of significance. If calculated value > tabulated value, null hypothesis is rejected and vice versa.

Example: Two hypothetical sample data is presented below. Test if the population exhibit identical distributions.

A	12	14	16	18	14	13	11	12	10	9	11	10	8	7	6	7	5	4	3
B	21	23	26	31	33	37	34	39	32	27	29	26	24	21	19	17	18	15	11

Solution: Null hypothesis H_0—The samples are from identical distributions. Performing required calculation through tabular format.

Chapter 23: Nonparametric Tests

A	B	Cumulative frequency-A	Cumulative frequency-B	Cumulative distribution function-A	Cumulative distribution function -B	Modulus of difference	D Maximum
12	18	12	18	0.063	0.038	0.026	
14	23	26	41	0.137	0.085	0.051	
16	26	42	67	0.221	0.140	0.081	
18	31	60	98	0.316	0.204	0.112	
14	33	74	131	0.389	0.273	0.117	0.117
13	37	87	168	0.458	0.350	0.108	
11	34	98	202	0.516	0.421	0.095	
12	39	110	241	0.579	0.502	0.077	
10	32	120	273	0.632	0.569	0.063	
9	27	129	300	0.679	0.625	0.054	
11	29	140	329	0.737	0.685	0.051	
10	26	150	355	0.789	0.740	0.050	
8	24	158	379	0.832	0.790	0.042	
7	21	165	400	0.868	0.833	0.035	
6	19	171	419	0.900	0.873	0.027	
7	17	178	436	0.937	0.908	0.029	
5	18	183	454	0.963	0.946	0.017	
4	15	187	469	0.984	0.977	0.007	
3	11	190	480	1.000	1.000	0.000	

(Cumulative distribution function is calculated in similar way, as done in the previous test)

Tabulated value of D for 'n = 15' at 5% level of significance = 0.338, which is greater than the calculated value of D = 0.117, the null hypothesis is accepted, that the samples have come from populations with identical distributions.

MANN-WHITNEY 'U' TEST FOR EQUALITY OF TWO MEANS

Mann-Whitney U test is used to test whether the means of two independent samples have significant difference or not. As an alternative to t-test, which also determines the significance of two sample means, Mann-Whitney U test does not assume or put a limiting condition, that the samples should belong to a normal population and have equal variance. The null hypothesis $H_0: \bar{x}_1 = \bar{x}_2$ and alternate hypothesis: $H_1: \bar{x}_1 \neq \bar{x}_2$, where $\bar{x}_1, \bar{x}_2$ are the two sample means.

Procedurally, U-test requires all the observations of the two samples be combined and ranked as one single group. The ranking sequence being smallest to largest and in case of tie, the average ranks are allotted. Subsequently the ranks are totalled for each sample and 'U statistic' is calculated by formulas given below. The lower of the two U values is considered as the U statistic and which is then compared with the tabulated U value. Contrary to most of other tests, in U test "null hypothesis is rejected if the calculated value is lower than the tabulated value for the specified level of significance".

$$U = n_1 n_2 + \frac{n_1(n_1+1)}{2} - R_1, \text{ and } U = n_1 n_2 + \frac{n_2(n_2+1)}{2} - R_2$$

Where, n_1, n_2 are the number of observations in each of the two samples and R_1 and R_2 are sum of ranks in each of the respective samples.

Example: Two new analytical methods of laboratory testing are introduced to reduce the turnaround time of testing. Following data pertains to the number of samples tested in one month after introduction through 20 chemists—10 each working through the two methods. Evaluate whether the two methods are equally effective or not.

Chemist	1	2	3	4	5	6	7	8	9	10
Method A:	150	146	165	138	142	159	153	171	159	147
Chemist	11	12	13	14	15	16	17	18	19	20
Method B:	138	174	166	164	157	161	141	159	169	149

Solution: Null hypothesis is to test that the two methods are equally effective, i.e mean laboratory test conducted through method A is equal to mean tests conducted by method B, i.e. $H_0: \bar{x}_1 = \bar{x}_2$. Preparing the desired table to calculate U statistic.

Chemist	Test done through method A:	Combined rank	Chemist	Test done through method B	Combined rank
1	150	8	11	138	1.5
2	146	5	12	174	20
3	165	16	13	166	17
4	138	1.5	14	164	15

Contd...

Contd...

5	142	4	15	157	10
6	159	12	16	161	14
7	153	9	17	141	3
8	171	19	18	159	12
9	159	12	19	169	18
10	147	6	20	149	7
	Mean $\bar{x}_1 = 153$	$\sum$ of Ranks $R_1 = 92.5$		$\bar{x}_2 = 157.8$	$\sum$ of Ranks $R_2 = 117.5$

Calculating U by the formula shown on previous page:

$$U_a = 10 \times 10 + \frac{10(10+1)}{2} - 92.5 = 62.5$$

$$U_b = 10 \times 10 + \frac{10(10+1)}{2} - 117.5 = 37.5$$

Lower of these is the U statistic, i.e $U_b = 37.5$

Tabulated value of U for n_1, n_2 (No. of observations) = 10 and at 5% level of significance = 23

As tabulated $U <$ calculated U at 5% level of significance, the null hypothesis is accepted, which implies that the two newly introduced methods are statistically equally effective.

WILCOXON-WILCOX TEST FOR COMPARISON OF MULTIPLE TREATMENT ON A SERIES

The Wilcoxon-Wilcox test is used to test the significance of the difference of multiple treatments applied to a series, through the use of 'Ranks'. If there are 'n' subjects who are subjected to 'K' treatments, then this test determines the significance of the difference of the treatments, like ANOVA. This is done by testing of the 'equality of the means' of the various treatments, while assuming that the subjects response to various treatments are independent (not affected) of each other. This test goes one step further than ANOVA, by allowing the identification of the treatments which are not having equal means and due to which of the treatments the null hypothesis is getting rejected. Though the test is suitable for qualitative (ranks), but it can also be applied to quantitative data which can be converted into ranks.

The null hypothesis established for this test is:

H_0: Mean of all treatments are equal or $m_1 = m_2 = m_3 = ... = m_n$ where m_i = various means

H_1: Mean of all treatments are not equal

Procedurally, a table is prepared with treatments in the columns and subjects in the rows. Ranks are given in each row, and column total are calculated for establishing the rank sum for each treatment. Thereafter next table 'difference of ranks' is established giving modulus of difference, when all treatments are compared with each other.

The tabulated value of Wilcoxon-Wilcox test statistics (for given sample size 'n' and given number of treatments 'k', at desired level of significance) is compared with the values of the difference in ranks table. If all the test statistic values are lower than the tabulated value, the means are considered equal and inferred that no statistical significant difference exists between the treatments. In case it is not so, then the null hypothesis is rejected and alternate hypothesis is accepted.

The pairs having unequal means get highlighted as these pairs are the ones where the calculated value is higher than the tabulated value.

Example: Below are ranks given by experts on treatments conducted on various subjects. Test if the multiple treatments have no significant differences among them.

Subject	Treatment			
	A	B	C	D
1	3	2	4	1
2	3	1	4	2
3	2	1	3	4
4	3.5	1	3.5	2
5	3	2	4	1
6	4	2	3	1

Solution: Let the null hypothesis be H_0: All treatments are equally effective and H_1: All the treatments are not equal. Preparing the table to calculate the rank sums

Subject	Treatment			
	A	B	C	D
1	3	2	4	1
2	3	1	4	2
3	2	1	3	4
4	3.5	1	3.5	2
5	3	2	4	1
6	4	2	3	1
Rank sum Σ	18.5	9	21.5	11

Creating Difference in Ranks Table

(Entry Cell $AA = |18.5-18.5| = 0$, Cell $AB = |18.5-9|$, Cell $AC = |18.5-21.5|$ and so on.

Treatment	Treatment			
	A	B	C	D
A	0	9.5	3	7.5
B		0	12.5	2
C			0	10.5
D				0

The tabulated value for $k = 4$ and $n = 6$, at 5% level of significance is given as 11.5. From the difference in rank table, one of the calculated values is higher than the tabulated value (highlighted cell), hence the null hypothesis is rejected, while alternate hypothesis is accepted. The highlighted cell denotes the difference between treatments B and C is significantly unequal, which has caused the null hypothesis to be rejected.

KRUSKAL-WALLIS RANK SUM TEST (H-TEST)

The Kruskal-Wallis rank sum test is used to test equality of means of several populations. It determines if the random samples could have come from populations with the same mean. The null hypothesis is like one-way ANOVA, but instead of variable values, this test is applied on the ranks. Whereas Mann–Whitney U test is applicable for two populations, this test can be used for more than two populations. If each sample has atleast five observations, the test statistic 'H' very closely follows χ^2 (Chi-square) distribution, and uses the same table for determining the critical values, with degrees of freedom (k-1), where k = number of treatments.

The procedure involved is assigning combined ranks to all the observations in all samples, with sequence being smallest to largest. The sum of rank for each sample is calculated and the following H statistic formula is applied.

$$H = \frac{12}{N(N+1)} \left(\sum_{i=1}^{k} \frac{R_i^2}{n_i} \right) - 3(N+1)$$

Where,

N = Total number of observations ($n_1 + n_2 + n_3 \ldots + n_k$)

$n_1, n_2, n_3 \ldots, n_k$ = Number of observations in each of k treatments

$n_1, n_2, n_3, \ldots, R_k$ = Sum of ranks for each treatment

The calculated value of 'H' is compared with the tabulated value of Chi-square for degree of freedom (k-1) and at desired level of significance. If calculated H value > tabulated Chi-square value, the null hypothesis is rejected and vice versa.

Example: Three batches of doctors are selected and imparted training on conducting virtual autopsy through digital imaging. Each batch is trained through a different method (A, B, C) using different softwares. Post-training, the doctors are given tests involving conducting of autopsies, and an overall score is given. Assuming similar skill levels of the doctors, test if the three methods of digital autopsy are equally effective or not.

Method A		Method B		Method C	
S. No.	Score	S.No.	Score	S.No.	Score
1	89	1	91	1	81
2	91	2	93	2	83
3	90	3	92	3	87
4	83	4	96	4	98
5	84	5	90	5	86
6	94	6	89	6	85
		7	85	7	88

Solution: Let the null hypothesis be H_0: All trainings are equally effective and alternate hypothesis H_1: All trainings are not equally effective. Combined rank table is prepared as given here:

Method A		Method B		Method C	
Score	Combined rank	Score	Combined rank	Score	Combined rank
89	10.5	91	14.5	81	1
91	14.5	93	17	83	2.5
90	12.5	92	16	87	8
83	2.5	96	19	98	20
84	4	90	12.5	86	7
94	18	89	10.5	85	5.5
		85	5.5	88	9
$n_1 = 6$	$\Sigma R_1 = 62.0$	$n_2 = 7$	$\Sigma R_2 = 95.0$	$n_3 = 7$	$\Sigma R_3 = 53.0$

Calculated H Statistics, by using the formula:

$$H = \frac{12}{20(20+1)} \left(\frac{62^2}{6} + \frac{95^2}{7} + \frac{53^2}{7} \right) - 3(20+1)$$

$$= 3.6068$$

The tabulated value (refer Chi-square table) at 5% level of significance and for (3 – 1) degrees of freedom = 5.99. As calculated value of H statistic < tabulated value, null hypothesis is accepted. There is no statistically significant difference between the three training methods of conducting autopsies.

FRIEDMAN TEST (TWO-WAY ANOVA)

Friedman test, with test statistic as F, is used to test whether the given samples have been drawn from the same population or not, by using the ranks. Similar to two-way ANOVA, but it does not assume normality and equal variance of the population. It is applied when the number of treatments is greater than three and each of the sample size pertaining to the each treatment is equal. The test statistic approximates Chi-square distribution for sample size greater than 5, hence, the tabulated values are referred from the Chi-square table for (k – 1) degrees of freedom and at desired level of significance, where 'k' is the number of treatments.

The calculation procedure would be clear from the example given below. In general sense, the null hypothesis is stated as H_0: The k treatments are equally effective, and the alternate hypothesis H_1: The treatments are not equal, which implies that either all of them are unequal or there is atleast one inequality present among all the treatments. If the tabulated value is greater than the F statistic calculated value, null hypothesis is accepted and vice versa.

The formula for calculating F Statistic is given as:

$$F = \frac{12}{nk(n+1)} \left(\sum_{i=1}^{k} R_i^2 \right) - 3n(k+1)$$

Where,

k = Number of treatments, n = Number of observations in each treatmen,

$R_1, R_2, R_3, \ldots, R_k$ = Sum of ranks for each treatment

The above formula is suitable if there are no tied ranks present in the data. For tied ranks, the formula is modified and presented after the below example on untied ranks.

Example: Six new diet combinations are being tested for effectiveness, with each diet being given to four study subjects. The feedback is taken on subjective as well as objective aspects and through predefined methods, the scores out of 100 are calculated for each respondent-method combination, as given below. Test if the treatments (diet) are same or differ.

Subject	Treatment					
	A	B	C	D	E	F
1	76	88	66	79	82	63
2	85	81	87	92	75	78
3	79	69	83	90	71	68
4	91	77	80	88	68	93

Solution: Let null hypothesis H_0: The treatments (diets) are equally effective, i.e. no significant difference exist among them. Accordingly alternate hypothesis would be that significant difference exists among the treatments and they are not equally effective.

Presenting the calculation in table format:

Subject	Treatment						Total ranks (Row)
	A	B	C	D	E	F	
1- Score	76	88	66	79	82	63	
Rank	3	6	2	4	5	1	24
2- Score	85	81	87	92	75	78	
Rank	4	3	5	6	1	2	24
3- Score	79	69	83	90	71	68	
Rank	4	2	5	6	3	1	24
4- Score	91	77	80	88	68	93	
Rank	5	2	3	4	1	6	24
Σ Rank	16	13	15	20	10	10	

Applying the formula:

$$F = \frac{12}{nk(n+1)} \left(\sum_{i=1}^{k} R_i^2 \right) - 3n(k+1)$$

$$F = \frac{12}{4 \times 6(4+1)} (16^2 + 13^2 + 15^2 + 20^2 + 10^2 + 10^2) - 3 \times 4 (6+1)$$

Solving, $F = 41$

Tabulated value of χ^2 at 5% level of significance, (6-1) = 5 degrees of freedom = 11.07. As calculated value of F > tabulated value, null hypothesis is rejected and alternate hypothesis is accepted. There exists, significant differences in the effectiveness of different diets.

Friedman Test: Case of Tied Ranks

In case of tied ranks, where ranks have been provided on basis of the average rank the observations would have occupied, the Friedman test statistic formula is modified and is as given here:

$$F_{Tie} = \frac{12 \sum_{i}^{k} R_i^2 - 3n^2 k(k+1)^2}{n^2 k(k^2-1) - n(\Sigma t^3 - \Sigma t)}$$

Where,
    k = Number of treatments,
    n = No. of observations in each treatment
    R_1, R_2, R_k = Sum of ranks for each treatment
    = Number of observations tied for a given rank

The below example would clarify the application of the formula,

Example: Evaluate if the treatments are significantly different or not in the below hypothetical data.

Group	Treatment					
	A	B	C	D	E	F
1	76	88	66	79	76	76
2	85	81	87	92	75	78
3	79	69	83	90	71	68
4	91	77	80	88	68	93
5	87	87	92	79	82	81

Solution: Null hypothesis H_0: No significant difference exists between the treatments. Alternate hypothesis would be that significant difference exists among the treatments.

Group	Treatment					
	A	B	C	D	E	F
1- Score	76	88	66	79	76	76
Rank	3	5	1	5	3	3
2- Score	85	81	87	92	75	78
Rank	4	3	5	6	1	2
	Treatment					
Group	A	B	C	D	E	F
3-Score	79	69	83	90	71	68
Rank	4	2	5	6	3	1
4-Score	91	77	80	88	68	93
Rank	5	2	3	4	1	6
5-Score	87	87	92	79	82	81
Rank	4.5	4.5	6	1	3	2
Total R_i	20.5	16.5	20	22	11	14

As there are tied ranks (highlighted cells), the suitable formula is applied, as stated below:

$$F_{Tie} = \frac{12 \sum_{i}^{k} R_i^2 - 3n^2 k(k+1)^2}{n^2 k(k^2-1) - n(\Sigma t^3 - \Sigma t)}$$

Total number of observations are 30, out of which 5 (3 no's and 2 no's) are involved in tie,

Hence $\Sigma t^3 = 3^3 + 2^3$ and $\Sigma t = 2 + 3$.

With these details, now applying the formula,

$$F_{Tie} = \frac{12(20.5^2 + 16.5^2 + 20^2 + 22^2 + 11^2 + 14^2) - 3 \times 5^2 \times 6(6+1)^2}{5^2 \times 6(6^2-1) - 5(3^3 + 2^3 - 5)}$$

$$F_{Tie} = \frac{672}{5100} = 0.132$$

Tabulated value of Chi-square for (k-1) degrees of freedom i.e. (6-1) d.f and at 5% level of significance is given as 11.07. The calculated value is lower than the tabulated value, hence the null hypothesis is accepted, i.e. no significant difference exists in the treatments.

TEST OF SIGNIFICANCE OF SPEARMAN'S AND KENDALL RANK CORRELATION

Other important nonparametric tests are 'test of significance of Spearman's rank correlation' and 'Test of significance of Kendall rank correlation.' These have been discussed already in detail in the chapter on 'correlation' and hence can be referred to from there. They are being mentioned here, so that the reader is aware of them being nonparametric tests.

McNEMAR'S TEST FOR PAIRED SAMPLES

McNemar's test is a nonparametric test and considered as an extension of Chi-square tests or a 'paired version of Chi-square.' Chi-square assumes independent samples and hence cannot be applied to data of the two groups which pertain to same subjects (paired data of two groups). In such cases, McNemar's test is used to evaluate the statistical significance.

Being a paired version, it involves common respondents observed in two conditions (e.g. before/after) in a crossover setting. There are two possible outcomes which are generally classified as yes/no or present/absent or success/failure, etc. For example, study of pain experienced before treatment (with the respondents stating yes or no), followed by responses taken from same respondents but after treatment (with respondents stating yes or no on pain experienced). Being a 2 × 2 contingency table there are four possible outcomes, namely:
1. Yes in both situations (before and after)
2. No in both situations (before and after)
3. Yes before treatment and No after treatment
4. No before treatment and Yes after treatment

McNemar's test determines the statistical significance between the probability of pairs which are different, i.e. cases as stated (3) and (4) above. Consider the below contingency table:

		After treatment		Total
		Yes/Present	No/Absent	
Before treatment	Yes/Present	a	b	a + b
	No/Absent	c	d	c + d
Total		a + c	b + d	N = a + b + c + d

McNemar's Test Statistic, under null hypothesis of the probability that the outcome present is same for the two conditions, is given as (with 1 degree of freedom):

$$\text{McNemar's } \chi^2 = \frac{(b-c)^2}{(b+c)}$$

McNemar's corrected χ^2 (if b or c are small or $10 < b + c < 25$) = $\frac{(|b-c|-1)^2}{(b+c)}$.

For critical values and inferences, the calculated test statistic is compared with the tabulated values of Chi-square and inferences are drawn on the lines of Chi-square tests. As can be seen it uses the count of frequencies from matched pairs of the nominal data and evaluates if the frequencies 'b' and 'c' occur in equal proportions. Though the test statistic above, mentions b and c, but it should be remembered that what is under consideration is the frequencies of the pair of categories, which are different. In case the row/column sequence is marked as no and yes, then the above formula needs to be suitably amended. In binary terms if yes/success/present is denoted as 1 and no/failure/absent as 0, then the formula relies on frequency of pair (0,1) and (1,0) and not (1,1) or (0,0). It should be further noted that in case the category change cases are less than 10, the test results may throw up inaccurate results.

COCHRAN Q TEST

Cochran Q test is an extension of the McNemar's test for matched (paired) samples. Whereas McNemar's test relates to two treatments, Cochran Q test provides method for testing the differences between three or more matched (paired) observation sets, with the limitation that the responses are binary in nature (dichotomous). If data is binary in nature Cochran Q test can be used instead of Friedman test. Cochran Q test should not be mixed up with Cochran's C test which deals with variance outliers.

The test has the limitations like responses need to be binary and from matched samples, with the respondents/subjects being independent and the sample size is sufficiently large. If 'n' is the number of respondents/subject/sample and 'k' is the number of treatments, then the sample size should be $(n \times k) \geq 24$.

The test statistic Q follows χ^2 Chi-square distribution with $(k - 1)$ degrees of freedom where 'k' is the number of treatments.

Under the null hypothesis that the proportion of success is same in all the groups and alternate hypothesis that the proportion of success is different in atleast one of the groups, the test statistics Q is given as:

Chapter 23: Nonparametric Tests

Subjects (i)	Treatments (j)			
	1	2	3	K
1	Y_{11}	Y_{12}	Y_{13}	Y_{1k}
2	Y_{21}	Y_{22}	Y_{23}	Y_{2k}
3	Y_{31}	Y_{32}	Y_{33}	Y_{3k}
4	Y_{41}	Y_{42}	Y_{43}	Y_{4k}
N	Y_{n1}	Y_{n2}	Y_{n3}	Y_{nk}

$$Q = \frac{(k-1)(kC - T^2)}{kT - R}$$

Where $C = \sum_{j=1}^{k}\left(\sum_{i=1}^{n} Y_{ij}\right)^2$

$T = \sum_{i=1}^{n}\left(\sum_{j=1}^{k} Y_{ij}\right)$ and $R = \sum_{i=1}^{n}\left(\sum_{j=1}^{k} Y_{ij}\right)^2$

The simplified version of the above test statistic, without use of Σ sign can be given as:

Sample	Treatments				Row total ΣR	Square of row totals $(\Sigma R)^2$
	T_1	T_2	T_3	T_k		
R_1	a	b	b	d	$\Sigma R_1 = (a+b+c+d)$	$(\Sigma R_1)^2$
R_2	e	—	—	—	ΣR_2	$(\Sigma R_2)^2$
R_3	f	—	—	—	ΣR_3	$(\Sigma R_3)^2$
R_4	g	—	—	—	ΣR_4	$(\Sigma R_4)^2$
R_n	h	—	—	—	ΣR_n	$(\Sigma R_n)^2$
Column total	ΣT_1	ΣT_2	ΣT_3	ΣT_k	Total ΣR (= say Y)	Total $\Sigma (\Sigma R)^2$ (= say Z)
Square of column total	$(\Sigma T_1)^2$	$(\Sigma T_2)^2$	$(\Sigma T_3)^2$	$(\Sigma T_k)^2$		
Total of squares of col. total	$\Sigma (\Sigma T^2)$ = (say X) $(\Sigma T_1)^2 + (\Sigma T_2)^2 + (\Sigma T_3)^2 + (\Sigma T_k)^2$				Number of samples = n	Number of treatments = k

(Note: $\Sigma T_1 = a+e+f+g+h$ and similarly so on)

$$Q = \frac{(k-1)(kX - Y^2)}{kY - Z}$$

Example: Each of the respondents was administered three different physical tests for testing lungs efficiency. Based on the standards established, the respondents performance in the three tests was termed either as success (denoted as 1) or failure (denoted as 0). Evaluate using Cochran's Q test if there exists any significant difference in the proportions of success in the three tests. (Observations are provided directly in the solution).

Solution: Preparing the appropriate table for performing the calculations:

Respondent	Test A	Test B	Test C	Row total	Square of row total
1	0	1	1	2	4
2	0	1	0	1	1
3	0	1	0	1	1
4	0	0	0	0	0
5	0	1	0	1	1
6	1	1	0	2	4
7	1	1	0	2	4
8	0	0	0	0	0
9	0	1	0	1	1
10	0	1	0	1	1
11	1	1	1	3	9
12	1	0	0	1	1
Column total	4	9	2	$\Sigma = 15 (= Y)$	$\Sigma = 27 (= Z)$
Square of column totals	16	81	4		
Sum of squares of column totals		$101 = X$			$k = 3$

With degree of freedom = (3-1) = 2, Calculating the Q statistic

$$Q = \frac{(k-1)(kX - Y^2)}{kY - Z}$$

$$Q = \frac{(3-1)(3 \times 101 - 15^2)}{3 \times 15 - 27} = 13$$

Tabulated value of Chi-square at 2 degrees of freedom and 5% level of significance is 5.99. As the calculated value of Q statistic is greater than the tabulated value, the null hypothesis is rejected, i.e. statistical significant difference exist among the proportion of success in the three tests.

The nonparametric tests discussed in this chapter are only a part of the armamentarium of this class of tests; there exist more such type of tests which are not as commonly used and hence not discussed. The tests mentioned here are most commonly used and can be applied in most of the research subjects. Nonparametric tests are particularly useful in cases where ranks are given instead of scores, and where the population distribution is not known or doubtful.

CHAPTER 24

Which Statistical Test to Use?

Statistical inferences rely on the appropriate statistical tests—application of improper test ruins the entire effort as the inferences would not be statistically valid. Understanding of descriptive statistics is relatively easier but inferential statistics is where the scope of faltering is much higher. The existence of more than a hundred statistical tests, make the situation even more confusing. The data types depend a lot on research question. With research questions being varied, it is natural to have different tests to handle this large variety. It is not possible to address all these numerous tests in this book and hence tests covered in this book in the various chapters are the ones which are widely used and can be performed in most research questions. This chapter provides the basic guideline on which test to use. The tests mentioned have been explained in preceding chapters, hence are not being detailed here. Similarly, definition of parametric and nonparametric tests has already been covered in the preceding chapters.

Recapitulating, parametric tests operate on the assumption of the data following normal distribution, whereas nonparametric do not assume this condition. Further, parametric tests are statistically more powerful than nonparametric tests and should be used as first choice, provided the data conforms to its assumption. Non-parametric tests provide less information and can be considered conservative. Whether to use parametric or nonparametric test, depends on:

- *Type of data:* If the data is nominal or ordinal, nonparametric test would be applicable, whereas for ratio and interval data parametric tests can be applied.
- *Distribution assumption:* Parametric test assumes normality (normal distribution) of the underlying population, whereas nonparametric tests do not make such specific assumption. It should be noted that the assumption is for the population from which sample is drawn and not the distribution of sample.
- *Assumed variance (homoscedasticity):* Parametric tests generally assume homogeneous population, whereas nonparametric tests do not assume this.
- *Independent and random samples:* Parametric tests assume the sample to be independent, whereas this assumption is not necessarily required for nonparametric tests.
- *Central value:* Generally mean is the central measure seen in data where parametric tests have been applied, whereas it is median in case of nonparametric tests.

- *Strength:* Parametric tests are considered statistically more powerful as they allow more conclusions to be drawn. Non-parametric tests though relatively simpler and weak, are less affected by outliers.

While deciding between parametric and nonparametric, the two very important aspects are normal characteristics of the population and the type of data. In addition, the amount of available data (sample size) is also important. If the sample size is sufficiently large then parametric tests are robust enough to provide appropriate results even though the population may not be properly normally distributed. Having stated this, caution needs to exercised if the large sample is derived from a population which is highly skewed or strongly deviated, as in that case nonparametric tests should be the only choice.

On the other hand, the challenge with small samples is that is they cannot be assumed to have close approximation to normal distribution and might not possess sufficiently close approximation of the population parameters. Inspite of these restrictions, at times and in certain cases, large samples may not be feasible; therefore small samples tests are required. Researcher should take a conscious decision before resorting to small samples as these are more prone to the sampling fluctuations.

Even though the sampling distribution might not be distributed normally, but in small samples tests, it is assumed that the population from which the sample is drawn does not markedly deviate from normal. More often than not, the interest in studying small samples lies in testing the hypothesis instead of estimation of population parameters. The generalizations from nonparametric tests should be done conservatively or observed with this view.

If required, necessary tests for conforming to normal distribution should be conducted for ensuring that the assumption of normality has been met before applying the parametric tests. Deciding whether to use parametric or nonparametric tests ensures overcoming an important hurdle in the process of inferential statistics. Most situations in research and data throw up a situation where the researcher is not sure whether normality condition has been violated or it has been satisfied. Whether a simple histogram suffices or a proper goodness of fit is used depends on the researcher's needs. The researcher has access to the background information to allow him to make an informed guess. The reasons for sticking to one or the other should be clear in design phase itself. The choice of test considered has a direct impact on the interpretation. Generally, it is witnessed that in biomedical sciences, most frequently used tests are t-tests among parametric tests and Wilcoxon Rank Sum Test or Mann Whitney U test. Researcher would do well to not to get influenced by the skewed usage of the tests, and should decide on the appropriate test based on the specific research requirements.

The choice of test to be used is necessarily be required to be established before the sample collection process is initiated, i.e. at the research plan and design stage itself.

SUMMARY OF STATISTICAL TESTS

Flow charts 24.1 and 24.2 provide the summary of the tests, followed by relevant comparative parametric and nonparametric tests.

Chapter 24: Which Statistical Test to Use?

Flow chart 24.1: Parametric tests

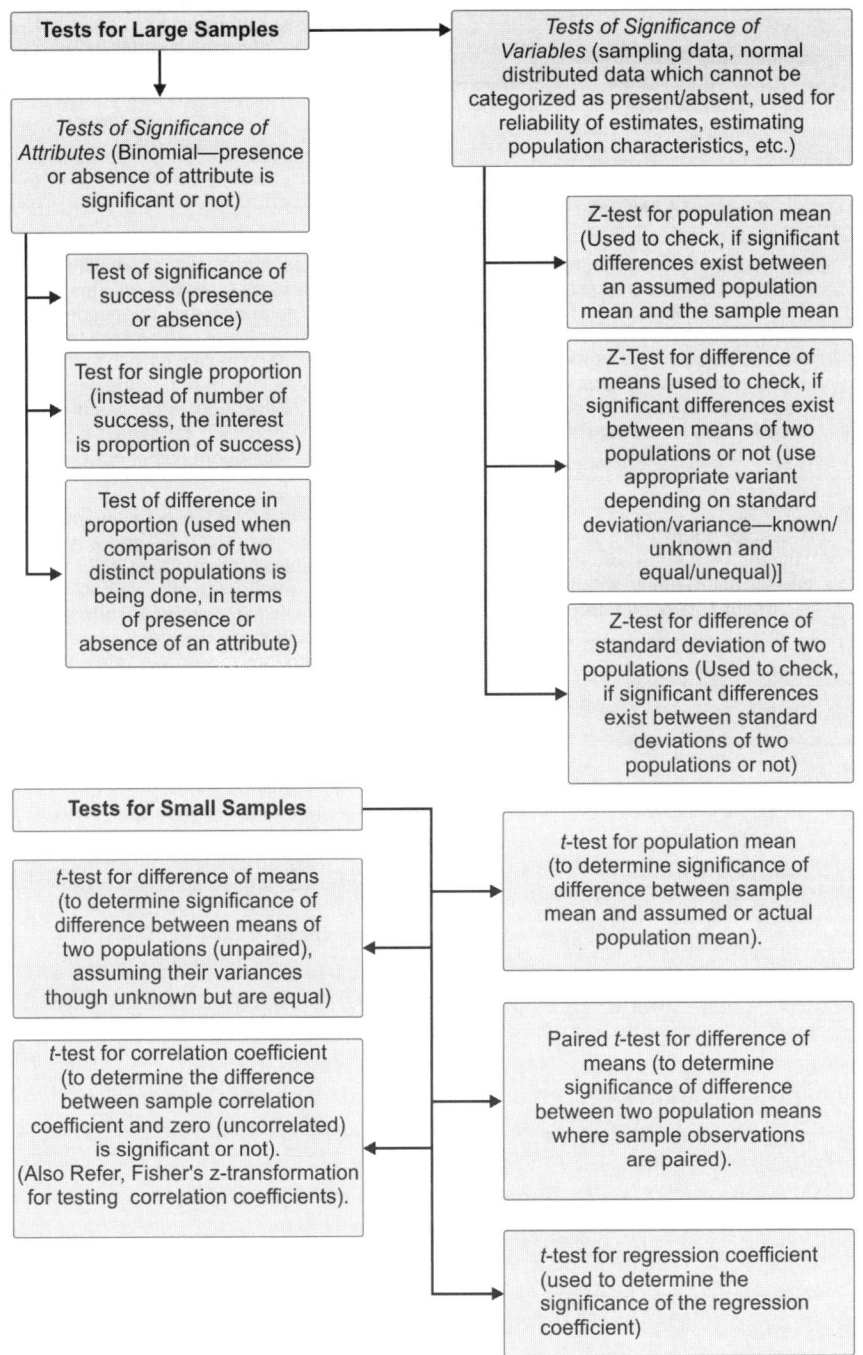

Contd...

Contd...

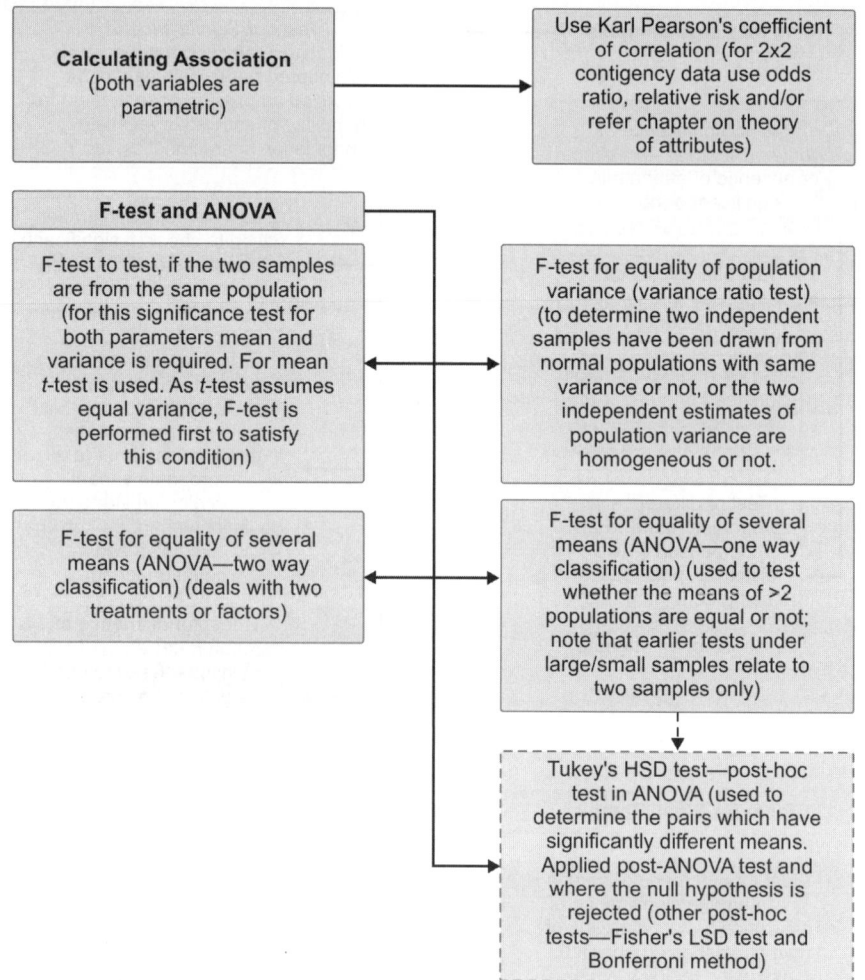

Chapter 24: Which Statistical Test to Use?

Flow chart 24.2: Nonparametric tests

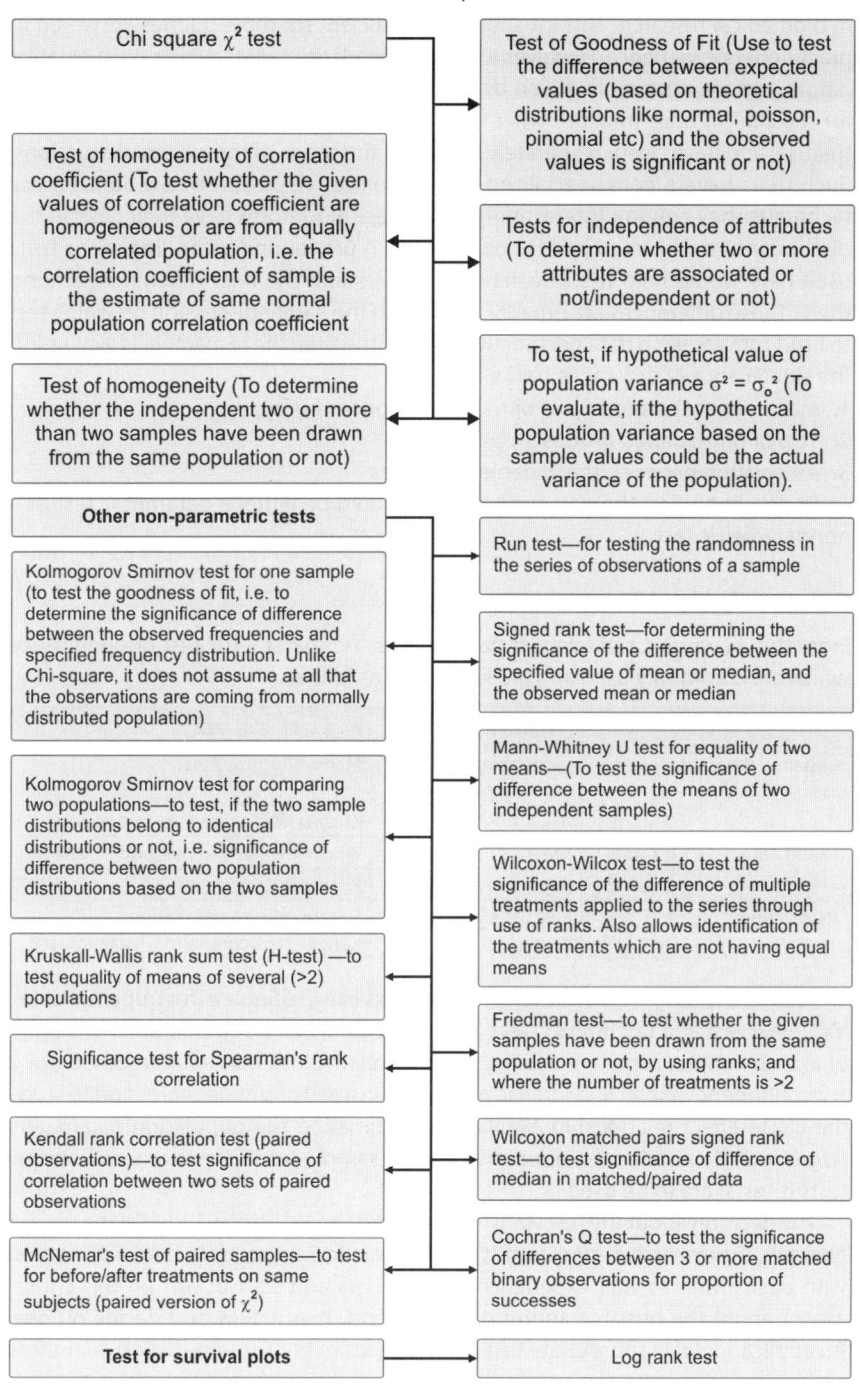

The listing of the tests is not exhaustive but covers most of the tests required in biomedical research. This list should be sufficient for most queries answered in practitioners' experience. However, a few advanced techniques, which are of iterative nature and are essentially driven by modern computing softwares, may be useful for the clinical trials (most of what is generally used even there is covered in this treatise) and have been deliberately left out of the scope of the current discussions. Such trials have a core specialized team in place which can throw light on the technique they employ. Interestingly, even these algorithms have their basis in the discussions incorporated in this book already. A proper understanding of the tests used here would help the researcher to understand the finer differences among them. These differences are broadly in line with the earlier discussion on which test to use. From research data perspective most of the time the researcher is evaluating the significance of difference between:
1. Mean/variances, etc. of two or more independent groups.
2. Two or more paired groups.
3. Association between the variable/attributes

Depending upon the data type, the use would be either a parametric test or a nonparametric test.

Commonly Used Tests

From this viewpoint, the relevant parametric or nonparametric test of significance which are commonly used and are suggested as follows:

		Parametric test	Nonparametric test
Unpaired data	2 groups	Large sample—Z-test Small sample—t-test	Mann-Whitney U test
	> 2 groups	ANOVA/F-test	Kruskal-Wallis rank sum test (H-Test)
Paired data	2 groups	t-test	Wilcoxon matched pair signed rank test
	> 2 groups	Repeated ANOVA	Friedman's test
Association		Pearson's correlation coefficient	Spearman rank correlation Kendall rank correlation—paired data

In case, the data is qualitative in nature, tests like Chi Square (for unpaired data), McNemar test and Cochran's Q test (for paired data) can be carried out. For degree of association the same are studied through relative risk/odds ratio method (for 2 × 2 contingency table) and logistic regression instead of simple regression to study the cause-effect relationship. Similarly for estimating the population parameters (1 group or 1 sample), the appropriate large sample test (Z-test) or small sample tests (t-tests) are to be used.

The decision about the test statistic to be used should be concluded at the time of finalizing the research design itself. It is a very practical approach to have a discussion with peer group as well experienced researchers and statisticians on this critical aspect about the possible appropriate tests and then review and decide on one. An application of inappropriate test statistic leads to improper statistical inferences and/or invalidates the research project, which needs to be avoided.

CHAPTER 25

Special Considerations about Inferential Issues in Biostatistics

An appropriate inference is the final outcome of the efforts and resources invested in research. Subsequent to handling successfully the contours of research process, which by itself was not simple with possible pitfalls at every step, the researcher arrives at the stage of drawing inference. A correct interpretation and inference of results is of prime importance as it allows other stakeholders, researchers etc to understand, evaluate, repeat or compare the results with their own. Though inference may sound straight-forward and easy, but in practice the use of inferential statistics tests to draw inference has its own challenges. Some of key questions which need answers can be (indicative list):

1. Was the research design used—appropriate and valid for the research question/objective?
2. Were there multiple research objectives and was the research design capable to handle them simultaneously?
3. Was the implementation or conduct of research thorough enough?
4. Were the measurements in order?
5. Did any possible bias creep into the study design or its measurements?
6. Was the data collection, input, output, analysis, etc free of typing/calculating errors?
7. Are the statistical methods and tests used appropriate?
8. Is the statistical significance significant enough and is it clinically significant?
9. Are the results significant enough to hold themselves in larger sample?
10. Do the results and the experiment itself have repeatability?

The list can go on, but what is important is to ensure that the study does not have material shortcomings. Thereafter, the results, inference and interpretations are logical and relatively uncomplicated. The researcher needs to evaluate and interpret the effort in a highly critical and stoically objective manner.

Research is carried out with specific objectives. The objective could be exploratory (implying study of different aspects of a problem), diagnostic (descriptive) or even analytical (hypothesis testing). Analytical research may further be divided into experimental and observational. The experimental studies are carried out by either Randomized or Non-Randomized Controlled Trials and generally aim at finding efficacy of interventions. On the other side, observational studies generally look at the causes of disease through Case Control (identifying risk factors) or cross-sectional studies (find prevalence of medical condition).

In certain research studies, inferences can be drawn straightaway and are uncomplicated. Due to underlying bias, inferences are not easy when the study involves cause-effect relationship, association of variables (exposure-disease) and where tests of statistical significance are involved. While assessing the true cause-effect relationship it is important to check for incidence of bias in measurements, presence of confounding variables and existence of spurious relationships. Similarly, in analytical studies, instead of completely relying on correlation coefficient, rate ratio and rate difference should also be considered. In statistical significance tests, the significance of the p-value should be examined before drawing inference. There could be more inferential issues, but these are major ones and are being considered here along with some methods to check for bias, control confounding and Bradford Hill Causality Criteria.

CAUSE-EFFECT RELATIONSHIP: IS IT REALLY TRUE?

The main aim of biostatistics is to draw out the patterns observed in the distribution and determinants of particular health conditions in the population under study. Once the data has been collected, collated, classified and analyzed, the next logical step is to make it applicable to the use of the population. The classic case of John Snow's Cholera experiment can be used to draw an analogy.

John Snow marked the home address of each cholera death in Soho in London during 1848–49 and 1853–54. An apparent association between the source of drinking-water and the deaths was noticed. Cholera deaths in districts with different water supplies were compared. This lead to the discovery that both the number of deaths and the rate of deaths were higher among people supplied water by the Southwark Company. Snow hypothesized about the communication of infectious diseases and suggested that cholera was spread by contaminated water. Improvements in the water supply could be made long before the organism responsible for cholera, *Vibrio cholerae* was even discovered by Filippo Pacini in 1865 or more famously by Robert Koch in 1884. John Snow had died in 1858.

All associations seen may not necessarily be leading to one or the other. A needle of suspicion raised should be put to rest beyond any doubt. At the same time, the needles should not pointedly cause unnecessary death and destruction because that will lead to loss of trust in the observations. Hence it is a fine line to tread. There must be sufficient evidence before we hang the determinant as a cause and at the same time an honest non-causal determinant should not be condemned. The questions to be addressed before the observed association between exposure and outcome is accepted to be a true cause-effect relationship are:

a. Can the observed association be explained by systematic errors or bias in selection, follow-up or information collection?
b. Can the differences between groups be explained by the distribution of another variable called confounder that was not measured or measured and not taken into account in the analysis?
c. Can the differences between groups be explained by chance and chance alone?
d. Is the observed association likely to be causal?

BIAS

Selection and measurement bias can be checked by critically examining the following:

Selection Bias

- Definition of study population clear or not?
- Inclusion and exclusion criteria have been laid or not?
- Attempts to keep low refusals, losses to follow-up, etc. made or not?
- *In case-control studies*:
  - Do controls represent the population from which the cases were picked?
  - Identification and selection of cases and controls influenced by their exposure status or not?
  - Can the controls actually be proved to not be subclinical or undetectable cases verifiably or not?
- *In cohort and intervention studies*:
  - Only exposure/ intervention differentiated the groups which were otherwise similar or not?
  - Follow-up adequate or not?
  - Follow up for all groups same or similar or not?

Measurement Bias

This requires assessment of the following questions, "Were:"
- Exposures and outcomes of interest clearly defined using standard criteria?
- Study subjects randomized to observers or measurements and method of randomization?
- Measurements as objective as possible?
- Clearly written standard operating procedures and protocols used to standardize procedures in data collection?
- Subjects and observers blinded to the study objectives or procedures and level of blinding?
- Observers and interviewers rigorously trained and user manuals made?
- Information provided by the patient validated against any existing records?
- The methods used for measuring the exposure(s) and outcome(s) of interest (e.g. questionnaires, laboratory assays) validated?
- Strategies present in the study design to detect and measure likely direction and magnitude of the bias?

CONFOUNDING

Confounding is a distortion or inaccuracy in the estimated measure of association that occurs when the primary exposure of interest is mixed up with some other factor that is associated with the outcome. The three conditions for confounding are:
1. The confounding factor must be associated with both the risk factor (independent variable) and the outcome (dependent variable).
2. The confounding factor must be distributed unequally among the groups being compared.

3. A confounder cannot be an intermediary step in the causal pathway from the exposure to the outcome.

A confounding variable (or confounding factor or confounder) is an extraneous variable in a statistical model correlating both the dependent variable and the independent variable. A potential confounder in biomedical research is a factor which is believed to have a real effect on the risk of the disease under investigation. It is related to both the subject and the outcome but the distribution of this variable affects the outcome under study. These factors may be directly linked to the disease (e.g. smoking and lung cancer) or maybe good proxy measures of direct unknown causes (e.g. age and social class).

Confounding Variables

Confounding variables according to their source can be classified as:
- *Operational confound*: It is due to choice of measurement instrument. This type of confounding occurs when a measure designed to assess a particular construct inadvertently measures something else as well.
- *Procedural confound*: It is due to situational characteristics. In a laboratory experiment or a quasi-experiment, this occurs when the researcher mistakenly allows another variable to change along with the manipulated independent variable.
- *Person confound*: These are inter-individual differences. This occurs when two or more groups are analyzed together (e.g. patients of different diseases), despite varying according to one or more other (observed or unobserved) characteristics (e.g. gender).

Confounding Control

Confounding can be dealt with:
- At the study design phase and/or,
- If the relevant data have been collected, in the analysis phase.

The approaches to control for confounding in the study design phase are:
- *Randomization:* The ideal method of controlling for confounders as it ensures that the distribution of known and unknown confounding variables will be similar in the compared groups. This requires a relatively large sample size to apply central limiting theorem. This method is available for experimental studies only.
- *Restriction:* This procedure limits study participation only to people, who are similar in relation to the confounder. It can work for known confounders only; and if confounder becomes known only after the commencement of the study then this technique is of no use.
- *Matching:* In this process, controls are selected in a manner that the distributions of potential confounders (e.g. age, sex, race or place of residence) are identical to that of the cases. In individual matching, this is done by selecting for each case one or more controls with similar characteristics (e.g. of the same age, sex); while in frequency matching, it is done by ensuring that as a group, the controls have similar characteristics to the cases. It is widely used in case-control studies

as it is an expensive technique in large cohort studies and unnecessary in large intervention studies. In small intervention studies like community trials, it is a common feature.

Confounding Control in the Analysis Phase Uses

- *Stratification*: The strength of the association is measured separately within each well-defined and homogeneous category of the confounding variable called stratum. For example, in case of age being a confounder, the association is estimated separately for each age-group; the results are then pooled using a suitable weighting to obtain an overall summary measure of the association. This can be adjusted or controlled for the effects of the confounder. This takes into account differences between the groups in the distribution of confounders. Standardization is an example of stratification.
- *Statistical modeling:* Sophisticated statistical methods like regression can be used to control for confounding. They are handy for simultaneously adjusting for various confounders. It is possible to control for confounders in the analysis phase only if data on confounders was collected and the extent to which it can be controlled for depends on the accuracy of this data. Random misclassification of exposure to a confounder leads to underestimation of the effect of the confounder and attenuates the degree of control of confounding. The association will persist even after the adjustment because of residual confounding.
- *Sample size calculations*: Sample size calculation at the design stage ensure that the study power (i.e. its sample size will be large enough) to detect the hypothetical effect or precision to quantify it is adequate. Multiple statistical tests on large data sets to determine significant associations should be used with caution and rather be deprecated. The probability of getting a statistically significant P value just by chance increases with the number of tests performed. If an unexpected association that was not hypothesized at the beginning of the study is found, it should be reproduced in studies specifically designed to test it, before it can be accepted as real. Statistical methods assess only sampling variation and cannot control for non-sampling errors like confounding or bias in the design, conduct or analysis of a study. That the study was not affected by any of these factors can never be ensured. Therefore, the findings should be studied in the context of all available evidence.

SPURIOUS RELATIONSHIP

A spurious relationship is a perceived relationship between an independent variable and a dependent variable. It may have been estimated incorrectly because the estimate failed to account for a confounding factor that had a bearing on the outcome or dependent variable and was related to both the dependent and independent variable. The incorrect estimation suffers from omitted-variable bias.

EFFECT MODIFICATION

Effect modification is said to occur when the magnitude of the effect of the primary exposure (independent variable) on the outcome (dependent variable) differs depending on the level of a third variable. Estimate of association in case of effect modification is misleading. It is better to examine the association separately for each level or stratum of the third variable. Effect modification is a biological phenomenon where exposure has a different impact in different circumstances. Multivariable methods and stratified analysis can be used to assess effect modification.

To identify these statistically the following analogy is used when subgroup analysis is done. In confounding, the measures of association in the subgroups differ from the crude measure of association but the measures of association across the subgroups tend to be similar. In effect modification, the measures of association in the subgroups differ from one another.

Therefore, all possible explanations should be considered before inference on cause-effect relationship is established. As discussed in previous pages, the observed association may exist due to bias, confounding or even pure chance. This implies that statistical association does not necessarily infer a causal relationship and conversely that absence of statistical association would not necessarily imply absence of causal relationship. Hence, instead of concluding proper causal association from a single component (statistical association), it is more appropriate to involve criteria that includes the strength of association as well consistency with other known facts or existing knowledge. The Bradford Hill criteria are widely used in epidemiology to evaluate whether an observed association is likely to be causal or not.

BRADFORD HILL CAUSALITY CRITERIA

In 1965, Bradford Hill mentioned criteria to be considered when assessing causality of an association.
- *Strength of association*: The stronger the relationship between the independent variable and the dependent variable, the less likely is that relationship to be due to an extraneous variable.
- *Temporality:* A cause should precede an effect in time or the independent variable should temporally precede the dependent one.
- *Consistency:* Multiple observations of the association reported by different observers under different circumstances with different measurement instruments support the credibility of the finding.
- *Theoretical plausibility:* A rational and theoretical basis for an association supports causality. However, it may not always be so. John Snow's observations preceded the isolation of the causative agent. A direct link still needs to be hypothesized to be tested by further research.
- *Coherence:* The association should be coherent with other knowledge to support causality. A cause-and-effect association is clearest when it does not conflict with what is known about the variables under study and there exist no plausible competing theories or rival hypotheses.
- *Specificity:* In an ideal situation, the effect should have one and only one cause. An outcome best predicted by one primary factor supports causality. However, that may not always be the case. In modern statistical techniques, there exist techniques to test for more than one cause but the causality association becomes weaker in such situations.

- *Dose response relationship*: A direct relationship between the risk factor (i.e., the independent variable) and the disease variable (i.e., the dependent variable) favours causal association especially if replicated over time.
- *Experimental evidence*: Related research based on experiments increases plausibility of a causal inference.
- *Analogy:* A commonly accepted phenomenon in one field is capable of application in another area.
- *Reversibility:* Removal or reduction of the exposure leading to a reduction of disease or risk of disease and reappearance of the dependent factor variable on repeat exposure to independent factor variable is clubbed under reversibility. It strongly favors causal association even though this may not occur in most instances.

Classically, described is the application of Hill's criteria to the case of smoking and lung cancer. All students have come across this example in their medical training. The lung cancer among smokers was higher than for nonsmokers (e.g. one study estimated that smokers are about 35% more likely than nonsmokers to get lung cancer) demonstrating the principle of strength of association. Smoking preceded the onset of lung cancer in most cases showing temporality. Different methods of studying the association, e.g. prospective and retrospective, case control and cohort studies among males and females produced similar findings showing consistency. Biological theory of smoking causing cumulative tissue damage over time resulting in cancer by tar products was biologically plausible. It is now coherent with the current knowledge about the biology and history of lung cancer. Lung cancer can be predicted from the incidence of smoking showing specificity of cause. It is not, however, the only source of the disease. A positive, linear relation between the number of pack years and the incidence of lung cancer is an example of dose response relationship. Experimental evidence in the form of cancer in the laboratory rabbits' ear tissue painted with tar supported presence of carcinogens in tobacco tar. Induced smoking in laboratory rats showed a causal relationship demonstrating an example of analogy and experimentation.

EXPOSURE-DISEASE ASSOCIATION

For association of variable correlation coefficient is determined. Information derived from the correlation coefficient does not include reliable and quantitative information about the risk or benefit of an observed association. In analytical studies, where the intent is to study the relationship between exposure and disease the basic measures considered are the rate difference, rate ratio, etc. These measures of association are routinely used to compare the frequency of diseases. These measures are broadly classified as absolute differences and relative differences or ratios. The absolute measures are generally preferred in public health matters and where preventive actions are required or being planned. The relative measures are generally used in studies involving cause and determinants with discrete outcomes.

The following table would help understand the calculations:

	Develops disease	Does not develop disease	Total	Risk rate (incidence rate)	Risk ratio (relative risk)
Exposed	a	b	a + b	$\dfrac{a}{a+b}$	$\dfrac{a}{a+b}$
Not exposed	c	d	c + d	$\dfrac{c}{c+d}$	$\dfrac{c}{c+d}$
Total	a + c	b + d	N = a + b + c + d		

The rate difference is a measure of the absolute difference between two incidence rates (e.g. incidence rate of lung cancer for the exposed minus the incidence rate for the unexposed in a cohort study).

Relative Risk or Rate Ratio

The relative risk or rate ratio (RR) is a relative measure of two rates. It is the ratio of the risk of occurrence of a disease among exposed to that among the unexposed. RR of 1.0 indicates no association and no increased risk. Any other value shows either increased or decreased risk. For example, a RR of 1.6 indicates a 60% increased relative risk of disease (possible causal association?); a RR of 0.7 indicates a decreased risk of 30% (possibility of protective association?). It is a better indicator of the strength of association than risk difference as it takes into account the baseline level of occurrence. It is used in assessing the likelihood of causality. Very large RRs signify that the relationship is unlikely to be due to chance. In case of small RRs a causal relationship may still be involved but this must not be accepted till other possible explanations have been ruled out.

ODDS RATIO

Statistically, odds of an event is the ratio of probability of occurrence of event to the probability of its non-occurrence. As denominator is same, this can be simplified as number of occurrence divided by number of non-occurrence. In a cohort study, the odds ratio (OR) measures the odds or chance of disease among the exposed divided by the odds of disease among the unexposed. The OR is essentially equivalent to the RR, if the disease is not extremely common. Odds Ratio is interpreted as: OR = 1 (exposure not related to disease); > 1 (disease positively associated with exposure); < 1 (disease negatively associated with exposure).

From the above table,

Odds that an exposed person develops disease = $\dfrac{a}{b}$

Odds that a not-exposed person develops disease = $\dfrac{c}{d}$

Therefore, odds ratio (OR) = $\dfrac{a/b}{c/d} = \dfrac{ad}{bc}$

Chapter 25: Special Considerations about Inferential Issues in Biostatistics

(The discussion preceding this related to cohort studies, but in case of case-control study only measurement is made of exposure and the disease together. There is no way of proving that exposure preceded disease. Therefore, the 2×2 contigency table is made as follows and the OR = odds that case was exposed/odds that control was exposed.

	Case	Control
Exposure	a	b
Not exposure	c	d
Odds ratio	$\dfrac{ad}{cb}$	

Whereas relative risk is calculated only for cohort study, odds ratio can be calculated for both cohort study and in case-control study. This has already been part of discussion elsewhere but this book is focusing on the statistical aspects only. The readers are well advised to look up books on epidemiology and clinical trials to further elaborate on the issue.)

ATTRIBUTABLE RISK

Attributable risk (AR) is the proportion of the outcome in exposed individuals that can be attributed to the exposure, i.e. a measure of excess risk attributed to exposure. This is useful in public health as it reflects the amount or percentage by which the risk of a disease or an outcome is reduced by elimination or control of a particular exposure. The number of people spared the consequences of exposure can be calculated by subtracting the rate of the outcome in unexposed from the rate in the exposed.

Attributable risk (AR) = (Incidence in exposed group–incidence in unexposed group)

Proportional AR = Attributable risk/Incidence in exposed group.

Population Attributable Risk

Population attributable risk (PAR) provides an estimate of the incidence of a disease in a population associated with, or attributed to the exposure or risk factor provided the association is causal. It is a measure of the relative importance of exposures for the entire population. PAR is calculated as follows:

Population attributable risk = Incidence in total population – Incidence in non-exposed.

It measures the proportion of reduction of incidence of the outcome in the entire population, if exposure were to be eliminated. Sometimes, it is distinguished into etiologic and excess fractions but unless otherwise specified, it should be taken to refer to excess fraction only.

Etiologic fraction is the proportion of cases in which the exposure played causal role. It is measured as:

$$\text{Etiologic fraction (EF)} = \frac{N_e - N_n}{N_e}$$

Where
N_e = Number of exposed individuals in a population who develop the disease.
N_n = Number of unexposed individuals in the same population who develop the disease.
Excess fraction is proportion of excess cases among exposed compared to the unexposed.

Statistically Significant and Clinically Significant

The statistical tests of significance discussed under various chapters are critical components of the research designs. Recalling, in most of the classical tests, if the calculated value of test statistic was less than the tabulated value at the given level of significance, then the null hypothesis stood accepted and vice versa. In broad sense, the inference being that the variations are due to chance and not due to the any acting forces, therefore null hypothesis remain valid. In terms of p-value terminology, the smaller the p-value, more significant would the result be. The test results (differences) would be statistically insignificant (accepting null hypothesis) if the p-value is more than the level of significance at which the test is being evaluated. Conversely, if the level of significance is considered at (0.05) 5%, then a p-value of ≤ 0.05 indicates result is real and not by chance resulting in rejection of null hypothesis. This brings us to the alternative hypothesis which is accepted by default if the statistical test of significance rejects the null hypothesis. Alternative hypothesis cannot be tested directly in the same study.

From the discussion on type I and type II errors in hypothesis testing, there is a real chance of result being false positive or false negative. At 5% level of significance, it correlates to the probability of one such outcome in 20 similar studies. Considering the number of research studies being carried out the world over and the possibilities that similar subjects are studied, the scope of erroneous statistical significance outcomes look plausible.

Therefore, even if p-value inference is significant, it is important to understand and critically question the impact and effects in a clinical setting. The researcher has to keep in mind the fact that the results are statistical in nature, which may turn out to be clinically insignificant. Clinically, 'effect size' is important. This is why the basic knowledge of the subject matter expert is important both in research design and in clinical import of the findings. To put it simply, both in writing and reading a paper the practitioner must always have adequate subject knowledge to draw meaningful conclusions from the studies. This is also important because all populations will have their unique characteristics. The doctor should be able to decipher the impact of the findings of a study on the population he works in or wishes to apply them on.

In addition to reporting p-value, many epidemiologists and researchers are moving towards using Confidence Intervals to report precision of the study results. Such a combined approach is probably the better way as compared to mentioning only the p-value.

CHAPTER **26**

Measurement and Error Analysis

Measurement is an integral part of research and statistics. Any data is a form of measurement of something or other. Objectives of research are achieved through measurement. Whether it is the measurement of a particle size or measurement of treatment effects, the baseline is measurement creates data which is analyzed to draw inferences. For generating useful data through research it is imperative to define clearly what is to be measured, ensure a valid and reliable tool to make that measurement, while having the necessary expertise to conduct, examine and investigate the measurement in order to draw inferences. Within these considerations, definition is important as through it the researcher can determine:
- Who or what is to be measured (attribute, characteristic, object, subject, etc.).
- Where is it to be measured (environment–random, controlled, etc).
- How and what tools and techniques are to be used for measurement and its analysis.

Therefore considering the criticality of measurement, it is appropriate to understand measurement theoretically.

Any measurement is open to errors. An error could be as simple and basic as the eye position while using a ruler to measure a line segment or advanced subjects like random errors in a process. Therefore, along with measurement, errors also need to be understood. This chapter provides the basic theory on measurement and error analysis, which would allow the researcher to comprehend the significance of these and possibly get a better insight about the research data.

MEASUREMENT

Measurement, "can be defined as a way of obtaining symbols to represent the properties of persons, objects, events or states, which symbols have the same relevant relationships to each other as do the things represented" (Adapted from Ackoff, Gupta and Minas, Scientific Method P179).

These symbols in most cases are numerals or can be converted into numerical form. This expanded definition of measurement is necessary because if measurement is simply defined as 'a number that shows the magnitude of something, usually referenced to some agreed standard' then such a definition does not include many situations like presence or absence of an attribute, or measurements in ranked form. Measurement can be construed as a process of assigning numerals to a characteristic or an attribute of an object to represent its qualities according to some rules. This allows comparability with other objects or events.

From the classical definition of measurement which considered it to be an estimation of ratios of quantities, to the representational definition where measurement was taken as the correlation of numbers with non-numerical objects, measurement theory has been evolving. The Information Theory of measurement further propounds inexactness of data which get measured by a range instead of a single value, i.e. all measurements are uncertain. Steering clear of the evolution of measurement theory, it is also well known that the modern system of measurement depends on seven base units and is called the International System of Units (in short 'SI Units' as abbreviated from *Systéme International d'Unités*).

Prior to SI Units systems other systems like FPS (Foot, Pound, Second), etc. were used in some countries but they have largely been superseded by the SI system. The SI Unit system comprising of measurement units like second (Time), metre (Length), kilogram (Mass), Ampere (Electric Current), Kelvin (Temperature), mole (Amount of substance/molar mass) and Candela (Luminous Intensity) is very true to Physical Sciences measurements. However, the fields like Social, Behavioral and Medical Sciences need measurements beyond these. This is because the study subjects in these fields of sciences are also of attributes, characteristics, treatments, etc. which do not have a standard unit. Thus these sciences use SI units wherever appropriate but also rely on additional tools to handle qualitative data.

LEVELS OF MEASUREMENT

The discussion above indicates that measurement covers a broad spectrum, but it would be possible that all forms of measurement may not necessarily be open and suitable to similar level of mathematical or statistical operations. For example, the length of a two line segment is stated as 5 cm and 10 cm. These can be added to state that the total length is 15 cm. Against this if measurement of an attribute is in ranked form, i.e. two ranks—5th rank and 10th rank, these cannot be added to derive 15th rank. But from definition both rank and length are measurements as they represent properties and have relevant relationship.

Measurement, depending on what it measures and the possibility to apply further statistical operations, etc. can be regarded as having four levels of hierarchy or Levels of Measurement, namely: Nominal, Ordinal, Interval and Ratio. These are also referred as primary scales of measurement. These scale or level of measurement help decide how to interpret data of the variable and allows the researcher to know what statistical tools are appropriate for the data.

Nominal

These are most basic scales where simply numbers are assigned (as tags or labels) to the study objects, like assigning roll numbers to the students of the class. It basically provides a unique and mutually exclusive 'numbers/names' to the attribute or qualitative variable. These measurements allow simple operation like counts, but no statistical operation like mean, etc. are possible. These are least informative and weakest in mathematical sense.

Ordinal

These are ranking scales or ordered data. The ranking or ordering is carried out on basis of specific attribute/characteristics of interest and in a specific direction. These scales include all information provided by nominal scale in an ordered form, with equivalent units getting equal ranks. Ordinal scale can be used for qualitative data, provides minimal information and has low statistical power. For example, the class roll numbers can be ranked on the basis of their scores in the examination to give rise to ordinal scale. Exam score being the characteristic and decreasing marks being the direction. It should be noted that marks may be decreasing but the ordinal scale is increasing (Rank 1,2,3, …). This is necessary as an increasing conversion preserves order. This scale allows few positional statistical tools to be used like Median, Quartile, etc. whereas mean cannot be used on ranked data. Apart from this, the other drawback of ordinal measurement is its inability to convey the quantum difference between any two or more ranks (like in above example: exam scores cannot be deduced from ranks and equal difference in ranks does not imply equal difference in actual scores). This implies that distance between ranks does not have any meaning, i.e. interval between the ranks is not interpretable.

Interval

Interval scales allow meaningful statements about differences between two characteristics. In such measurements, proper unit of measurement exists but zero point may be arbitrary and does not signify absence. For example, temperature of 60°F and 70°F indicate a difference of 10°F, which can be compared with the difference between 30°F and 40°F. This measurement is still statistically powerful and is used for quantitative data. The drawback of this scale is that ratio of two interval scale values are not interpretable. A temperature of 80°F does not imply temperature twice as hot as compared to 40°F.

Ratio

These measurements are statistically strongest as they allow statistical operations. These scales have an absolute zero which is meaningful, i.e. a zero which signifies absence of the attribute or characteristics. Further the scale values reflect equal ratios. For example length of 4 cm is twice of 2 cm. Further this measurement can be converted into mm, inches (other units of measuring length) but their ratio remains same as 2:1. Further, 0 cm implies nil length.

Interval and Ratio are used for quantitative variables, while nominal and ordinal are used in qualitative variable. Accordingly, significance of nominal and ordinal scaled data is tested through nonparametric tests, while parametric tests can be utilized for interval and ratio scaled data.

EVALUATING MEASUREMENT

In research, the aim is to find the truth or the true value of the variable by measuring various objects, subjects, events, etc. which the researcher believes would be representative of the true value. Contrary to expectations, measurements are more often than not, less than perfect. This does create situations where best available data is used but whose validity is still weak. Such situations may still be in the range of tolerance in certain nonmedical research subjects, but in medical studies significant consideration is to acquire valid and reliable measurements which are either without or with minimal bias and errors.

In order to assess and evaluate measurements about their validity and reliability, the concepts of accuracy and precision of measurements also need to be understood.

Precision

Precision implies how close two or more measurements are to each other. Precise measurement may or may not be accurate. For example a person weighs 57 kg but the weighing machine returns the weight as 55 kg every time weight is taken. This implies that the machine measurement of weight is precise but it is inaccurate. Therefore, precision is independent of accuracy. The following representation **(Fig. 26.1)** differentiates between more and less-precise measurements.

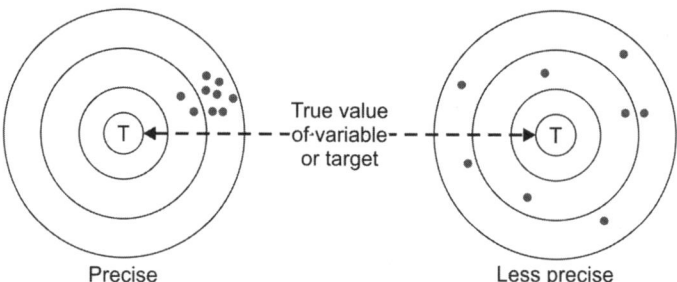

Fig. 26.1: Precision

Considering this involves two or more measurements, Repeatability and Reproducibility of the measurement becomes significant. Repeatability implies same tools of measurement are applied by the same experimenter and to measure the same subject or object, whereas Reproducibility implies same tools of experiment applied on same subject/object but by different experimenter.

From the perspective of statistical sampling, precision also depends on the sample size and the efficiency of the sample as representative of the true value. Precision of measurement can be assessed through Standard Deviation, Coefficient of Variation or Kappa Statistic. Lack of precision in measurements introduces random errors which result in reduction of confidence interval, thereby widening the gap between the estimated and true value of the variable or association.

Accuracy

Accuracy implies how close two or more measurements are to the true value. It reflects conformity to the truth. It can be stated as the closeness of agreement between a test result or measured result and accepted reference value or true value (**Fig. 26.2**).

It is considered as a function of systematic error. Less accuracy indicates increased systematic error which results in a biased estimate. This bias does not get influenced by repeating measurements or by increasing samples size. In case of comparative measurements if accuracy is lacking in both variables then it tends to either camouflage or overstate the true difference between them.

The graphical representation in **Figure 26.3** highlights the difference between the concepts of accuracy and precision. It should be noted that accuracy at times is also stated as 'Trueness.'

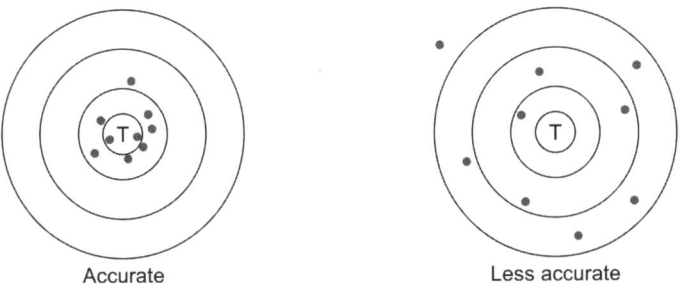

Fig. 26.2: Accuracy

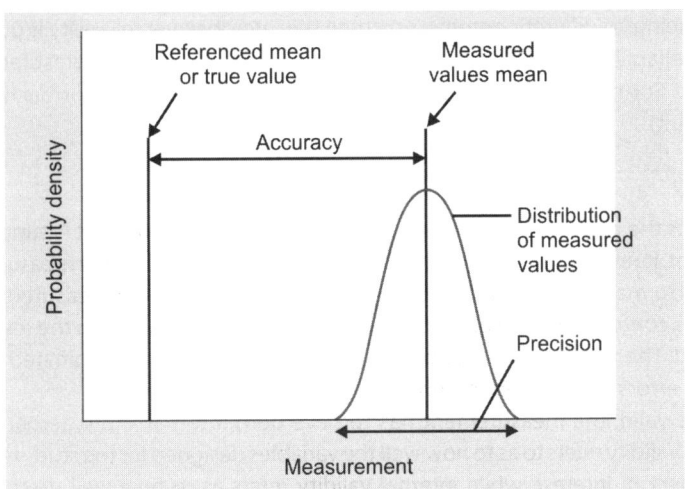

Fig. 26.3: Concept of accuracy and precision in relation to true value of variable. Closer the measured mean to true value greater is the accuracy

Reliability

Reliability refers to 'the extent to which an experiment, test, or measuring procedure yields identical results on repeated trials'. It implies comparison of repeated measurements indirectly reflecting on the precision and is affected by the random errors. Reliability is a condition where a measurement process produces consistent scores (given an unchanged measured phenomenon) over repeated measurements, i.e. consistency and trustworthiness of the measurements. Though reliability is important and necessary to prove validity of the test, but it is not a sufficient condition in this regard.

The components of reliability are:

Test-Retest Reliability

When a researcher administers the same measurement tool multiple times, follows the same research procedures, etc. the measurements should give the same results each time if there has been no change in whatever is being measured. Measurement of the piece of pencil using a vernier caliper will have high test-retest reliability.

Inter Item Reliability

This is applicable where multiple items are used to measure a single concept. There should be agreement between all items used to measure the attribute, e.g. the worse score should be associated with higher signs or symptoms or poorer quality of life.

Interobserver Reliability

Interobserver reliability refers to the extent to which different observers using the same measure get equivalent results. If different observers, using the same instrument to score the same thing, display matching scores then the interobserver reliability is good. For example in noncontact tonometry in automatic mode the same scores are obtained after training by different optometrists then the interobserver reliability is good.

The reliability of measurements is judged by methods like Test-Retest Technique, Multiple Forms method, Inter-Rater or Split half Reliability (Cronbach's alpha coefficient).

Validity

Validity is the degree to which a variable actually represents what it is supposed to represent. It reflects the extent to which a measurement, test or study measures what it claims to measure or what the experimenter thought is being measured. Validity indicates conformity to accuracy as it relates to the comparison with the referenced standard. The validity improves once the systematic errors are eliminated and the random errors are reduced thereby increasing accuracy.

To be valid, the measurement has to have both internal and external validity. External Validity refers to as to how well the variables designed for the study represent the subject of interest, while internal validity refers as to how well do the actual measurements represent these variables.

A valid measure should satisfy the following criteria:

Face Validity

This is the minimum requirement—if a measure cannot satisfy this criterion, then the other criteria become redundant. This criterion is an assessment of whether the measure appears to measure the concept it is intended to measure on the face of it. Offering assistance to a patient would meet the criterion of face validity for helping or kindness. However, measuring height to assess the educational status of an individual does not meet the face value of validity.

Content Validity

Content validity refers to the extent to which a measure adequately represents all facets of a concept or study subject. Consider a series of questions in the McMonnies' Index used in dry eye, e.g. age, contact lens use, treatment for dry eye, presence and duration of symptoms, etc. If there were other kinds of common indicators that are clubbed with dry eye that were not included in the index, then the index would have low content validity since it did not adequately represent all facets of the concept.

Criterion-related Validity

Criterion-related validity is empirical based and relates to the correlation between the instrument, as indicator of specific trait or behavior and external criteria. It means that the criterions that are used to measure the trait actually are useful for extrapolation or generalization. For example, performance on a driving test correlates well with safe driving ability.

Construct Validity

Construct validity refers to the extent to which a measure is related to other measures as specified by theory or previous research. This becomes useful when the criterion related validity is not easily available. Does a measure stack up with other variables the way it is expected to? Clinical observations had shown that people with low self-esteem often had depression. To establish the construct validity of the self-esteem measure, the researchers showed that those with higher scores on the self-esteem measure had lower depression scores, while those with low self-esteem had higher rates of depression.

RELIABILITY V/S VALIDITY

Though these terms are connected but as discussed above they are different. Any measurement which is non reliable will not be considered valid, i.e. a measurement has to be reliable to be termed as valid. At the same time, this condition is insufficient as nonvalid result may also be reliable as reliability is to get identical measurements. The following pictorial representation of earlier stated concepts would make the inter-relations more clear. This pictorial representation Reliability vs Validity (**Fig. 26.4**) here is a theoretical construct for the sake of explanation.

In research, it is difficult to state the difference between poor and good because the study is an attempt at unearthing the unknown. Therefore, this is a relative environment and needs to be treated and inferred appropriately. What is accurate

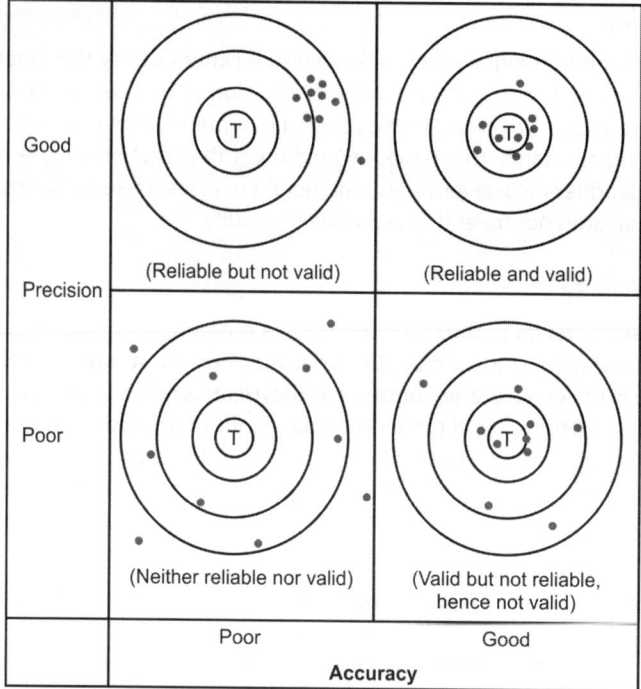

Fig. 26.4: Reliability and validity

about variable in past, may look of poor accuracy in today's context. For example, the unit of measurements has become more accurate as it moved from a simple ruler to Vernier caliper and then to a screw gauge. Though for sake of explanation, the words 'poor' and 'good' have been used above, it would be well suggested that the reader should consider these as 'less' and 'more'. Further, the noting above 'valid but not reliable hence not valid' implies that by theoretical definition the measurement is valid as it relatively closer to true value, but as it lacks reliability and therefore it is considered as 'Not Valid'. Apart from this, it is also worthwhile to note that in the current scenario truth is different from facts. Truth here is reality behind facts. The data which gets reflected through measurements, studies, etc. are the facts, which may or may not reflect truth. As more and more facts are discovered, truth is uncovered, which may be different from earlier understanding of the subject or builds upon the existing truth. The goal of every researcher is to find methods to find facts which mirror the truth.

ASSESSMENT OF VALIDITY: SPECIFICITY AND SENSITIVITY

Many of the epidemiology studies pertain to dichotomous subjects like Positive or Negative, Case or Non-Cases, Exposed or Not Exposed, etc. The validity of such experiments is analysed by denoting them as Positive or Negative firstly from the test/survey results and secondly from the actual (true) or referenced state. Accordingly four probable situations arise from this, as shown in following matrix.

Results comparison- true state and test results		Actual (referenced) condition	
		Positive	Negative
Test measurements results	Positive	Corrects—true positive	Incorrect—false positive
	Negative	Incorrects—false negative	Correct—true negative

The matrix inspired above is called a contingency table. A contingency table is a table in a matrix format displaying the frequency distribution of the variables. This term was first used by Karl Pearson in "On the Theory of Contingency and Its Relation to Association and Normal Correlation" in 1904. Nowadays, these tables are used in survey research, business intelligence, engineering and scientific research. They provide a basic picture of the interrelationships between two variables and can help find interactions between them.

Continuing from our discussion above the four probable situations can be stated as:
1. The Actual state is Positive and Tests are Positive—Correct measurement—True Positive.
2. The Actual state is Negative and Tests are Negative—Correct measurement—True Negative.
3. The Actual state is Positive but tests are Negative—Incorrect—False Negative (The test results are negative, which is false hence denoted as False Negative)
4. The Actual state is Negative but tests are positive-Incorrect- False Positive (The test results are positive, which is false hence denoted as False Positive)

From the preceding contingency table, various important statistical paradigms applied in medicine can be derived.

Various biostatistic paradigms derived from analysing validity in dichotomous subjects:

		Actual (Referenced) condition		Marginal total		
		Positive	Negative			
Test results	Positive	Correct-true positive (TP)	Incorrect-false positive (FP) (Type I error)	(TP+FP)	Positive predictive value (PPV) $= \dfrac{(TP)}{(TP+FP)}$	False discovery Rate (FDR) $= \dfrac{(FP)}{(TP+FP)}$
	Negative	Incorrect-false negative (FN) (Type II Error)	Correct-true negative (TN)	(FN+TN)	False omission rate (FOR) $= \dfrac{(FN)}{(TN+FN)}$	Negative predictive value (NPV) $= \dfrac{(TN)}{(TN+FN)}$
Marginal total		(TP+FN)	(TN +FP)	(TP+FP+ FN+TN)		
		Sensitivity $= \dfrac{(TP)}{(TP+FN)}$	Specificity $= \dfrac{(TN)}{(TN+FP)}$		Accuracy $= \dfrac{(TP+TN)}{(TP+FP+FN+TN)}$	

Contd...

Contd...

False negative rate $(FNR) = \dfrac{(FN)}{(TP+FN)}$	False positive rate $(FPR) = \dfrac{(FP)}{(TN+FP)}$
Positive likelihood ratio $(LR+) = \dfrac{\dfrac{(TP)}{(TP+FN)}}{\dfrac{(FP)}{(TN+FP)}}$	Negative likelihood ratio $(LR-) = \dfrac{\dfrac{(FN)}{(TP+FN)}}{\dfrac{(TN)}{(TN+FP)}}$
Odds ratio (OR +) = $\dfrac{(LR+)}{(LR-)} = \dfrac{(TP)(TN)}{(FP)(FN)}$	

The concepts of Sensitivity and Specificity are universally used in diagnostic tests. The basic theory behind research methods for hypothesis testing also derives from contingency tables. All aspects of the rubrics are chopped and brought together for the understanding of practitioners. Of course, they are discussed differently and at greater lengths in community medicine texts. This is a concentrated abstract and the reader can refer to the previously mentioned texts for further reading.

Sensitivity (True Positive or Recall Rate)

A sensitive test detects the proportion of true cases. It is the percentage of actual positives which are identified as positive by the test measurement, i.e. this statistic measures true positive. Thus, sensitivity is

$$\text{Sensitivity} = \frac{\Sigma \text{True Positive}}{\Sigma \text{Condition Positive}} = \frac{\Sigma TP}{\Sigma TP + FN}$$

It measures the ability of the test to identify presence of a condition correctly. It refers to the probability of getting a positive test result given that patient has the medical condition. It is the ability to detect a case. High sensitivity test will have low type II error rate.

Specificity (True Negative Test)

Specificity is the percentage of negatives which are identified as negative, i.e. these statistic measures true negatives.

$$\text{Specificity} = \frac{\Sigma \text{True Negative}}{\Sigma \text{Condition Negative}} = \frac{\Sigma TN}{\Sigma TN + FP}$$

It measures the ability of a test to exclude a condition correctly. It is probability of a negative test given that the subject is healthy. It is also known as 'ability to detect a non-case.' A test with high specificity will have a low Type I error rate.

Sensitivity versus Specificity

From above definitions:
a. If a procedure correctly identifies 46 out of 50 persons having medical condition/ characteristic (True Positives), its Sensitivity = 46/50 = 0.92 or 92%.

b. If this procedure correctly identifies 43 out of 50 persons as not having any medical condition/characteristics (True Negatives), its Specificity = 43/50 = 0.86 or 86%

If the criteria for identification are made stricter, it may result in lesser False Positive but the test may become less sensitive. Similarly relaxation of criteria may reduce False Negatives, but renders the test less specific.

The bigger limitation of sensitivity and specificity is the fact that they do not take in account the correct identification (true positive or true negatives correctly identified), which may happen purely due to random behavior of variable.

Positive Predictive Value (Precision)

Positive predictive value (PPV) is the probability that subjects testing positive on the screening test truly have the condition.

$$PPV/Precision = \frac{\Sigma \text{ True Positive}}{\Sigma \text{ Test Outcome Positive}}$$

$$= \frac{\Sigma TP}{\Sigma TP + FP}$$

False Discovery Rate = FDR
It is the corollary of the above. Those who are falsely test positive out of all testing positive.

$$FDR = \frac{\Sigma \text{ Total Falsely Positive}}{\Sigma \text{ Test Outcome Positive}}$$

$$= \frac{\Sigma TP}{\Sigma TP + FP}$$

$$= 1 - \text{Precision/PPV}$$

Negative Predictive Value

Negative predictive value (NPV) is the probability that subjects with a negative screening test truly do not have the condition.

$$NPV = \frac{\Sigma \text{ True Negative}}{\Sigma \text{ Test Outcome Negative}}$$

$$= \frac{\Sigma TN}{\Sigma TN + FN}$$

False Omission Rate = FOR

$$FOR = \frac{\Sigma \text{ False Negative}}{\Sigma \text{ Test Outcome Negative}}$$

$$= \frac{\Sigma FN}{\Sigma FN + TN}$$

Accuracy (ACC)

This is not covered in standard epidemiology texts as it is more of a statistical concept. This refers to the proportion of sum of all true positive and true negative tests to the total tests done for the condition. It reflects the correctly identified truth in the study.

$$\text{Accuracy} = \frac{\Sigma \text{ True tests positive and negative}}{\Sigma \text{ All tests or population}}$$

$$= \frac{\Sigma \text{ True Positive + True Negative}}{\Sigma \text{ All tests or population defined in out matrix}}$$

$$= \frac{\Sigma \text{ TP + TN}}{\Sigma \text{ TP + FN + TN + FP}}$$

Fallout, False Positive Rate

$$\text{False positive rate (FPR)} = \frac{\Sigma \text{ False positives}}{\Sigma \text{ Condition negative}}$$

$$= \frac{\Sigma \text{ FP}}{\Sigma \text{ FP + TN}}$$

False Negative Rate (FNR)

$$\text{False negative rate (FNR)} = \frac{\Sigma \text{ False negative}}{\Sigma \text{ Condition positive}}$$

$$= \frac{\Sigma \text{ FN}}{\Sigma \text{ FN + TP}}$$

Positive Likelihood Ratio (LR+)

It is the probability of a person with the condition testing positive divided by the probability of a person without the condition testing positive. It has a relevance for positive test only. A likelihood ratio > 1 (more than 1) indicates that the test result is likely to be associated with the condition. Usually useful diagnostic tests will have positive likelihood ratios which will be much higher than unity. A likelihood ratio of <1 (less than 1) shows a negative connotation showing that it is more likely to be associated with absence of the condition. Tests in which the likelihood ratios are 1 or close to 1 will obviously have little use practically. When applied in clinical practice the pretest probability is multiplied by the LR to get the post-test probability of the condition.

$$LR+ = \frac{\text{True Positive Rate}}{\text{False Positive Rate}}$$

$$= \frac{\Sigma \text{ TP } \Sigma \text{FP + TN}}{\Sigma \text{ TP + FN } \Sigma \text{ FP}}$$

The rule of the thumb is that LR of 1, 2, 5 and 10 represent 0, 15, 30 and 45 percent change in pre-test probability. In other words the following table is true.

Likelihood ratio	Change in post-test probability
0–1	Decrease the probability of presence of the condition
0.1	– 45%
0.2	– 30%
0.5	– 15%
1	0% or no change
>1	Increase the probability of presence of the condition
2	15%
5	30%
10	45%

Since mostly sensitivity and specificity is what most of the medical practitioners are usually aware of, it is well to remember that there is an important relationship that exists which allows LR+ calculations.

$$LR+ = \frac{Sensitivity}{1 - specificity}$$

Negative Likelihood Ratio (LR–)

$$LR = \frac{FNR}{TNR} = \frac{False\ Negative\ Rate}{True\ Negative\ Rate}$$

$$= \frac{\Sigma FN}{\Sigma TP + FN} \frac{\Sigma TN + FP}{\Sigma TN}$$

The same discussion as above holds. To put it simply, it is the probability of a person with the condition testing negative divided by the probability of a person without the condition testing negative. Similarly,

$$LR-- = \frac{1 - sensitivity}{specificity}$$

If this sounds intimidating for bedside practice, take heart. Most of the research suggests that physicians do not use these calculations in clinical practice, as when many of them did so they often made errors. Practitioners presented with sensitivity and specificity, likelihood ratio or graphic of the likelihood ratio were likely to interpret the results not very differently using any of the methods. The importance of these ratios lies in the fact that they give us the Odds Ratio or the Cross Product Ratio. Remembering the ones for common tests in bedside practice and consulting the book or the net for the oddball is not a bad practice when looking at Likelihood Ratios.

Odds Ratio (OR+)

$$\text{OR of Diagnostic Test} = \frac{LR+}{LR-}$$

$$= \frac{\Sigma TP \, \Sigma FP + TN \, \Sigma TP + FN \, \Sigma TN}{\Sigma TP + FN \, \Sigma FP \, \Sigma FN \, \Sigma TN + FP}$$

$$= \frac{\Sigma TP \, \Sigma TN}{\Sigma FP \, \Sigma FN}$$

= Gross Product Ratio of Diagonals

Those who remember their community medicine tutorials would remember the diagonals in the epidemiology contingency tables on tests. Odds ratio is important because it is amenable to logistic regression and for the old reason of case control studies where it gave us a measure of occurrence of the outcome of interest (disease) given exposure to the variable of interest (trait or attribute). Thus, it helped determine whether a particular exposure was a risk factor and determine the magnitude of this effect on outcome in a given population. It is discussed in greater detail elsewhere.

To recapitulate, any good test should increase true negatives and true positives. Unfortunately that does not happen in the real world. False Positives tend to distort the proportion values of true positives, especially in cases of rare disease/conditions. In rare occurrences, True Positives are themselves limited and any False Positive will tend to increase their percentages to much higher values. In such situations, repeat measurements or additional tests help improve the study and research measurements. In clinical practice, in such situations it is better to revisit the findings before making the rare diagnosis. Remember the cliché, "When you diagnose rare diseases you will be rarely right, when you diagnose common diseases you will be commonly right."

ERRORS IN MEASUREMENT

As far as the laws of mathematics refer to reality, they are not certain; and as far as they are certain, they do not refer to reality

—*Albert Einstein*

All experiments involving measurement will have uncertainty as measurement cannot be entirely accurate and are termed as Experimental Errors. Finding an estimate about the deviation of unknown, true value of the quantity from the measured value is an important principle in research. The estimation of these deviations is an uncertainty analysis which historically has been referred as error analysis.

Experimental errors are different from common mistakes and miscalculations. Mistakes and miscalculations can be corrected or improved by repeating the experiment in a correct manner, whereas experimental error is the inherent error which cannot be eliminated by repeating the experiment. As stated previously, experimental errors are classified as 1) Systematic Errors and 2) Random Errors. The sum of these is the total error, i.e.

Total Error = Systematic Error + Random Error

Systematic Errors

Systematic errors arise from mistakes in the measurement scheme. These are repeated at each measurement in the same direction. Thus, the result differs from the correct result systematically in the same direction at each measurement, i.e. they affect the accuracy of the measurement.

Systematic errors are difficult to detect and analyze statistically, and if detected they result in refinement of measurement methods or techniques. Systematic errors do not average to zero no matter how many measurements are repeated. If the systematic error is discovered, a correction in the data is possible. For example, if the scale reads 2 mm more in each reading then that can be subtracted from the final results. There is no prescribed way to find systematic error and it needs to be worked out at the design stage itself. All the possible sources of error in a given measurement should be listed and then, if possible with small experiments the effect of each of them can be quantified.

These errors are broadly classified as:
- Instrumental, related to instrument or scale (instrument calibration, etc.). These include
  - *Zero error*: (Offset or zero setting error) where the instrument does not read zero when the quantity to be measured is zero.
  - Multiplier error: (Scale factor) error where the instrument consistently reads changes in the quantity to be greater or less than the actual changes.
- Observational, related to error in measurement method (Example: Parallax Error)
- Environmental, related to the surroundings of the experimental space. (Temperature affects viscosity or other physical parameters of many liquids, metals, etc.)
- Theoretical or due to the error in conceptual framework or assumptions in the analysis.

Systematic errors can be attempted to be controlled by Blinding. In single blind study the subjects are not aware of the intervention that they are getting. Blinding eliminates the differential systematic bias which may affect one group in a particular direction. Measuring unobtrusively and using dummy measurement or interventions can also reduce the measurement error. When people know they are being observed they may behave differently leading to a situation where one may get false results. So it is important that when observations are made they should not affect the normal working.

Apart from above mentioned classification, Systematic Errors or Bias, especially in behavioral or epidemiological studies do creep in due to observer, subject, information and selection. The goal of a good experiment should be minimization of systematic errors to least value possible, as they affect the mean of the sample, and hence potentially more serious than random errors.

Random Errors

Random errors affect precision of the measurement and are caused by unknown and unpredictable variations in the measurement process. Random errors are two sided as they generally occur in either direction and are inherent in any process. Due to two-sided nature they offset or cancel out to some extent, and can be reduced by repeating the measurements. Sources of random errors cannot be always identified. Possible sources could be environmental (unpredictable fluctuations, mechanical vibrations of equipment or other nonmutable environmental factors), observational (where the reading falls between the least count leading to uncertainty in interpolating between the smallest divisions), or due to subject being measured being prone to random fluctuations. e.g. labile blood pressure.

Random errors are nonsystematic in nature as they contribute to the variability and do not influence the sample averages.

Random errors usually follow a Gaussian normal distribution and can be analyzed statistically. Reduction of random errors improves precision and can be done by increasing sample size, improving sampling procedures or even by using statistically stronger analytical methods and tests. In terms of sample size, it should be noted that the sample size should not be increased indefinitely as the error will reduce but will reach a limiting value. Beyond such a level the cost of increasing the sample size would not in proportion to the improvement in measurement.

Apart from the sources of both systematic and random errors, inaccuracies can come in reporting the observations too. The two concepts of significant figures and propagation of errors would highlight this.

Significant Figures

Researchers need to appropriately report the number of meaningful digits which imply the error in the measurement. The accepted convention is that only one uncertain digit is to be reported for a measurement. For example, if the estimated error is 0.04 m, the report of result would be 0.95 ± 0.04 m, not 0.948 ± 0.04 m. If the experimental instrument is not equipped to measure up to three decimal places but only up to two decimals, it does not make sense to report it to three decimal places as this value would not be replicable. It is always better that the mathematically calculated value is considered important to the level it can be measured by instrument, in terms of the least count.

Zero is a significant figure if the zero has a non-zero digit anywhere to its left, otherwise it is not. For example, 7.00 has 3 significant figures while 0.0007 has only one significant figure. Thus, 7.0005 has 5 significant figures. A number like 700 is not well defined as it can be represented as 7×10^2 with one significant figure or as 7.00×10^2 with 3 significant figures.

Absolute and Relative Errors

While reporting accuracy related errors, they can be reported as an absolute or a relative form.

Absolute error is the uncertainty in the measured quantity and has same units as the quantity itself. Mathematically it can written as

= Observed Value - True Value or $= X_0 - X_T$

Relative error (or fractional error) is the ratio of the absolute error to the quantity measured. The relative error is usually more significant than the absolute error. Relative errors are dimensionless. Mathematically

$$= \frac{\text{Observed Value} - \text{True Value}}{\text{Observed Value}} \text{ or } \frac{X_0 - X_T}{X_0}$$

Propagation of Errors

Initial measurements are subjected to mathematical and statistical calculations before meaningful information is developed. This involves combining the measured quantities with some formulae or functions or relationships. The errors in the measured quantities are random and in either direction leading to some cancelling out. But if the variables are independent (quantity being measured is larger than it really is while another quantity is still just as likely to be smaller or larger) then error theory shows that the uncertainty in a calculated result (the propagated error) surfaces and can be obtained as below:

The general formula for the propagated error is given by:

$$(\Delta f(x_1, x_2, K))^2 = \Sigma \left(\frac{\delta f}{\delta x_1} \right)^2 (\Delta x_1)^2$$

A simple corollary of the above is used more commonly to show the propagated errors in z due to errors in variables x and y. The errors in constants a, b and c are assumed to be negligible. Only for a few cases it is as shown below in order to introduce the researcher to the propagation concept.

Function	Propagated Error
$z = ax \pm b$	$\Delta z = a\Delta x$
$z = x \pm y$	$\Delta z = \sqrt{((\Delta x)^2 + (\Delta y)^2)}$
$z = cxy$	$\frac{\Delta z}{z} = \sqrt{\left(\left(\frac{\Delta x}{x}\right)^2 + \left(\frac{\Delta y}{y}\right)^2\right)}$

Blunders in Measurement: Short Note

A blunder is an outright mistake. It is usually not described in statistical texts. It is being described here because it is not too uncommon to come across such mistakes or 'bloopers'. Wrong value, misread scale, forgotten digit in recording all constitutes a blunder. They may or may not stick out like sore thumbs if we make multiple measurements. Another method of control is supervision or peer review where one person checks the work of another. The reason for mentioning this here is that blunders should be excluded in the data analysis. Otherwise the quality of interpretation is severely compromised. It is never too late to correct for a blunder.

Understanding of concepts of measurement, assessment of validity and error analysis allow the researcher to ensure that appropriate control is exercised while handling measurements. Further it helps in understanding the observations in a better manner thereby avoiding mistakes either while researching himself or evaluating research data of others.

Section 7

Other Statistical Concepts Frequently Used in Research

- Vital (Demographic) Statistics and Life Tables
- Time Series Analysis
- Interpolation, Extrapolation and Forecasting Methods

Section 7

CHAPTER 27

Vital (Demographic) Statistics and Life Tables

Demography is the scientific study of the population of humans, their social statistics such as births, deaths, fertility, etc. It studies the changes experienced by the human population in its composition, distribution, and various other parameters of interest. It can also be described as the study of the human population for its static and dynamic components. Static components include sex, race, etc. whereas dynamic components can include mortality, fertility, population growth, etc. Vital statistics as a branch of statistics, studies the population, its existing landscape and the patterns of demographic changes occurring in it.

In this chapter the word *population* and other demographic variables would imply human population and variables, unless otherwise specified.

Popularly known as Vital Statistics, because it studies the vital events, like birth, marriage, death, etc. in the life of the population, it is a challenging subject due to it constant dynamic behavior, its size and its direct influence on the lives of millions of people worldwide, as many governmental welfare and health schemes are based on the vital statistics of the population. Due to importance of the demographic studies, international bodies such as World Health Organization (WHO), United Nations Organization (UNO), etc. are very well involved along with respective Governments, in this important field of human statistics.

BASICS OF VITAL STATISTICS

The importance of vital statistics is immense, both at individual and Governmental level. Some of the uses include:
- Helpful in Public Administration and acts as a basis for Public Welfare and Health Schemes
- Starting point in retrospective epidemiological studies
- *Life table construction*: Basis of Life Insurance sector and other insurances too.
- Indicates the status of the population, cohorts and period studies.
- Allows futuristic population projections, estimates, etc.
- At individual level, as its offshoot-documents such as birth certificate, death certificate, etc. are legally necessary; many other activities which are to be carried out by the individual mandatorily require these documents.

The entire vital statistics is based on the challenging task of data collection, its correctness (verification) and timeliness, especially in a country like India. The census conducted every 10 years does provide great deal of insights but cannot be the only source for collecting the vitals. Further, due to dynamics of the subject, the gap of ten years forces the researchers to use forecasting, interpolation and extrapolation techniques for the data of intervening years. The sources of data for demographic studies, in India are:

Sources of Data

The main demographics pertain to birth, death, mortality and reproductive rates of the populations; hence the main sources for data for these are being discussed here. Lot of other specific demographics pertaining to human population is possible, but at this discussion level, the main concern is restricted to certain vital statistics and not every statistics of the human population. Accordingly in India, the main sources of the data are:

- *Civil registration system*: Certain vital events have to be regularly and mandatorily recorded on their occurrence. These include births, deaths, marriage, divorce, etc. The compulsory recording is mandated through the law, in order to provide legal status to the records and certificates issued from this system. These are accomplished through the civil administrative set-up. In India the birth and death recordings are covered under *The Registration of Births and Deaths Act of 1969*.
- *National Sample Survey Organization (NSSO)*: In 1961 census, NSSO tried to provide estimates of birth and death rates, through one time retrospective recall survey. These estimates were found to be unreliable, that alternative mechanism was required. Though still there are many other demographics which are dealt through NSSO sample surveys.
- *Sample registration system (SRS)*: This system was established in 1960's, and works on a dual mechanism. There exists a system of continuous recording of births and deaths at localized (village) level part-time recorder, followed by six monthly retrospective surveys, which verifies the events at field level. These provide a reasonable level of reliable data. Though as per the Act such events are to be compulsorily reported, but reporting falls short due to various reasons. The births and deaths, occurring in hospitals do get reported, but events occurring outside the fold of such organized set-ups, do not get fully reported. Hence gaps exist between the data collected through this system and the data available through the Civil Registration Process, with the figures from registration being lower than SRS. The difference between Registration figures and SRS has considerably narrowed down as the literacy rates have improved and more awareness has been created by the Indian Governmental and Government machinery. Between the birth rates and death rates, the difference is much more in the death rates. Collectively with the census, SRS offer best available tools for estimating the various demographic rates. Many insurance organizations use life tables on sample registration system.
- *Analytical data*: As actual surveys are not possible every year, lot of data is generated by using statistical analytical tools. These are created by specific private or Government organizations and for their specific purposes.

The above mentioned sources of data are supplemented through various other information sources, like different ministries which allow the researchers to get more details about the population statistics.

Notations

The general terminology used in vital statistics is as given below:

- P = Population size
- P_x = Population aged 'x'
- P_o = Population at the start of period
- P_t = Population at any given point of time
- B = Births
- D = Deaths
- E = Emigration
- I = Immigration
- M = Mortality

Measures of Vital Statistics

From the above notations, the Population of a given geographical area is given by the following simplistic equation.

$$P_t = P_o + (B - D) + (E - I)$$

The values of B, D, E and I pertain to the time duration encapsulated between P_t and P_o.

The data about population is evaluated in many ways. Few underlying concepts are necessary for appropriate understanding of the different measures. These concepts are:

Absolute and Relative Measures

Certain values of the population are always considered in absolute number, like the population itself; whereas most of the measures are relative in nature and denoted as 'per 1000', 'per 100,000', etc. The advantage with this standard notation is that it is easy to communicate, easy to understand and does not allow the scope of confusion, which can creep if fractions, which have been solved to different levels by different people (3/39 =1/13= 21/273, etc.), are used.

Rate, Ratio and Proportion

In the measurement of demographics, the values are mentioned as a rate (For example, Crude Birth Rate) or Ratio (Sex ratio) or proportion.

Ratios are used for static measurements and indicate measurement at a given *point of time*. It expresses the relationship in magnitude between two factors by dividing one with the other.

In comparison 'Rates' (in vital statistics) may be considered as special case of ratios indicating population is changing all the time, i.e. dynamic or sense of movement (difference between growth ratio and growth rate). In other words, it denotes the occurrence of the vital event in a population during a given *period* of time. Rates could be *Crude* or *Specific*, with crude rates related to entire population and specific rates being specific to specific population groups.

Proportion is the relationship in magnitude of one to the whole. For example, in a given sample of 10 people, 3 are Females and 7 are Men, the ratio of Men to Women would be 7:3; ratio of Women to Men would be 3:7, whereas proportion of Men in the sample would be 7/10. In proportion the numerator value gets included in the denominator.

With these basics, the measurements of vital statistics can be studied. For various measurements, formulae are provided and are self-explanatory, hence not detailed out.

Note:

1. Mean population during the period or year =

$$\left(\frac{\text{Population at start of given period or year} + \text{Population at end of given period or year}}{2} \right)$$

2. The formulae are stated for period, but most of them are calculated for a period—one year basis, accordingly, if applicable, these should always be considered for one year unless otherwise specified.

Fertility Rates

Crude Birth Rate

$$\text{Crude birth rate} = \frac{\text{Livebirths during the period}}{\text{Mean population during the period}} \times 1000$$

Crude birth rate in India as per 2011 census was 22.5 per 1000 people.

Specific Fertility Rate

Fertility: Fertility denotes the actual bearing of children, i.e. livebirths. Cases of still birth, fetal death, abortions, etc. are not included in fertility. It should not be confused with fecundity, which is a measure of physical ability to reproduce, whereas fertility is the actual reproduction of an offspring. On the opposite, infecundity relates to lack of capacity to reproduce.

Specific fertility rates are established, as women in all age groups do not have the same reproductive capacity. Each of these rates is for specific conditions, which should be kept in mind while dealing with these.

Specific Fertility Rate (Age Specific Fertility Rate)

$$= \frac{\text{Livebirths occurring to females of specified age groups during the period}}{\text{Mean female population in that specified age group during the period}} \times 1000$$

General Fertility Rate

The drawback of crude birth rate is that it wrongly assumes that the entire population of males and females of all age groups is capable of reproducing. To arrive at better results, general fertility rate is calculated which only considers livebirths in relation to female population in the reproductive age group. The reproductive age group is considered as 15 years to 49 years.

$$\text{General fertility rate} = \frac{\text{Livebirths during the period}}{\text{Mean female population between ages 15-49 during the period}} \times 1000$$

For example: General fertility rate (India, 2011) = 88 per 1000.

Total Fertility Rate

It is more specific than general fertility rate (GFR), which is *Number of births per 1000 women in the reproductive age*. Against this total fertility rate denotes the total number of children expected to be born to women through their reproductive span. This measure indicates fertility conditions in the population, as it provide the number of livebirths, females are expected to give during their entire reproductive years. It is the total sum of all the *age specific fertility rates*.

Total fertility rate = Σ Specific fertility rate for all age groups
Total fertility rate (India, 2011) = 2.6

Reproductive Rates

Gross Reproduction Rate

This measure considers the fact that only female births would later conceive and add to the population hence the formula is given as:

$$\text{Gross reproduction rate} = \frac{\text{Number of female births}}{\text{Total number of births}} \times \text{Total fertility rate}$$

It is a better measure of fertility than total fertility rate, as it considers only female births.
Gross reproduction rate (India 2011) = 1.2

Net Reproduction Rate

All female births are not expected to be able to survive till the end of their reproductive age group, due to incidence of deaths. Gross reproduction rate overlooks the mortality rate which is experienced by the population. Net reproduction rate is an improvement on gross reproduction rate in this aspect, and represents the replenishment rate of the population.

$$\text{Net reproduction rate} = \frac{\Sigma(\text{Number of female births} \times \text{survival rate})}{100}$$

Mortality Rates

Crude Death Rate

$$\text{Crude death rate} = \frac{\text{Number of deaths during the period}}{\text{Mean population during the period}} \times 1000$$

Crude death rate (India, 2011) = 7.3 per 1000 population

Age Specific Death Rate

$$\text{Age specific death rate} = \frac{\text{Number of deaths in the specific age group}}{\text{Mean population during the period of that specific age group}} \times 1000$$

Specific death rate can be calculated for any specific parameter or cause of the population, if the deaths specific to the parameter and the total population related to that parameter is known. For example, sex specific death rate, etc.

Infant Mortality Rate

It is a case of age specific death rate, where the specific age is <1 year, i.e. number of livebirths which are not able to survive till the age of 1. Though it is called as Rate, but if the formula is seen, it is actually a ratio.

$$\text{Infant mortality rate} = \frac{\text{Total number of deaths in population of age <1 year during the period}}{\text{Total number of livebirth during the period}} \times 1000$$

Infant mortality rate (India, 2011) = 48 per 1000 livebirths.

Neonatal Death Rate

As a sub group of infant mortality rate, in neonatal death rate, deaths occurring within 1 month of livebirth are considered.

$$\text{Neonatal mortality rate} = \frac{\text{Total number of deaths in population of age <1 month during the period}}{\text{Total number of livebirths}} \times 1000$$

Post Neonatal Mortality Rate

This measure considers the death of an infant after 1st month and before 1 year.

$$\text{Post neonatal mortality rate} = \frac{\text{Total number of deaths in population age group 1 month} - 1 \text{ year}}{\text{Total number of livebirths in the given period}} \times 1000$$

Infant, Neonatal, perinatal mortality rates are given due importance, as the chances of deaths during this period are similar to old age deaths. A high rate possibly indicates desirable improvement in the efficiency of the health programs and medical facilities.

In certain countries even fetal deaths are measured. The categories under which fetal deaths are classified are:
a. Early fetal death pertains to <20 weeks of pregnancy,
b. Intermediate fetal death relates to death between 20th week and up to 28th week (both inclusive), and

c. Late fetal death accounts for deaths in and after 28th week. On basis of this, fetal death ratio is given as $\dfrac{\text{Fetal deaths}}{\text{Number of births}}$

Maternal Mortality Rate

This measure signifies the maternal death due to maternity condition and calculated as per the below given formula. Maternal condition death is considered, if the death occurs during pregnancy or within 6 weeks of delivery and does not include death caused due to other factors like accident or accidental.

$$\text{Maternal mortality rate} = \dfrac{\text{Number of death due to maternity condition}}{\text{Total female population in reproductive age group 15-49}}$$

Maternal Mortality Ratio

This is different from the maternal mortality rate, as it expressed as the measure of deaths due to maternal cause per 100,000 births

Maternal mortality ratio = Death due to maternity causes per 100,000 livebirths.
Maternal mortality ratio (India, 2009) = 212 per 100,000 livebirths.

Other Rates/Ratios/Measures

Natural Increase Rate

It measures the natural increase in population, i.e. due to births and deaths; and excludes immigration and emigration from the calculation.

$$\text{Natural increase rate} = \dfrac{(\text{Number of birth} - \text{Number of deaths}) \text{ in a given period}}{\text{Mean population of that period}} \times 1000$$

Net Migration Rate

It measures the net effect of migration on the population. Migration is the geographical movement of the people and involves shift of residence.

$$\text{Net migration rate} = \dfrac{(\text{Immigration} - \text{emigration}) \text{ in a given period}}{\text{Mean population of that period}} \times 1000$$

As a subset of this, immigration rate is calculated as:

$$\text{Immigration rate} = \dfrac{\text{Total number of immigration/arrivals in a given period}}{\text{Mean population of that period}} \times 1000$$

$$\text{Emigration rate} = \dfrac{(\text{Number of emigration/departures}) \text{ in a given period}}{\text{Mean population of that period}} \times 1000$$

Vital Index of the Population

It is ratio of the livebirths and the deaths in the population during the period. Population increase or decrease can be inferred from it. If the ratio is >1, it implies population is increasing, if it is <1, then the population is experiencing decline and if it is =1, then population is stagnant. This measure is for natural increase or decrease as it does not consider migration. Further a consistent ratio of = 1 continuously for many years, does not imply that population would continue to remain stagnant. The change could be due to the possibility of alteration of survival rates, or the fertility rates might get altered as it is possible that more male children are being born, etc.

$$\text{Vital index} = \frac{\text{Total number of livebirths in the given period}}{\text{Total number of deaths in the given period}}$$

Age Dependency Ratio

Any given set of population would have economically productive population and nonproductive population. This measure indicates how much population is dependent. The age groups of <15 years and >64 years are considered as dependent age groups, i.e. population between 15 years and 64 years is considered as economically productive population.

$$\text{Age dependency ratio} = \frac{\text{Population age <15 years + Population age >64 years}}{\text{Population between 15 years and 64 years of age}} \times 1000$$

In the numerator, if only population below 15 years is being considered, then it is called as *Child Dependency Ratio*, and if only population age of more than 64 years is considered, then it is called as *Old Age Dependency Ratio*.

Sex Ratio

It is the number of females in a population per 1000 males. In Indian context the sex ratio was more than 950 before independence and fell down to 927 in 1991, though it has slightly increased after that to 940 in 2011.

$$\text{Sex ratio} = \frac{\text{Total number of female population}}{\text{Total number of male population}} \times 1000$$

Literacy Rates

Literacy rates are the measure of literates in the population. An increase in literacy tends to improve other parameters of the demography. Literacy rate can be crude literacy rates and effective literacy rates. Crude literacy rates consider the number of literates in the total population, whereas effective literacy rates consider the same, but for population with age higher than 7 years.

$$\text{Crude literacy rates} = \frac{\text{Total number of literates}}{\text{Total population}} \times 1000$$

$$\text{Effective literacy rate} = \frac{\text{Total number of literates in 7 years + age group}}{\text{Total population in 7 + years age group}} \times 1000$$

Life Expectancy

It is the measure of age that a newborn is expected to survive till. It does not signify that every newborn today will reach the life expectancy age, but only indicates on an average a new born is expected to survive till this age. This will be dealt in detail while discussing Life Table.

Population Change

As stated in the initial part of the topic, the current population can be denoted as:
$$P_t = P_o + (B - D) + (E - I)$$
From this equation, population added per unit of time can be denoted as:
$$\text{Population growth} = \frac{P_t - P_o}{t} = (\text{say}) f$$

If the population is experiencing linear growth, then it can be denoted as $P_t = P_o + ft$. But this condition, in which the population continues growing at equal amounts (an absolute value), does not occur frequently. Population growth rate tends to behave in exponential manner. In such a situation the exponential growth can be denoted by the exponential equation $P_t = P_o e^{rt}$, where t is time, and r = growth rate.

In this exponential equation if $P_t = 2P_o$ i.e. doubling of population, then at various assumed growth rates, the time taken for the population to double itself can be calculated. Accordingly, at a growth rate of 1.0, the population doubles in nearly 70 years, at growth rate of 2.0 it takes approx 35 years, and at a growth rate of 3.0, it takes on 23 years for the population to double.

The demographic statistics are crucial and indicate to a large extent the development of the society. In developed economies, the life expectancy is relatively higher, mortality rates are generally lower, and sex ratio's closer to being equal, the literacy rates very high and nonproductive population being low. The biggest challenge developed economies face is with the fertility rates going down. Certain economies are even experiencing negative population growth and old age dependency ratio is on rise due to aging population. Due to use of statistical techniques, these have been identified and the respective governments are working towards rectifying the situation.

Example: For the following sample values, calculate specific fertility rate, general fertility rate, total fertility rate, gross reproduction rate and net reproduction rate. (Hypothetical values):

Section 7: Other Statistical Concepts Frequently Used in Research

Age group	Number of women	Total number of livebirths	Of which, female birth	Survival rate of female births
15–24	183874	12546	6210	0.934
25–34	231543	35498	17430	0.898
35–44	247651	24951	12124	0.854
45–49	235564	4287	2214	0.832

Solution: Calculating the various results, through tabulation and calculation as shown below:

Age group	Number of women	Total number of bivebirths	Specific fertility rate
(i)	(ii)	(iii)	(iv)
15–24	183874	12546	68.23
25–34	231543	35498	153.31
35–44	247651	24951	100.75
45–49	235564	4287	18.20
Total	898632	77282	

Specific fertility rate (SFR) = (12546÷183874) × 1000 = 68.23 and so on
General fertility rate =

$$\frac{\text{Total number of livebirths}}{\text{Total number of women in age group } 15-49} \times 1000$$

$$= \frac{77282}{898632} \times 1000 = 85.999$$

Total fertility rate: For calculating the total fertility rate, the table is reproduced: SFR is multiplied by the interval, because the fertility rate for the age group is in *per year* basis and the women would experience 'i' number of years, before the specific fertility changes.

Age	Specific fertility rate (SFR)	Number of years	SFR × i (Interval of age group)
(i)	(ii)	(iii)	(iv)
15	0	0	0
25	68.23	10	682.3
35	153.31	10	1533.1
45	100.75	10	1007.5
50	18.20	5	91.00
Total			3313.9

Total fertility rate (TFR) = Σ(SFR × i)/1000 = 3313.9/1000 = 3.31
Gross reproduction rate =

$$\frac{\text{Number of female births}}{\text{Total births}} \times \text{TFR} = \frac{37978}{77282} \times 3.31 = 1.63$$

Net reproduction rate:

Age group	Number of women	Total number of livebirths	Of which, female-birth	Survival rate of female-births	Specific fertility rate per women	SFR x survival rate
15–24	183874	12546	6210	0.934	0.034	0.0315
25–34	231543	35498	17430	0.898	0.075	0.0676
35–44	247651	24951	12124	0.854	0.049	0.0418
45–49	235564	4287	2214	0.832	0.009	0.0078
'i' = 10					Total	0.1488

Only female births are considered as only they can produce offsprings, therefore in this case:

$$SFR = \frac{\text{Female births in the age group}}{\text{Total women population in that age group}}$$

$$= \frac{6210}{183874} = 0.034$$

and therefore net reproduction rate = Σ(SFR × Survival rate) × i = 0.1488 × 10 = 1.488.

Lexis Diagram

Lexis Diagrams are graphical representation of certain population parameters. These graphs have time on x-axis and age on y-axis as a standard notation. The lines called as Life Lines, start from x-axis and move continuously in upwards diagonal direction (left to right) signifying the changing age along with changing time (years). The line ends in a cross 'x' signifying death at an age/time point. A line ending is dark filled circle, indicates exit from the population (emigration) and a hollow outlined circle at the starting point indicates immigration. These help in easy understanding of the certain parameters. Being a specific actuarial statistics concept, the idea is to mention the concept here so that the reader is aware of it in general.

LIFE TABLES

Life tables are a comprehensive tool to describe parameters like mortality, life expectancy, etc. of a given population. These are used widely in Life Insurance Industry and though they are complex looking but are easy to understand. A life table summarizes the mortality experience, survivorship and provides a concise measure of life expectancy at varying ages.

Before the discussion on Life Tables and their construction is considered, certain terms and topics need to be broached:

Cohort group: In vital statistics, cohort group is a group of persons, who experience the same statistical events in their lives. Consider a group of children born in year x. These set of children will pass through the same mortality rates through their lives and have same life expectancy at the time of birth, but it does not imply that all of them will survive till that age.

Mortality rates: Measures such as crude death rate, age-specific death rate, infant mortality rate and maternal mortality rate have already been discussed in the earlier part of this chapter. Building on the same, mortality can also be viewed as *Failure of Life* or *Failure to Survive*. As an offshoot of the Life Tables, these tables can be built for other subjects (nonhuman population, etc.), where such *failure* occurs and measured in age or consistent time interval. This is subject to condition that starting point (birth, success) and ending point (death, failure) are clearly and without any ambiguity, can be established on specific points of time (just like age).

Apart from the life tables for the entire population, life tables are also constructed separately for both male and female population. Life tables, assume that the data available on births, deaths and population are accurate and the population would continue experiencing the same mortality rates. Beyond these, life tables can be constructed for specific purpose, in which case the deaths caused due to the specific parameter is considered.

Types of Life Tables

There are two types of Life Tables:

Cohort Life Table (Also called Generation Life Table or Dynamic Life Table)

This is used to represent the overall mortality rates of a certain cohort population. These are developed by following a birth cohort through their entire life time, until all the persons of group have died. Based on the data collected all along the actual lives of the group, the mortality rates are calculated, and life tables generated. This implies that if the last member survives till the age of 100 years, then these tables would be generated after 100 years of birth cohort, which is not a practical situation. Further at the time of constructing these tables, they will be rendered as past, historical data, and with a realistic possibility of alterations of the demographic rates during this long duration. Therefore these life tables will be at best of limited use and hence these cohort tables are rarely used.

Period Life Table (Also called Static Life Table)

These tables are based on the mortality experience of a cohort, based on the current age-specific mortality rates for the various age-groups. Instead of following the lives of the cohort as in cohort life table, period life tables assume that the those being born today, will experience the current age-specific mortality rates, as being experienced by the different ages (or age groups) today, through their life span, and accordingly will die in concurrence with these. To have standard notation, the life tables are calculated on the basis of 100,000 newborns as the starting population and thereafter the population keeps on reducing as per the mortality rates until none of these survives. It is generally assumed that any given life table is *Period life table*, unless otherwise mentioned.

Life Table Construction

The life tables have standard notation/column headings, etc. These along with the description are given below:

S. No.	Column heading	Description
1	$x, x+n$	x denotes (Completed) age (0, 1, 2, ...), or age interval x to $x+n$. If all values from 0 to 100 are stated, then the interval is 1 year. If age groups are given then the interval is the difference between start year and end year of the age group.
2	$_nq_x$	The proportion of population at each age that is alive at the start of the year but die before end of the year, i.e. the conditional probability of dying in the interval 'x' and '$x+n$' after surviving till age 'x'. It is calculated from the actual mortality rates of the population and is the most critical column of the table. All other columns are calculated from values in this column.
3	l_x	Denotes the number of people alive at the start of the age interval. First entry for age 0 is called the *radix*. At entry age 0, it denotes the size of the population at start, usually taken as 100,000. Provides the probability of survival for age 'x' which is equal to l_x/l_{x-n}, if Radix is set to 1. If radix is in absolute number then this denotes the number of survivors at the given exact date.
4	$_nd_x$	Denotes number of deaths between the age interval 'x' and '$x+n$'. Adding up all these values will result in the radix value as everyone dies with time. If the values is equated to 1, then it reflects the probability of dying between age interval 'x' and '$x+n$.'
5	L_x	Denotes number of persons lived between exact ages 'x' and '$x+n$'. It is assumed that deaths occur throughout the year (evenly spread) therefore average population for the year is considered and calculated as $\left(l_x - \dfrac{d_x}{2}\right)$. If age group is being considered then the value needs to be multiplied by the age interval to arrive at total person-years.
6	T_x	Denotes the total number of years lived by all the persons collectively of a particular age/age group, i.e. cumulative sum of all L_x. Differently stating, it is the total number of years, which the group is collectively expected to live from the age 'x' until all of them die.
7	e_x	Denotes the average remaining lifetime of a person who is alive at age 'x'. At e_0 the value reflects the *life expectancy at birth* or *number of years a person is expected to live*, and which is often quoted in the census results. $e_x = T_x/l_x$.

Other Important Points

In the notations above, there is mention of presubscript and postsubscript. The presubscript in $_nd_x$, 'n' denotes the width of the interval and postsubscript 'x' denotes the starting point of the interval. If data pertains to all ages from 1 to 100 with 1 year as interval then presubscript may be omitted.

The notations without presubscript are exact ages, hence no presubscript is necessary.

The assumptions while constructing life table include: (a) no migration effect on the population, (b) mortality rates to be experienced are same as being experienced today by the population, and (c) underlying data is accurate.

The Life tables contain all the years from 0 to 100 or 105/110 on a yearly basis, or in age group basis. Many applications of the life tables rely on age groups instead of all ages. This can be done without much loss of accuracy, but it is always preferred that the age group 0–1 should be treated separately. Next age group should be age 1–5, and thereafter at equal intervals like 5–10, 11–15 and so on. This is so as the mortality rate in 1st year after birth is relatively very high, and drops considerably in the second year and thereafter keeps on reducing albeit slowly, until starts increasing again in aged population.

Life Tables: Exhibit and Calculations Explained

The raw data on which the period life tables are based are the population at each age and age specific deaths. Based on these, the mortality rates are calculated. It is assumed that the observed age-specific mortality rate will remain same and will still be experienced by the population, and on this basis the life tables are created. To facilitate comparison and standard understanding, the population is taken as 100,000 to start with, i.e. radix = 100,000. The mortality rates if expressed in per 1000 basis are converted to per 100,000.

Given below is an example of a life table. Life tables which contain the details for all the individual years from 1 to 100+, with 100+ being the last age, i.e. x = 0, 1, 2, 3,… 100+, such life tables are usually called *complete life tables*. Life tables in which instead of all the ages, the ages are represented through groups, such life tables are called *abridged life tables*. For example, 'x' in such cases can have interval of 5 years like 6–10, 11–15, 16–20 and so on.

To explain the calculation, consider a small extract of hypothetical *Complete life table*.

Life Table Calculations—Noting's

n = 1 year (The interval in the above example is 1 year as no grouping has been done).

q_x = *Conditional probability*: It is assumed that the deaths occur uniformly throughout the year, but in real situation it does not happen uniformly. Therefore for calculating the probabilities the mid-point of the year is considered. a_x is the factor for proportional distribution of deaths and is assumed at 0.5. Only in the age bracket 0–1, this factor is higher than 1, because within this bracket most of the deaths occur within the 1st month itself. Assuming (for this example only) that 50% deaths occur in the first month and rest of the deaths, are evenly spread out, the weighted mean is 0.7272, which is considered in this example. This factor can be fine tuned with the knowledge of infant mortality rates, etc. Sometimes it is considered as high as 0.85 for age 0–1 and 0.60 for age groups 1–5. This factor is more helpful when instead of all ages, age groups are being considered. In those cases the formula undergoes changes and is dealt separately.

Complete Life Table: Extract

The complete life table (Extract) is listed in following table:

Age	Proportional distribution of deaths during interval	Total population	Total deaths	Age specific mortality rate	Conditional probability of dying in the interval	Number of persons alive at start of age interval	Number of persons dying during the age interval	Equated to 100,000 Number of years lived		Average (Mean) number of years life remaining at start of interval
								Persons alive during the age interval 'x' and '$x+n$'	Person-years lived after age 'x' by the cohort	
x	a_x	P_x	D_x	$m_x = \dfrac{D_x}{P_x}$	$q_x = \dfrac{m_x}{1+a_x m_x}$	$l_{x+1} = l_x - d_x$	$d_x = l_x m_x$	$L_x = l_x - a_x d_x$	$T_x = L_x + \ldots$	$e_x = \dfrac{T_x}{l_x}$
0	0.72	1298734	74638	0.05747	0.055164	100000	5747	95690	7378549	73.79
1	0.50	1758921	38734	0.02202	0.021782	94253	2076	93215	6898654	73.19
2	0.50	1687548	18845	0.01117	0.011105	92177	1029	91663	6717532	72.88
3	0.50	1599721	9765	0.00610	0.006086	91148	556	90870	6587435	72.27
—	—	—	—	—	—	—	—	—	—	—
—	—	—	—	—	—	—	—	—	—	—
98	0.50	274789	142254	0.51768	0.411239	11860	6140	8790	15823	1.3
99+	0.50	89745	73000	0.81342	0.578241	5720	4653	7032	7032	1.2

L_x = All values are calculated as per the formula, except the last age, which is open ended. Being open ended, it is not possible to estimate how many years will the person remain alive in future. For this last group the value of L_x is estimated by using the equation $= l_x/m_x$.

T_x = Total number of years lived in the given age interval and all subsequent age intervals. This is calculated as $= L_x + L_{x+1} + ...$ until last age values of L. Can also be calculated as reverse $T_x = L_x + T_{x+1}$.

Abridged Life Table

The earlier example listed all the ages with $n = 1$. In lot of uses of life table, the ages can be clubbed together to form age groups, without losing any significant accuracy. Such tables are called abridged life tables. The construction of such tables is similar to the earlier example, with slight modifications to incorporate the age interval and its impact on the calculations. An example for the understanding is given in table on next page:

Formulae used in constructing the given life table are similar to previous life table. For ease of references all formulae are being given again:

x = Age; n = interval (different intervals for different age groups are valid).

a_x = indicates distribution of death within interval, death is assumed to be spread evenly except for age groups <1 and 1–4.

P_x = Total population of the age group and D_x = Total number of deaths pertaining to the age group.

$$m_x = \frac{D_x}{P_x}$$

$$q_x = \frac{n \times m_x}{(1 + n \times a_x \times m_x)}$$

(where n = interval, in above example three different intervals are there, i.e. 1, 4 and 5 years)

$$L_x = n(l_x - a_x d_x)$$

Chapter 27: Vital (Demographic) Statistics and Life Tables

x	a_x	P_x	D_x	m_x	q_x	l_x	d_x	L_x	T_x	e_x
<1	0.85	1543765	17645	0.011430	0.011320	100000	1132	99038	7230300	72.30
1–4	0.60	5653728	7648	0.001353	0.005393	98868	533	394192	7131262	72.13
5–9	0.50	8689645	3465	0.000399	0.001992	98335	196	491184	6737069	68.51
10–14	0.50	9356473	1900	0.000203	0.001015	98139	100	490446	6245885	63.64
15–19	0.50	9765437	2135	0.000219	0.001093	98039	107	489929	5755440	58.71
20–24	0.50	10276483	2654	0.000258	0.001290	97932	126	489345	5265511	53.77
25–29	0.50	10573822	3763	0.000356	0.001778	97806	174	488594	4776166	48.83
30–34	0.50	9745637	4021	0.000413	0.002061	97632	201	487657	4287571	43.92
35–39	0.50	10027649	8764	0.000874	0.004360	97431	425	486092	3799915	39.00
40–44	0.50	9653874	9654	0.001000	0.004988	97006	484	483820	3313823	34.16
45–49	0.50	9183784	12764	0.001390	0.006925	96522	668	480939	2830003	29.32
50–54	0.50	8769836	17543	0.002000	0.009952	95854	954	476883	2349064	24.51
55–59	0.50	7976453	43992	0.005515	0.027201	94900	2581	468045	1872180	19.73
60–64	0.50	5432876	76387	0.014060	0.067914	92318	6270	445917	1404135	15.21
65–69	0.50	4765329	135273	0.028387	0.132529	86049	11404	401733	958218	11.14
70–74	0.50	4087643	365489	0.089413	0.365389	74645	27274	305038	556484	7.46
75–79	0.50	3453678	543872	0.157476	0.564961	47370	26762	169946	251447	5.31
80+	0.50	2546832	643980	0.252855	1.000000	20608	20608	81501	81501	3.95

Calculations for l_x, T_x, e_x and d_x remain same as in previous example.

Considering mortality rates and other population parameters being dynamic, while Period Life Tables are static, these tables need to be updated at regular frequency, and for which the researchers, generally rely on the data sample registration system (SRS) for updating in between the census years. For many applications of life tables and for policymaking, it may not be possible to wait for the census results, hence SRS provides appropriate data for the intervening period.

Uses of Life Tables

Life tables have more applications than being just indicator of population characteristics. Specific life tables for specific causes of death or death among population having specific attributes can be constructed which have immense use in the insurance industry. In-depth analysis on mortality rates, impact on life expectancy at various ages, future population estimates and many more are possible through the use of life tables.

SURVIVAL ANALYSIS AND KAPLAN-MEIER CURVES

The understanding of life tables can be extended to survival analysis, which not only involves survival data relating to death, but considers occurrence of any event on a time scale. Such situations include time on one scale and any event which can be considered as a factor of time, like time interval in recurrence of disease, time taken for de-addiction, relapse of addiction, time taken for drug to be effective or time survived post an critical operative procedure. In short, survival analysis deals with the evaluation and analysis of the data on time taken for an event to occur and aims to establish models based on these.

Kaplan-Meier Curves are the graphical plots representing the survival probabilities, after duly accounting for censored entries and can be referred as *graphical method of describing survival characteristics*.

To explain the concept of survival analysis and establishing the Kaplan-Meier Curves, consider a hypothetical situation. A sample of 100 cancer patients is randomly selected and studied until death occurs. During the study information on some patients could not be tracked due to various reasons. The study reveals the following data. Starting month is considered as baseline time period, i.e. 0 and time of event is the time elapsed, in months from the starting time.

Time of event (Month)	Number of deaths	Cases with number information
0	0	0
1	1	0
2	0	3
5	8	0
6	0	1
9	10	0
10	18	1
12	14	3
14	16	4
17	19	2
Total	86	14

The months that are not appearing in the list, like 3rd, 4th, 7th, 8th etc. indicate neither an event (death) nor a case of *no information (censored)* was reported in those months.

Censoring: It is the term denoting the cases where the survival time (or death point) is not known accurately. This can be due to many reasons including patients withdrawing, drop outs, shifting to another location, missing count, etc. The reason could be anything, but the important aspect is, that it is not known for sure that those patients died (or the event occurred to them) at that time point.

From the above table, it can be inferred that at the start of 1st month, all 100 patients were surviving. At the start of 2nd month, 99 patients were surviving (because 1 patient died during the 1st month). At the start of 5th month, only 96 patients were known to be surviving as prior to that 1 patient had died and for 3 patients data was censored. Before the Kaplan-Meier curves are discussed, the graphical representation of the raw data also needs to be understood **(Fig. 27.1)**.

Time is plotted on the x-axis, and the *study start point* could be either shown as *y*-axis or a time point on *x*-axis (as shown above). If *y*-axis is taken as the start point, then all lines originate from *y*-axis and not from left of it, i.e. no entries with negative '*x*' value'. The end point (study end), is the point of time at which the data collection is stopped. The study start and end point are shown as dotted vertical lines.

Each plotted line on the graph, represents a patient or subject being followed up. The lines ending in '*x*' indicates that the event (death) has occurred for the subject at that point of time. The first line (from top) indicates that the subject was surviving at the time when the study ended. The end point shown as '//' indicates censored data i.e. no information was available about the subject/patient beyond that time point. A dotted line after the censored indicator, implies that after the study ended it was found out that the subject was surviving and survived till the time point marked '*x*'. The notation for censored data and event occurrence (death) shown as '*x*' and '//' here are generally used, but use of other symbols or indicators is also seen, like *filled circle* and *outlined circle*. Accordingly the legends on those graphs indicate such notations.

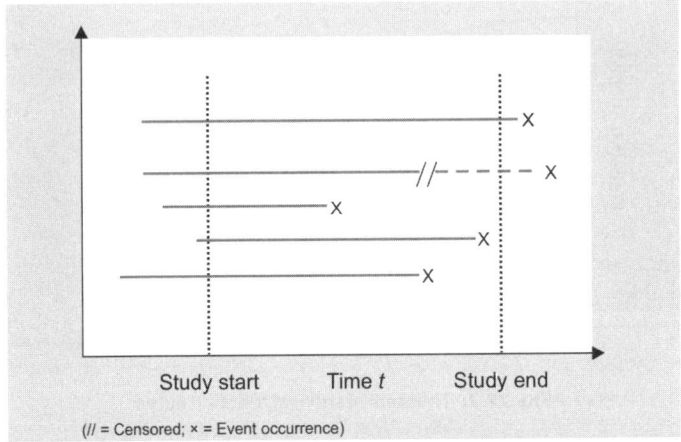

Fig. 27.1: Graphical representation of raw data of life lines

This representation method provides an initial bird's overview about the survival data.

For further analysis, of the example, the following points needs to be considered:
a. Survival Function S(t) = Probability that a subject survives longer than time 't'
b. Probability of surviving until any point = Cumulative probability of each preceding period.
c. The entire group (population or sample or study subjects) is at risk at all points of time.
d. The deaths (event) occurs at a specific point of time (say minutes or seconds), but using minutes or seconds is not practical to use for most of studies. Hence more realistic time frames are considered such as days, weeks, months or year, etc. Even though the event (death) can occur during any time within the period, but it is considered to have occurred at the specific time point 't'. This results is the Kaplan-Meier Curves becoming stepped type instead of a smooth curve. Theoretically the survival function curve **(Fig. 27.2)** appears as shown on next page, but in practice the Kaplan-Meier curve is stepped one, as shown in the following pages.

The Survival Analysis Notations

The notations used are:
t_0 = Failure time
m_0 = Number of failures (events or death)
q_0 = Number of censored cases
n_0 = Number of subjects survived till that point
$S(t)$ = Survival function = Probability $(T > t)$.

With this understanding, the earlier example can be reproduced and updated to conduct the survival analysis, as shown below, with explanations after the table.

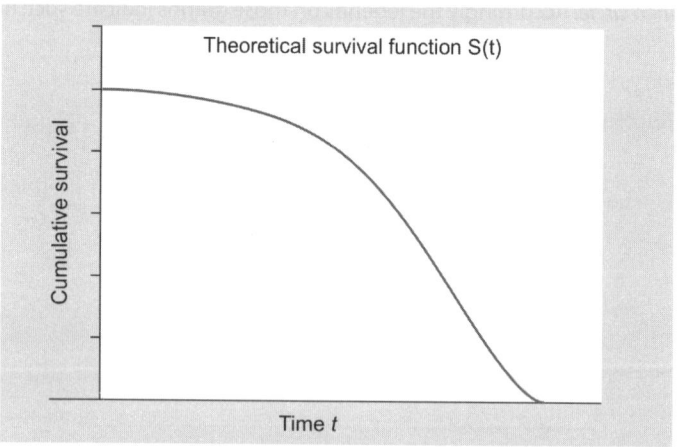

Fig. 27.2: Theoretical survival function curve

Chapter 27: Vital (Demographic) Statistics and Life Tables

Original given data			Survival analysis					
Time of event (Month)	Number of deaths	Cases with number information	t_0	m_0	q_0	n_0	$\dfrac{n_0 - m_0}{n_0}$	$S(t)$
0	0	0	0	0	0	100	1.0000	1.0000
1	1	0	1	1	0	99	0.9899	0.9899
2	0	3	2	0	3	96	1.0000	0.9899
5	8	0	5	8	0	88	0.9091	0.8999
6	0	1	6	0	1	87	1.0000	0.8999
9	10	0	9	10	0	77	0.8701	0.7830
10	18	1	10	18	1	58	0.6897	0.5400
12	14	3	12	14	3	41	0.6585	0.3556
14	16	4	14	16	4	21	0.2381	0.0847
18	19	2	18	19	2	0	0	0
Total	86	14	Total	86	14			

Explanation

t_0: In the original data *time of event in months* is irregular in nature. While *ordered data* has not been created in this example, but if required it can be done. Data ordering implies converting data into a regular time period. Regular time periods facilitate comparison if there are more than one curves on the same subject. If ordering is being done, by clubbing time of events into regular period, then suitable clubbing should be carried out in other columns too.

n_0: The number of subjects survived till that point of time. The number of deaths and censored numbers are happening during the period are deducted from the number of subjects who survived until the end of previous period. For example, the starting sample size was 100. In 1st month 1 died, hence the people surviving till the end of 1st month is (100 − 1) = 99 subjects. The censored cases are also deducted.

Method of treating of censored cases should be specifically noted from the above table. Censured cases reduce the number of subjects/patients surviving (denominator), but do not influence the numerator, where only those cases are considered for which the event has occurred.

Survival function $S(t)$—Probability of survival until that point is the cumulative probability until that time and is calculated as (example for value t_0 = 5) = 0.9091 × 0.9899 = 0.8999, i.e. survival rate for that period multiplied by the probability of survival until preceding period.

When these survival function value are plotted with time, it is known as Kaplan-Meier Curves **(Fig. 27.3)**. These plot help to estimate population curves from the sample observations. They indicate the censored data by crosses on the curve line.

These curves can be established in situations of drop-outs (censored), data collection at different intervals of time and can be compared for same study subjects. For comparing two or more Kaplan-Meier Curves, Log-Rank Test is most popular tool as a test of significance, where no assumption regarding the distribution of survival times is required. Log-Rank Test, tests for no difference in the probability of an event at any time point, between the populations.

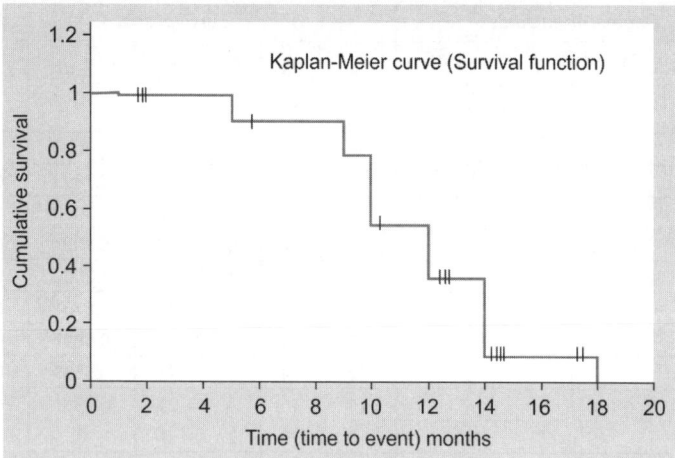

Fig. 27.3: Kaplan-Meier curve

The precision of survival function is higher when more censoring happens on the left side (initial period) as compared to right side (later time period), due to the simple reason that more death like events occur in the later periods. Though it depends a lot on subjects and events, too much censored data may not elicit proper results, hence needs to be monitored.

LOG-RANK TEST

Mantel-Haenszel Log-rank Test or simply Log-rank test, a non-parametric test is a test of significance between two or more Kaplan-Meier Curves or survival distributions of two samples. It is widely used in clinical trials to test the new treatments with old or controlled treatments. It compares the two or more survival curves under the null hypothesis that there is *no difference* in the two survival curves or survival times at all time points covered in the study. This test helps to determine the difference being statistically significant or not.

The comparison is done by comparing the hazard functions of the two groups at each observed event time. The test statistic is given as:

$$X^2_{logrank} = \frac{(O_1 - E_1)^2}{E_1} + \frac{(O_2 - E_2)^2}{E_2}$$

where O_1 and O_2 = Observed number of events in 1st and 2nd groups respectively at 't' time, and

E_1 and E_2 = Expected number of events in 1st and 2nd groups respectively and corresponding to the observed values at 't' time.

The expected values are calculated as:

$$E_i = \sum \frac{d_i\, n_i}{n}$$

where d_i = Number of deaths or occurrence of event recorded at time t.

n_i = Number of individuals under observation or surviving of that group at that point of time.

Chapter 27: Vital (Demographic) Statistics and Life Tables

n = Total number of individuals under observation or surviving in both groups combined.

The test statistic follows Chi-Square, i.e. χ^2 at $(N - 1)$ Degree of Freedom, where N = Number of groups being compared.

The below example would make the calculations and test application easy to understand.

Example: Given below is the (hypothetical) observed data on two groups on their survival during the 12 months period. In each group 25 individual were monitored. Test if the survival curves are same or not for the two samples.

Sample 1

Time	1	2	3	4	5	6	7	8	9	10	11	12
Deaths	1	0	2	1	3	1	0	1	1	2	4	1
Censored	0	0	0	0	1	0	1	0	0	1	2	0

Sample 2

Time	1	2	3	4	5	6	7	8	9	10	11	12
Deaths	0	1	2	4	1	3	1	2	0	0	2	2
Censored	0	0	1	0	1	0	0	0	0	0	1	1

Solution: Null Hypothesis H_0: The two survival curves have no difference.
For calculations, necessary table is to be prepared as shown:

Time t	Sample 1 O_1			Sample 2 O_2			Combined (1+2)			E_1	E_2
	n_1	d_1	c_1	n_2	d_2	c_2	n	d	c		
1	25	1	0	25	0	0	50	1	0	0.500	0.500
2	24	0	0	25	1	0	49	1	0	0.490	0.510
3	24	2	0	24	2	1	48	4	1	2.000	2.000
4	22	1	0	21	4	0	43	5	0	2.558	2.442
5	21	3	1	17	1	1	38	4	2	2.211	1.789
6	17	1	0	15	3	0	32	4	0	2.125	1.875
7	16	0	1	12	1	0	28	1	1	0.571	0.429
8	15	1	0	11	2	0	26	3	0	1.731	1.269
9	14	1	0	9	0	0	23	1	0	0.609	0.391
10	13	2	1	9	0	0	22	2	1	1.182	0.818
11	10	4	2	9	2	1	19	6	3	3.158	2.842
12	4	1	0	6	2	1	10	3	1	1.200	1.800
Total		17			18			35		18.334	16.666

Explanation on calculation table:
- The values in columns—time t, d_1, c_1, d_2 and c_2 are as given in the question above.
- d = Number of death in the time t (can be generalized as number to whom the event has occurred)
- c = Number of censored items in the time t

- Columns of n_1 and n_2 have been calculated based on the starting number of individuals being monitored/sample size, in this case both given as 25. Value for each succeeding time t has been calculated by reducing the number of deaths and censored items, from the number of individuals at risk in the previous time t.
- Column under combined are total of sample 1 and sample 2 values under respective headings.
- The column pertaining to *death* are the *observed values* to whom the event (death) has occurred, and hence denoted as O_1 and O_2.
- Expected values are shown in columns marked as E_1 and E_2 for samples 1 and 2 respectively.
- Expected values have been calculated by the formula given earlier, i.e. $E_i = \sum \dfrac{d_i n_i}{n}$
- Expected value for first entry in column $E_1 = (1 \times 25) \div 50 = 0.50$ and so on. (Highlighted).
  Expected value for last entry in column $E_2 = (2 \times 6) \div 10 = 1.20$ and so on. (Highlighted in table for ease of understanding).

Calculating the test statistic:

$$\chi^2_{\text{logrank}} = \dfrac{(O_1 - E_1)^2}{E_1} + \dfrac{(O_2 - E_2)^2}{E_2}$$

$$\chi^2_{\text{logrank}} = \dfrac{(17 - 18.334)^2}{18.334} + \dfrac{(18 - 16.666)^2}{16.666}$$

$$\chi^2_{\text{logrank}} = \dfrac{(-1.334)^2}{18.334} + \dfrac{(1.334)^2}{16.666}$$

Solving, $\chi^2_{\text{logrank}} = 0.204$

Tabulated χ^2 at $N - 1$ Degree of freedom, i.e. $(2 - 1) = 1$, at 5% level of significance = 3.841

As calculated value of test statistic < tabulated value, statistically there is no evidence to reject the null hypothesis. Therefore, null hypothesis is accepted and there is no significant difference between the two survival curves.

As an alternative to the Log-rank test, there are number of tests for comparing survival experience of two samples. These include tests such as Wilcoxon test, Peto test, Flemington Harrington Test, etc. All these are primarily variants of Log-rank test as they provide weights to the different failure times. The Wilcoxon test gives more importance to the deaths in the earlier period as compared to later period. The Log-rank test gives equal importance to failure (death) at all times, i.e. equal weights or no weights. Due to this any differences in the initial part of the survival curve have higher chance of getting detected through Wilcoxon test. Still, most of the times, even with different weights the results inferred are similar. Therefore, as a standard measure, Log-rank test, having wider acceptability should be the first choice for determining the difference between survival curves being significant or not.

CHAPTER 28

Time Series Analysis

Time Series is defined as 'a set of data of a variable recorded at various points of time, in chronological order.' Analysis of such a series to decipher it and also to use it for forecasting its future values or the demand in future is the core of Time Series Analysis.

From the definition, time series is a set of values of a variable at a given point of time. The underlying basis of time series is that, even if the variable is not a function of time, but a function of multiple factors, it is assumed that all these known and unknown factors, routinely combine in various complex interactions through the course of time, to provide the outcomes, that they can be considered as a function of time. For example, it is common knowledge that rainfall occurs due to many various reasons and not because it is the month of June or July. The underlying factors consistently combine every year to provide rainfall. And similarly these factors would continue to impact the variables in a consistent manner, so as to allow forecasting. Forecasting, based on historical data, may or may not be true and has its own pitfalls, but it provides enough pointers towards the possible values which the variable will take, if there are no different forces acting upon it. The study of the time series is greatly helpful to understand the past behavior of the variable, which then can be utilized for forecasting. In epidemiology, mention is often made of the expected occurrence of the disease at that given point of time, e.g. malaria incidence is lower in winters than summers in India and this is showing a rising trend with peaks every few years. Time series analysis is an efficient tool for resource planning. The forecasts may or may not come true; and the signal to noise ratio may help decide how reliable the series of forecasts may be on an extended period.

Any set of values of a variable plotted against time in a chronological order will exhibit four basic types of components: trend, seasonal, cyclical and random or irregular. Consider the quarterly sales (in millions of rupees) of a pharmaceutical company:

	Years						
	2009	2010	2011	2012	2013	2014	2015
Jan - Mar	42	46	51	48	60	79	120
Apr - Jun	44	48	49	61	75	96	143
Jul - Sep	49	56	63	66	72	88	174
Oct - Dec	53	57	85	74	69	106	167
Total	188	207	248	249	276	369	604

Visual inspection of data indicates the annual sales have consistently grown. The organization appears to have had a reasonable success to start with; had practically no growth in one year and thereafter appeared to enter an exceptional growth phase. The factors could be many, may be initial time was a learning curve for the organization, or the organization introduced a new technology after market research which allowed it exceptional growth in later years. Beyond the realms of the organization itself, there could be other factors at play like the industry or economy had experienced restricted growth, which impacted the sales of the organization. It may also be due to government and policy decisions, like imposition of antidumping duties on competitive products or removal of restrictions on production capacities, etc. Time series analysis is primarily not concerned with these as a primary concern, but to decipher if there is any pattern or patterns at play in the variable, i.e. 'Sales.'

Once the patterns in the data are deciphered, the secondary step is to correlate the subjective reasons behind patterns, and can be undertaken to understand the data in more depth. This second step, though not a core part of time series analysis, but still is critical for proper assessment, understanding and provides immense help in forecasting.

TIME SERIES COMPONENTS

A time series can be divided into the following components:
- Trend (T)–long-term direction, also sometimes referred as secular trend.
- Seasonal component (S)
- Cyclical component (C)
- Random component or irregular variations (I).

Trend

Trend is the long-term direction or behavior of the data. Long-term, does not signify any specific number of years, decades or months but depends on the matter under study. For example, for population growth studies of tigers, the long-term could be a factor of many years like 7–10 years, but for a lac insect, it may be just 18–24 months, as the life cycle of lac insect is just six months. Therefore, trend is not based on numbers of years, but more logically the period should be such that it includes atleast two cycles of the variable under study or has enough points on the time scale that it brings out the true essence of the variable under study. Beyond this it does not matter, if the time period is of 10 years, or 2 years or few minutes. Generally speaking, if the variable grows or diminishes in linear manner, then chances are that its trend values will get recognized over a larger time horizon, as compared to any variable which changes in parabolic manner or exponential manner. In case the variable behaves in exponential manner, then its trend values can be assessed in much shorter time frame.

Seasonal Component

Seasonal component is the component of the time series, which shows itself as short-term variations in the data, which are repetitive in nature. These variations could be due to seasons (sale of ice creams in summers is always high, water-borne diseases in rainy seasons), due to festivals (white goods experience higher sales around festival season), due to school holidays (travel industry experience extra rush and sales in summer holidays and to some extent in winters). Usually in business parlance, these are referred as peak season or lean season.

Cyclical Component

Cyclical component relates to the component which reflects the overall business cycles experienced by the industry and economy. The cyclical nature is more prominently seen in basic industries like steel, metals, coal, mining, housing, etc. In ecological studies and dietetics again there may be such cycles which may last from a few months in the latter to many years in the former. It is very difficult to understand these cycles as they do not have a fixed duration (how many years does it last), frequency (how many time they will occur in a given period of time) or intensity (the downturn experienced in late nineties and the decline in 2008 due to global meltdown had different intensities). Therefore, it is difficult to separate them from the series.

Random Component

Random (irregular) component refers to the erratic movements or sudden prominent deviations in data series. These are caused by unexpected, unpredictable, abnormal events and move the data point significantly from its couple of neighboring points on its either side. For example, such irregular components are generally caused by natural calamity, mass hysteria, etc.

These four components are either considered to have a multiplicative effect or additive effect on the time series. Mathematically.

$$y = T \times S \times C \times I \text{ or } y = T + S + C + I$$

From the equations it can be inferred that the multiplicative effect treats the four components as related and impacting each other, while the additive effect consider these as independent components, not related to each other. Though random component, by its definition could be independent of other, but the rest three components do have an association. There exists no definite answer as to which of these two equations should be used, but generally speaking multiplicative model is preferred.

Ideally, a time series should be evaluated in both objective and subjective manner. The variations and the factors causing them should be understood in light of available information. Adjustments in data should be carried out if there is strong and valid reason. If required two or more trend lines can be established, if there exists a fundamental shift due to external factors especially imposed by the environment or the Government, e.g. emergence of resistant strains of microorganisms would lead to increase in the morbidity or a price control order issued by the Government would result in abrupt decline in sales value, though quantity may remain same.

It has to be kept in mind that the methods which will be discussed to determine the trends, etc. serve as guide to interpret the data series, but at the same time to decode the series to higher extent, subjective treatment can also be carried out, if necessary. This subjective treatment is in shape of subjective explanation of factors which caused the aberration, and hence allows enhanced knowledge of the series, which can be utilized in forecasting.

In time series analysis, the next key question is how to determine and measure the components of time series.

MEASURING THE COMPONENTS OF TIME SERIES

Trend

Measuring trend allows making the first statement about the data series. Though from the graph, the direction of trend can be seen, but in very slow growing or declining series, a graph may be error prone. Trend line can be established by semiaverage method, where the series is split into two equal halves (for odd number of data points, the central data point is ignored) and their means are calculated. These means are plotted on the graph at central point of each respective half, and joined. The line thus, formed is called the trend line based on semiaverages method. The limitations of semiaverage method lies its assumption that the series is progressing only linearly and being a simple mean, it gets unduly affected by extreme values which might have been caused by random variations.

Note: In time series analysis, the word 'average' implies 'arithmetic mean'. It is pointed out in the chapter on 'measure of central tendency' that statistical difference between average and mean, but historically time series analysis has used the word average instead of mean. Therefore, 'moving averages' calculations imply 'Mean' as the measure of average and this fact needs to be noted for this chapter/topic.

The standard methods for measuring trend are:
1. Moving averages method and
2. Curve fitting method.

Moving Averages Method

Moving averages method tends to smoothen out the seasonal fluctuations and are able to provide the trend of the series. Moving averages method is very suitable if the data exhibits a linear trend, and does not perform well in data having exponential variations. Because this method does not generate a mathematical equation, forecasting is not possible through this method.

While calculating the moving averages, the first step is to decide the period of moving averages. If it is taken as 3-year moving average, then simple mean of the first three values of the series is calculated and placed in alignment with the second value. The next value is derived by removing the first value and adding the fourth value, and calculating their simple mean and placing in line with the third value, and so on.

In case, a moving average of even number is to be calculated (say a 4-year moving average), then the values placed at center would get placed in between the second and the third period. This would be difficult to plot on the graph, therefore another column is added as two-year moving averages of the four year moving averages already developed. The two-year moving average will bring back the average to the central period which can be plotted in the graph. For example, the two years moving average, for the four years moving averages will get back the value in alignment with the third period. This process is also called as 'Centering.'

Sample calculations of three quarters moving average and four quarters moving averages:

Year/Qtr	Sales	3- Period		Four periods moving total	Four periods moving averages	Two period moving averages (Centering)
		Moving totals	Moving average			
2009-I	42	–	–	–	–	–
'09-II	44	135 (42+44+49)	45 (135÷3)	–	–	–
'09-III	49	146 (44+49+53)	48.7 (146÷3)	188 (42+44+49+53)	47 (188÷4)	47.5 [(47+48)÷2]
'09-IV	53	148 (49+53+46)	49.3	192 (44+49+53+46)	48 (192÷4)	48.5 [(48+49)÷2]
2010-I	46	147 (53+46+48)	49	196 (49+53+46+48)	49	49.9
'10-II	48	150 (46+48+56)	50	203 (53+46+48+56)	50.8	51.3
'10-III	56	161 (48+56+57)	53.7	207 (46+48+56+57)	51.8	–
'10-IV	57	–		–	–	–

In the above sample calculations, the word 'quarter' and 'period' has been interchangeably used. In regular terminology, the moving averages are referred as 'yearly moving averages' but the wording depend on the period of the data series. If the data is on 'daily basis' then it is called 'daily moving averages', and if it is in monthly basis, then it is referred as 'monthly moving averages'.

The key question is what does calculation of moving averages do to the data series. As it is an average of the moving totals, it smoothens out the movements of the values by reducing the impact of extreme fluctuations. There is no single value which is impacting the trend fully, but many values situated next to each other which are impacting the trend. The higher the period of moving average, the lower is the impact of an individual value. As the period (order) of the moving average increases, number of calculated points available for plotting, are reduced. As seen from earlier example, in three periods moving average, the first and last values are lost. In case of 4 period moving averages, first two and last two values are lost, which is similar to five period moving average.

The period of moving average which needs to be considered is the critical decision required while calculating the moving averages. If the data points are limited, then the higher order moving averages will not bring out proper trend.

Curve Fitting Method

The data series can transverse in many ways. It can show itself as a straight line, parabolic curve, an exponential curve, or many other types. In the moving averages method, the limitation existed because of an absence of mathematical equation which could denote the trend line, thereby reducing its usage. In curve fitting methods, this drawback is eliminated as the trend component is stated as a mathematical equation, and can take a shape of a curve or a straight line. Depending upon the type of curve, various methods are available for curve fitting. These options can give rise to errors if a wrong curve is fitted. The choice of the curve to use depends upon the data behavior.

Least Squares Method (Linear Trend)

Assume a data series 10, 23, 32, 46, 58, 71. If the differences between each value and its succeeding value show a nearly constant difference, then it indicates a linear growth/degrowth trend. In this case the differences are 13, 9, 14, 12 and 13. For such a data, the method of least squares would be the most appropriate method. Least squares principle is the underlying basis of the regression equation too, and the trend line through this method would be straight line. As in regression analysis, this line, i.e. trend line would satisfy the conditions (i) sum of the deviation of individual values from the trend line would be zero and (ii) sum of the squares of these deviations (from the trend line) would be minimum (least) possible. The second condition implies that there is no other trend line which can be drawn for this data, which has lower 'sum of squares of deviations from the line' than the trend line based on least squares.

The trend line based on least squares method is mathematically denoted as:

$$y = a + bx$$

where y = dependent variable,
x = independent variable,
a = y – intercept and
b = slope of the line – change in value of the variable y when x changes one unt

Graphically the equation can be denoted as in **Figure 28.1**.

For calculating the two constants, a and b the following equations are required to be solved, where y is the value of dependent variable at a given point of independent variable 'time' x.

$$\Sigma y = Na + b\Sigma x$$

$$\Sigma xy = a\Sigma x + b\Sigma x^2$$

For example, consider the following data, for which through the least squares method, the curve fitting shall be done:

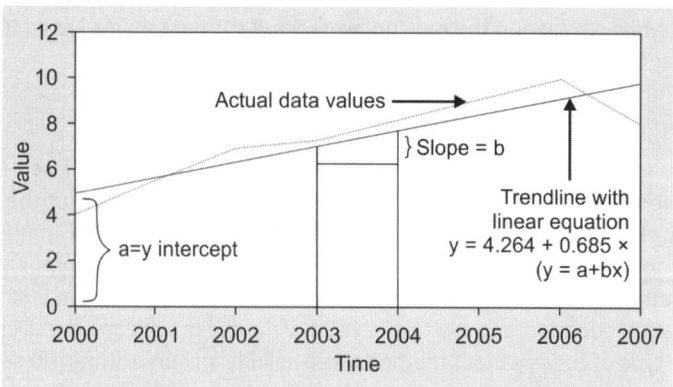

Fig. 28.1: Sample trendline by least squares method

Years	Value (y)	x	xy	x^2
2000	4.0	1	4	1
2001	5.5	2	11	4
2002	6.9	3	20.7	9
2003	7.3	4	29.2	16
2004	8.1	5	40.5	25
2005	9.0	6	54	35
2006	10.0	7	70	49
2007	8.0	8	64	64
N = 8	Σy = 58.8	Σx = 36	Σxy = 293.4	Σx^2 = 204

To reduce the calculations, the time period is changed for its origin and instead of starting point, it can be taken as the mid value of the series when the number of values is odd, but for understanding sake this example contains the simpler and straight forward way.

Using the two equations and placing the required summation values:

$$\Sigma y = Na + b\Sigma x \Rightarrow 58.8 = 8a + 36b$$
$$\Sigma xy = a\Sigma x + b\Sigma x^2 \Rightarrow 293.4 = 36a + 204b$$

Solving these equations for a and b, we get the desired values of a = 4.26 and b = 0.685, hence the trend line by the method of least squares can be denoted as y = 4.26 + 0.685x. Based on this equation, forecasting can be also estimated by equating x = 9, which will mean year 2008. The forecast will be at best a mathematical estimate (The same principle is the basis of regression analysis too).

Parabolic Trends

Least squares method is suitable only when the trend follows a linear change. In case the trend is not changing uniformly in absolute terms, the method of least squares would not be appropriate to determine the trend. Consider a data series as given below:

Years	Values (y)	First order differences	Second order differences
2000	8		
		12	
2001	20		4
		16	
2002	36		4
		20	
2003	56		5
		25	
2004	81		2
		27	
2005	108		4
		31	
2006	139		2
		33	
2007	172		

On calculating the first differences, the values do not reflect a behavior exhibited previously, i.e. somewhat similar difference between each succeeding value (linear trend), but on the contrary, the observation of first difference indicate that they are increasing slightly more than previous increase in absolute terms (not in percentage terms but in absolute terms). Therefore, second differences are calculated. The second differences indicate a pattern where the absolute growth is approximately +4 over and above the previous increase. Such a case where the second differences indicate a relationship, will signify a quadratic equation in the series and a second degree parabola would be the choice of the curve of the trend. The mathematical equation of a parabola is written as:

$$y = a + bx + cx^2$$

The constants a, b and c are determined by solving the equations:

$$\Sigma y = Na + b\Sigma x + c\Sigma x^2$$
$$\Sigma xy = a\Sigma x + b\Sigma x^2 + c\Sigma x^3$$
$$\Sigma x^2 y = a\Sigma x^2 + b\Sigma x^3 + c\Sigma x^4$$

Parabolic trend curve based on sample values is shown below in the chart. The linear chart is also shown for comparative understanding of the treatment of the same data by two different methods **(Fig. 28.2)**.

It can be seen for such relationships, the method of least squares provides a linear fit, but a parabolic curve fits better and indicates the trend in a much better manner.

Exponential Trend

Many a times, the values in a time series change with percentage growth instead of absolute growth. Due to this compounding effect the values reach a very high level in a much shorter duration. Such behavior cannot be properly studied through linear or parabolic curves, but instead exponential curve is required. These curves are represented by equation:

$$y = ab^x$$

To solve such an equation, its Logarithm value is required and given as $Log\ y = Log\ a + x\ Log\ b$, and the values of '$a$' and '$b$' can be derived from the equations:

$\Sigma Log\ y = N Log\ a + Log\ b\Sigma x$ and $\Sigma x. Log\ y = Log\ a\Sigma x + Log\ b\Sigma x^2$

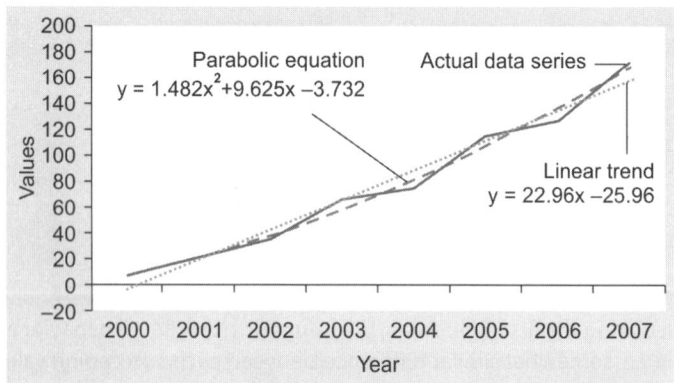

Fig. 28.2: Trendlines using parabola and least squares method

Consider the below time series:

Years	Values	Years	Values
2000	189	2008	2225
2001	255	2009	2781
2002	330	2010	3059
2003	462	2011	3885
2004	632	2012	5012
2005	861	2013	6215
2006	1145	2014	7645
2007	1660		

Within a span of 15 years, the value has increased tremendously. The graph of this series is depicted in **Figure 28.3**.

The chart above provides all the trend lines developed through different methods. In this chart, the least squares method will appear to pull all the values towards itself, smoothening out all variations. The moving averages method seems to work with slight lag. The parabolic curve tends to be close to actual values but tends to increase at a slower rate. In the middle section it can be seen that the parabola has been slightly higher at the time, when the growth rate fell, but the moment growth rate was back, the parabolic curve again fell behind. The exponential curve tends to overstate at the end.

From this chart, it becomes clear that no method is universal, which can handle all types of time series. In such a situation, the best way forward is to draw the graph and do a visual inspection. Take the first order differences to see if any linear trend shows up or not. If not, calculate the second order differences to see the series conforms to second degree parabolic curve and increase with similar absolute values. If not check the percentage increase in each succeeding value to check if there is any exponential growth seen. If no conclusion can be reached by any of these, then two possibilities exist. Plot all the curves with the help of a computer program (excel) and check the

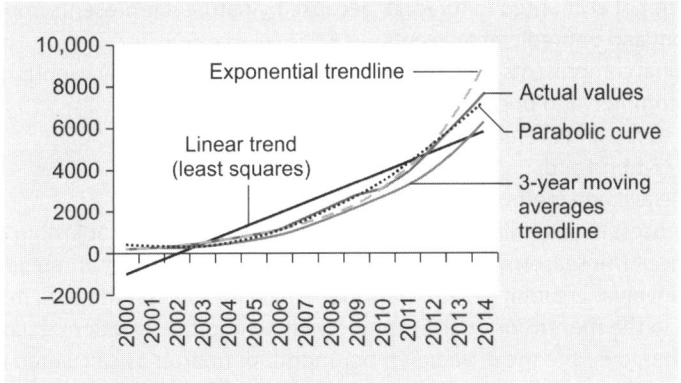

Fig. 28.3: Sample trend curves through various methods but for same data series

best fit (mostly the result would be second degree parabola) or use the most popular method, i.e. method of least squares.

The fundamental requirement for any time series is the availability of enough time values for proper analysis. Insufficient data would lead to errors in the analysis. In earlier times, too much data was also a problem, but with the advent of computing devices, large amount of data series are easily handled, therefore there should be no aversion to utilising large amount of data. But at the same time, if any tectonic shift has happened in the underlying factors or results, then it is better to split the data around the shift point, so that proper analysis can be undertaken and values on either sides of the shift point do not unduly influence each other.

At times, the graph of a data series appears to be an exponential curve in the first half and then, despite showing growth, tends to appear as a parabolic curve, i.e. growing at a increasing rate initially, followed by 'growth at a decreasing rate' in later half. Such series are a subject of growth curves known as Gompertz curves. These curves being very mathematical are outside the scope of this text.

Once trend of the time series has been identified, the next step is to check the data for seasonal variations.

Seasonal Component

In regular business, sales, admissions and economic cycles, seasonal component present themselves routinely and regularly. Though, for different variables the seasonal component would vary in its frequency in a year or two years or can even occur twice or more times in a year. For example, for real estate or automobile business the seasonal impact is seen around Navratras, twice a year. It is possible that the seasonal impacts may occur with different intensities too.

In the multiplicative model of time series $y = T \times S \times C \times I$, the seasonal component is an another major component which needs to determined. For forecasting, these two components—trend and seasonal, are necessary as cyclical would automatically get included through these. If the cycle is small (very frequent), it will get included in the trend and if it is too long, it would not impact the forecast as forecast is supposedly should be conducted for maximum next two or three periods. Irregular component would not get accounted in forecast, because by nature it represents impacts due to random and unpredictable events.

Seasonal components can be determined in the time series by various methods, but the common and popular methods are:
1. Ratio to moving average method,
2. Ratio to trend and
3. Simple average methods.

The process of determining the seasonal component and then adjusting the trend line without the seasonal impact is called as Deseasonalization of the Series. This process involves creation of seasonal index which is based on the data movement through in the months or quarters of the year, to find the variations each month/quarter has, over the mean value for per month or quarter as calculated from the year totals. The procedure to derive the seasonal components through the various methods is given below:

Ratio to Moving Average Method

In this method, each month's values are expressed as a ratio of its corresponding 12-month centered moving average. In case the series is in quarterly basis, then 4-quarterly centered moving averages are considered. By considering 12 months moving average or 4 quarters moving average, it is expected that all trend and cyclical components are removed from the data. The process involves calculating the average in a particular month or quarter, spread over many years, which eliminates the irregular variations also. Thus, the data series is left with seasonal component only. This method is very widely used for determining the seasonal component despite the fact that due to 12 months moving averages, the loss of values at the beginning and end occurs. This is ignored, especially, if the data pertains to many years, and the seasonal component would anyway, get reflected through the remaining data.

Ratio to Trend Method

This method involves calculation of the trend through the least squares method. Once the linear equation is established, all values are recalculated from the linear equation. The next step involves calculating the ratio of the actual values to the calculated values (from trend linear equation), multiplied by 100. The average for each month is calculated over the many years for which the data is available to arrive at the monthly or seasonal values. Though a good method, it is not able to eliminate the cyclical impacts which a moving averages method is able to do.

Simple Averages Method

This method involves calculating the marginal total for the months spread over various years and thereafter calculating the percentage of the monthly averages. The total of these percentages would be 1200 if it is a monthly series or 300 if it is a quarterly series. It is the simplest method, but is not able to avoid the trend and cyclical patterns from influencing the analysis.

In all the above methods, any logical adjustments can be carried out prior to analysis, if it is known that the value was highly influenced by an irregular variation. This improves the analysis, but should be done not by discretion but by pure, valid logic; and therein lies the difficulty.

All the calculation methods would be clear from the following example:

Example: Calculate the seasonal indices for the following quarterly time series by all the three methods discussed. Also calculate the linear equation and forecast the year 2016, quarter-wise values.

	Years							
	2009	2010	2011	2012	2013	2014	2015
Jan - Mar	42	46	51	48	60	79	81
Apr - Jun	44	48	49	61	75	96	97
Jul - Sep	49	56	63	66	72	88	106
Oct - Dec	53	57	85	74	69	106	126
Total	188	207	248	249	276	369	410

Section 7: Other Statistical Concepts Frequently Used in Research

Solution: 1. Seasonal indices by simple averages method:
Preparing the table for calculating the Indices by simple averages method:

Years								Quarterly total-7 years	Quarterly average	Seasonal index
	2009	2010	2011	2012	2013	2014	2015			
Jan -Mar	42	46	51	48	60	79	81	407	58.14	83.62
Apr - Jun	44	48	49	61	75	96	97	470	67.14	96.56
Jul - Sep	49	56	63	66	72	88	106	500	71.43	102.72
Oct -Dec	53	57	85	74	69	106	126	570	81.43	117.10
Total								1947	278.14	400
Average								486.75	69.54	100

The seasonal Index value is ratio of quarterly average divided by average of quarterly average, multiplied by 100, i.e. $(58.14 \div 69.54) \times 100 = 83.62$.

From above, a seasonal impact can be seen in the above series, in the form of maximum sales in the last quarter and minimum sales in the first quarter every calendar year.

2. Seasonal indices by ratio to trend method:
As this method involves trend, the trend values are required to be calculated by least squares method based on the mean quarterly sales after converting the data into a yearly series instead of quarterly.

Years	Annual sales	'x'	Mean quarterly sales (y)	'xy'	'x²'
2009	188	1	47	47	1
2010	207	2	51.75	103.5	4
2011	248	3	62	186	9
2012	249	4	62.25	249	16
2013	276	5	69	345	25
2014	369	6	92.25	553.5	36
2015	410	7	102.5	717.5	49
N = 7	Σ = 1947	Σ = 28	Σ = 486.75	2201.5	Σ = 140

$\Sigma y = Na + b\Sigma x \Rightarrow 486.75 = 7a + 28b$ and
$\Sigma xy = a\Sigma x + b\Sigma x^2 \Rightarrow 2201.5 = 28a + 140b$

Solving these equations for values of a and b, the calculated values of $a = 33.18$ and $b = 9.09$, i.e. the linear equation is $y = 33.18 + 9.09x$.

As the quarterly data is being considered, the yearly trend equation has been based on quarterly mean values calculated from actual given values. The process has been (a) add all quarterly values to arrive at yearly total. (b) Calculate mean quarterly value. (c) Apply the least squares method on these quarterly values to get the annual trend line. The reason being that if the yearly total values are considered the trend line will be same but on a larger scale, i.e. 4 times, which needs to be reduced subsequently to bring it at par with the data periods. The summary of calculations done so far:

Years	Annual actual sales	Mean actual quarterly sales (y)	Mean quarterly sales (based on linear equation derived from least squares)	Annual sales derived from least squares
2009	188	47	42.27	169.07
2010	207	51.75	51.36	205.43
2011	248	62	60.45	241.79
2012	249	62.25	69.54	278.14
2013	276	69	78.63	314.50
2014	369	92.25	87.71	350.86
2015	410	102.5	96.80	387.21
N = 7	$\Sigma = 1947$	$\Sigma = 486.75$	$\Sigma = 486.75$	1947.00

From this equation, it is inferred that the yearly change is 9.09 and therefore quarterly change would be (9.09 ÷ 4) = 2.27, the value used in preparing the above table.

Based on the linear equation, the mean quarterly sales as derived from the Trend Line has been calculated by giving x the values = 1, 2, 3 and so on, in the linear equation derived earlier. From these quarterly averages, annual values based on trend line have been calculated by multiplying the derived values by 4.

Calculating individual quarterly values: The mean quarterly value of 42.27 pertains to the central quarter of the year. 'Central quarter of the year' is the period that would fall, half in the second quarter (i.e. 16th May–30th June= 45 days) and half in third quarter (1st July–15th Aug = 45 days), therefore the quarterly value as per trend-line, pertaining to 2nd quarter would be 42.27 − (2.27 ÷ 2) = 41.13 and for 3rd quarter = 42.27 + (2.27 ÷ 2) = 43.41. As the values now are in sync with the quarter, rest of the values can be calculated by subtracting or adding 2.27 to subsequent quarters. The following table gives the quarterly values derived from the trend line, for the entire period.

Quarterly values derived from the trend:

	2009	2010	2011	2012	2013	2014	2015
Jan -Mar	38.86	47.95	57.04	66.13	75.22	84.31	93.40
Apr - Jun	41.13	50.22	59.31	68.40	77.49	86.58	95.67
Jul - Sep	43.41	52.50	61.59	70.68	79.77	88.86	97.95
Oct - Dec	45.68	54.77	63.86	72.95	82.04	91.13	100.22
Total	169.08	205.44	241.80	278.16	314.52	350.88	387.24

Calculating ratio-to-trend: The actual value of the quarter is expressed as a ratio to the value derived from the trend and multiplied by 100 (i.e percentage to the trend values). For example, the first value (Jan-Mar 2009) would be stated as (42 ÷ 38.86) × 100 = 108.08. The rest of the values and further calculations are as below.

Actual quarterly values as a percentage of trend value and other calculations:

	2009	2010	2011	2012	2013	2014	2015	Total	Mean %age	Adjusted %age
Jan-Mar	108.08	95.93	89.41	72.58	79.76	93.70	86.72	626.19	89.46	89.36
Apr-Jun	106.97	95.57	82.61	89.18	96.78	110.87	101.39	683.37	97.62	97.52
Jul-Sep	112.89	106.67	102.30	93.38	90.26	99.04	108.22	712.77	101.82	101.71
Oct-Dec	116.03	104.07	133.11	101.44	84.11	116.32	125.72	780.80	111.54	111.42
Total	443.96	402.25	407.42	356.58	350.92	419.93	422.06	2803.12	400.45	400.00

The percentage values are calculated against all quarters. The total of each quarter spread over different years is calculated to find the mean percentage. These mean percentages are added up. As the data was in quarterly basis, this should add up to 400 (1200 in case it was monthly data series). In case it does not, then the mean percentages are adjusted by equating the total to 400, i.e. (89.46 ÷ 400.45) × 400 = 89.36. These final values provide the seasonal variations on the trendline.

Forecasting for next four quarters: As per the yearly linear trend equation $y = 33.18 + 9.09x$, the value for year 2016, by equating $x = 8$ (2015 was year 7, therefore 2016 would be year 8 or $x = 8$) is calculated as 105.89. As the trend line was based on quarterly values, the total value for year 2016 = 105.89 × 4 = 423.56. Based on the trend line the quarterly values for 2016 are given in below table:

Quarter	2015 trend based quarterly values	From the trend line: Projected value for year 2016 = 423.56	Seasonal component	Forecast 2016 sales value (Quarter wise)
Jan-Mar	93.40	Projected quarterly value = 105.89	89.36%	94.62
Apr-Jun	95.67		97.52%	103.26
Jul-Sep	97.95		101.71%	107.70
Oct-Dec	100.22		111.42%	117.96
Total	387.24		400.00%	423.54

Multiplying the mean quarterly value derived from trendline with the seasonal component, the forecasted values are derived, and these include the impact of seasonal variations as deduced from the data.

3. *Seasonal indices*: Ratio to moving averages method:
   As the data is quarterly, calculating with 4-period moving averages and other subsequent analysis.

   Calculation sample for last column: actual value as a percentage of moving average-
   = (49/47.50) × 100 = 103.16 (as appearing on the table on the next page as the first entry)
   The next step is to prepare the seasonal indices, as shown below:
   Seasonal indices table (derived from the table of moving averages):
   Tabulating the Percentage values (last column of next table) in quarter-wise period tabulated format and doing the final calculations to arrive at the seasonal components.

Chapter 28: Time Series Analysis

Years	Quarter	Value	4-Period moving totals	4-Period moving averages	Centered 2-year average	Actual value as a percentage of moving average
2009	I	42			---	---
	II	44	188	47	---	---
	III	49	192	48	47.50	103.16
	IV	53	196	49	48.50	109.28
2010	I	46	203	50.75	49.88	92.23
	II	48	207	51.75	51.25	93.66
	III	56	212	53	52.38	106.92
	IV	57	213	53.25	53.13	107.29
2011	I	51	220	55	54.13	94.23
	II	49	248	62	58.50	83.76
	III	63	245	61.25	61.63	102.23
	IV	85	257	64.25	62.75	135.46
2012	I	48	260	65	64.63	74.27
	II	61	249	62.25	63.63	95.87
	III	66	261	65.25	63.75	103.53
	IV	74	275	68.75	67.00	110.45
2013	I	60	281	70.25	69.50	86.33
	II	75	276	69	69.63	107.72
	III	72	295	73.75	71.38	100.88
	IV	69	316	79	76.38	90.34
2014	I	79	332	83	81.00	97.53
	II	96	369	92.25	87.63	109.56
	III	88	371	92.75	92.50	95.14
	IV	106	372	93	92.88	114.13
2015	I	81	390	97.50	95.25	85.04
	II	97	410	102.5	100.00	97.00
	III	106			---	
	IV	126			---	

The process at this step is similar as explained under ratio-to-trend method:

Quarter	Years							% Total	Mean %	Adjusted seasonal index
	2009	2010	2011	2012	2013	2014	2105			
I	-	92.23	94.23	74.27	86.33	97.53	85.04	529.63	88.27	88.41%
II	-	93.66	83.76	95.87	107.72	109.56	97.00	587.57	97.93	98.09%
III	103.16	106.92	102.23	103.53	100.88	95.14	-	611.85	101.98	102.15%
IV	109.28	107.29	135.46	110.45	90.34	114.13	-	666.95	111.16	111.34%
							Total	2396.00	399.34	400

Comparison of the seasonal components from the three methods:

	Ratio-to-moving averages	Ratio-to-trend method	Simple averages method
Jan-Mar	88.41%	89.36%	83.62%
Apr-Jun	98.09%	97.52%	96.56%
Jul-Sep	102.15%	101.71%	102.72%
Oct-Dec	111.34%	111.42%	117.10%
Total	400%	400%	400%

It can be seen that the Simple Averages Method (using Mean) is influenced by the extremes and hence has bigger seasonal variation range. The results from Ratio-to-Moving Averages and Ratio-to-Trend are close to each other, as they are able to avoid impact from the random and cyclical influences to a large extent. Hence, these two methods are preferred.

If the series exhibits nonlinear trend, then other curves can be used, instead of moving averages or least squares methods. If the extreme values are unduly having high variations, then 'Median' instead of 'Mean' can be used a measure of central tendency, though in above calculations Mean has been used for easier understanding of the calculations and concept.

Cyclical and Random Components

In any data series, these components are very difficult to identify until subjective knowledge and understanding is available about the variable under study. Mathematically, though certain methods have been tried, but they are not able to identify these very accurately. In previous century, the cyclical variations were reasonably clear and had differing but ample wavelength and amplitudes that identification was still possible, and so could be superimposed on the trend. But after turn of the century, the dynamics of business has seen immense speed and technological changes. Due to the cumulative effect, currently being witnessed, the business cycles are difficult to decipher. Concepts like Double Dips, etc. make the understanding bit more complex. Therefore, it can be stated that for cyclical and irregular components, subjective knowledge of subject matter expert can be considered and thereafter decision should be arrived as to whether the data values need any adjustments or not. This is to emphasize that reading patterns requires in-depth subject knowledge. This is true of any data series be it business or medicine. Business models also depend on a multitude of factors like those seen in medicine and hence it makes more sense to look for patterns in the data generated by the institutions. This is important for the milieu of today where more knowledge will be generated in coming years from linkages and forward chaining of the data being generated in healthcare institutions of today.

Time series is a powerful tool, having use in practically all scientific and economic activities. With use of computers, it has become easier to evaluate and analyse voluminous data with efficiency, speed and accuracy, but necessary caution needs to be exercised about the amount of data to be studied, the method to be used and the curve to fit. If proper objectivity is maintained, alongwith subjective adjustments wherever strongly valid, it can help researcher to use time series to forecast with reasonable probability the immediate succeeding values. Though this concept is regularly used in business matters, but it can be applied to all fields including medicine and medical administration.

All disease control programs and their effectiveness, study the incidence of diseases over the years, etc. are all subjects which can be evaluated through time series. The possible use of time series is too broad and can be applied in lot of studies, where a data can be considered as a dependent variable of time factor.

CHAPTER 29

Interpolation, Extrapolation and Forecasting Methods

Interpolation refers to estimating the intermediate values of a given data. Consider the following example:

Year   :   2000 2002 2004 2006 2008 2010 2012
Market size (₹ mn) : 750   810   ?   900   970   1050 1200

The data above is for a Market Size of a product in ₹ Millions, and based on actual survey conducted every alternate year. Values for all the years in which survey was conducted are available, except year 2004 and due to this gap, it leads to a restricted usage of this data. This data can be filled in either by some intelligent guess or through some scientific method. Apart from the missing value, at times there is a need to know the value of the variable during the intervening period, say 2009. For such situation, interpolation is the statistical technique to estimate the missing values.

Interpolation is the estimation of the intermediate value between two extreme points of a series, by use of various statistical methods. Contrasted to interpolation, extrapolation refers to the estimation of values outside the two extreme points of the series. In the example above, estimating of any value for any year between 2000 and 2012 would require use of interpolation technique, whereas estimating any value for year prior to 2000 and post 2012 would entail use of extrapolation technique. The results arrived at from the use of interpolation and extrapolation, are most 'likelihood estimators' and not the actual values.

The unavailability of data could be due to multiple reasons like data loss, data collection not carried out, flawed data collection, etc. Interpolation technique provides systematic approach to estimate the intermediate value, but does not obviate the researcher's prerogative to fine tune the estimate further, on basis of additional information and knowledge about the event or the factors influencing the event. For example, interpolation of data on say medical equipment imports, the value derived by interpolation technique is say, 'x' but the researcher is aware that for that year, to restrict medical equipment imports and give a fillip to domestic industry, the Government had imposed additional import duties as a temporary measure; hence the researcher can reduce the calculated estimate suitably. Though generally, such subjective adjustments should be avoided, unless there is a strong evidence to support this adjustment.

Chapter 29: Interpolation, Extrapolation and Forecasting Methods

TERMINOLOGY

To understand the terminology used in interpolation and extrapolation, consider the following data on demand over the years. The year being an independent variable (x), and demand being the dependent variable (y).

Year (x)	Demand (y)		'x'	'y'
2004	1500		a	U_a
2006	1700		$a+h$	U_{a+h}
2008	2000	Representing these	$a+2h$	U_{a+2h}
2010	2300	values in interpolation $\Rightarrow$	$a+3h$	U_{a+3h}
2012	2700	terminology	$a+4h$	U_{a+4h}
2014	3100		$a+5h$	U_{a+5h}

1. *Arguments:* Values of independent variables are called arguments ('x' values above)
2. *Entries:* Values of dependent variables are called entries ('y' values above)
3. $y = U_x$ signifies that y is dependent on 'x'
4. First argument is equated as = 'a' i.e. Year 2004 = 'a'
5. 'h' = equidistant finite difference between the arguments, i.e. 2006 − 2004 = 2 = h (also referred as 'interval differencing')
6. 'y' entries are denoted as U_a. If h = 2, and a = 2004, then U_{a+3h} is equal to (2004 + 3 × 2) = 2010, i.e. it is the entry for year 2010.
7. Thus difference operator Δ (delta) can be defined as

$$\Delta U_x = U_{x+h} - U_x \text{ where } x = a, a+h, a+2h \ldots$$

To explain the concept of difference operator, consider the previous example and creating a 'difference table" for the values:

Argument year (x)	Entries demand (y) = Ux	1st order difference ΔU_x	2nd order difference $\Delta^2 U_x$	3rd order difference $\Delta^3 U_x$	4th order difference $\Delta^4 U_x$	5th order difference $\Delta^5 U_x$
2004 = a	1500 = U_a					
		200				
2006 = a + h	1700 = U_{a+h}		100			
		300		−100		
2008 = a + 2h	2000 = U_{a+2h}		0		+200	
		300		+100		−400
2010 = a + 3h	2300 = U_{a+3h}		100		−200	
		400		−100		
2012 = a + 4h	2700 = U_{a+4h}		0			
		400				
2014 = a + 5h	3100 = U_{a+5h}					

First order difference: $(U_{a+h} - U_a)$
    1700 − 1500 = 200;
    2000 − 1700 = 300;
    2300 − 2000 = 300 and so on.
Second order difference:
    300 − 200 = 100;
    300 − 300 = 0 and so on.
Third order difference:
    0 − 100 = −100;
    100 − 0 = 100;
    0 − 100 = −100

Fourth order difference:
100 − (−100) = 200;
−100 − 100 = −200
Fifth order difference:
−200 − 200 = −400

The first set of calculations to determine Δ from the U_x values is called the first order of difference. If Δ is again carried out on values of the first order difference, the resulting values are called second order differences. This operation is to be carried out until only one value is left. Δ along with its power is the symbol used to denote the order of the difference. Hence Δ^3 would signify third order difference and not cube of the difference.

METHODS OF INTERPOLATION

There are various methods by which the intervening values can be calculated, but these methods are based on assumptions, namely:
- The data, within reasonable accuracy can be expressed as a polynomial function.
- No large variation exists in the missing dependent variable value for which the interpolation is being sought. As the interpolation techniques derive the missing value from available values, any extreme variation due to special forces which might have occurred in the period of missing values, will not be captured and impact the results.

Graphic Method

It is the simplest method and involves plotting the available values on *xy*-graph, and then joining them with a smooth curve. Thereafter, for finding any interpolated value, a perpendicular is drawn from the point for which value is required, to the curve and then again a perpendicular from the point on curve to the other axis. This method can also be used to find reverse values, i.e. missing value is in the independent variable, while dependent variable's value is known. Therefore from graphic method the following questions can be answered:
a. What was the demand in year '*x*', with '*x*' being between the two extreme years?
b. In which year was the demand equal to a specific level?
c. What was the demand and which year was it, for any specific point of interest on the curve?

Though easiest, this method fails when data has large number of arguments or large value of entries. Large number of arguments will result in a large graph paper or elimination of certain arguments to make it fit in a reasonable size, resulting in a bigger scale. Similarly, if the values of the entries are also large, it will result in a bigger scale. In both these situations, the bigger scale will result in loss of accuracy. If a graph uses a scale of 1 mm = 10 units and another, 1 mm = 1000 units, the accuracy levels would differ. In other words, with increase in values the least count tends to increase, resulting in decrease in accuracy. The other drawback of this method is its subjectivity, as the plotting of smooth curve would vary in individuals. Though this drawback has been largely eliminated due to software plotting the curves. Due to its limitations and because it is not a popular method, further discussions are not required.

Parabolic Curve Fitting Method

Parabolic curve method is based on the algebraic theorem, if there are 'n' distinct points, then one and only one polynomial curve with degree less than or equal to (n – 1) passes through these given 'n' points. This curve of nth degree, passing through (n + 1) points is represented by

$y = f(x) = a + bx + cx^2 + dx^3 + ex^4 \ldots nx^n$ where $a, b, c, d, e \ldots n$ are constants.

Once the equation is known then the same can be used for obtaining any other intermediate value.

Example: Based on the following data, find the value of 'y' when 'x' = 7

x:	2	4	6	8
y:	9	12	16	20

Solution: The total number of arguments n = 4, hence a curve based on polynomial degree of (n – 1) would pass through all the points.

Further as y is dependent on x, i.e. $y = f(x)$, the polynomial curve equation would be $y = f(x) = a + bx + cx^2 + dx^3$. In this equation substituting the values provided of x and y:

$y = f(2) = a + 2b + c2^2 + d2^3 \Rightarrow a + 2b + 4c + 8d = 9$
$y = f(4) = a + b4 + c4^2 + d4^3 \Rightarrow a + 4b + 16c + 64d = 12$
$y = f(6) = a + b6 + c6^2 + d6^3 \Rightarrow a + 6b + 36c + 216d = 16$
$y = f(8) = a + b8 + c8^2 + d8^3 \Rightarrow a + 8b + 64c + 512d = 20$

Solving the above algebraic equations, we get the results as follows:

$a = 8 \quad b = \dfrac{-1}{6} \quad c = \dfrac{3}{8} \quad d = \dfrac{-1}{48}$

Using these in the above equation, to derive the value of y for x = 7,

$y = 8 + \left(\dfrac{-1}{6} \times 7\right) + \left(\dfrac{3}{8} \times 7^2\right) + \left(\dfrac{-1}{48} \times 7^3\right) = 18.06$

The main drawback of this method is involvement of large calculation, which cannot be done easily through manual process, but require computing devices especially for higher degree polynomials.

Newton's Forward Interpolation Method

Newton's forward interpolation method, for equal intervals is given as:

$U_{a+xh} = U_0 + {}^xC_1 \Delta^1 U_0 + {}^xC_2 \Delta^2 U_0 + {}^xC_3 \Delta^3 U_0 + \ldots + {}^xC_n \Delta^n U_0$

where U_0 = first value of the respective order of difference, a = first argument, U_{a+xh} = value of independent variable for which dependent variable is to be calculated, h = equal interval difference between arguments, and

$x = \dfrac{\text{Value to be argument for which interpolated value is desired} - \text{first argument}}{\text{Internal difference of the arguments}}$

In order to make the formula clear, consider the example below

Example: The OPD registrations (in '000) of a hospital in a year are given below. Estimate the number of registrations for the month of April.

Month:	Jan	Mar	May	Jul	Sep	Nov
Number of registrations: (in '000)	20	23	27	28	31	34

Solution: Preparing the 'difference table' for the above given data, by equating Jan = 1, Mar = 3, May = 5 and so on. Therefore, value to be interpolated, i.e. April = 4

Month (x)	Number of Reg'n (U_x)	ΔU_x	$\Delta^2 U_x$	$\Delta^3 U_x$	$\Delta^4 U_x$	$\Delta^5 U_x$
1	20					
3	23	3				
5	27	4	1	-4		
7	28	1	-3	5	9	-16
9	31	3	2	-2	-7	
11	34	3	0			

Newton's forward interpolation formula is

$$U_{a+xh} = U_0 + {}^xC_1 \Delta^1 U_0 + {}^xC_2 \Delta^2 U_0 + {}^xC_3 \Delta^3 U_0 + \ldots + {}^xC_n \Delta^n U_0$$

'Value of argument' for which 'entry value' needs to be calculated = 4, i.e. $a + xh$ = 4 or U_4

U_0 = First arguments for each order difference
h = Equal interval difference = 3 − 1 = 2
a = First argument = 1

Therefore, $(a + xh) = 4 \Rightarrow x = (4 − a)/h = (4 − 1)/2 = 3/2 = 1.5$, i.e. $x = 1.5$
Applying the Newton's formula:

$U_4 = 20 + {}^{1.5}C_1(3) + {}^{1.5}C_2(1) + {}^{1.5}C_3(-4) + {}^{1.5}C_4(9) + {}^{1.5}C_5(-16)$
$= 20 + 1.5 \times 3 + 0.375(1) + (-0.0625)(-4) + 0.0234(9)$
$\quad + (-0.0117)(-16)$
$= 20 + 4.5 + 0.375 + 0.25 + 0.211 + 0.187$
$= 25.52$

Hence, the number of registrations done in April are estimated to be 25.52 (in '000).

Newton's Backward Difference Interpolation

Newton backward difference formula, generally used for interpolation of values nearer to the end of the arguments, is given below (Operator ∇ denotes backward order difference)

$$U_{a+xh} = U_0 + x\nabla U_0 + \frac{x(x+1)}{2!}\nabla^2 U_0 + \frac{x(x+1)(x+2)}{3!}\nabla^3 U_0$$
$$+ \ldots + \frac{x(x+1)(x+2)(x+n-1)}{n!}\nabla^n U_0$$

where U_0 = last value of the respective (∇) order of difference, a = last argument, U_{a+xh} = value of independent variable for which dependent variable is to be calculated, h = equal interval difference between arguments, and

Chapter 29: Interpolation, Extrapolation and Forecasting Methods

$$x = \frac{\text{Value to be argument for which interpolated value is desired} - \text{last argument}}{\text{Internal difference of the arguments}}$$

Consider the previous example: If the need this time is to find the OPD registrations in the month of October, Newton's backward difference method is used. Reproducing the earlier table,

Month (x)	Number of Reg'n (U_x)	ΔU_x	$\Delta^2 U_x$	$\Delta^3 U_x$	$\Delta^4 U_x$	$\Delta^5 U_x$
1	20					
		3				
3	23		1			
		4		-4		
5	27		-3		9	
		1		5		-16
7	28		2		-7	
		3		-2		
9	31		0			
		3				
11	34					

And applying Newton's backward difference interpolation formula:

$$U_{a+xh} = U_0 + x\nabla U_0 + \frac{x(x+1)}{2!}\nabla^2 U_0 + \frac{x(x+1)(x+2)}{3!}\nabla^3 U_0$$
$$+ \ldots + \frac{x(x+1)(x+2)(x+n-1)}{n!}\nabla^n U_0$$

Value of argument for which entry value needs to be calculated = 10, i.e. $a + xh$ = 10 or U_{10}

U_0 = Last arguments for each order difference.
h = Equal interval difference = 3 − 1 = 2
a = Last argument = 11

Therefore, $(a + xh) = 10 \Rightarrow x = (10 − a)/h = (10 − 11)/2 = -1/2 = -0.5$ i.e. $x = -0.5$

Applying the Newton's formula:

$$U_{10} = 34 + (-0.5)(3) + \frac{(-0.5)(-0.5+1)}{2!}(0)$$
$$+ \frac{(-0.5)(-0.5+1)(-0.5+2)}{3!}(-2)$$
$$+ \frac{(-0.5)(-0.5+1)(-0.5+2)(0.5+3)}{4!}(-7)$$
$$+ \frac{(-0.5)(-0.5+1)(-0.5+2)(-0.5+3)(-0.5+4)}{5!}(-16)$$

$$= 34 + (-0.5)(3) + 0 + \frac{(-0.5)(0.5)(1.5)}{3!}(-2)$$
$$+ \frac{(-0.5)(0.5)(1.5)(2.5)}{4!}(-7) + \frac{(-0.5)(0.5)(1.5)(2.5)(3.5)}{5!}(-16)$$

$= 34 - 1.5 + 0 + 0.125 + 0.273 + 0.437$
$= 33.33$ (in '000)

As a general rule, the Newton's forward difference method is used if the required missing value lies in the top half of the data and Newton's backward difference method is used if the missing value lies in bottom half of the data.

The drawback of both these methods, is the need to have equal intervals. To overcome this limitation, Newton's divided difference method is used where unequal intervals exist.

Newton's Divided Difference Method

For unequal interval difference in arguments, Newton's divided difference method is used. Divided difference is denoted as Δ and given as (DDD-Divided Difference Delta):

$$DDD\ U_a = \frac{(U_{a1} - U_{a0})}{(a_1 - a_0)}$$

Where, the ordinary difference between the entries is divided by the difference in their arguments. The order of DDD is ($n-1$), where 'n' is the number of arguments.

The interpolation formula of Newton divided difference is given as:

$$U_x = U_0 + (x - x_0)\ DDDU_0 + (x - x_0)(x - x_1)\ DDD^2\ U_0 + \ldots +$$

where U_0 = First divided difference for respective order, a = First entry, x = Argument value for which entry is to required to be calculated.

The following example would help understand the application of the formula:

Example: Given are the values of independent variable 'x' and their corresponding dependent values 'y'. Interpolate the value for $x = 18$.

| x: | 12 | 15 | 17 | 21 | 26 |
| y: | 196 | 400 | 900 | 800 | 1000 |

Solution: The given values of independent variable do not have equal intervals. Therefore, Newton's divided difference formula is required to be used. Preparing the divided difference table for the given values:

x	y	$DDD\ U_x$	$DDD^2\ U_x$	$DDD^3\ U_x$	$DDD^4\ U_x$
12	196				
15	400	= (400 − 196)/(15 − 12) = 68	= (250 − 68)/(17 − 12) = 36.4	= $\frac{(-45.8 - 36.4)}{21 - 12}$ i.e. −9.13	= $\frac{4.81 - (-9.13)}{26 - 12}$ = +0.996
17	900	= (900 − 400)/(17 − 15) = 250	= (−25 − 250)/(21 − 15) = −45.8		
21	800	= (800 − 900)/(21 − 17) = −25	= (40 − (−25))/(26 − 17) = 7.2	= $\frac{(7.2 - (-45.85))}{26 - 15}$ i.e. 4.81	
26	1000	= (1000 − 800)/(26 − 21) = 40			

Using the Newton's divided difference formula:

$U_{16} = 196 + (16 − 12)(68) + (16 − 12)(16 − 15)(36.4) + (16 − 12)(16 − 15)(16 − 17)(−9.13) + (16 − 12)(16 − 15)(16 − 17)(16 − 21)(+0.996)$

$U_{16} = 196 + 272 + 145.6 + 36.52 + 19.92$

$U_{16} = 670.04$

Lagrange's Method

Lagrange's method is a very useful method and can be applied to unequal argument values (interval differences). The calculation of interpolation through Lagrange's method is done by the following formula:

Chapter 29: Interpolation, Extrapolation and Forecasting Methods

$$U_x = \frac{(x-a_1)(x-a_2)\ldots(x-a_n)}{(a_0-a_1)(a_0-a_2)\ldots(a_0-a_n)} \times U_{a_0} + \frac{(x-a_0)(x-a_2)\ldots(x-a_n)}{(a_1-a_0)(a_1-a_2)\ldots(a_1-a_n)} \times U_{a_1}$$
$$+ \ldots + \frac{(x-a_0)(x-a_1)\ldots(x-a_{n-1})}{(a_n-a_0)(a_n-a_1)\ldots(a_n-a_{n-1})} \times U_{a_n}$$

where, x = argument value whose entry value has to interpolated, a = argument values (a_0 first argument, a_1 second argument and so on) and U_a being their respective entries.

The formula would be clearer to understand from the following example:

Example: Given independent variable = x and dependent variable = y having following values. Interpolate the value for $x = 6$.

x:	1	2	7	8
y:	14	15	25	24

Solution: Given $y = f(x)$, denoted $y = U_x$, the values are stated as:

X	$y = U_x$
1	14
2	15
7	25
8	24

Using Lagrange's formula,

$$U_6 = \frac{(6-2)(6-7)(6-8)}{(1-2)(1-7)(1-8)} \times 14 + \frac{(6-1)(6-7)(6-8)}{(2-1)(2-7)(2-8)} \times 15$$
$$+ \frac{(6-1)(6-2)(6-8)}{(7-1)(7-2)(7-8)} \times 25 + \frac{(6-1)(6-2)(6-7)}{(8-1)(8-2)(8-7)} \times 24$$

$U_6 = -2.67 + 5 + 33.3 - 11.43$
$U_6 = 24.2$

The interpolated value for $x = 6$, by Lagrange's method is 24.2.

Lagrange's Inverse Interpolation

In 'inverse interpolation', the objective is to find the value of independent variable for a given dependent value, i.e. finding the argument value, for a corresponding given value of entry. As it is opposite to the interpolation methods, hence referred as inverse interpolation. The Lagrange's inverse interpolation formula is given as:

$$x = \frac{(U_x - U_{a_1})(U_x - U_{a_2})\ldots(U_x - U_{a_n})}{(U_{a_0} - U_{a_1})(U_{a_0} - U_{a_2})\ldots(U_{a_0} - U_{a_n})} \times a_0$$
$$+ \frac{(U_x - U_{a_0})(U_x - U_{a_2})\ldots(U_x - U_{a_n})}{(U_{a_1} - U_{a_0})(U_{a_1} - U_{a_2})\ldots(U_{a_1} - U_{a_n})} \times a_0 + \ldots$$
$$+ \frac{(U_x - U_{a_0})(U_x - U_{a_1})\ldots(U_x - U_{a_{n-1}})}{(U_{a_n} - U_{a_1})(U_{a_n} - U_{a_2})\ldots(U_{a_n} - U_{a_{n-1}})} \times a_n$$

Example: Find the argument value *x* for given entry value *y* of 40, for the following data:

| x: | 7 | 9 | 12 | 14 |
| y: | 21 | 29 | 37 | 42 |

Solution: Denoting the dependent variable *y* as U_x, and restating the data:

x	y = U_x
7	21
9	29
12	37
14	42

The value of *x* which is needed to be calculated is for $U_x = 40$.

This being a case of inverse interpolation, Lagrange's inverse interpolation method needs to be used. Applying the formula:

$$x = \frac{(40-29)(40-37)(40-42)}{(21-29)(21-37)(21-42)} \times 7$$

$$+ \frac{(40-21)(40-37)(40-42)}{(29-21)(29-37)(29-42)} \times 9$$

$$+ \frac{(40-21)(40-29)(40-42)}{(37-21)(37-29)(37-42)} \times 12$$

$$+ \frac{(40-21)(40-29)(40-37)}{(42-21)(42-29)(42-37)} \times 14$$

$x = 0.172 - 1.233 + 7.838 + 6.431$

$x = 13.21$

Hence, the value of *x* for given *y* = 40 is 13.21.

EXTRAPOLATION AND FORECASTING

The Newton's and Lagrange's methods of interpolation can be utilized for extrapolation too. Extrapolation aims to derive the value of dependent variable outside the given data values, which could be in past or future, based on the relationship as witnessed in the data available.

Therefore, Extrapolation can be considered as a form of 'forecast', but there are various other statistical tools used for forecasting. Forecasting can be stated as an estimation of future values in a systematic and logical manner. Though future can never be predicted precisely and accurately, but forecasting techniques provide possible situations, based on past behavior of values. These techniques are routinely applied in the various fields especially in commercial activities, economics, hazard analysis, etc.

Forecasting involves understanding past values, knowing the changes that have occurred in the past, reasons and factors that affect change. There are various methods used in forecasting, with some of them given below:

1. Extrapolation and 2. Regression analysis.

Both these techniques have been already dealt separately in this book.

3. *Econometric models:* With the advent of computers, the econometric models have grown in usage. These models consider as many factors as required and possible, to arrive at suitable analysis. As an extension, these models are very useful in simulation and scenario building. Development of such models does require adequate data to ensure that correct relationships are established among various factors.

4. *Causal models:* These models express mathematically the cause-effect relationship of a particular process flow.

The methods of forecasting are not being discussed in detail as the application of these is more prominent in other fields like business, economics, etc. In biomedical sciences, there is now an effort to develop forecasting methods in the form of risk factor analysis; but this is still a work in progress. In years to come we may be looking at increasing use of these techniques in medicine.

Section 8

Statistical Concepts for Medical Administration

- Index Numbers
- Statistical Quality Control
- Decision Theory

CHAPTER 30

Index Numbers

Index numbers is a concept whose results are very commonly heard and used by most of the people, without realizing it. BSE Sensex or NSE Nifty, CPI (Consumer Price Index) and WPI (Wholesale Price Index), IIP (Index of Industrial Production) are common index numbers seen every day.

There are many meanings of the word "Index" in a dictionary, including "something used to point out, provide indication, a reference point directing towards bulk of values or numbers." Index number as a concept in statistics provides an indication of the amount of change in a phenomenon over a defined period of time.

Consider the below table giving the number of surgeries conducted by a hospital per year, over the 5-year period.

Year	Surgeries conducted	Percentage change over 2011	Indexed
2011	1200	0	100
2012	1464	22	122
2013	1296	8	108
2014	1536	28	128
2015	1728	44	144

Apart from the two columns mentioned, two additional columns are also shown depicting (a) How much is the percentage change in surgeries conducted in each year in relation to the starting year 2011 and (b) Indexed = If value for the starting year is equated to 100, then what would be the value for the rest of the years (calculated as $[(1464/1200) \times 100] = 122$). Upon closer examination, both these additional columns are similar, as from indexed column the percentage change can be derived by subtracting the base year value from the value of the required, i.e. for Year 2012, the change is 122–100 = 22 (or 22/100 × 100 = 22%).

The indexed column is nothing, but the 'Index Number' and the series of index numbers, i.e.100, 122, 108, 128, etc. is known as the index series.

Spiegel defined index number as a "Statistical measure designed to show changes in a variable or a group of variables with respect to time, geographic location or other characteristics such as income, etc."

Index number measure changes in a variable or a group of variables over a defined period of time. It is a relative measure of change, in the subject of interest, or a ratio of changes occurred or in simplistic term—a quantitative value measuring growth or degrowth of a phenomenon, at specific intervals in a defined period of time.

From all the above various explanations and meanings of index numbers, the important points that emerge about constitution of index numbers are:

- Is it for a single variable or for a group of variables?

Index number of a single variable is called a univariate index and is the simplest form. When an index number is calculated for a group of variables, it is called a composite index. When the consideration is for a group of variables, it becomes pertinent to decide what all variables are to be included, will they have equal or unequal weights and what kind of average (simple, weighted or geometric mean, etc.) would be used to arrive at their central value.

- Against what the values being 'equated to' or are being 'based' on?

In the previous example, we considered the starting year as the year against whose corresponding value, the values of all other years were equated to, i.e. if value of year 2013 is 108, it meant that against year 2011, which is equated to 100, there has been a change of 8 units (which equal 8/100 × 100 = 8% in this case). The year which is considered as the benchmark year and whose value is the base for calculating all other values is called the Base Year. All measurements of change are done against the Base Year. Which year to be considered as base year, depends upon the creator of the index number, its utility, purpose, etc. There is no fixed rule about fixing of base year, but generally it is not considered a too back a date/year in the past data.

- What is being studied for quantum of change? Price, quantity, value, etc.?

The subject of interest also is a consideration, as it determines what type of data would be required to be collected, from where it would be collected, etc.

KEY DECISIONS IN CREATING INDEX NUMBER

Objective

The index numbers are created for a specific purpose and have the variable or variables included in it, on the basis of the specific objective. The BSE Sensex (base year 1978–79 = 100) or Nifty are specific index numbers which pertain to the market capitalization of specifically included listed companies having different weights and serving as a benchmark index of the stock market. This index cannot be used to measure, say, consumer price index.

Objective determines for whom is the Index relevant, what variables (and, sometimes, what sub-variables within these variables) are required to be included, what are the sources of data for these variables, how the data is to be collected, how frequently is the Index number to be generated [daily (stock exchange), monthly (manufacturing index), quarterly (index based on quarterly results of companies), yearly, etc.] what would be the weights (Importance) given to each variable, what measure of central tendency of the variable would be applied (arithmetic mean or geometric mean, etc.), how the Index number would be calculated (Method of calculation) etc. As generally Index Numbers are ongoing series which need to be updated periodically, it has to be decided who would monitor, maintain and update the index number in-future periods.

In years to come standardization and inventorization of the medical services is expected to push up the need for concepts of comparison-like index numbers. These are easier for non-bioscience lay people to understand also.

Reference Period

As stated above, the index numbers are expressed or equated to specific year which is the reference year. The year which is the reference year and on whose value are all other values are expressed as, is called the 'Base Period or Base Year.' It is generally denoted as period/year 0, but may not necessarily be the first period/year. Depending on the method of calculating the index year, reference year can be the current (latest) year or first year, or any other period as decided by the creator of the index. It can also keep on changing on the basis of pre-decided methodology, similar to moving averages. As all other periods are based on this, the base period should be as close to normal behavior of the variable under consideration, to the maximum extent possible. This will ensure that non-normal periods either positive or negative will readily show up in the index numbers. In case a non-normal period is considered as a base period, it will either reduce or amplify the index series. This can be explained by the following example:

Year	Sales	Index number (Base year)		
		2010	2011	2012
2010	270	100	123	71
2011	219	81	100	58
2012	380	141	174	100
2013	295	109	135	78
2014	310	115	141	82
2015	330	122	151	87

The given table brings out the disadvantages in considering non-normal period as base year. It can visually seen that if the non-normal peak side value (year 2012 – value 380) is considered, it reduces the change in value in other years, whereas if a non-normal lower side value (year 2011–value 219) is considered, it amplifies the changes in value during the other periods. While considering price-based index number, a too old or outdated base period may result in error, due to the fact that money has a time value attached to it. A Rupee today is not equal to a Rupee one year down the line, due to inflation or vice versa if deflation is being experienced in the economy. Many of the Governmental Indices were updated for their base period of 2004–05 to new base year 2011–12 around 2014–15.

For index numbers within an index series to be truly comparable, it has to ensured that the basket of products, i.e. variables being considered are preferably same or reasonably similar in terms of their quality, maintaining consistent unit of measurement (UoM) for the specific products, using a unit of measurement (UoM) which is appropriate to products usage or its consumption (it is not appropriate to use 'square millimeters' or 'square kilometers' as UoM for consumption of Cloth, or 'Metric Tonnes' for food and vegetables at Retail Level). It may be argued that in index numbers, the interest lies in the quantum of change, therefore, UoM should not have an impact. From the formula of index number, this perspective may be true, but what is being overlooked is the fact that in most of the commodities, price is a function of quantity (larger quantum of purchase reduces the prices, due to supplier's willingness to offer discount). This may sometimes not be true also (like in case of electricity, household cooking gas where extra consumption entails extra cost, either due to non-subsidy or higher tariff for large unit consumers). In such a scenario, choosing a reasonable unit ensures better reflection of actual state of affairs by the index.

Once the 'Objective' and all its associated decisions, are firmed up, along with the decision on base year has been finalized, the next step lies in deciding the method which needs to be employed for calculating the index number.

Method of Calculating Index Numbers

As mentioned in the earlier section of this chapter, index number can either be based on weights (relative importance) of each variable or without assigning any weights to the variable (no weights are assigned but in actual sense, it is a situation where all the variables have equal weights).

The standard notations applicable to index number calculations methods and formulae are:

P = Prices, $\quad Q$ = Quantity, $\quad V$ = Value, i.e. $P \times Q$

Subscript '0' = Base year, i.e. P_0 = Price in base year; Q_0 = Quantity in base year, etc.

Subscript '1' = Current year, i.e. P_1 = Price in current year; Q_1 = Quantity in current year, etc.

P_{01} and Q_{01} = Index number based on prices and quantities respectively and N = Number of variables.

Note: Though in terminology, words 'Base Year' and 'Current Year' have been used which permits easier understanding of the concept, but in reality the base 'year' could be any defined starting point of the period, i.e. for monthly index, it would be read as base month, for biannual, it would be the Base Biannual Period and so on.

The broad classification of index number is based on 'Weights' or 'No Weights'. In case of no weights, the prices are expressed as a factor of base-year price of the variable, multiplied by 100. In case weights are assigned to the variables, the key issue that arises is what is to be considered as weight. The weight could be the quantities in the base year or quantities in the current year or any other value which is appropriate. On the basis of which year quantities are being considered as 'Weight,' different methods of construction of Index Numbers have been developed. The following chart would allow easy understanding of the various Index Number construction methods.

Price Index Number: Types and Construction

Each of these methods has its own advantages and disadvantages. Index numbers without weights are too simple and can be applied only where all the variables are to be given equal importance or the index is for a single variable **(Flow chart 30.1)**.

Index Numbers Without Weights

Simple Aggregative Method

It is the simplest method, but has the drawback of getting unduly influenced by the units of measurement. For example: In case an Index number is desired for precious metals—Gold and Silver, the Gold prices are generally expressed on per 10 g basis and Silver prices are on per Kg basis.

Chapter 30: Index Numbers

Flow chart 30.1: Index number construction

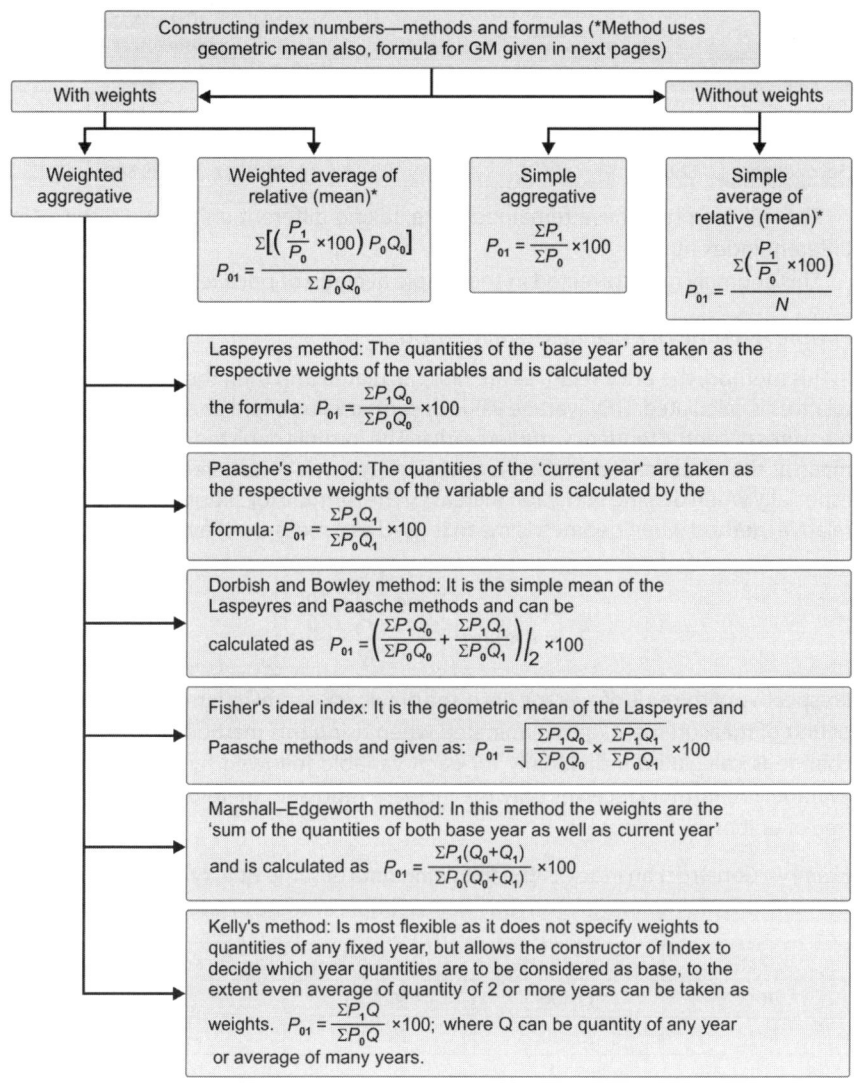

Consider the calculations below:

	Typical case calculations				Calculations after changing unit of measurement		
	Gold (/10 g)	Silver (/Kg)	Total	Index	Silver (/10 g)	Total	Index
Base year	25,000	42,000	67,000	100	4,200,000	4,225,000	100
First year	26,750	38,000	64,750	96.64	3,800,000	3,82,6750	90.57
Second year	29,500	41,000	70,500	105.22	4,100,000	4,129,500	97.77

Though the prices have remained same, taking different units have resulted in different Index numbers.

This anomaly gets corrected in the simple average of price relative method.

Simple Average of Price Relative Method

In this method, the price relatives are first calculated and then the average of these relatives is calculated. This average value or central value could be any of the possible measures of central tendency studied earlier. The formula depicted in the chart is with mean as the measure of average. As geometric mean is considered as best measure especially when dealing with ratio increases, the formula for simple average of price relative method when geometric mean is used as a measure of average, is given as:

$$P_{01} = \text{Antilog} \left\{ \frac{\sum \log \left(\frac{P_1}{P_0} \times 100 \right)}{N} \right\}$$

Irrespective of the type of average used, the drawback of previous method on account of unit of measurement, gets eliminated when using this method, as the quantum change is calculated individually for each variable followed by calculating their average, the formula receives only the increase ratio and, therefore, this method is free of unit of measurement.

Example: Construct an index of the following data of same quality product, through the methods of simple aggregative as well as simple average of price relative method:

Year	Price of fridge	Price of TV	Price of audio system
2014	13,400	43,500	12,000
2015	15,100	39,000	16,000

Solution: Creating the table for the given data:

Prices	2014 (P_0)	2015 (P_1)	Price relative (P_1/P_0) × 100	log of price relative
Fridge	13,400	15,100	112.69	2.0519
Television	43,500	39,000	89.66	1.9526
Audio system	12,000	16,000	133.33	2.1249
Total (Σ)	68,900	70,100	335.68	6.1294

- Index number by simple aggregative method
  - $P_{01} = \Sigma P_1 / \Sigma P_0 \times 100 = 70100/68900 \times 100 = 101.74$

- Index number by simple average of price relative
  - Using mean as the average

$$P_{01} = \frac{\sum\left(\frac{P_1}{P_0} \times 100\right)}{N} = 335.68/3 = 111.89$$

  - Using geometric mean as the average

$$P_{01} = \text{Antilog}\left\{\frac{\sum \log\left(\frac{P_1}{P_0} \times 100\right)}{N}\right\} = \text{Antilog}\,(6.1294/3)$$

$$= \text{Antilog}\, 2.0431 = 110.40$$

The difference between the indices shown above derived from the two methods is reasonably high. This explains why the simple aggregative method is not used frequently. Additionally the limitation of the 'Price Relative Method' lies in the fact that it gets highly influenced by changes in the extreme values. This method uses percentage to arrive at the index number, and percentage change is higher if base is low and appears smaller when base is high. For example, if an additional product is added to the above data, with a price of ₹ 2000 in 2014 and ₹ 5000 in 2015, but sparingly used, the Index Value by using simple average (Mean) of price relative becomes 146.42—a very high increase for a very low priced product. Hence, the need for weighted Index becomes essential.

WEIGHTED INDEX NUMBERS

Simple Aggregative Methods

These are listed in the chart, and the difference lies in the weights to be used, i.e. which year quantities are considered as the weights for calculating the Index number. None of these methods is universally applicable and accordingly it is left to the judgment of the creator of Index to decide which one to use, though Laspeyres and Kelly are frequently used.

Laspeyres Method

This method uses the base year quantities as weights. It is very widely used, but has the drawback if the base year becomes too distant. Considering the price-demand relationship in economics, an increase in price may lead to reduction in demand, but in Laspeyres method, the weights remain same even if the demand has fallen in the following years due to price increase. Theoretically, this creates an upward bias in this index number. If the index is related to some variables which do not subscribe to this theory, for such variables it is an index number of choice. This method allows easy comparison of the change values over the entire life of date and index numbers once created need not to be recalculated, until the base year itself is being changed.

Paasche's Method

In this method the Index numbers are generated by using the current year quantities as weights. It is opposite of Laspeyres and creates a minor downward bias. The index derived through this method would be close to Laspeyres, until high fluctuations have been experienced by the variables in the intervening periods. Though the calculations required during creation of this Index are similar to Laspeyres, maintenance and sustaining this Index number requires a great deal of effort. As current year quantities are taken as weights, and current year will keep on changing, this implies that the index number has to be recalculated each year. This restricts its usage.

Dorbish and Bowley's Method

Considering both base year and current year quantities, as weights, have their own benefits and limitations, Dorbish and Bowley's method expresses index number as the arithmetic mean of the Laspeyres method and Paasche's method. This method also requires recalculating the index every year, as the index number derived from Paasche's method based on current year quantities, will change every year. Therefore, it is used in a limited manner.

Fisher's Ideal Index

Considering that geometric mean is theoretically the best measure of ratio changes, instead of using arithmetic mean as in Dorbish and Bowley's method, Fisher's method entails use of geometric mean of the Laspeyres and the Paasche's Index. This method is also referred to as an ideal index due to its property of considering both base year and current year quantities as weights, using geometric mean which is better than arithmetic mean and it satisfies both time reversal test and factor reversal test.

Time reversal test and factor reversal test were two tests developed by Professor Fisher. Time reversal test specifies the condition that the Index number should be such that if its base and current year are reversed, the percentage change should remain same. Elaborating, assume with 2014 as base, the index number for 2015 stands at 120, i.e. 20% increase. Reversing, for the same set of data, if index number for year 2014 is calculated, keeping the base year 2105 (i.e. index number for 2015 = 100), then the index number so derived for year 2014 should be $83\frac{1}{3}$. If 20% increase is calculated on this index number ($83\frac{1}{3} \times 1.2 = 100$), it equates to 100 which is the indexed value for 2015. From both perspectives the percentage change is 20%. Time reversal test specifies an index number to conform to this. Out of the many methods listed in the chart- Fishers index, simple average (Geometric Mean) of price relative, aggregated with fixed weights, Marshall-Edgeworth Method and Weighted (Geometric Mean) of price relative satisfy this condition.

Similarly, factor reversal test suggests that interchanging of price and quantity in the formula should not change the percentage (like the time reversal test), where the base period is interchanged. Only Fisher's Index satisfies this condition.

Theoretically, these tests look significant, but in practice, it is not feasible to have consistency as desired by these tests. Fisher's Ideal Index may be considered theoretically ideal, but in practice it is rarely used. Two main reasons are (a) It involves Paasche index number, which needs calculation and restatement every year, hence the value of Fisher's index also needs to be calculated and restated every ever and (b) Calculation of geometric mean is relatively more difficult compared to calculation of arithmetic mean.

Marshall-Edgeworth Method

In this method, the weights are the sum of the quantities of base year and current year. As it involves current year figures, it also is not comparable with earlier results, until recalculation and restatements are done every year.

Kelly's Method

This method of constructing index numbers does not specify which year quantities are to be considered as weights. The weights can be an average of many years, or it may be any specific year not necessarily the base year. The advantage of Kelly's method lies in its flexibility and consistency of calculation which do not require recalculation or restatement every year of past index numbers.

Weighted Average of Relatives

This method is similar to the unweighted simple average of price relative, but with weights attached. The weights are the value of the base year, i.e. $P_0 Q_0$. In this method, arithmatic mean as well as geometric mean can be used as the average. The equation while using arithmatic mean, is already stated in the earlier chart, while the equation using geometric mean is given here:

$$P_{01} = \text{Antilog} \left[\frac{\Sigma \left\{ P_0 Q_0 \log \left(\frac{P_1}{P_0} \times 100 \right) \right\}}{\Sigma P_0 Q_0} \right]$$

Considering base year value (V_0) = Price (P_0) × Quantity (Q_0) = $P_0 Q_0$, the above equation can be restated as:

$$P_{01} = \text{Antilog} \left[\frac{\Sigma \left\{ V_0 \log \left(\frac{P_1}{P_0} \times 100 \right) \right\}}{\Sigma V_0} \right]$$

Price relatives have the advantage of not allowing the unit of measurement to unduly affect the index number.

Example: By using various weighted methods, develop the price index number for the purchase of fruits for the hospital kitchen. All quantities are in tons and prices in ₹/Kg.

Fruit	2014		2015	
	Price	Quantity	Price	Quantity
Apple	85	36	94	38
Pear	52	12	60	13
Banana	40	44	42	43
Watermelon	35	6	38	4
Orange	70	40	80	40
Guava	55	12	50	14

Solution: Placing the given data in a desired form as given below.

Fruit	2014		2015		P_0Q_0	P_1Q_1	P_0Q_1	P_1Q_0	$P_0(Q_0+Q_1)$	$P_1(Q_0+Q_1)$
	Price (P_0)	Qty (Q_0)	Price (P_1)	Qty (Q_1)						
Apple	85	36	94	38	3060	3572	3230	3384	6290	6956
Pear	52	12	60	13	624	780	676	720	1300	1500
Banana	40	44	42	43	1760	1806	1720	1848	3480	3654
Watermelon	35	6	38	4	210	152	140	228	350	380
Orange	70	40	80	40	2800	3200	2800	3200	5600	6400
Guava	55	12	50	14	660	700	770	600	1430	1300
Total (Σ)					9114	10210	9336	9980	18450	20190

Solving for various methods of developing the index number:
- Weighted aggregative methods

$$\text{Laspeyres method} = P_{01} = \frac{\Sigma P_1 Q_0}{\Sigma P_0 Q_0} \times 100 = \frac{9980}{9114} \times 100 = 109.50$$

$$\text{Paasche method} = P_{01} = \frac{\Sigma P_1 Q_1}{\Sigma P_0 Q_1} \times 100 = \frac{10210}{9336} \times 100 = 109.36$$

Dorbish and Bowley method =

$$P_{01} = \frac{\left(\frac{\Sigma P_1 Q_0}{\Sigma P_0 Q_0} + \frac{\Sigma P_1 Q_1}{\Sigma P_0 Q_1}\right)}{2} \times 100 = \frac{\left(\frac{9980}{9114} + \frac{10210}{9336}\right)}{2} \times 100 = 109.43$$

Fisher's method=

$$P_{01} = \sqrt{\frac{\Sigma P_1 Q_0}{\Sigma P_0 Q_0} \times \frac{\Sigma P_1 Q_1}{\Sigma P_0 Q_1}} \times 100 = \sqrt{\frac{9980}{9114} \times \frac{10210}{9336}} \times 100$$

$$= \sqrt{1.19753} \times 100 = 109.43$$

Marshall-Edgeworth method =

$$P_{01} = \frac{\Sigma P_1 (Q_0 + Q_1)}{\Sigma P_0 (Q_0 + Q_1)} \times 100 = \frac{20190}{18450} \times 100 = 109.43$$

Another weighted aggregative method-Kelly's method will not be useful as only two year values are available and both have been considered either under Laspeyres or Paasche methods. The index numbers created by all the methods are very close to each other, due to presence of very stable values of price and quantity, and absence of any extreme values or fluctuations. For comparison sake, the index number through simple aggregative method for above data, on calculation = 108.01.
- *Weighted average of price relative methods:* For calculating the index numbers through this method, the additional table columns are required in line with the formula. Preparing the table again:

Fruit	2014		2015		P_0Q_0	$P = P_1/P_0 \times 100$	$(P_1/P_0 \times 100) \times P_0Q_0$	$Log(P_1/P_0 \times 100)$	$P_0Q_0 \times log(P_1/P_0 \times 100)$
	Price (P_0)	Qty (Q_0)	Price (P_1)	Qty (Q_1)					
Apple	85	36	94	38	3060	110.59	338400	2.0437	6253.749
Pear	52	12	60	13	624	115.38	72000	2.0621	1286.78
Banana	40	44	42	43	1760	105.00	184800	2.0212	3557.293
Watermelon	35	6	38	4	210	108.57	22800	2.0357	427.5003
Orange	70	40	80	40	2800	114.29	320000	2.0580	5762.377
Guava	55	12	50	14	660	90.91	60000	1.9586	1292.681
Total (Σ)					9114	644.74	998000		18580.38

By using mean as the average =

$$P_{01} = \frac{\sum\left(\left(\frac{P_1}{P_0} \times 100\right) P_0 Q_0\right)}{\sum P_0 Q_0} = \frac{998000}{9114} = 109.50$$

By using geometric mean as the average:

$$P_{01} = \text{Antilog}\left[\frac{\sum\left\{P_0 Q_0 \log\left(\frac{P_1}{P_0} \times 100\right)\right\}}{\sum P_0 Q_0}\right]$$

$$= \text{Antilog}\left[\frac{18580.38}{9114}\right] = \text{Antilog } 2.0386 = 109.20$$

QUANTITY AND VALUE INDEX NUMBERS

All the index numbers stated so far are related to the prices and hence called as price index number. Similarly if the index number is to be developed for quantity, then the same methods can be used by taking the prices as the weights. Instead of P_{01}, which denoted the price index number, quantity index numbers are denoted as Q_{01} and value index number as 'V_{01}.'

The methods for calculating quantity index number can be inferred from price index formula by interchanging price with quantities. For Laspeyres formula for quantity index number would become as $Q_{01} = (\sum Q_1 P_0 / \sum Q_0 P_0) \times 100$ and similarly for other methods.

The value index numbers are simple to calculate by using the formula $V_{01} = (\sum V_1 / \sum V_0) \times 100$, where V_{01} = Value index number for 0 = Base period and 1 = Current period, i.e. V_1 = Value in current period and V_0 = Value in the base period, and with value being calculated as price × quantity.

CHAIN INDEX NUMBERS

The index numbers methods discussed so far had either weights or no weights. If weights were attached, they were on the basis of a fixed base. One the biggest challenge facing constructing of Index numbers is the constant alteration in the basket of the items constituting the Index. From the 'objective' perspective of the index, the alteration and changes due to environment, technology shifts, consumer behavior, etc. need to be captured in the index. For example, in radiology division, the shift from X-ray films to digital storage, the quantum leap in the imaging devices with very high resolution, 3-D images and in color, has happened over the last decade. If there was an index pertaining to cost of running the radiology division over the years, either these changes would have been ignored from the index number or included. In both the events, the Index number would not have showed the actual status, unless the weights were modified.

Apart from this, in most of the commercial, economical and business sections, the speed of change propels the management to become greatly interested in the current year's values in relation to the last 2–3 years values (and not much older data) and especially, performance with respect to the previous year. Values prior to a couple of years, are practically historical. Chain index numbers, to a large extent, fill this need of comparing immediate previous year and flexibility to change the constituent items. This is achieved by constantly changing the base from a fixed period to a period immediately preceding the period for which the index is being calculated. For example for year 2014, the base year can be taken as 2013 and then for 2015, the base year would be 2014. To guard against any spikes or decrease in values in a year, especially if the variables experience a biannual cycle, 2-year moving averages can also be considered as the base year for weights.

The chain index is calculated by first calculating the link relative (current year price expressed as a percentage of previous year), then multiplying this by the chain index of the previous year and dividing the product by 100. The starting year chain index (base year) is taken as 100. For a single variable data series chain index would be same as the simple unweighted aggregative method. The below example would help understand the creation of a chain index number:

Example: Considering the following data of 5 years, on annual cost of school fees segregated in fixed charge, tuition fees and transport/ other expense. Create a chain index for the data set.

	2011	2012	2013	2014	2015
Fixed charge	16000	18000	20000	22000	24000
Tuition fees	36000	38000	41000	42000	44000
Transport/other	25400	30000	31200	32400	34200

Solution: Preparing the data in table, with necessary columns for link relatives, etc.

Chapter 30: Index Numbers

Year	Fixed charge	Tuition fees	Transport etc.		Link relative table			Total-link relative	Link relative average	Chain index number
					Fixed	Tuition	Transport			
2011	16000	36000	25400	⇑	100	100	100	300	100	100
2012	18000	38000	30000	⇑	112.50	105.56	118.11	336.17	112.06	112.06
2013	20000	41000	31200		111.11	107.89	104.00	323.00	107.67	120.66
2014	22000	42000	32400	⇑	110.00	102.44	103.85	316.29	105.43	127.21
2015	24000	44000	34200		109.09	104.76	105.56	319.41	106.47	135.44

- Calculation details:
  - The year 2011 has been taken as base year, hence the index number for 2011 is taken as 100.
  - For the link relative value for fixed charge for year 2012 = (18000/16000) × 100 = 112.5 and for year 2013 = (20000/18000) × 100 = 111.11, and so on for each entry.
  - Link relative for the year are added up to give total link relative
  - Average of link relative is derived by dividing the total link relative by the number of link relatives in there.
  - Chain index number for base year is taken as 100. For succeeding years, it is calculated as:

Chain index number

$$= \frac{\text{Period's Link Relative} \times \text{Previous Period's Chain Index}}{100}$$

i.e. chain index for year 2012 = (112.06 × 100)/100 = 112.06; and for year 2013 = (107.67 × 112.06)/100 = 120.66; and so on.

BASE SHIFTING AND SPLICING

Due to continuous changes occurring, a need does arise to shift the base year to a more recent one. This is warranted on many factors including the main ones like: (a) The real value of money gets affected due to inflation or deflation, affecting the prices and quantities over a considerable period of time, (b) the consumption behavior shifts and articles consumed previously may not be much relevant in current periods—for example, coal used for locomotive or for household cooking is not largely valid today, or (c) technological shift bringing highly significant price variations on either side—for example, the computers and laptop prices have continuously reduced, while speed and performance have increased. Technological shift can cause price increase too, without corresponding increase in quality. For example, spares of old models of industrial equipment or machinery etc. are more expensive than they were earlier.

While the series needs to be upgraded to account for these shifts, but it is also not worthwhile to lose all historical data of the Index. When the need is to bring the base closer to current year, in that case, all previous indices are restated as percentage of the new base. For example, the Index numbers for five years are 100, 110, 130, 142 and 156; and the base has to be shifted to fourth year, then the values are restated as (100/142) × 100; (110/142) × 100 and so on. The restated series becomes 70.4, 77.5, 91.6, 100 and 109.9. This process is called 'Base Shifting.' This is generally undertaken when the constituent items included in the Index have not changed too much, but the 'real value of money' (in finance terminology—'Time Value of Money') has considerably changed and accordingly the quantities may have also undergone reasonable change. From the mathematical perspective, base shifting is possible only if the index number method used to generate the index number satisfies 'Circular Test'. However, in practice most of the methods provide reasonable consistency of data for comparison to be done of the previous data.

Against shifting the base, the other concept is of 'Splicing' index number, which arises generally due to shift in constituents of the index number. The items which were originally present in the index may have become obsolete, withdrawn, technologically outdated, consumer taste or consumptions pattern may have changed. In such cases, the importance of index has not reduced, but need is felt to upgrade the constituents of the index number. Consider an example: A small hospital had created an Index of its Purchases. In 1980s when the index was established, it included items like X-ray Films, Sterilizing Equipment for Injections, large amount of paper sheets, file holders, paper weights, etc. If today's, purchases are evaluated, there has been a complete shift in the purchases being done—no X-ray films, instead small discs, reusable glass hypodermic injections to disposable syringes, etc. In such situations, a new index needs to be created.

While creating the index with new constituents or weights, the old historical data is also desired, in order to maintain consistency of the series. In such situations, for the year in which the index shift happens, two index numbers are created. One as per the earlier items, weights and other parameters; and second as per new items, weights, etc. The new reconstituted index will have its value as 100 as it will be its starting year, whereas the index derived on the basis of earlier items and weights will have some value, say 'x.'

Index number for all subsequent years will have the calculations based on revised parameters, whereas for the previous years, the indices will be restated. The restatement would be done by equating the two index numbers calculated in the changeover year, i.e. $x = 100$; and then expressing all previous index values on the basis of this. For example, consider an index series of four years—100, 110, 125 and 120. For same objective, a new series was created in the fourth year and its values are 100 and 102 for the fifth year. This new series will start with 100 as the base year index. For restating the previous series, the common year values are equated, i.e. previous year index numbers are calculated as a percentage of the changeover year's index number, i.e. $(125/120) \times 100 = 104.2$; $(110/120) \times 100 = 91.7$ and $(100/120) \times 100 = 83.3$. Consolidating the index series for the five years becomes—83.3, 91.7, 104.2, 100, 102.

INDEX NUMBER TESTS

Apart from the time and factor reversal tests discussed earlier, the tests include circular test and unit test, which are outlined below.

Circular Test

- Consider three Index Numbers:
  - Index number for year 'z' constructed with base year 'y',
  - Index number year 'y' constructed with base year 'x', and
  - Index number year 'x' constructed with base year as 'z'.

In the first two instances, indices, 'z' is based on 'y', and 'y' is based on 'x'. Against this, if there was a direct calculation done for generating index for year 'x' with 'z' as the base year; then, as per circular test, the value of index number calculated directly, should be same as derived when calculated through base 'y'. Consider the following example:

Year	Price	Simple aggregative index		
		Base year 2013	Base year 2014	Base year 2015
2013	75	100	83	58
2014	90	120	100	69
2015	130	173	144	100

Mathematically, $\dfrac{120}{100} \times \dfrac{144}{100} \times \dfrac{58}{100} = 1$ if it satisfies the circular test. In case the equation $\neq 1$, then circular test is not satisfied. This test is only satisfied by simple aggregative (without weight), fixed weight aggregative method and geometric mean of price relative methods.

Unit Test

This test specifies that the formula for developing the index number should not be influenced by the units of measurement of the constituents. All tests conform to this condition, except simple (without weight) aggregative.

CONCLUSION

Index number, as a concept is very useful and can be made applicable to any data series. For medical field, it can be used to establish say index of noncommunicable diseases, index of cost of operating an hospital, index for number of patients treated, index of productivity of the hospital, in which the constituents could be department wise cases handled, including laboratory and radiology divisions. These may find increasing use with the corporatization of health care and the advent of investment in the medical tourism.

In many cases, each of the variables included in the index is a function of its own constituent variables. Weights need to be provided at the sub-division level and again at the main level. This can be done any number of times, but trade-off has to be managed as each further level increases the cost and resources of managing the index in future.

If while calculating proper attention is given to the objective and detailed notes on: what products/items are being included, why are they being included, their quality parameters, replacement products and their price—quality matrix, data collection techniques to be employed, base year, what weights to be employed, which calculation methodology to be used, how will it be updated and at what frequency, etc. are well-documented, it will ensure the reliability and applicability of index number at a much higher and more productive level.

CHAPTER 31

Statistical Quality Control

Statistical quality control, as the name suggests, pertains to monitoring, maintaining and improving quality of a process or an activity through the use of statistical techniques. This concept relies on the fundamentals of sampling and principles of normal curve, and builds upon them to provide the necessary tools for achieving the quality objective.

INTRODUCTION

In an ever increasing competitive world, every provider of goods and services wishes to deliver good and consistent quality; and every consumer expects to receive the same. There is an inherent difference between good quality and consistent quality. 'Quality' as a word has a positive bias in psychological terms, as 'a quality product' so often seen on product labels, invariably is construed as a good product, but in objectivity, 'quality' is fundamentally a 'specification level' of a product or service, which may be superior, good or inferior/poor.

A superior quality alongwith its specifications and related information is what the research and development or technical departments of an organization create, but 'consistent quality' depends on the manufacturing process, process efficiency, raw material quality, skilled personnel, etc. To ensure quality control, most of the manufacturing organisations and some service providers have quality control departments to check the quality at various stages/levels of the process. Compared to these quality control activities, statistical quality control is more scientific, objective and reliable in its approach to measure, monitor, maintain and improve the quality. Introduction of statistical quality control does not obviate the need to have the quality control department, but on the contrary it depends on quality control department to provide it with data on quality issues, which can then be studied by statistical tools to arrive at appropriate improvements.

A product or service quality could falter due to nonassignable causes or assignable causes. Nonassignable causes are random occurrence and can be attributed to minor imperfections of the process. The minor imperfections of the process, most of the time, cancel out, but sometimes they do not, which impacts the quality. Against this, assignable causes could be due to raw material, labor skills, machine wear and tear, error in process flow, etc. Statistical quality control focuses on studying and controlling the assignable causes. Considering that even nonassignable causes get known in the monitoring of statistical quality control, and as a byproduct, it is prudent to study these nonassignable causes and monitor them to the extent possible, to reduce their occurrence.

Statistical quality control (SQC) relies on sampling, in situations where it is not possible for 100% inspection. In many processes, quality control has become 'in-line' through use of advanced automation solutions and software's for inspection. These devices at various stages of the production process inspect the product and accept (pass through) or reject (eject) them on the basis of predefined parameters. Some processes have even in-line rectification systems, which automatically take corrective actions to bring the parameters within established specifications, without stopping the process. Though such, auto-rectification works, in selected product manufacturing processes but it cannot be applicable to every product.

The SQC does not deal to replace these automated systems, as these are primarily 'inspection devices.' On the contrary, role of SQC is to evaluate the process so that minimum rejection happens. These inspection devices complement SQC as with their help, inspection levels can be made 100%, making available large amounts of data, which can be thus studied through SQC tools. Medical facilities are by definition service industries but they use a lot of products to deliver the service. Hence, the practitioners as healthcare administrators need to understand these tools if they are used in their organizations and indeed encourage their use.

Statistical quality control tools can be applied at process level, i.e. during manufacturing or at end of process, i.e. final inspection. At process level, the aim of SQC is to determine and ensure that the process is in 'control' through the tool known as 'process control charts.'

CONTROL CHARTS

Broadly stating, these are graphical representations of the actual measured values and their frequencies in the process, plotted against the desired values or specification along with permissible tolerances **(Fig. 31.1)**.

The average line, also called the "central line (CL)" is the specified standard of the process, while upper control limit (UCL) and lower control limit (LCL) are control limits or the tolerances which are acceptable as per the specifications. Any measured value lying outside the control limits need to be checked for assignable reasons which are leading to rejections. Even within the control limits, the points are evaluated for existence of patterns, streaks of similar value, points continuously lying on one side, or any another evidence which doubts the randomness of the process. This is due to the assumption that all processes have randomness and if such randomness does not exist, it indicates potential problems. Statistical quality control, along with many other activities, forms the basis of many quality improvement initiatives like corrective action and preventive action (CAPA), etc.

The control limits are set up or based on any of the parameters like past data, current production practice, specification established by the technical department. The central line is first established based on any of these parameters. Thereafter, the LCL and UCL are established, and generally taken as ± 3 sigma (± 3σ) to the central

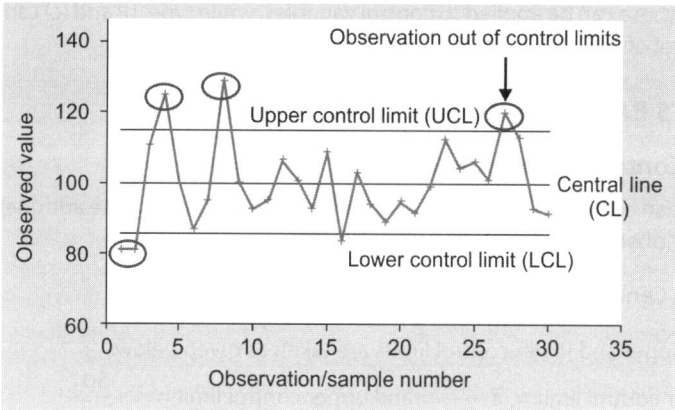

Fig. 31.1: Sample control chart

line. This is because, as per normal curve, ± 3 sigma covers 99.73% values and anything beyond this range would not be a normal occurrence, which needs to be evaluated.

It would be worth noting that the 'control limits' should not be considered as 'specification limits'. Control limit can be referred to as the 'process capability' whereas 'specification limits' refer to 'design parameter' established by product development and technical teams. In current business environment, it is expected that the process has to perform as per the design and as such the specification parameters can be taken as control limits. These limits can also be influenced by the organization's own standards, industry benchmarks, technical standards installed by the industry/government/international standards, but these standards being not a subject under consideration are not being dealt here.

TYPES OF CONTROL CHARTS

Control chart of variables and control chart of attributes are the two types of control charts.

Control Chart of Variables

These charts are used when the product features to be controlled, can be measured or quantified, like dimensions, tensile strength, elasticity, etc. and have a unit of measurement.

Control Chart for Attributes

Control chart for attributes are used where the characteristic, which needs to be controlled cannot be measured, but can be inspected for its presence or absence like product's visual esthetic look, flavor, odor, etc.

To incorporate these possibilities, the different charts are:
(a) Based on actual measurements
(b) Based on number of defectives
(c) Based on defects per unit.

All of these can be applied to control variables, while type (b) and (c) can be used for control of attributes.

CHARTS BASED ON ACTUAL MEASUREMENTS

Mean Control Chart ($\bar{x}$ Chart)

In the mean control chart ($\bar{x}$ chart), the central line is equated to the arithmetic mean ($\bar{x}$) of the observations or samples, i.e.

$$\text{Central line} = \frac{x_1 + x_2 + x_3 + \ldots + x_n}{n} = \frac{1}{n}\sum_{i=1}^{n} x_i = \bar{x}$$

The upper and lower control limits are taken as given below:

$$\text{Lower control limit} = \bar{x} - \frac{3\sigma_{\bar{x}}}{\sqrt{n}} \text{ and upper control limit} = \bar{x} + \frac{3\sigma_{\bar{x}}}{\sqrt{n}}$$

where $\bar{x}$ = arithmetic mean of the samples, n = number of samples, and

$$\sigma_{\bar{x}} = \frac{\sum_{i=1}^{n}(x_i - \bar{x})^2}{(n-1)}.$$

At times, instead of calculating $\sigma_{\bar{x}}$ by the above given formula, it is based on the range (maximum – minimum value), which can provide an estimate of $\sigma_{\bar{x}}$. If the control limits are based on range, then $\sigma_{\bar{x}}$ is given as:

$$\sigma_{\bar{x}} = \frac{\bar{R}}{d_2}, \text{ where } d_2 \text{ depends upon } n \text{ and its values provided in}$$

the d_2 tables.

Accordingly the control limits can be restated as:

$$\text{UCL or LCL} = \bar{x} \pm \frac{3\bar{R}/d_2}{\sqrt{n}} \text{ i.e.} = \bar{x} \pm \frac{3\bar{R}}{d_2\sqrt{n}}$$

In the above equation, if $\frac{3}{d_2\sqrt{n}}$ is taken as A_2, then the control limits become:

$$\text{UCL} = \bar{x} + A_2\bar{R} \text{ and } \bar{x} - A_2\bar{R}$$

(Values of A_2 are given in the tables in appendix)

Example: Randomly five samples, each containing 5 pieces were collected from the production line. Develop the mean ($\bar{x}$) control charts for the process based on the sample observations given below:

Sample	Sample observations				
A	10	8	13	11	10
B	7	9	8	12	11
C	8	7	7	8	9
D	9	6	14	9	12
E	12	8	9	10	14

Solution: Based on the given values, tabulating the data and carrying out the necessary calculations:

Sample	Observations					Σ	Sample mean (x)	Sample range	(x − x̄)	(x − x̄)²
	1	2	3	4	5					
A	10	8	13	11	10	52	10.4	5	0.76	0.578
B	7	9	8	12	11	47	9.4	5	−0.24	0.058
C	8	7	7	8	9	39	7.8	2	−1.84	3.386
D	9	6	14	9	12	50	10	8	0.36	0.129
E	12	8	9	10	14	53	10.6	6	0.96	0.921
							Σ = 48.2	Σ = 26		Σ = 5.072

Mean of all samples $\bar{x} = 48.2/5 = 9.64$
Calculating mean of range $\bar{R} = 26/5 = 5.2$
Central line = Mean of sample means = 9.64
UCL = $9.64 + A_2\bar{R} = 9.64 + 0.577(5.2) = 9.64 + 3.00 = 12.64$
(A_2 values from tables for $n = 5$)
LCL = $9.64 − A_2\bar{R} = 9.64 − 0.577(5.2) = 9.64 − 3.00 = 6.64$
Calculating the limits by $\sigma_{\bar{x}}$ method, i.e the formula is

Control limits = $\bar{x} \pm \dfrac{3\sigma_{\bar{x}}}{\sqrt{n}}$; and $\sigma_{\bar{x}} = \dfrac{\sum_{i=1}^{n}(x_i - \bar{x})^2}{(n-1)}$

Calculating $\sigma_{\bar{x}} = 5.072/(5 − 1) = 5.072/4 = 1.268$

Therefore, the control limits would be given as = $9.64 \pm \dfrac{3(1.268)}{\sqrt{5}}$

= 9.64 ± 1.70.

Hence, upper control limit = 9.64 + 1.70 = 11.34 and lower control limit = 9.64 − 1.70 = 7.94.

It can be observed that by using the approximation method of σ (by using range and A_2), the control limits are not very precise but provide an indication, whereas using the proper sigma calculations, the results are more precise. Nevertheless, these approximations are used whenever immediate estimates are essential. The control limits have been mentioned as 3σ, but depending upon the process, product, quality requirements, etc. they can be fixed at 2σ, or 1σ or any number which is logically evident and appropriate. Naturally 2σ or 1σ limits would be more stringent than 3σ limits.

The **Figure 31.2** for 'mean control charts' for the given example, by both methods stated above would clarify the relative stringency of these methods. The range control chart is provided in the graph and pertains to the next example.

R-Chart

R-chart shows the dispersion or variability of the process. It complements the $\bar{x}$-chart. In case of R chart, the central line is equal to $\bar{R}$, where R is the range of the sample and the control limits are given as $\bar{R} \pm 3\sigma_R$; where σ_R is the standard error of the range. Alternatively the control limits in R-chart can also be calculated using the formula:

$$UCL_R = D_4\bar{R} \text{ and } LCL_R = D_3\bar{R}$$

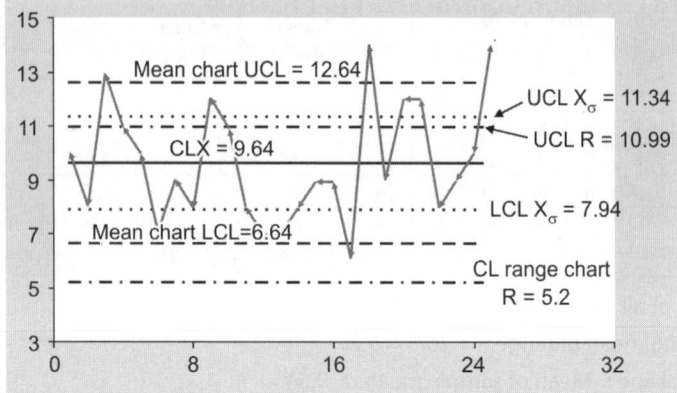

Fig. 31.2: Mean and range control chart

Note:
CL X = Central line of mean chart.
- - - Mean chart *LCL* and *UCL* control limits based on $A_2\bar{R}$ method (In this example, this method shows there are four observations outside the control limits).
...... *LCL* X_σ and *UCL* X_σ are the control limits mean control chart, based on $\sigma_{\bar{x}}$ values. (In this example, this method indicates ten observations are outside the control limits)
-.-. The central line range and *UCL* range are the range control limits (R-chart) as explained in the next part.

(Values of D_3 and D_4 can be taken from tables provided in appendix for given sample sizes n).

In general sense R chart is preferred for small sample size.

Example: Prepare R chart for the previous examples value:

Solution: Representing the data in tabulated form:

Sample	Observations					Sample range
	1	2	3	4	5	
A	10	8	13	11	10	5
B	7	9	8	12	11	5
C	8	7	7	8	9	2
D	9	6	14	9	12	8
E	12	8	9	10	14	6
						$\Sigma = 26$

$\bar{R} = 26/5 = 5.2$
$UCL = D_4\bar{R} = 2.115 \times 5.2 = 10.998$
$LCL = D_3\bar{R} = 0 \times 5.2 = 0$

The R chart indicates the dispersion of the values and is evaluated alongwith x-chart. If the values of R charts are outside the limits, then it is advisable to rectify the process on basis of these before proceeding to evaluate the x-chart.

Standard Deviation Control Chart (σ Chart)

In standard deviation chart, the central line is equal to the standard deviation (σ) of the data and the control limits are:

$$UCL = B_2\sigma \text{ and } LCL = B_1\sigma.$$

Like the constants A_2, D_4, etc. the values of constants B_2 and B_1 for given 'n' are provided in the control chart tables (see appendix).

In case the standard deviation (σ) of the process is not known, then the limits can be calculated as:

$$UCL = B_4\bar{s} \text{ and } LCL = B_3\bar{s};$$

where $\bar{s}$ = mean of the sample standard deviation = $\Sigma\sigma/n$

Standard deviation chart and range charts are used together as both indicate the dispersion of the variable. For immediacy and ease of calculation range chart is preferred, whereas for better accuracy the standard deviation chart is better.

'C-CONTROL CHART' FOR NUMBER OF DEFECTS PER UNIT

These charts are used where potential of occurrence of defect is large, but actual occurrences are small, e.g. an unwanted black dot in newsprint roll. The central line of the control chart is denoted as:

$\bar{C}$, where C is the number of defects per unit and $\bar{C}$ is the mean of these defects (per unit basis) of many such samples.

The control limits of C chart are: $UCL/LCL = \bar{C} \pm 3\sqrt{\bar{C}}$.

C-control chart require uniform sample sizes, otherwise p-charts are used **(Fig. 31.3)**.

Example: The number of defects in 10 rolls of newsprint is given below. Develop the control limits.

Roll no.:	1	2	3	4	5	6	7	8	9	10
No. of defects:	1	3	2	4	2	1	2	3	2	1

Solution: For the above given values $\bar{C}$ = avg. no. of defects per roll = 21/10 = 2.1.

Therefore, $UCL = \bar{C} + 3\sqrt{\bar{C}} = 2.1 + 3(\sqrt{2.1}) = 2.1 + 3(1.45) = 2.1 + 4.35 = 6.45$ and $LCL = 2.1 - 4.35 = -2.25 = 0$, as it cannot be negative.

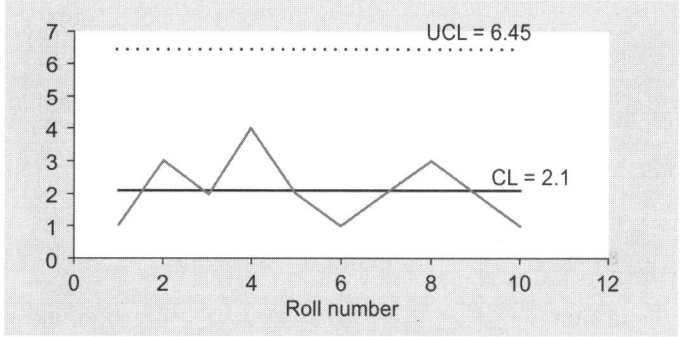

Fig. 31.3: C-control chart for number of defects per unit

Section 8: Statistical Concepts for Medical Administration

P-CONTROL CHART (FRACTION DEFECTIVES OR PROPORTION OF DEFECTS)

P-control chart **(Fig. 31.4)** takes into consideration percentage of defectives per sample. Such treatment helps in situation where sample sizes vary.

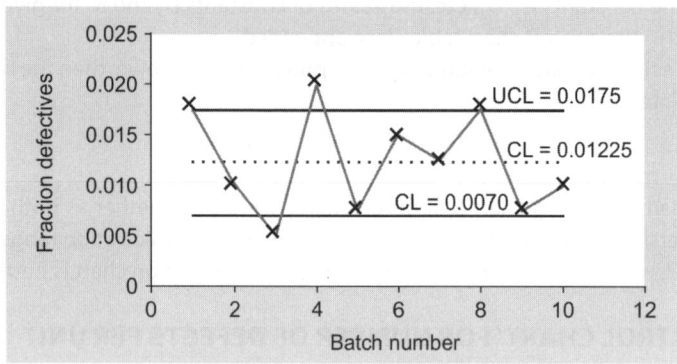

Fig. 31.4: P-control chart for fraction defectives

The central line for P chart and the limits are as given:
Central line (CL) = $\bar{P}$, where P is no. of defects and $\bar{P}$ is average percent defectives, i.e. $\dfrac{\text{Number of defectives}}{\text{Number of units}}$

The control limits are UCL and LCL = $\bar{P} \pm 3\sqrt{\dfrac{\bar{P}(1-\bar{P})}{n}}$

Example: A packing and filling operation in a printing ink manufacturing plant operates on batch processing. In each batch 400 cans are filled and packed. Defect occurs when the filled cans do not have label affixed on its correct position on the can. The data for ten batches is given below. Establish P-chart for the operation.

| Batch No: | 1 | 2 | 3 | 4 | 5 | 6 | 7 | 8 | 9 | 10 |
| No. of defectives: | 7 | 4 | 2 | 8 | 3 | 6 | 5 | 7 | 3 | 4 |

Solution: Presenting the provided data in tabular format, along with the fraction defective calculations:

Batch no.	No. of defects	Fraction defects
1	7	=7/400 = 0.0175
2	4	0.0100
3	2	0.0050
4	8	0.0200
5	3	0.0075
6	6	0.0150
7	5	0.0125
8	7	0.0175
9	3	0.0075
10	4	0.0100
	Σ = 49	

$n = 400 \times 10 = 4000$

Calculating the central line and the limits:

Central line $\bar{P} = \dfrac{P}{N} = \dfrac{49}{4000} = 0.01225$

$$UCL = \bar{P} + 3\sqrt{\dfrac{\bar{P}(1-\bar{P})}{n}}$$

$$= 0.01225 + 3\sqrt{\dfrac{0.01225(1-0.01225)}{4000}}$$

$$= 0.01225 + 0.0052$$
$$= 0.0175 \text{ (rounded)}$$

and $\quad LCL = \bar{P} - 3\sqrt{\dfrac{\bar{P}(1-\bar{P})}{n}} = 0.01225 - 0.0052 = 0.0070$

Hence, the P control limits are: $CL = 0.01225$, $UCL = 0.0175$, $LCL = 0.0070$.

Analyzing the given data in relation with the control limits established, it may be noted that two points (Batches) lie outside the control limits, i.e. where the number of defects in the batch is 2 (Lower than LCL) and 8 (Higher than UCL). The operations team can investigate these two points on immediate basis, along with introducing measures to reduce the number of defects per se.

ACCEPTANCE SAMPLING

Acceptance sampling relates to the sampling inspection done by the purchaser of goods in order to decide whether to accept or reject the supply of goods. This is necessary as it is not feasible to conduct a 100% inspection of goods. Therefore, the purchaser decides on an acceptance number, which is the threshold limit of the number of defects, beyond which, if the supplied material contains, it stands rejected.

If a decision to accept or reject the supplied material is based on a single sample plan, it is called as 'single sampling plan.' It is termed as plan, because it may involve multiple samples to be drawn from the supplied batch. For example: A hospital receives 1000 bedsheets against its purchase order. It may establish a single sampling plan as follows:
- Total supplied batch = N = 1000 (No. of bedsheets supplied)
- Sample size = n = 25 (No. of bedsheets to be checked for quality)
- Acceptance level = c = 2 (If the sample contains not more than 2 defective bedsheets, it is accepted, and if it contains more than 2 defectives, it stands rejected).

The main drawback of single sampling plan is that it appears to be biased towards the purchaser of the goods. The acceptance level is established by the purchaser, the inspection happens mostly at the purchaser's location and is conducted by the purchaser without any representative from the supplier's being present. All and all, it appears a little bit unconvincing or sounds unfair to the seller. To establish some level playing field to both purchaser and supplier, some buying organizations operate on 'double sampling plan' for deciding on acceptance or rejection of the supply.

In double sampling plan, a second sample or sample set (as it may contain many samples) is drawn, but decision to accept or reject is based on combined results of first and second sample sets.

Continuing the previous example used in single sampling plan, the notations for double sampling plan are:

N = Total size (1000 bedsheets)
n_1 = First sample size (25 bedsheets)
c_1 = Acceptance level for first sample (2 bedsheets)
n_2 = Second sample size (50 bedsheets)
c_2 = Acceptance level for second sample (assume 5 bedsheets)

Explanation: From the supply of 1000 bedsheets, a sample of size 25 bedsheets is to be taken as first sample. If this sample contains defective (c_1) i.e. ≤ 2, then the supply is accepted. If this sample contain defectives greater than first acceptance level (c_1) but less than or equal to second acceptance level (5 in this case), then second sample of 50 bedsheets needs to draw for quality testing. If the defectives in total combined sample size, i.e. $n_1 + n_2$ is less than or equal to c_2, the supply is accepted, i.e. number of defective in sample sizes (25 + 50) is less than or equal to 5, will result in acceptance, otherwise the supply is rejected. In case the number of defectives in first sample is higher than the acceptance level of second sample (c_2), then the supply is rejected without pursuing the testing of second sample.

Depending on the need, necessity or any relevant factor, the sampling plan can be extended further to give rise to multiple sampling plan. Generally, double sampling plan suits most of the situations, but in case the supply size N, is very large, either triple sampling plan may be resorted to or the sample size can be increased.

Whether it is single, double or multiple sampling plan, the key consideration is that all these are based on sampling techniques and hence open to inherent limitation of sampling errors creeping in. Similar to hypothesis testing, in acceptance sampling it may also happen that an acceptable supply gets rejected or rejectable supply gets accepted. The first instance where an acceptable supply get rejected in called as 'producer's risk' and denoted by 'α', and the second instance, where a supply gets accepted inspite of it of inferior quality is termed as 'consumer's risk' and denoted by 'β'.

The probabilities of accepting a defective lot can be calculated through use of Poisson distribution, as it is assumed that population size is large (that is why sampling is being done), and probability of occurrence of defective is very small. These are characteristics of Poisson distribution. For understanding, consider the following example:

Example: Consider a single sampling plan, consisting of N = 2000, n = 120 and c = 1. What is the probability of accepting a supply containing (a) 1% defectives and (b) 0.5% defectives.

Solution: Given total supply Size N = 2000, sample size n = 120 and acceptance level c = 1. As this follows Poisson distribution, using its probability formula $P(r) = \dfrac{e^{-\lambda}\lambda^r}{r!}$ by representing the data in tabular form.

Situation	No. of defectives in the sample size	P(0)	P(1)	Total probability P(0) + P(1)
1% defective	(1/100) × 120 = 1.2	0.301	0.361	0.662
0.5% defective	(0.5/100) × 120 = 0.6	0.549	0.329	0.878

Using the Poisson formula with $\lambda = 1.2$ in first case and 0.6 in second, i.e.
Calculations for 1% defective $P(0) = e^{-\lambda} = e^{-1.2} = 0.301$
$P(1) = e^{-\lambda}\lambda = e^{-1.2} (1.2) = (0.301)(1.2) = 0.361$
Similarly for 0.5% defective $P(0) = e^{-0.6} = 0.549$
$P(1) = e^{-0.6}(0.6) = 0.329$

Probabilities till the acceptance level are calculated, as beyond this level, the supply anyways would be rejected.

The probability that a supply of 2000 size and containing 1% defective items would get accepted is 66.2% (the total defective in the entire lot = 1% of 2000 = 20)

While the probability for the similar supply but having only 0.5% defectives would get accepted is 87.8% (the total defective in the entire lot = 0.5% of 2000 = 10). As the absolute number of defectives is half of previous 1% defective, the probability increases but does not reach 100%. It will tend towards 100% if the total number of defectives in the supply tends to zero, or the number of defectives goes lower than the acceptance level c.

For further understanding, assume that in the same example, the number of defectives in the lot was 0.05% (i.e. 0.05% of 2000 = 1). This results in the number of defective becoming equal to the acceptance level. Calculating the probability for such a situation, the results obtained are as follows:

Number of defectives in a sample size of 120 = (0.05/100)*120 = 0.06.
$P(0) = e^{-0.06} = 0.9418$
$P(1) = e^{-0.06} (0.06) = 0.0565$

Total probability = 0.9983, i.e. 99.83%. This value is higher than those for 0.5% and 1%, which indicates higher probability of acceptance as it contains less number of defectives. Statistically the probability will become 100% when the absolute number of defects reaches zero because $e^{-0} = 1$, i.e. 100%.

Irrespective of whether it is a loss for producer or consumer, all the acceptance sampling quality tests are time consuming, use resources and have cost attached. An ideal situation is where the supplier does the outgoing quality inspection as per its own standards and also any other test, according to any special needs of the customer. The detailed quality inspection report is shared with the consumer, who on the basis of this report accepts the supply. Such instances of supplier-buyer partnerships are common in Japanese business environment, and also being practiced by very few but certain sections of Indian businesses.

Statistical quality control and acceptance sampling can be used in most of the commercial activities including medical field. With medical facilities getting institutionalized, such tools come in handy to establish standards and monitor performance of services to patients, as well as in buying process.

CHAPTER **32**

Decision Theory

Our lives revolve around decision making—right from mundane things like which shirt to wear to important decisions involving investments and taking official decisions. Fundamentally, decision making involves selecting the appropriate option among the multiple alternatives available. Though it sounds simple, decision making is one of the most important and complex aspects of our lives. Where we stand today in our lives is the sum total or the result of the choices and decisions we took at every stage of our lives. Life presented us with situations and circumstances, which upon evaluation resulted in alternatives and from which we opted for one of them or at times even none. Sometimes situations present *Hobson's Choice,* i.e. only one option to choose but then a *no decision* at times is also a decision.

What treatment method needs to be followed? Is surgery to be conducted now or should it be postponed? These are also decisions. Amongst others, decisions are influenced by the knowledge, information, personality, experience of the decision maker. Statistically, there are tests of hypothesis which can help in decision making, but they cannot be used as quantitative data, may not be always available, number of causing factors may be multiple and the economic consequences of decision cannot be captured in these tests, etc. With medical field now expanding very fast and being run as corporate, lot of medical professionals are also involved in the management of the hospitals. Accordingly, apart from medical-related decisions a medical practitioner needs to take decisions on investment of new equipment, their payback, operating costs, etc. Hence, the chapter on decision theory has been incorporated, so as to allow the medical professionals to get introduced to the subject of scientific process of decision making.

The decision theory is a systems approach for evaluating, selecting with the help of quantitative tools, an alternative out of the various alternatives available, for achieving an objective. Every decision relies on multiple factors which could be controllable or uncontrollable, and known or unknown. Considering that there is time and cost involved in gathering data, decision maker should be aware of how much additional time or cost is necessary and how much accuracy can this bring to the decision, while making decisions.

Before the methods of decision making can be discussed, it is pertinent to understand the terminology involved in the theory of decision making.

NOTATION/TERMINOLOGY

Space of Alternatives

The list of all the various alternatives available for selection as a decision is called the Space of Alternatives or Activity Space. For deciding an X-ray machine, all the brands available would conclude as the activity space.

State Space (Outcomes)

Every decision has an outcome, which depends on various factors affecting it. For example, a decision to invest in new efficient equipment may result in higher price (to recover the cost), better quality, improved customer service as waiting time may reduce, higher cost of maintenance or vice versa. These outcomes of one decision may or may not have equal probabilities, but all these possible outcomes are collectively called as 'State Space' or 'Nature Space'.

The 'Outcomes' should not be construed as 'results' of the decision but the possible factors that will determine the 'pay-off' or 'quantitative result' of the decision.

For example: An organization wants to open a new hospital in a city. As a norm, it builds a 50 bed tertiary care hospital, if expected demand is low, a 200-bedded secondary care hospital, if the expected demand is reasonable and a 500-bedded hospital, if it expects demand in the area to be significant.

Explanation: In this example, the action space or alternates is the 'kind of hospital.' The pay-off quantitative result depends upon the level of demand. Thus 'level of demand' is the 'Outcome' or 'States of Nature.'

It cannot be predicted with accuracy, the 'state of nature' that will occur; hence, it is not possible to establish accurately the exact result of the decision. But, it is possible to attach probability or likelihood of occurrence of a particular 'state of nature.' In the example above, additional information can be stated as the probability of significant demand is 0.4, medium demand is 0.5 and low demand is 0.1. This becomes the basis of the next terminology.

Value or Pay Off

Considering that there are 'x' number of alternatives and 'y' states of nature. Combining these, results in $(x \times y)$ consequences. These consequences can be stated in terms of their economic value, profit, cost, loss, etc. This value is known as 'Pay Off' of the specific combination of a decision and the specified outcome.

Risk versus Certainty versus Uncertainty

In case, all the alternatives and outcomes are accurately known, then pay off can be calculated with certainty, and therefore these situations have neither risk nor uncertainty.

In case, all the alternatives are known but the outcomes or states of nature are not known for certainty, but based on current or past statistical data/behavior, some probability can be attached to each state of nature, then such a situation is called 'Risk.'

In case, there is little or no statistical evidence about the states of nature, then it is termed uncertainty.

Decision Criteria

From same set of alternatives and states of nature, different decisions are possible. Which alternative gets selected depends on the objective. Therefore, a decision criteria needs to be specified and accordingly the pay off table is developed. The decision criteria could be expected monetary value or expected opportunity loss or something else. A sample pay off table is given here:

Sample pay-off table		Alternatives for size of hospital		
		50 bed	200 bed	500 bed
Outcome/States of Nature	Low	P_{11}	P_{12}	P_{13}
	Medium	P_{21}	P_{22}	P_{23}
	Significant	P_{31}	P_{32}	P_{33}

P_{11} = Pay off/Expected value of exercising alternative 1 in outcome/State of nature 1, and so on.

Opportunity Loss

Opportunity loss is defined as the difference between the highest possible value for an outcome and the value of alternative selected, or in other words, the profit foregone while selecting an alternative.

Example: A perishable goods manufacturer wants to decide production level, but is faced with fluctuating demand. The possible options for production given are 50, 60 and 75 tonnes per month, whereas demand is expected to be 40, 60 and 80 tons/month. The cost of production is ₹ 100/ton and sales price is ₹ 150/ton. Develop a Pay-off Table for expected value and for an expected opportunity loss.

Solution:
- The alternatives provided are production level of 50, 60 and 75 tons/month.
- The states of nature/outcomes are demand levels of 40, 60 and 80 tons/month.
  Production cost/ton = ₹ 100   Sales revenue/ton = ₹ 150

Pay-off table		Alternatives: Production level (tons/month)		
		50	60	75
Outcome: Demand (Tons/month)	40	1000	0	−1500
	60	2500	3000	1500
	80	2500	3000	3750

Calculation P_{ij} = (Sales Quantity × Sales Price) − (Production Quantity × Production Cost)

On basis of this calculating all the pay offs:
- P_{11} : (40 × 150) − (50 × 100) = 6000 − 5000 = 1000
- P_{12}: (40 × 150) − (60 × 100) = 6000 − 6000 = 0
- P_{13}: (40 × 150) − (75 × 100) = 6000 − 7500 = −1500

- P_{21}: (50 × 150) – (50 × 100) = 7500 – 5000 = 2500
  (Demand is higher, but production is a constraint)
- P_{22}: (60 × 150) – (60 × 100) = 9000 – 6000 = 3000
- P_{23}: (60 × 150) – (75 × 100) = 9000 – 7500 = 1500
- P_{31}: (50 × 150) – (50 × 100) = 7500 – 5000 = 2500
- P_{32}: (60 × 150) – (60 × 100) = 9000 – 6000 = 3000
- P_{33}: (75 × 150) – (75 × 100) = 11250 – 7500 = 3750

Expected opportunity loss (EOL) table: As EOL is the difference between maximum pay-off of an alternative and the alternate selected for a given state of nature/outcome. In the given case, the maximum pay-off for first outcome—demand at 40 tons/month, the pay offs for the three alternatives are 1000, 0 and –1500. Therefore the maximum pay off is for P_{11}. Hence, EOL for P_{11} = 1000–1000 = 0; EOL for P_{12} = 1000–0 = 1000 and EOL for P_{13} = 1000–(–1500) = 2500, as shown below:

EOL table		Alternatives: Production level (tons/month)		
		50	60	75
Outcome: Demand (Ton/month)	40	0	1000	2500
	60	500	0	1500
	80	1250	750	0

Explaining EOL: Assume that the state of nature turns out to be 40 tons/month. For this demand level, maximum pay off is given by alternate: 50 tons of production. In such a scenario, if the decision had been 60 tons, then there would have been no profit no loss (Pay Off =0), but it would have lead to 'Loss of opportunity" to earn ₹ 1000, had alternative (50 tons production) been opted. Hence, EOL is not an actual loss, but more so a loss of opportunity to have earned more.

DECISIONS METHODS

As outlined previously, decisions are made within three conditions, viz. certainty, risk and uncertainty. Each of these conditions requires different treatments, with the final decision depending on the objective.

Decision Under Certainty

These decisions are easiest as all alternatives, as well as states of nature are known and quantified. Pay offs can be calculated with accuracy with these details and appropriate decision can be made. Such situations are very seldom seen.

Example: Three X-ray machines in a hospital have a per day operating cost of ₹ 10000, ₹ 15000 and ₹ 20000. The number of patients each machine can handle per day is 600, 700 and 800 respectively. On a given day only one operator is available and patients registered for X-ray are 620. Which machine should the operator use to maximize profit, given that the hospital keeps profit margin of ₹ 200 for each X-ray test.

Solution: Let the three machines be denoted as M1, M2 and M3. The pay off table can be prepared as below:

	M1	M2	M3
Operating cost/day (₹)	10000	15000	16000
Capacity/day	600	700	800
Demand/day	620	620	620
Profit/unit (₹)	200	200	200
Expected gross profit (₹)	1,20,000	1,24,000	1,24,000
Less operating cost (₹)	−10000	−15000	−16000
Pay off	1,10,000	1,09,000	1,08,000

Hence, if profit maximization is the objective, then Machine M1 must be used, even though it having marginally lesser capacity than the expected demand. It will entail postponing 20 patients X-ray for the next day, but statistically this is the best option for profit maximization. In case, customer service is the objective, then turning back 20 customers would not be an ideal situation. In that case use of Machine M2 would be the appropriate decision, as the pay-off is better than M3.

Decision Under Uncertainty

Such situations are highly prone, as statistically no evidence is available about the states of nature. Decisions in this situation do not have scientific basis and manifest themselves as pure guess work, being highly influenced by the decision maker's knowledge and psychological traits.

Decision Under Risk

This situation presents itself where the states of nature are not accurately known, but based on statistical evidence, probabilities can be attached to the states of nature and pay-off table is created.

Creation of pay-off table is an important aspect. After its creation, arriving at a conclusion and decision is relatively simpler. In practical situations, the decision process most likely to be taken is to use the statistical tools to objectively narrow down the alternatives to two or three; and then decide after fine tuning and minor adjusting these alternatives for any state of nature which could not have been quantified or have not been considered. For example, in the previous example, there are off-beat chances that few patients may be unregistered but may require X-ray service, then the best decision is M2, because it needs only 5 more patients to have pay-off equal to maximum pay-off ₹ 1,10,000.

Accordingly, there are many ways to look at the decision making. Though there is no fixed rule to decide which alternate to select, but some principles exist, which can guide the decision maker towards optimal decision.

PRINCIPLES OF DECISION MAKING

Maximin Principle (Maximize the Minimum Pay-off)

It involves finding the minimum pay-offs for each of the various states of nature against an alternative, and then selecting the specific 'alternative-state of nature' combination which has maximum pay-off among these minimums. In the earlier example, on perishable goods' demand and production, if Maximin Principle is applied, the minimum pay-offs for the an alternative is ₹ 1000 for 50 ton Production level, 0 for 60 ton production level and –1500 for 75 ton production level. Choosing the maximum out of these results in selection of ₹ 1000, hence the combination of (Alternative) Production level 50 tons and when expected (State of nature) Demand is 40 tons, as the Decision. This principle tends to have a pessimistic and cautious approach toward decision making.

Minimax Principle

This principle is applied to the Opportunity Loss Table instead of Pay-off Table. It directs to observe all opportunity loss values over various states of nature and select the maximum opportunity loss for each state. Next step is to decide the alternative corresponding to the nature state which has minimum opportunity loss. Applying this principle, to the same example, maximum opportunity loss for alternative 50 ton production level is ₹ 1250, and for 60 ton = ₹ 1000 and for 75 ton = ₹ 2500. As a next step, selecting the minimum of these, i.e. ₹ 1000 and the alternative pertaining to this values becomes the decision.

Maximax Principle

As the name suggests, this involves selecting the maximum value for each alternative among various states of nature, and then selecting the maximum of these. It considers an optimistic approach and relatively aggressive stance of decision making.

Bayes' Principle

As per Bayes' principle, pay offs are calculated considering the assigned probabilities to the occurrence of the various states of nature. The assigned probabilities to the states of nature should be non-negative number and the sum of these probabilities assigned to various states of nature should be 1. The basis of these probabilities could be statistical data analysis of past behavior (Objective) or the confidence of the decision maker (Subjective).

According to Bayes' Principle, the best alternate is the one which has highest expected pay-off, considering probabilities of each state of nature against each alternate. The pay-offs derived on basis of the probabilities are also known as 'Expected Monetary Value' and the criteria 'EMV Criteria.'

Within Bayes' Principle, the decision arrived, either maximum pay-offs or minimum opportunity loss, is called as 'Prior Analysis.'

Example: Continuing with the same example and incorporating additional data on probabilities, hypothetically based on past behavior of demand/state of nature, the table is reproduced. Calculate pay offs and decide the best option.

Solution: Reproducing the pay-off table along with the probabilities of various states of nature is as below:

Section 8: Statistical Concepts for Medical Administration

Pay off table		Alternatives: Production level (tons/month)		
Demand	Probability	50 (A1)	60 (A2)	75 (A3)
40 Tons	0.4	1000	0	–1500
60 Tons	0.3	2500	3000	1500
80 Tons	0.3	2500	3000	3750

On basis of this, calculating the expected pay off for each alternative
- EP (A1) = 1000 x 0.4 + 2500 x 0.3 + 2500 x 0.3 = 1900
- EP (A2) = 0 x 0.4 + 3000 x 0.3 + 3000 x 0.3 = 1800
- EP (A3) = –1500 x 0.4 + 1500 x 0.3 + 3750 x 0.3 = 975.

On basis of this, Alternate A1 is selected as it has maximum expected pay off.

The decision maker can either accept or reject this alternate derived from prior analysis, or he may review the decision by probing more about the probabilities. In this example, the pay-off of EP (A1) and EP(A2) are close enough for further investigation to be done.

Perfect Information

The process of calculating the expected pay off on basis of probability is an optimal decision and not a perfect decision. For a perfect decision, perfect probabilities are needed. Terming the perfect probabilities, as a 'Perfect Predictor', and assuming that a perfect predictor exists, then on basis of such predicted probabilities, the expected pay offs can be maximized or loss can be minimized. For deriving a perfect predictor, perfect information is required. The expected pay off when perfect information is known is denoted as Expected Pay-off of Perfect Information (EPPI). Availability of EPPI converts the decision from Risk to Certainty.

Logically, then the gap between the expected value (EP), which provides optimal pay off and the EPPI, is due to the information available. This difference is called Expected Value of Perfect Information (EVPI),

$$i.e.\ EVPI = EPPI - EP\ or\ EVPI = EPPI - EMV_{Max}$$

The objective of this analysis, which is also referred as 'Preposterior Analysis' is to determine whether additional data to improve the accuracy of the decision is desirable or not. Additionally, it allows determination of the maximum cost which can be incurred for gaining this additional data and information. There is no reason to spend more than EVPI in order to convert the EP into EPPI, as thereafter the additional cost would reduce the pay off to that extent.

Continuing with the same example, wherein in the previous stage the EP had been calculated, being reproduced below, along with the expected pay off (EP).

Pay-off table		Alternatives: Production level (tons/month)		
Demand	Probability	50 (A1)	60 (A2)	75 (A3)
40 Tons	0.4	1000	0	–1500
60 Tons	0.3	2500	3000	1500
80 Tons	0.3	2500	3000	3750
Expected pay off (EP)→		1900	1800	975
EP_{Max}		Alternate A1 (EP = 1900)		

Calculating EPPI for these values: The maximum pay-off decision, provided the state of nature is known, i.e. if the demand turns out to be 40 Tons, the best alternative would be A1 (giving maximum payoff), similarly if the demand turns out to be 60 tons, the best alternative would have been A2, and if the demand had been 80 Tons, the best alternate would have been A3. Hence, the EPPI is calculated as the sum of 'product of the highest pay-off for each state of nature and its probability', i.e.

EPPI = $1000 \times 0.4 + 3000 \times 0.3 + 3750 \times 0.3$
= ₹ 2425

Therefore, EVPI for these values = EPPI − EP_{Max} = ₹ 2425 − 1900

EVPI = ₹ 525, which implies that the maximum amount that can be spent for gathering more information is ₹ 525, as any expense beyond this level, would result in reduction of the pay off.

Posterior Analysis

Assume as a next step to the ongoing illustration, the decision maker decides to research further, in order to improve the state of nature probabilities and gathers additional information. In posterior analysis, the new information is combined with the prior probabilities and the resulting new probabilities are then called as 'Posterior Probabilities' and the pay offs calculated from these are called 'Posterior Pay-offs.'

Calculating posterior pay-offs: Continuing with the same example, based on further research on demand forecast, the reliability statistics on forecasted demand are as follows:

Conditional probability table		Level of demand		
		40 Tons	60 Tons	80 Tons
States of nature- Demand	40 Tons	0.70	0.20	0.10
	60 Tons	0.20	0.60	0.20
	80 Tons	0.10	0.30	0.60

From the above table, it is inferred that the probability = 0.70 is the probability that while the actual demand is 40 tons (from state of nature), the research also points out the demand to be 40 tons. Similarly, the probability figure 0.30 signifies that while the actual demand is 80 tons, the research points to a demand of 60 tons. Indirectly, it reflects the reliability of the probabilities.

From the above conditional probability table, joint and marginal probabilities are calculated, as follows:

Joint and marginal probability table		Alternatives: Production		
Demand	Prior probabilities	50 Tons	60 Tons	75 Tons
40 Tons	0.4	0.28	0.08	0.04
60 Tons	0.3	0.06	0.18	0.06
80 Tons	0.3	0.03	0.09	0.18
Marginal probabilities		0.37	0.35	0.38

Joint probability = Prior probability × Corresponding conditional probability = $0.4 \times 0.7 = 0.28$, similarly $0.4 \times 0.2 = 0.08$, and so on.

From the joint and marginal probability table, posterior probability table is calculated, as follows:

$$\text{Posterior probability} = \frac{\text{Joint probability}}{\text{Marginal probability}}$$

(Example = 0.28/0.37 = 0.76)

Posterior probability table	Alternatives: Production		
Demand ↓	50 Tons	60 Tons	75 Tons
40 Tons	0.76	0.23	0.143
60 Tons	0.16	0.51	0.214
80 Tons	0.08	0.26	0.643
	1.00	1.00	1.00

Based on the posterior probabilities, the expected pay off are calculated for each estimate, as: Reproducing the pay-off table established initially as these values would be used to calculate the expected pay off (posterior).

Pay-off table		Alternative: Production level (tons/month)		
Demand	Probability	50 (A1)	60 (A2)	75 (A3)
40 Tons	0.4	1000	0	−1500
60 Tons	0.3	2500	3000	1500
80 Tons	0.3	2500	3000	3750
Expected pay-off (EP) →		1900	1800	975
EP_{Max}		Alternate A1 (EP = 1900)		

Sum of products of each alternative with posterior probabilities against each alternative, as given here:
- 1000 × 0.76 + 2500 × 0.16 + 2500 × 0.08 = 1360
- 1000 × 0.23 + 2500 × 0.51 + 2500 × 0.26 = 2155
- 1000 × 0.14 + 2500 × 0.21 + 2500 × 0.64 = 2286
- 0 × 0.76 + 3000 × 0.16 + 3000 × 0.08 = 720
- 0 × 0.23 + 3000 × 0.51 + 3000 × 0.26 = 2310
- 0 × 0.14 + 3000 × 0.21 + 3000 × 0.64 = 2571
- −1500 × 0.76 + 1500 × 0.16 + 975 × 0.08 = −600
- −1500 × 0.23 + 1500 × 0.51 + 975 × 0.26 = 1395
- −1500 × 0.14 + 1500 × 0.21 + 975 × 0.64 = 2518

Representing these in tabulated form in an expected posterior pay-off table:

Expected posterior pay-off table	Alternatives: Production		
Demand ↓ (State of Nature)	50 Tons (A1)	60 Tons (A2)	75 Tons (A3)
40 Tons	1360	720	−600
60 Tons	2155	2310	1395
80 Tons	2286	2571	2518

Analyzing the preceding table, it can be inferred that if demand is 40 tons, then choosing alternate A1 (50 tons) is the best decision, similarly, A2 for demand at 60 tons and A2 for demand at 80 tons. The best pay off is for A2 at ₹ 2571, hence, alternative 60 tons should be considered as the best decision.

DECISION TREE ANALYSIS OR TREE DIAGRAM

The diagrammatic representation of the alternatives, their possible outcomes or states of nature, and the associated probabilities along with expected pay-offs is called the tree diagram. It is particularly useful in cases where the complexity of the action space or outcome, makes it difficult to represent the data in a matrix format. The symbols used in tree diagram are standardized and are mentioned here:
- *Event*: Represented by 'Circle,' an event denotes situation which can have one or more outcomes. In case, it has only one outcome or state of nature, then it becomes a certainty. Expected monetary value is used at every event.
- *Decision box*: Represented by a square box, it denotes the alternatives.

Methodology

1. List all the alternatives and events properly and systematically. The tree develops in chronological order, in terms of events and decisions. Any assumption should be recorded as a note separately.
2. Apply the appropriate probability values to the branches and sub-branches. Expected values are to be calculated and final pay-offs are listed at the end of each branch.
3. Locate the branch having largest expected value to arrive at the EP_{Max} and trace the branch to reach the alternative which needs to be selected.

The main advantage of the decision tree is its flexibility to simulate changes in events and outcome in much easier than other methods.

Example: A medical professional is planning to invest in starting a clinic. For this four sites have been shortlisted and based on a survey, the situation can be described as here:

Site	Investment size (in lac rupee)	Expected return (R) and its corresponding probability (P)					
		R1	P1	R2	P2	R3	P3
A	100	10%	0.2	7%	0.3	2%	0.5
B	90	8%	0.4	6%	0.4	3%	0.2
C	70	10%	0.3	5%	0.7		
D	60	15%	0.5	10%	0.5		

In addition to these, he has the option to invest the amount of ₹ 70 lacs in a fixed deposit and earn an interest of 8.5% pa. Use tree diagram to arrive at the best option maximizing the return on investment in one year's time.

Solution: From decision tree analysis **(Fig. 32.1)**, site *D* has maximum expected pay off of ₹ 7.5 lacs and hence should be selected as the decision of investment.

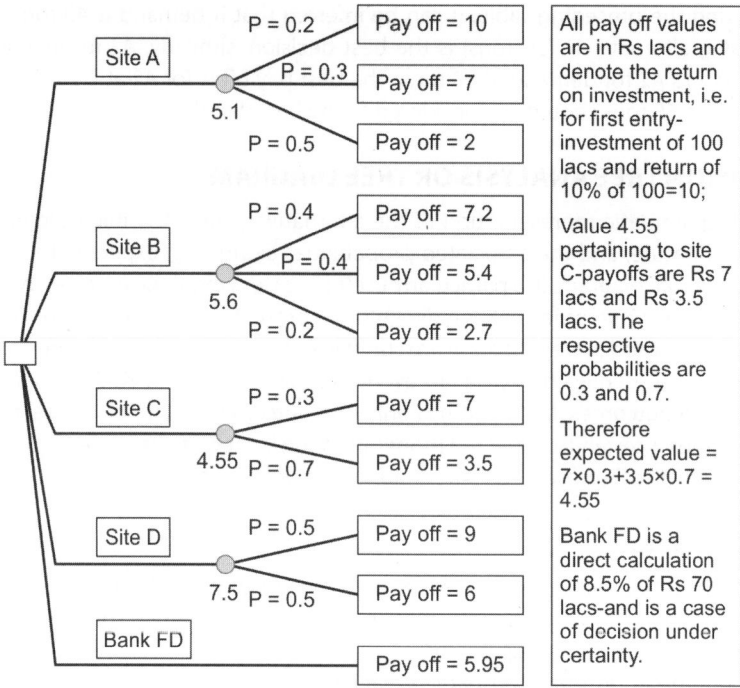

Fig. 32.1: Decision tree analysis

The ease of handling decision tree makes it a preferred choice for usage. The example shown is a simplistic approach, whereas the tree diagram can be extended to as many levels as desired, by following the same pattern of depicting the nodes and calculating the probabilities.

In conclusion, theory of decision making does not provide decisions, but it greatly helps to narrow down the alternatives, as well as highlight the relative difference among the alternatives. This allows the decision maker to maintain objectivity in the decisions and still remain flexible towards subjectivity and incorporate it, if required to finalize the alternative as a decision.

Section 9

Report Writing and Applying for Grants

- Research Report Writing: How to do?
- Applying for Grants

CHAPTER 33

Research Report Writing: How to do?

Research is an exploration into the grey areas or unknown aspects of a subject. New knowledge so created needs to be communicated to the stakeholders and the community at large. This can be done through a research report or a paper. Science depends on unbiased, objective and succinct observations, comprehensive analysis of the results arising from these observations, and precise transmission of facts and ideas. Preparation of a comprehensive, comprehensible and concisely written research report is an essential part of any valid research effort. Interim reports are nowadays required by ethics committees and data safety monitoring boards to determine if the results merit an early termination of a given trial. Therefore, the investigator should be prepared to write and file his reports at a short notice if required.

This chapter outlines how to write research reports in order to bring out the essence of the project and help the progress of scientific research. It also briefly discusses the pitfalls, which need to be taken care of and suggestions on presentation of tables, figures, etc. No universal format can be framed for report writing and these are dictated by the institutions to which the report is submitted depending on the purpose for which the report is submitted. For example, the thesis dissertation format for a postgraduate degree for one University may differ from the PhD format of another one. However, the format can still be generalized to contain the important sections which remain common to all types of reports. The 'organization of the research report' delves in the topics which are necessary and should be included in the report.

ORGANIZATION OF THE RESEARCH REPORT

Scientific research reports usually follow the method of scientific reasoning—problem definition, hypothesis creation, conceiving of experiment to test the hypothesis, experimentation or observation, analysis of results and conclusions.

This framework is consistent with the following organization of a research report:
- Title
- Abstract
- Introduction or background
- Experimental details or theoretical analysis

- Results
- Discussion
- Conclusion and summary
- References.

Title and Title Page

The title should reflect the content and the emphasis of the study. It should be as short as possible. But it should include essential key words. Key words are asked for separately also. But having key words included in the title ensures better search ability in these days of web crawlers and search engines. The analytics of search engines keeps changing to better reflect the content required by the question asked. Having the keywords in the title ensures that these changes have lesser effect on the ranking of the research paper.

Title should be informative as well as readable, for example "Association of self-administered aspirin with development of bleeding tendency in patients of dengue fever without shock." It should have a context and arouse interest if possible. A title such as "Bleeding Patient in Ward 20" is not informative and definitely not reportable. The author's name should be written on a separate line after the title, followed by the authors' affiliations (e.g. Department of Ophthalmology, Hindu Rao Hospital, Delhi), the date and the cause of the report (e.g. In partial fulfillment of a Diplomate of National Board under the supervision of Professor S Gavaskar, May, 2015). The name(s) and address(es) of all the authors should be included on the title page itself. Sometimes the review boards and journals give their own guidelines for reports. In such cases those instructions need to be complied with before the editorial review can begin. A title page with all information such as the number of words, tables, figures and photographs; alongwith the details of authors and intellectual inputs of each of them, is not a bad first starting point for most articles.

It is best for all the above to appear on a single cover page. Sometimes a certificate confirming the originality of the work is required to be appended after the title page by the academic institutions and other review agencies.

Acknowledgments and table or list of contents may be added along with the preface pages.

In all sections of the report separate paragraphs should be used to highlight each important point. This requirement is done away with only in the abstract. The flow of the argument should be in logical or chronological order. Present tense should be resorted to in the introduction or background section to describe the facts which are already established or form the bedrock of the work. For example, 'the sky appears blue.' On the other hand, past tense should be used to describe results of specific studies especially your own. For example, 'When an intravitreal injection of 1.25 mg of bevacizumab was given, the retinal thickening decreased significantly below the baseline.'

One should remember that different Universities have their own guidelines for submitting research reports. The current discussion is general in nature and broadly outlines the commonly used conventions. Always consult the guidelines of the University or the journal to which the report is proposed to be submitted.

Abstract

The abstract should describe the topic, its scope, the principal observations and the conclusions in the briefest possible words. While it is placed at the head of the report, ideally it should be written last. This helps to reflect the content of the report accurately. The length of abstract is usually prescribed. Most commonly, it is not expected to exceed 200–250 words. It is important to stick to this limit. Long abstracts lose the eye of the reader. When the abstract is to be edited, remember that 'brevity is the soul of wit' —or the lack of it. Always remove the superfluous adjectives and adverbs or anything which does not convey relevant information to the reader. 'The method performed very well in the hands of the investigators' conveys little and wastes words. 'The results from the current method were better than (or as good as) the current gold standard (quote of the tests used)' conveys scientific temper.

The effort should be to summarize the study, focus on the results and major conclusions. Relevant quantitative data should be included briefly. The abstract may be structured or unstructured. In the former, it follows the pattern laid down and mentions the subsection separately. In the latter, it is usually a single concise paragraph or two and sub-headings are not given.

The abstract should be capable of standing on its own meaning thereby that it should be capable of being read alone. Therefore, reference to any other part of the report, its figures or tables or appendices is not desirable. Long winding introductory or explanatory statements should be avoided. If there is nothing else to say, it is better not to say it. Keeping it short and terse manages to get the reader to go to the manuscript directly. The abstract should be written in past tense because it describes the work already done by the researchers.

The primary objective of any abstract is to communicate the essence of the study or report. The reader can judge the relevance of the report from this and can then decide whether to read the full report or not. In primary literature, the abstract serves as a source of indexing terms and key words to be used in information retrieval.

Introduction or Background

"A good introduction is a clear statement of the problem or project and why you are studying it." (Dodd JS, Ed. *The ACS Style Guide*; American Chemical Society: Washington, DC, 1986). That sums up the reason for writing this section. The rationale behind conducting the study in its current format should be given. The following points should preferably be covered:
- The research question and its relevance to science or to the knowledge of the subject
- Suitability of the experimental model and objective
- Experimental design and specific hypothesis
- Furtherance of knowledge expected from the outcome.

The idea is to write herein the appropriate background information. This does not mean that the researcher tries to write all that he knows about the subject. This is a common mistake that new writers make in this section. The focus should be on including only the relevant information and a little more. The nature of the problem and the reason for interest in the specific area of the problem should constitute the opening paragraphs. The background information on the problem and previous efforts with proper literature citations should be included in this section. The researcher should build the objectives of the current project and attempt to establish a clear relationship between the current project and the scope and limitations of the previous work. The reasons for undertaking the project and the approach used, are better understood in the context of limitations and scope of earlier attempts.

Experimental Details or Theoretical Analysis or Conceptual Framework

Conceptual framework is not required for every study. It is required only for studies which use concepts from basic science and apply it into the specific knowledge area, where little work has been done earlier. This is a description of what was done for the study. It elaborates laboratory recordings, describes procedures, techniques, instrumentation, special precautions, user manuals, checklists and so on. The details should permit other experienced researchers to repeat or duplicate the work and obtain comparable results. In other words, the research should be repeatable and should allow other researchers to verify the results with similar findings. Most of the times, this is condensed into materials and methods which indeed form the core of all that is contained in this section. Sufficient details about theoretical or mathematical analysis to enable derivations and numerical result checking should also be included without making it too 'busy' or difficult to read. Computer programs and algorithms from the public domain, if used, should be cited. Newer computer programs or algorithms as well as proprietary platforms, if used, should be described in outline ideally.

If the section becomes too lengthy and detailed, some parts of it can be included in appendices so that the conceptual flow of the report remains uninterrupted. This is not commonly used by new authors. The nature of the project and the discretion of the writer decide whether this method is utilized or not.

The objective is to document all materials and general procedures.

Writing a Materials and Methods Section

Materials: This describes the materials used for the research. It is not a bad idea to include all the parameters and the methods of measurement initially. Then for the sake of economy, the superfluous ones can be dropped. Commonly found supplies need not be included. Use of any specific type of equipment, reagent or culture from a particular supplier should be mentioned clearly. The methodology used should be reported in detail. Sometimes the user manuals describing each procedure are included in the appendices. But usually, commonly used procedures are not described beyond mention of the equipment and the technique used. The inclusion and exclusion criteria should be mentioned and described in unambiguous terms. Most authors use third person passive voice and past tense in this section. It is advisable to use normal prose and complete the sentences. The use of headlines or incomplete sentences is not advisable. Reporting methods in chronological order in a narrative form are usually not as effective as presenting it under headings devoted to specific measurement methods or groups of procedural methods.

Any statistical techniques used should find a mention alongwith the software used. If special sampling techniques are used, then the rationale for using them should be mentioned. This underlines the importance of knowing the statistical methods for making valid decisions and drawing scientifically correct conclusions.

Results

Relevant data, observations and findings should be presented with tables and figures wherever required. Raw data should be included in a master-chart appended to the end of research report. In shorter forms of the report, like a research paper, this is not required but it comes in handy. Analyzed data can be presented as a figure, graph, table or less commonly in narrative prose. Presentation of data should not be repeated. If it has been presented in a table, the table should be referred to and it should not be described in prose again. The most effective method to convey the data or the information should be used. To do so succinctly and clearly, usually figures score over tables, and tables are preferable to straight text. However, this is just a rule of the thumb and does not hold true in every situation. Sometimes, figure or graph is not appropriate or the data may be more presentable in narrative form. This is especially so if the number of observations is few. The final approach to this is a contingency approach and the researcher gets a hang of this as he goes along. It is not a bad idea to rely on the peers or the faculty advisor or the editor for the advice. The writer does not have the objective bird's eye view that an unconnected reader or even a co-author has.

The results should be described in continuity. The relationship of groups of observations and thus, each section of converted data should dovetail from one to another and thereby contribute to the overall study. The description should be concise. Rather than just putting a table and going on to the discussion it may be more worthwhile to say, 'To test the null hypothesis that high altitude has no effect on the ocular structures, we measured the nasal bridges, the inner and outer canthal distances, and the inter pupillary distances in different age groups. Table 1 lists the measurements in the high altitude group and Table 2 in the control group.' This can be followed by the complete table with appropriate title and headings. The table should be able to stand alone. Ideally all graphs, tables and converted data should go into the body of the report after the methods and before the discussion. Sticking graphs or other data onto the back of the report is resorted to when they are printed or prepared on different (read glossy) paper and is often resorted to by some researchers. It is a less than ideal situation because it interrupts the flow of the report. Photographs stand on a different footing. They are equally useful inside the report or as an appendix. Sometimes it makes sense to put in appendix any extra observations that do not form part of the main analysis. However, this is not the method that beginners should use. It is often difficult to determine the relative importance of the information to the readers. Addition of the master-chart allows secondary analysis to be carried out by other fellow researchers. Nowadays the log of the statistical analysis is also submitted for peer review if asked for. Therefore, it is not a bad idea to have the log saved in the primary analysis and refer to it when required.

Conclusions should not be drawn in the results section. Many writers make this mistake as they feel this saves space or prevents duplication of prose or appears logically flowing into the conclusion. Data interpretation should be kept for the discussion and discussion alone. The results section is for what was observed and it should be irrefutable.

Discussion

The "Results" and "Discussion" sections are interrelated and it can be tempting to combine them. This can be done and is indeed accepted by some journals also. For the sake of traditional teaching and wisdom, it makes sense to keep them separate, as indeed they have been for decades.

Analysis and interpretation of the results is the cornerstone of the report and the research work. Data should be interpreted in the discussion. The most important decision involves the hypothesis or hypotheses—whether each one of them is supported, rejected, or it is not possible to make an inferential decision with confidence. Draw the conclusions that can be made, and then suggestions on modifications or improvements to the experiment or study to properly test the hypothesis(es) should be made. It is common to find interesting observations being left out in the cold by the label 'inconclusive' and no further elaboration. It has been repeatedly pointed out by the editors that 'the paradox of the reportable' forces them to drop negative reports. Sometimes negative reports are as important as the positive ones. So the attempt should be to draw a conclusion even in cases bordering the significantly different and mention it. This is the area where further research would profitably be directed. The move away from reporting just '$p < 0.05$ or $p > 0.05$' and towards mentioning '$p= 0.06$ or whatever the numerical value' in most biomedical journals is a welcome step in this direction. Explanation for observations made should focus on the underlying mechanisms, if possible. It is a good idea to distinguish data from the current project from published information from studies. Again the principles for reporting remain the same—work done by specific individuals should be reported in the past tense and the generally accepted facts and principles, in the present tense. It is important to comment on whether the experimental or study design adequately answered the hypothesis and, if it did, then whether it was properly controlled or conducted. The strengths and limitations of the current effort should be discussed with dispassionate clarity. If the experimental design does not answer the hypothesis, then it is important to mention the further courses of action emanating from the current study. The best studies open new avenues of research. Addressing the questions that remain unanswered and the new questions thrown up by the current project, make the researchers' effort more valuable to the community. After all, an expert's considerable time and energy time gets invested in the project or endeavor towards new knowledge. If the intention is to guide further research, then this is the place to suggest any new hypothesis and thus studies to further address the main research question. Creativity, conjecture and speculation are all permitted only in this place.

Conclusion and Summary

This is often a dilemma. Since conclusions have already been drawn in the discussion then inclusion in a separate section is a repetition. A separate section outlining the main conclusions of the project is most appropriate if conclusions have not already been stated in the 'Discussion' section. However, even if that has been done, it is not a bad idea to encapsulate the main conclusions alongwith a brief summary of the work in this section. Suggestions for future work may also be included. In reports of long or complex studies a paragraph summarizing the main features of the report, three quarters of the objectives, the findings and most of the conclusions (especially the main ones) are very desirable. Most of the readers jump from the abstract to this section and it makes sense to give a complete outline here to arouse the interest of the reader.

The last paragraph of text in publication manuscripts traditionally is devoted to acknowledgments. In research reports or theses, acknowledgments usually follow the title page.

References

Literature references are collated at the end of the report and cited in one of the many formats available. It is best to follow the format prescribed by the institution to which the report will be submitted. Formats should not be mixed. This is a common anomaly seen in reports. All references should be checked against original literature. A reference that the researcher has not read should not be included. It is commonly seen that people copy from other bibliographies, often with hilarious or even perilous consequences. The researcher is responsible for the accuracy of the citation.

A citation contains basic information to identify or locate a specific publication (book, article, video, etc.) or url (universal resource locator). Citations are used in print indexes, catalogs and databases to identify listed resources. Their use in research papers, articles and books to reference quoted text or a source that has been used as an authority makes them indispensable for replicable research.

Parts of Common Citations

Books
- Author(s) or editor(s)
- Book title
- Edition
- Publication date
- Publisher and place of publication

Articles
- Article title
- Author(s)
- Page numbers
- Periodical title
- Publication date
- Volume and issue
- DOI (Digital Object Identifier)

The citation gives credit to individuals for their creative and intellectual works utilized to support the research. It can also be used to locate sources and combat plagiarism. The citation style dictates the information included in a citation and the manner in which it is ordered. The punctuation and other formatting differentiate different citation styles. There are many different citation styles for research. The style may be dictated by the discipline, the institution, the journal or the funding agency involved. Commonly used citation styles include:
- APA (American Psychological Association) style used by Education, Psychology, and Sciences
- MLA (Modern Language Association) style used by the Humanities
- Chicago/Turabian style generally used by Business, History, and the Fine Arts
- The physical, natural, and social sciences use the Author-Date or "Harvard" system and Author –Number Vancouver system or Vancouver reference style system usually.
- Specialty styles like ACS (American Chemical Society) Style, CSE (Council of Science Editors), AMA (American Medical Association)

Vancouver Author–Number system and "Harvard" Author-Date system are commonly used in medical literature. The researchers need to consult faculty and find out requirements of their specific course or visit the journal website and look at 'instructions for authors' in case of journal articles. Online citation managers are available to keep track of references with tools such as Mendeley and Endnote. They import citations from different sources and article databases to automatically integrate them into the research paper and bibliography.

Mendeley and EndNote are reference managers. They enable online bibliographic database searches, reference organization and formatting of bibliographies. Online citation managers include the following:

EasyBib–The Free Automatic Bibliography and Citation Maker

Zotero–A free, easy-to-use Firefox extension helps to collect, manage, cite, and share research resources from the web browser.

NoodleTools–NoodleBib is a comprehensive and accurate MLA, APA, and Chicago/Turabian bibliography composer with fully-integrated note-taking.

BibMe–The fully automatic auto-filling bibliography maker is an easy way to build up a bibliography.

Citation Machine–*Allows researchers to cite their information* sources.

Some reports may not require references and there "no references were consulted," should be stated.

MANUSCRIPT PREPARATION FOR JOURNALS

The first and foremost thing to remember is that each journal has its own set of 'instructions for authors'. It must be read and adhered to. The article type should be chosen before starting on the endeavor of writing the paper.

The personal computers, laptops and cloud store all data and results. Word processing tools like Word on Windows and Pages on Apple make manuscript preparation and revision very easy. A number of spread sheets and graphics software are available like Excel, Lotus123, Numbers, etc. They allow numerical data to be graphed and mathematical equations to be represented. Any technical writer should be conversant with these tools. Online resources to guide the budding writer also abound. Unless the researchers are too weak in the language, it is probably best to use one's own presentation style with inputs from the faculty supervisor or a peer who has good command over the language. Professional help is often available at a price at academic institutions.

All manuscripts should be checked for spelling and grammar either manually or automatically with programs. Careful proofreading is essential before taking the final version for printing and binding. A distinctive style of writing can be developed by using own words and framing grammatically correct short sentences. Long-winding sentences are often used by experts but they can lose the interest of the reader. Many times they convey a different meaning to different readers. It makes more sense to use smaller and concise sentences.

Designing Tables

Tables are used to present results concisely and clearly. Data analyses begin by making basic tables, graphs and summary statistics to describe the data. They must be presented sensibly and be self-explanatory. A table should have a title that gives a clear indication of its contents. The presentation should not be confusing. The reader should be able to understand precisely what is tabulated. Column and row headings, though brief, should be self-explanatory. The source of the data must be put at the end so that original context and source are conveyed. Units of measurement are frequently forgotten. They should always be mentioned. Totals and percentages wherever required must be included. They indicate the quantity of data in the table and allow comparisons. In percentages, the base for the percentage should be indicated. If in complex tables percentages are displayed without indication of base, the reader is uncertain how or in which way the percentages total to become hundred percent. Misinterpretation in such cases is possible and should be avoided. Subtotals adding to 100% show the reader how the percentages were arrived at and make comparisons succinct. Approximation, assumption and omission, if any, should find a mention in the footnotes. This should be explained in the text if it is not possible to put a small abbreviated reason in the footnote itself. Exclusion of observations from the table for whatever reason should be mentioned with their number clearly. Tables score over text to present information concisely and clearly though with a degree of complexity added to it. Often this technique is necessary for saving words and space. However, over-compression of information into a single table should be avoided. Two or three simple tables are often less 'busy' and more readable than a single large and complex one.

The number of observations on which the summary is based should always be reported. For binary responses (0 or 1) the percentage of either 0 or 1 should be reported and used. Avoid using both percentages together because it tends to confuse the issue. If median is used as the measure of central tendency of a quantitative distribution then the lower and upper quartiles (or the range) should be quoted as well. If mean is used, then the standard deviation should be quoted as well.

Graphs

Every graph should be accompanied with a self-explanatory legend. It should aid the reader in understanding the data or what is called data visualization. It should not be cluttered with too much detail. The axes need to be clearly labeled and units of measurement indicated. The reader should be able to understand precisely what is being illustrated and the units in which the measurement was made. Scales are extremely important. They should either start at zero or clearly show a break in the axis. Usually the explanatory (exposure) variable is displayed along the horizontal axis and the response (outcome) variable along the vertical axis. Three-dimensional graphs are also coming into use now but readers usually visualize data less easily in three dimensions. Graph type used is dictated by best representation of the data and ink consumption without information-loss.

Effective technical writing skills prepare the researcher for his job of communicating his results to the scientific community and the public at large. The idea of writing is to communicate or to convey the message using a medium. This is not a literary activity. The writing should be precise, concise and succinct. Written and oral communication skills are universal qualities in professionals. Developing them requires effort and capability. However, with practice, it is not a difficult skill to acquire.

Quotes

In literature, history and social sciences quotations are necessary for a full discussion of the subject. In science, there is very rarely a need for a direct quotation. This does not make a nice impression on the reviewer especially if the writer does not understand the concept and the context in which the quote is being used. The original ideas take the paper farther than any long-winded superfluous quotations. This is not a literary treatise. It should be comprehensive and concise. It is a good practice to try to chop the unnecessary words in all manuscripts because this hones the skills to get to the point directly.

Grammar Use

Use of the wrong verb, adverb or tense breaks the flow of thought and is difficult to read. It irritates the reader and reflects poorly on the researcher's writing skills. The reader may get confused about what is already known and what was newly discovered in the actual study described in the report. As a rule, past tense is used to describe events that have happened. Such events include procedures done and the results observed. Present tense is used to describe the generally accepted facts.

Proofing

Incomplete sentences, redundant phrases, misspellings and other oversights show shoddy or lazy work. After being through with the draft, ensure that it is proofread properly, not once but a few times. Otherwise the consequences can sometimes be harsh. It is not uncommon to find good work being affected by tawdry editing. While the authors get away with this in articles in the editorial review itself but in case of thesis or research reports, there is often no revision. This casts a shadow on the researcher and his work. Spelling of the scientific names, names of people, names of procedures, references and their layout are common mistakes picked up at this stage. If it is missed, then the mistake goes into the original manuscript with its attendant problems and consequences. Misspelling and grammatical errors leave a poor impression and can be embarrassing for the researcher.

Splitting of tables over more than one page is a common mistake that gets picked up after the printing and binding is over, leaving the applicant or the writer flummoxed. The reader finds it difficult to read orphaned headings on the first page and the data starting from the next one or spilling out onto the next page with no headings. It becomes difficult to read through such documents. Numbering of pages is important because it is often seen that pages have been bound out of sequence making it difficult to understand the content or the context of the observation or the discussion. When someone reads the report, the content and context has to be communicated to that reader-be he a reviewer or a reader.

Anecdotal Information

In older literature anecdotes and insights could find a way in. The writer would actually tell the story of the investigation process. This made interesting reading. But with the volume of publication nowadays and the modern writing ethic, papers omit such information today. The readers do not have the time or the patience to get to the point by wading through side material. The publication costs are high and irrelevant information is not published. Even reviewers are not kind with superfluous information which does not contribute to the solution or the discussion.

Irrelevant Information

Unnecessary background is another example of irrelevant information. For example, when an application for studying the use of bevacizumab in diabetic macular oedema is made, it is not appropriate to start from Banting and Best. A common mistake made very often is the introduction of a disease or condition which does not contribute to the question why the study should be done. In writing everything the researcher knows or has read about the subject, he loses the interest of the reader and wastes his own time- often with disastrous consequences. It makes sense to write a clear introduction with specific points in mind and include the references in background information.

Subjectivity and Use of Superlatives

Scientific writing makes very little use of adjectives and superlatives. It differs from the writing of fiction, opinion pieces and scholarly literature in social sciences. Objectivity is paramount. If it cannot be verified, repeated and quantified by an unbiased observer, then it is not reportable usually. Subjectivity refers to feelings, opinions, etc. For example, "We felt that the procedure was more difficult than bare sclera technique," is not a good report. The reviewer or another researcher would not like to risk time and resources on the basis of someone's feeling no matter who the writer may be. On the other hand, if this is objectively cited like, "The time taken for the current technique at mean $=1440 \pm 24$ seconds was significantly more than the bare sclera technique (mean = 1024 ± 20 seconds) [p=0.03, *t* test]. The technique was rated to be more difficult to perform by different surgeons with a mean of 3.54 ± 0.46 for the current technique and 2.44 ± 0.42 for the bare sclera [p = 0.03, *t* test](Five point Likert Scale 1 = Most Easy to do, 5 = Most Difficult to do). The complications encountered were more in latter than in the former." This is more informative and objective. It is comprehensive and unambiguous. This can be used by another researcher or the reader.

Superlatives or adjectives like huge, terrible, unfortunate, wonderful, incredible, exciting, etc. should have no place in scientific writing. An elephant is bigger than a horse which is bigger than a mouse. 'Big or small' tumor has no meaning unless quantified. Similarly, 'we believe' is not exactly a rosy phrase to be used. Presentation of the evidence with suggestion of strong support for a line of action or thought is the method of science and must be presented in the report that way. Beliefs, values, morals and attitudes should remain in the background only. It is wrong to state that a particular result was expected except when some hypothesis is propounded.

Evidence or Proof

Scientific proof needs to be robust and the process extremely rigorous. It should be proved repeatedly and no other conclusion should be possible. It is unlikely for any single experiment to prove or disprove a particular research problem in totality. For any result to be accepted, it must be confirmed independently by different observers in different samples. A single set of observations just proves that the data may strongly support a position or allow rejection of a hypothesis, but it is not a proof. That has to be more rigorous and evidence has to be interpreted in its own space using the rubrics it uses.

Inaccurate Words and Phrases

It is about homonyms. The observation cited is, "Changing pH had the following 'affect' on the subject." Now, 'affect' is a verb and 'effect' is a noun. What happens to the subject is an 'effect' and the temperature change 'affects' the subject. Affect can be taken to mean the change in attitude alone also.

The word 'data' is plural. The singular is 'datum' which is rarely used. Therefore, the correct use is, "the data were skewed in both groups" and not "the data was skewed in both groups". However, it may be some time before this usage becomes familiar. And the word 'datas' unfortunately, does not exist. There may be two sets or more sets of data but they should not be referred to as 'datas.'

Superficiality of Discussion

The purpose of a discussion is to interpret the results. Many times it is just used to express the results in a slightly different way by the writers. It is a travesty. It does not contribute to the paper or the report. At the same time there should be no discussion in the results section. A superficial discussion ignores mechanisms or fails to explain the results and the mechanisms. It, therefore, fails the purpose of the paper completely. From the discussion, the reader should be able to form an opinion on the reasons of getting a particular specific. Such explanations may or may not turn out to be correct later. However, credit for posing a reasonable explanation will always be given to the researchers when the paper gets cited in later communications.

Common Mistakes in Reporting Results

Converted data refers to data that have been analyzed, summarized or tabulated and presented in a manner that the information pertinent to the objectives of the study is presented.

Raw data refers to the proforma of the individual cases, individual observations, chart records, etc. and these usually find a place in the appendices. Repetition should be avoided and if graphs have been plotted, then tables should not present the same information again or vice versa, unless there is a compelling reason for doing this (which is very unlikely). The caption and legends of tables and figures should be self-explanatory and referral to the body text should not be required. This is usually never the case. Use of appropriate number of decimal places is important and it should be used to report measured or calculated values. The number of decimal places and/or significant figures reflects the degree of precision of the original measurement. Once the concept of precision of measurement is understood, then the quoted values speak for themselves. Since the number of significant figures used reflects the level of precision of the measurement or calculation and there is no need to talk of 'about' or 'approximate' when discussing the results. A special mention of this aspect has been made to draw the readers' attention to the chapter on accuracy and precision. This is a weakness in most medical studies or their reports especially by the younger or newer authors.

Special attention should be paid to reporting results and data visualization by the new authors or writers. Figures are graphs and other pictures that represent data. They should be numbered in sequence. Tables are differently numbered from figures and sequentially again. For example, a paper with two graphs, a reproduction of a segment of chart record and two tables will have Figures 1, 2, and 3, and Tables 1 and 2. This is of importance when submitting to electronic submission systems. Many times people upload figures and tables in images and this causes problems to the sub-editors at the time of publication.

The Significance of 'Significance'

A statistically significant difference simply means that the data analysis threw up a low p value and thus the probability that the difference was due to sampling error (random error) alone is remote. If the sample size is adequate and statistical significance is not achieved, the investigator should conclude that the null hypothesis is supported. Thus he reports that there is no significant difference. Lack of a significant statistical difference does not mean that the result itself is insignificant. A finding, for example, that there are no intrinsic differences in intra-ocular-pressures among different racial groups would be a very significant finding. This means that "clinical significance" is to be differentiated from "statistical significance". On the other end of the spectrum is the case of statistical difference being there but there is no clinical significance. For example, the difference in the nose bridge width between those living at sea level compared to those living above 3000 meters in a tribal population in the Himalayas was reported by an author. The difference between the two was too small to have a bearing on any clinical parameter. Thus significance has two aspects to be considered—statistical and clinical. Therefore, it needs to be interpreted properly and then reported with diligence and care. The reader also needs to differentiate between the two when reading a paper.

The propensity to reject a study as inconclusive just because no statistically significant differences were found needs to be deprecated. This is actually misunderstanding of the scientific method. A well-designed experiment will result in support for the null hypothesis in this situation. The catch here is to interpret what is encountered and draw conclusions from it. This should not be affected by expectation or prior knowledge. Even if the researcher is at variance to the previous knowledge he needs to understand that this may be due to a multitude of factors and he needs to explore the findings as they have been observed. The purpose of experimental and observational methods in science is to discover the truth. It is not to make the data conform to posterior knowledge. Indeed if that is what happens then it supports the prior knowledge in that population also. Thus, whatever be the results, they should be reported as they have been observed and valid conclusions should be drawn from them.

In the last, it has to be understood that sufficient time is required for satisfactory completion of reports. It is important to revise the draft and get it peer reviewed or reviewed by the faculty or someone who understands the nuances as well as the importance of the report, before its submission.

Therefore, report writing activity should be planned properly. Adequate time should be budgeted for carrying it out. The researcher should acknowledge and appreciate that all research guidelines or methodology always mention report writing as a proper activity in the end, which needs to be considered as an integral and important part of research. It is not just a support activity.

CHAPTER 34

Applying for Grants

The grants or funding for biomedical research is available from government and nongovernment agencies. Some of these agencies are international in scope while others fund research within the country. In India it is not uncommon to find the research being done in pursuance of one degree or another. In this case, the funding comes from the parent institution supporting the infrastructure for the study or award of the degree. The newer regime of regulation of the trials and ethical clearances has ensured that the studies for the award of degrees should be of lower risk than the cutting edge research. This has led to an impetus being given to other forms of research now. The current chapter focuses on these people who would like to secure funding for their research.

The funding agencies can be broadly classified as:
- National and International agencies
- Commercial and noncommercial agencies (including charities/foundations, etc.)
- Government and nongovernment agencies.

In India, a host of research funding agencies are available and they can be approached for research work in the specific areas they cater to. It is not a bad idea to visit the websites of these sources to get an overview of the opportunities available. Some of the funding agencies are listed below.
- Tata Institute of Fundamental Research
- Science and Engineering Research Board (SERB)
- Petroleum Conservation Research Association (PCRA)
- Naval Research Board (NRB)
- National Board for Higher Mathematics (NBHM)
- Various Government Ministries/Departments such as Water Resources, Urban Development, Small Scale Industries, Rural Development, Railways, Health and Family Welfare, Environment and Forests, Defence, Chemical and Fertilizers, Biotechnology, Atomic energy, etc.
- Various Government Organizations and bodies such as Indian Space Research Organization (ISRO), Indian National Science Academy (INSA), Indian Council of Social Science Research (ICSSR), Indian Council of Medical Research (ICMR), Indian Council of Agricultural Research (ICAR), Defence Research and Development

Organization (DRDO), Council of Scientific and Industrial Research (CSIR), Board of Research in Nuclear Sciences (BRNS), Atomic Energy Regulatory Board (AERB), Aeronautics Research and Development Board (ARDB) etc.

Apart from Indian agencies, the researcher can approach International and National Institutes of different countries, for example, National Institute of Health (NIH), USA, National Institute of Allergy and Infectious Diseases, National Institute of Neurological Disorders and Stroke (NINDS), etc.

Regional grouping agencies like European Commission, Research Councils like The Royal Society in UK or University or hospitals' internal schemes can also be evaluated for scope of funding. Beyond these industry associations, private companies and nonprofit trusts/foundations too support a lot of research, even though they may be specific to their own area of objectives and operations.

Irrespective of the funding organization, the application for grants should be specific and professional in approach. This chapter provides general guidelines on key issues which need to be considered while seeking funds for research.

PREPARING FOR APPLICATION PROCESS

Whatever be the area of interest, the agencies have a basic set of rules and a given set of instructions to be followed for funding applications. The instructions are given on their websites. Information documents are available in these agencies. They may differ from one to another and the researcher must familiarize himself with the agency he is applying to.

The following discussion is a description of the important issues to be addressed in general for all funding agencies as mentioning each one separately is quite out of the scope of this chapter.

Any detailed project report is a major undertaking and that of a research proposal is more so because of the diverse elements to be incorporated into the report. All information should already have been collected before the writing is started to ensure that there is continuity in the document and time is not wasted. Knowledge of the area of research, choice of a creative idea, information about the instructions from the agency and a passion for the research problem solution are good starting points. The focus when writing the report should be to put the information across effectively without too much decoration and use of hyperboles. Things should be stated as they are. The review is done by peers—knowledgeable investigators about the research area. A lack of knowledge gets a bad review. The reviewers need to be convinced about the importance of the problem, the soundness of the hypothesis and the background, logic of the objectives, validity of the construct, feasibility of the study, potential problems and proper plan of statistical analysis of the data.

For peer review the following criteria are usually zeroed in on:
- Significance
- Approach
- Innovation
- Investigator
- Environment.

The Basic Questions in the Review Process revolve around:
- Intellectual quality and merit of the question being attempted to be answered by the proposal.
- Potential impact of positive and negative findings of the proposal.
- Creativity in the project or novel ideas springing forth from it; probability of new data and concepts emerging or confirming the existing knowledge.
- Validity of the hypothesis valid along with the supporting evidence.
- Assessment of the logic of the objectives and aims.
- Assessment of the standard operating procedures in terms of appropriateness, adequacy and feasibility.
- Qualification, competence, credentials and experience of the investigators.
- Facilities and organizational climate of the institution for research especially in the given field.

GROUND REALITY: THE BACKGROUND CHECK

Full awareness of the competition in the field gives the writer an edge. This may allow the tweaking of the area and objectives of research to the advantage of the applicant. The applicant must possess resources required to compete, both intellectually and financially. Often this may not be easy and multiple agencies may be approached to build research. Knowing strengths, weaknesses, opportunities and threats (SWOT) is an important first step. This SWOT analysis should be done at the individual and then at organizational level with reference to the core research area. There are opportunities for collaboration in the field. However, it is important to document and understand that researcher and collaborators are qualified and trained for the research. Known laboratories, hospitals or mentors help the researcher find a niche area which helps the research sail through the funding ocean. A mentor is a good idea for those who are not grounded in the area of research at the time of the funding.

The priority areas keep changing and it is important to be armed with a list of institutes supporting research in the selected area of research. Often institutions have experienced faculty and extramural program staff, which can be of immense help. A discussion with them on the subject area can sort out the gray and black areas in the objectives before embarking on the journey. Grant applications of successful grantees will give enough information to orient the research in the successful direction or if such orientation is not possible then the agency itself should be changed. This does not mean that an application should not be made; it simply means that one should not put all hopes in a place where the probability of success is limited or nonexistent. It is not uncommon to find a rejected proposal from one agency getting the grant from another. All comments and critiques received should be viewed constructively. They help to refine the proposal.

A good grant application needs a strong hypothesis. It should be important to the field. It should be testable and repeatable. The rationale behind the hypothesis should be stated clearly and be unambiguous. It should be based on current knowledge and should answer some gap in the current knowledge. A good hypothesis should further the knowledge and understanding of biologic processes, diseases, treatments and/or preventions. A successful proposal is usually driven by one or more hypotheses. Mere advances in technology and fishing expeditions lacking scientific basis do not find favor with agencies. Emphasis should be made on the hypothesis in the abstract and the aims section of the research plan.

A strong grant application must be on a creative, exciting, and funding-worthy subject with clear objectives. A rigorous, well-defined experiment or observation method should support the objectives set out in the plan. The information should be clear and unambiguous in a language free from spelling or grammatical errors. It should follow guidelines detailed in the grant application agencies' instructions. If experienced grantees are available then they may help critique and improve the application which should be adequately proofread before submission. It is important to apply to the correct level of funding section in terms of time, finance and expertise required.

MAIN APPLICATION: CONTENTS AND PITFALLS

The main sections in an application may vary from one to the other agency but broadly most proposals will require the following:
- Face or title page
- Description or abstract
- List and description of performance sites
- List of key personnel
- List of contents
- Detailed budget for initial budget period
- Budget for entire proposed period of support
- Biographical sketch
- Personal data
- Supporting documents
- Resources
- Research plan
- Appendix
- Personnel report.

The page limitations for the application type must be kept in mind. Reviewers appreciate comprehensive but concise proposals. Font size, page size, margins and spacing requirements are mentioned by the agency. Hard-to-read type and formatting are not good for the application's success. The active voice should be preferred over the passive voice. For example, *We will construct a pocket in the skin*, should be preferred over *A pocket in the skin will be constructed*. It is easier to read and comprehend. Related knowledge, ideas and information should be kept close together and preferably keep one main idea in one paragraph. Small sentences are better than long ones. Simplify and breakup long sentences and paragraphs. The goal is communication and literary merit needs to be kept subservient to the goal of getting the idea across. Redundant words, phrases and sentences should be chopped mercilessly. Editing and proofreading before submission should be thorough. Typographical errors, grammatical mistakes, omitted information, dissonance from text in figures and tables can defeat an application. A sloppy application conveys a lazy or incompetent or disorganized researcher.

A good research plan is important for success in peer review. It should be based on a strong hypothesis. It needs to have coherence and logical direction. The sections of the plan should be related to the keystone and be well organized around it. A good grant application asks questions about biological mechanisms. The timetable mentioned should be feasible. The reviewers can usually spot an overambitious timeline. Specific aims and methods should answer the hypothesis to be tested.

The project aims should be highly focused on answering the hypothesis. The general purpose or major objectives of the problem are stated in the starting and then the specific aims are outlined. All objectives should relate directly to the hypothesis. In case of multiple hypotheses there should be specific aims for each one. Alternatives to the hypothesis should be outlined and reasons for selection of the current one explained. Objectives should be verifiable by the review committee. Specific aims are not the same as long-term goals.

BACKGROUND AND SIGNIFICANCE OR INTRODUCTION

Statement of significance should be brief and direct. Innovation and novelty of the research should be highlighted. Hypothesis and research should remove gaps in knowledge in the field. The proposal with background information about the research field should justify the proposed research.

The literature section should demonstrate the grasp and knowledge of the investigators in the area of research. The difference between successful applications and the unsuccessful ones is that the latter usually miss this opportunity to reveal awareness of gaps or discrepancies in the field. Familiarity with unpublished work often through personal contacts scores beyond the systematic review of published work. The next logical stage of research beyond current application may be outlined for making a pitch for continuing work.

PRELIMINARY STUDIES/PROGRESS REPORT

Preliminary studies/progress report help build reviewers' confidence that the researcher is conversant with the technologies, understands the methods and can interpret results. Preliminary data should support the hypothesis and the feasibility of the project. They should expound how the early studies support the hypothesis and how they could be expanded in scope or size. Critical appraisal of the results and looking at alternative meanings demonstrates that the problem has been considered at length. Preliminary data may consist of own publications, publications of others, unpublished data or any combination thereof. If any manuscripts have been submitted for publication they may be included.

RESEARCH DESIGN AND METHODS

Research design and methods should be detailed and describe the design, construct, concepts and procedures. The stated aims should be addressed by the methods which may be innovative or well established. For the former it is important to point out how the existing proven methods have been modified and how they are advantageous to the proposed research. Colored charts, graphs and photographs may spruce up the application. Many people also place a copy of the item in an appendix, noting this in the body of the text. Important figures should not be placed only in the appendix. Place reduced figures in the body of the application with enlargements in the appendix. The research plan must be a self-contained note. In many agencies the study section members receive only black and white photocopies of the original application and thus all the hard work may be undone. Assigned reviewers will receive originals of the appendices if multiple copies have been submitted.

The current choice of approach(es) as opposed to others should be justified. In case of standard approaches or gold standards nothing much is required by way of explanation except citations. In case of nonconventional approaches, the advantages over a conventional one should be explained. The innovative procedures should be valid, repeatable, and feasible in the setup described and the researcher should have the qualification as well as competence required for the procedures. Potential problems and their proposed alternatives reassure the reviewers that the researcher is aware of the potential limitations. It is not a bad idea to look at the limitations of all alternative approaches. Usually the reviewers are looking for quality answers to the research question and more rigorous the approach the greater is the chance of success. Being detailed and specific improves the chances of success. Nothing should be left to assumption. When funding is involved, the reviewer would like to know exactly which bacterium, medicine, concentration, test and technique the researcher proposes to use. It helps them understand the costs involved and the eye on details demonstrates that the researchers are conversant with the methods. The proposed model should address the research question adequately. A lacuna here kills the entire effort. One should revise, review and revise till it is certain that no stone has been left unturned.

RESULTS

The results are important to interpret. The applicant should show knowledge of the value and limitations of the expected results based on current knowledge and paradigms. Generalizability and repeatability of the results is important—the conditions under which the data supports or contradicts the hypothesis and the extent to which the interpretation of the results can be applied outside the study population. Researchers' ability to interpret results based on the current knowledge of the subject usually favorably impresses the reviewers. A good statistical plan can be of immense help. Description of proposed statistical methods and definition of criteria for evaluating the success or failure of a specific test for the data proposed to be collected demonstrates a level of understanding and commitment which is usually rewarded with favorable reviews.

MONITORING AND EVALUATION

Monitoring and evaluation is important for some agencies. It is not a bad idea to have some targets to accomplish and their timelines. These are called milestones. This may be against an industrial yardstick like benchmarking or against the academic requirement or deadline. Expected or apprehended potential delays should be outlined in the proposal so that the organizational climate is available for the reviewers. Sometimes this becomes important when extensions are applied for. However, it is better to make only a short mention of this in the application. Most reviewers already would be aware of the institutional affectations so focus on the processes of delay in the scientific endeavor itself may go a long way. If collaborators would be providing tissue, samples, etc. then letters from the sources should be incorporated in the application. Informed consent process, relevant publications and supporting data should be incorporated. Well-designed tables and figures with informative titles, properly labeled axes and legends described in the text help gain communication superiority. Prior publications in the proposed methods and area are always a plus.

PROTECTION OF HUMAN SUBJECTS AND INSTITUTIONAL REVIEW BOARDS

Protection of human subjects and the assurance about this protection are the responsibility of the principle investigator. The applicant and his institution must ensure that the conventions are followed. An undertaking to this effect is insisted upon by institutions. Awards cannot be made until assurances are on file. If the proposed research does not involve human subjects this should be indicated by noting *Not applicable*. It should never be left blank. Anyone reading the application understands immediately that the section has not been missed but been answered in the negative.

In case of human subjects or samples from human subjects enough information must be given so that reviewers do not have to query the applicant. The proposal must be certified by institutional review board (IRB) or institutional ethics committee prior to funding (unless exemption is applicable as above). IRB approval is not mandatory at the time of application. But it is not a bad practice to apply only after the approval has been received. Some agencies especially in the US may require a Human Subjects Assurance. This is obtained by registration with the Office for Human Research Protections (OHRP). This office provides 'leadership in the protection of the rights, welfare, and well-being of subjects involved in research conducted or supported by the US Department of Health and Human Services (HHS).

All studies involving drugs or devices which are not mere observation studies only or involve interventions require registration with Clinical Trials Registry-India (CTRI). The Clinical Trials Registry- India (CTRI) is hosted at the ICMR's National Institute of Medical Statistics (*http://nims-icmr.nic.in*). It is a free and online public record system for registration of clinical trials in India launched on 20th July 2007 (*www.ctri.nic.in*). From 15th June 2009 trial registration with CTRI is mandatory. Editors of Biomedical Journals of major journals have agreed to publish only registered studies. A trial involving human participants having any intervention such as drugs, surgical procedures, preventive measures, lifestyle modifications, devices, educational or behavioral treatment, rehabilitation strategies should be registered with the CTRI before enrollment of the first participant. IRB approval and Drug Controller General of India's approval (if applicable) is essential for trial registration in the CTRI. Multi-country trials with India as a participating country should also be registered in the CTRI even if they have been registered in an international registry. CTRI maintains details of Indian investigators, trial sites, Indian target sample size and date of enrollment. Regular updates of the trials are uploaded to the CTRI. After a trial is registered, all updates and changes are recorded and available for public display.

BIBLIOGRAPHY

Literature should be thoroughly and adequately cited but not to excess. The publications quoted do not need to be exhaustive but should include relevant ones in the proposed area of research. Failure to refer relevant published research can sound the death knell of an application especially if this shows that the proposed approach had already been attempted or the methods are deemed to be inappropriate for the research questions posed.

CONTRACTS

Consortium/contractual arrangements should be described in full along with the relevant documents for any arrangements made in the proposed research plan. The roles of individuals or organizations should be noted along with reference to any letters from them included in the application. Letters should be from authorized personnel and indicate that the individual/s or organization/s have understood its role in the consortium or contractual arrangements. Addition of consultants after due diligence/careful selection can add credibility to the application and greatly improve its quality.

ABSTRACTS

Abstracts should be carefully written as the secretarial staff and research officers use abstracts and titles for assignment of the application to institute or peer review panel or specific reviewers. Getting to appropriate primary reviewers increases the probability of success. The abstract is best written after the entire writing work is finished. It should be clear, concise and complete. It should be a summary of the project within the word limit. The background, hypothesis, objectives, reasons for importance of the plan, innovations in plans and methods for accomplishment of goals and key takeaways should be highlighted.

TITLE

Title should be specific and preferably reasonably detailed. There is often a word limit and character limit which includes spaces between words. If specified the running title should also be proposed.

BIOGRAPHICAL SKETCHES

Biographical sketches should be utilized to showcase the knowledge, skills and abilities of the key staff and consultants of the project. Reviewers give weightage to proper experience with the proposed techniques. That is the reason collaborators and key personnel should be chosen with care. The reviewers look at the bio-sketches to garner information about reliability of investigators. Now some agencies have started modular grant and application process, it is now advisable to include the aims of all past and current related research along with related publications of key personnel. The order of collection of data should start with principle investigator and include all the key personnel. The following details should be included:
- Name, title and designation.
- Educational qualifications—degree(s), institutions, year conferred, and field(s) of study, memberships of professional bodies and trainings.

- Roles in other relevant current or past research.
- Employment history in reverse chronological order—dates, places, nature of position, professional experience, honors.
- Publications in chronological order, titles and complete references (preferable to include all authors)
- List of staff—professional and nonprofessional along with estimates of the effort for each person.

BUDGET

An estimated budget shows that the investigator has experience with the subject and is aware of the current ground conditions in the area of research. This request is assessed for being realistic and justified. It is important to keep the aims and methods of the project in mind when drawing up the budget. It is prudent to draw up a budget after the research plan has been drawn up and have a good idea of costs.

Budgeting has to be done prudently—Just enough—neither too much, nor too little. Significant deviation in either direction is taken as an indication of lack of understanding of the scope of the proposed work or worse still as an error of understanding of the ground conditions in the field. Expensive equipment should be requested only if necessary and should have a valid justification. Equipment already listed in the resources section and available with the researcher's institution does not get listed in budgeting unless there is an outlay that is required for say maintenance or upgrade which the parent is not financing. In such a situation a suitable explanation should be incorporated. Discrepancies in budget are looked at very minutely and liberal pruning of excess demands is done normally if there is any error of judgment or intention. Some agencies may have salary caps or limit to payments handed out as salaries. It is important to refer to such limits and stick to them otherwise it may lead to problems later.

The release of funds is done either as lump sum or in instalments or modules. Even in case of the latter all expected modules should be budgeted for in the main grant application itself at the outset. When all demands are explicitly made in the application it is unlikely that the reviewers or subsequently agencies will refuse to support fluctuations in annual demands if they are already budgeted. When it comes to deviations from demanded budget or cost overruns it is a different story and it becomes difficult to justify the increase unless the ground conditions have drastically changed. This indicates that either the departure was unforeseen which should be justified by the circumstances, or it shows that there was a lack of application at the outset.

FORMATTING AND INSTRUCTIONS TO APPLICANTS

The agency's guidelines and instructions should be followed completely. Formatting is strictly enforced. The applications are usually returned if the page or word limits are exceeded, or an improper font or font size is used, or if the contents do not follow the laid down norms and procedures. Proofreading and editing should be thorough. Unless the application is perfect it makes no sense to submit it. If the work is not ready, then, delaying to the next receipt date is a better idea than doing a rush job and regretting it later. Reviewers expect the research plan to be organized exactly as described in the guidelines –they are reviewing many applications of the same nature. Label sections exactly as is indicated in the instructions: For example, (A) Aims, (B) Background and significance, (C) Materials and Methods, etc. once the draft is ready then independent peer review at the researchers' end can yield rich dividends. Colleagues in the field who are experienced and have a track record of successful grants or are themselves reviewers are the *'go to'* people. Value the critique received as *the more critical they are, the better are the chances of improving the application*. They can identify problems before the application is sent in which is better than learning about them after the review.

COVER LETTER

A cover letter with the title of the application, very brief description of the research proposed and the program or request for applications (RFA) or request for proposals (RFP) submitted to make the life of the ministerial staff easier and ensures that the application is not misdirected. Some agencies allow applicants to refer to or in no case refer to a specific reviewer. This does not mean that it will definitely go to the same person but many times this may be acceded to if the area of interest is also the area of specialization of the reviewer. Not sending to a specific reviewer is a protection in case of conflicts in the area of research with the investigator. So, if there is such a situation then the investigator gains by mentioning it at the outset and then the agency usually gives less weightage to a negative review from such a reviewer.

A cover letter is used to introduce the subject and identify people who should not review your application because of potential conflict of interest (e.g. competitor or peer with a long-standing scientific disagreement). The reasons for avoiding specific reviewers should be clearly stated (e.g. conflict of interest). Short lists are more welcome than long ones. Some agencies may allow researchers' to prefer some reviewers. In such a situation the cover letter is a nice place to use the opportunity to nominate the reviewer with whom the team feels there is a higher likelihood of success.

Please remember that a number is assigned for correspondence by the agency to every application. If it is quoted in every stream it becomes easier to search for the same.

DECISION PROCESS EXPLAINED

Several factors come into play when the issue of which proposals to fund are decided- application's percentile ranking derived from its priority score, the outcome of peer review, the relevance of the proposed project to its mission, the availability of funds and funding priorities of the agencies.

The usual process is that primary reviewers write critical appraisal before the agency's funding meeting and hand them over to be circulated in the funding board members. One or more members serve as readers to identify the strengths and weaknesses of each application. At the meeting applications are discussed and evaluated. Any schema like yes/no or Likert Scales or rating scores can be used. In discussion differences of opinion are explored. Sometimes the extreme scores are removed and the rest evaluated or there may simply be resort to a mean or median score.

Applications of clinical research with no significant and substantial scientific merit, or with inadequate protection against risks to human subjects may not be recommended for consideration. In case of lack of adequate information, the review group maybe unable to make an adequate determination of the scientific merit of an application and request the application to be deferred to a later review date. In this time the information from the applicant or from outside sources may be sought. Therefore, it is imperative to include all relevant details in the primary application itself. However, it must be borne in mind that too much information or very *busy* applications also do not do too well because the key points may become buried in the deluge. Therefore, the method of *satisficing* i.e. satisfactory and sufficient is an ideal that cannot be over preached—meaning neither too much nor too less.

PAY LINES

A pay line is a funding cutoff point. All applications with percentiles better than the pay line are usually funded whereas those worse than the pay line, except some high-priority applications at the pay line margin, would remain unfunded.
- Pay line is a budget management tool which is dependent on the precise amount of funds available to the agency and gets usually revised through the financial year.
- It is usually used for deciding research project grants only. Many other funding opportunities are usually not within the ambit of the pay line for example, training and career awards. Tweaking around them may be an option if the application falls outside the priority area and then secondary funding from that agency is the alternative.
- Some agencies have no fixed pay line and indeed the pay line or cutoff may vary through the financial year. Therefore, it is wise to find out how the agency tackles the applications. Some agencies keep the applications for a fixed period like say six months or some quarters after the application fails to make grade. Others are not so generous. A nonfundable percentile in one agency may be funded in another. Looking at the reviewers comments and improving for the next round is often the key to successful creativity if one does not make the cut in the first attempt. If an institution declines to pay an application, another institution may agree to take on primary assignment and commit the funding.

Second Level Peer Review or revision meeting is often conducted for a small number of high-priority grants at the pay line margin. These would not have made the grade if the strict criteria were applied but since these are in areas of vital importance or in priority areas they may enjoy funding. Many of these committees are driven by the personality of the opinion leaders in the group. However, the decision on funding is usually a quasi-scientific one wherein the scores are interpolated with the intangible judgment of the members. Sometimes even unexpected results are seen but that is usually the exception rather than the rule.

OUTCOMES

Based on the review the following outcomes are possible:
- Funding approval.
- Transfer to another institution which may fund it or recommendations to do so. This gives an edge to the application in the other institution.

- Kept for later decision when funds are not immediately available but would be expected from savings in the review in usually second quarter (read September in India).
- File closed and funding denied.

The decisions are differently worded but fall into the four broadly stated above. These decisions are communicated to the applicants. The directors or the staff communicates the decision to the applicants in due course. An electronic communication like email or sms after the meeting is well appreciated by the applicants in the interim. Ideally, the summary statements should include the reviewers' write ups, a summary of the deliberations of the committee or scientific grant group, the average priority score, changes recommended in the budget and other comments. Agencies should and usually do have an appeal process. In an appeal, an applicant should outline the reason/s to believe that the review process was flawed. Flawed should mean errors due to reviewer conflict of interest or personal bias and not mere differences in scientific opinion. Appealable errors should be reported to the appellant authority or the person whose name and contact information appear on the face page of the communication. If possible a phone call to the office of the person is enough to set the process in motion if provisions for it exist. Even if they do not, there is always the benefit of correction of systemic failures.

COMMON PROBLEMS ENCOUNTERED

The common reasons for failure are:
- Lack of significance to the scientific issue being addressed.
- Lack of original or new ideas.
- Proposal of an unrealistically large amount of work or scientific rationale not valid.
- Project too diffuse or superficial or lacking focus.
- Proposed project a fishing expedition lacking solid scientific basis or a solution looking for a problem.
- Hypothesis or background data unreliable or alternative hypotheses not considered.
- Descriptive experiments without a specific hypothesis.
- Technology driven rather than hypothesis driven proposals or a method in search of a problem.
- Rationale for experiments not clear.
- Direction or sense of priority not clearly defined or jumbled American pie.
- Insufficient methodology description to convince reviewers the investigator knows what he or she is doing highlighted by non- enunciation of potential problems and pitfalls.
- The proposed model/s does not address the proposed questions or invalid conclusions (i.e. proposing to study T-cell gene expression in a B-cell line).
- The proposed experiment does not include all relevant controls or problem of measuring against whom.
- Innovative proposal not backed by enough preliminary data.
- Preliminary data underlines unfeasibility of the project.
- Inexperienced investigator/s (i.e. publications or appropriate preliminary data) especially in the proposed techniques or has no collaborator with the experience.

- Lack of literature references causing reviewers to think that the applicant either does not know the literature or has purposely neglected critical published material.

If the application is not selected for getting funded, it may not be a bad option to revise the application and resubmit. This gives an opportunity to address reviewers' concerns. Many applications succeed on the second or even third submission if the permission to resubmit exists. Usually there may be a cap on the number of times resubmission will be permitted. However, some projects are plagued by what is called the *systemic sin of never succeeding* and these include but are not limited to the following:
- Lack of new or original ideas.
- Ill-defined, superficial, lacking, unfocused hypothesis unsupported by preliminary data.
- Unsuitable or defective methods not capable of yielding appropriate high quality results.
- Inappropriate design, instrumentation, poor timing or conditions for data collection.
- Inadequate expertise or knowledge of field or too little time to devote to the project on part of investigator or principle collaborators.
- Limited access to patients, resources or research facilities
- Philosophical issues, e.g. the reviewers not convinced about the significance of the work.
- Hypothesis not sound or not supported by data presented.
- Work has already been done. Nothing more to report.
- Methods proposed not suitable for testing the hypothesis.

Other common problems amenable to correction are:
- Poor writing.
- Insufficient information or background data.
- Unconvincing significance.
- Insufficient discussion of problems expected and alternatives available.

REVISION OF APPLICATION

In case of failure to secure funding remember Robert Bruce won on the seventeenth attempt and Stephen King had countless rejection slips for his greatest masterpiece. Even Amish Tripathi was rejected forty nine times before he got published. Data shows that persistence pays. Many proposals have gone through repeated modifications before being accepted for funding.

To revise the application for resubmission the summary statement should be read carefully, the concerns identified, the office coordinator's comments taken and someone experienced in grantsmanship but not involved in the initial application asked to review the application, summary statement and revision plans. The response should be careful and the critique taken constructively. An introduction with point wise reply to the comments and suggestions of the reviewers is a good first step. The applicant is not duty bound to accept every comment of the reviewers but they must all be responded to with clear identifiers so reviewers can easily find where new data or revised plans have been added. Some authors use a table with two columns—one pertaining to the observations of the reviewers and the second a reply to it. A bar in the margin is another good way to show the revisions. Highlighting new sections with indenting, bracketing, underlining, or change of type can help convey the message to the reviewers clearly. In case of disagreements careful delineation of the reasons should be supported with reasons and additional information with references if needed. A positive tone goes a long way and an angry rebuttal gets nowhere. A summary statement is not meant to be an exhaustive critique and the peer review comments are the go to places for answers. Sometimes the changes may be radical and new problems may be introduced leading the researcher into a catch 22 situation alike *Scylla and Charybdis problem*. Accepting limitations imposed by the methods and the demonstration that this current course is the best *bounded rationality* approach is often enough to win the reviewers over as they are aware of the research process and its potential pitfalls.

VOILA! MANNA FROM HEAVEN: THE FUNDING

In case of selection for funding, the agencies usually discuss when the award is to start and the funding level of project. Additional information may be required to be submitted, e.g. updated information on budget, certification of institutional approval of human subjects, other support or information, and the timelines for disbursal as well as follow-up information. It is important to send these as soon as possible. Upon satisfactory completion of all requirements, the Notice of Grant Award stating the amount of funding approved along with the details for the current financial year, funding committed for future years, start and end dates, and the terms and conditions of the award is sent to the investigator and his parent institution.

Most agencies inform the investigators the items that can be paid for by the agency in the grant. In most cases, direct (project-specific) costs plus the facilities and administrative costs (F and A costs or indirect costs) are covered. This varies from country to country and depends on the type of institutions. This must be adhered to in the budget as well as when incurring expenditure. Some agencies just require utilization certificates while others will want an audited report. Keeping computerized inventory and accounts is always a winning formula because this saves a lot of space as well as paperwork later.

RECORDS AND ACCOUNTING

Once the funding is received it is imperative to use it for the specified purposes and maintain meticulous records. The voucher, receipts, invoices and records should be kept in properly labeled file folders. They should be catalogued. Accounting softwares are easily available in the market to provide adequate records which generate reports like income and expenditure report, balance sheet, fund flow statement, etc. Well kept records assure the agencies about the integrity and work ethic of the researcher. This reputation comes into play when subsequently one applies again or acts as a collaborator in a research project.

Appendix

Oxford Centre for Evidence-based Medicine (CEBM) 2011: Levels of Evidence

List of Tables

1. Random Number Table
2. Binomial Coefficients
3. Binomial Probabilities
4. $e^{-\lambda}$ Values for Calculating Poisson Probabilities (for $0 < \lambda < 1$)
5. Area under the Standard Normal Curve (Values for 0 to + z)
6. Area under the Standard Normal Curve (Values for $-\infty$ to –z)
7. Area under the Standard Normal Curve (values for $-\infty$ to + z)
8. Critical Values of 't'
9. Critical Values of Chi Square χ^2
10. Critical Values of 'F'
11. Factors for Construction of Control Charts
12. Studentized Range—Critical Values
13. *Pearson's Correlation Coefficient* (r): Critical Values
14. *Spearman Rank Correlation Coefficient* (ρ): Critical Values
15. *Run Test*: Acceptance Region for Values of 'R'
16. *Sign Test*: Critical Values of 'T'
17. *Kolmogorov-Smirnov Test*: Critical Values of 'D'
18. *Wilcoxon-Wilcox Two-sided Test*: Critical Values
19. *Mann-Whitney Test (U Test)*: Critical Values of 'U'
20. *Critical Values of* τ: Kendall Rank Correlation Significance Test
21. *Wilcoxon Signed Rank Test 'W'*: Lower and Upper Critical Values
22. *Common Logarithms Table*
23. *Antilogarithms Table*

OXFORD CENTRE FOR EVIDENCE-BASED MEDICINE 2011 LEVELS OF EVIDENCE

Question	Step 1 (Level 1*)	Step 2 (Level 2*)	Step 3 (Level 3*)	Step 4 (Level 4*)	Step 5 (Level 5*)
How common is the problem?	Local and current random sample surveys (or censuses)	Systematic review of surveys that allow matching to local circumstances**	Local non-random sample**	Case-series**	n/a
Is the diagnostic or monitoring test accurate? (Diagnosis)	Systematic review of cross-sectional studies with consistently applied reference standard and blinding	Individual cross-sectional studies with consistently applied reference standard and blinding	Non-consecutive studies, or studies without consistently applied reference standards**	Case-control studies, or "poor or non-independent reference standard**	Mechanism-based reasoning
What will happen, if we do not add a therapy? (Prognosis)	Systematic review of inception cohort studies	Inception cohort studies	Cohort study or control arm of randomized trial*	Case-series or case-control studies, or poor quality prognostic cohort study**	n/a
Does this intervention help? (Treatment Harms)	Systematic review of randomized trials or n-of-1 trials	Randomized trial or observational study with dramatic effect	Non-randomized controlled cohort/follow-up study**	Case-series, case-control studies, or historically controlled studies**	Mechanism-based reasoning
What are the common harms? (Treatment Harms)	Systematic review of randomized trials, systematic review of nested case-control studies, n-of-1 trial with the patient you are raising the question about, or observational study with dramatic effect	Individual randomized trial or (exceptionally) observational study with dramatic effect	Non-randomized controlled cohort/follow-up study (post-marketing surveillance) provided there are sufficient numbers to rule out a common harm. (For long-term harms the duration of follow-up must be sufficient.)**	Case-series, case-control, or historically controlled studies**	Mechanism-based reasoning

Contd...

Contd...

Question	Step 1 (Level 1*)	Step 2 (Level 2*)	Step 3 (Level 3*)	Step 4 (Level 4*)	Step 5 (Level 5*)
What are the rare harms? (Treatment Harms)	Systematic review of randomized trials or n-of-1 trail	Randomized trial or (exceptionally) observational study with dramatic effect			
Is this (early detection) test worthwhile? (Screening)	Systematic review of randomized trials	Randomized trial	Non-randomized controlled cohot/follow-up study**	Case-series, case-control, or historically controlled studies**	Mechanism-based reasoning

*Level may be graded down on the basis of study quality, imprecision, indirectness (study PICO does not match questions PICO), because of inconsistency between studies, or because the absolute effect size is very small; Level may be graded up, if there is a large or very large effect size.
**As always, a systematic review is generally better than an individual study.

How to cite the Levels of Evidence Table?

OCEBM Levels of Evidence Working Group*. "The Oxford 2011 Levels of Evidence".
Oxford Centre for Evidence-Based Medicine. *http://www/cebm.net/index.aspx?0=5653*

*OCEBM Table of Evidence Working Group = Jeremy Howick, Iain Chalmers (James Lind Library), Paul Glasziou, Trish Greenhalgh, Carl Heneghan, Alessandre Liberati, Ivan Moschetti, Bob Phillips, Hazel Thornton, Olive Goddard and Mary Hodgkinson

RANDOM NUMBER TABLE

9	96	61	366	580	645	3971	4341	3967	43018	67518	36992	74529
66	12	41	540	533	370	1269	1180	3594	95506	48205	30601	80287
40	64	55	758	319	227	5965	4895	9668	56451	94432	53365	96596
2	69	5	300	735	705	7963	1437	6713	79516	67278	83997	44161
76	33	52	698	227	352	9353	2529	3780	95907	51201	75904	67032
99	1	93	937	289	877	2395	2089	1512	30875	70066	47365	49613
74	41	53	861	305	939	5147	7882	3172	41364	64262	53475	66294
42	1	30	963	890	339	7581	4721	8825	90841	55381	17576	43780
16	97	66	301	604	923	9227	7825	8816	52530	46563	15950	14854
15	98	10	974	742	562	8100	7554	4161	82290	71821	95762	75568
62	67	75	962	688	847	6305	2967	2062	99422	23418	82162	62622
29	27	77	337	772	223	8291	1048	6967	64674	77251	78970	48540
72	64	43	393	371	953	4277	2417	7771	39586	96866	31064	86479
66	42	59	178	151	913	6127	2566	3078	44416	54077	57090	62971
53	55	16	177	534	919	8260	4824	5051	55978	37654	77690	27930
47	24	37	565	440	207	4150	6953	2318	64915	76571	37766	69762
40	62	95	670	727	113	4642	3533	3446	10745	72120	14353	17510
86	81	1	327	339	662	8194	8854	2913	66274	97898	74918	66530
55	76	68	440	755	948	8474	4866	6545	58122	82597	39380	95919
49	32	18	883	453	160	4294	3216	3642	54514	65021	15537	37902
96	29	66	870	480	737	5714	6074	5851	50556	84379	46154	60540
64	31	70	434	443	364	7835	8005	1517	53406	84870	20775	70383
29	22	69	988	428	810	5740	8335	3772	96001	88237	12779	11931
76	23	33	263	989	190	2899	8281	1489	92687	41801	46316	65394
70	54	41	826	659	547	3326	6395	5853	52309	68583	24469	73777
48	84	33	333	643	835	9496	1848	9073	64396	85500	80037	84995
45	56	90	606	637	121	7096	9385	9546	13577	68530	52406	97931
35	71	9	995	635	829	4307	5364	1037	31685	94230	77713	60028
2	87	49	683	521	915	4844	6186	3426	72112	59320	57385	27118
47	69	66	230	919	751	7373	1240	8420	94299	45903	44331	94994
5	54	92	274	279	298	1434	2527	3427	96434	90227	48955	12782
75	37	5	948	351	192	4267	3502	8010	48611	92910	36990	42202
1	47	73	391	960	513	5390	6543	8625	37274	24246	60169	83404
32	4	63	803	716	442	3491	4675	7858	34810	75182	93200	95345
11	73	38	533	619	317	8677	9873	3390	28419	10643	45719	63012
58	36	51	840	661	320	9093	1186	4844	15756	23976	47726	61324
37	11	77	502	102	561	5987	9681	8920	88030	35251	72475	42166
88	12	90	595	113	430	3354	5847	2904	50424	55904	80067	49424
74	50	21	184	641	371	4730	7248	1826	96394	99774	98426	52012

BINOMIAL COEFFICIENTS $\left[{}^nC_r = \dfrac{n!}{r!\,(n-r)!} \right]$

n \ r	0	1	2	3	4	5	6	7	8	9	10	11	12	13	14	15
0	1															
1	1	1														
2	1	2	1													
3	1	3	3	1												
4	1	4	6	4	1											
5	1	5	10	10	5	1										
6	1	6	15	20	15	6	1									
7	1	7	21	35	35	21	7	1								
8	1	8	28	56	70	56	28	8	1							
9	1	9	36	84	126	126	84	36	9	1						
10	1	10	45	120	210	252	210	120	45	10	1					
11	1	11	55	165	330	462	462	330	165	55	11	1				
12	1	12	66	220	495	792	924	792	495	220	66	12	1			
13	1	13	78	286	715	1287	1716	1716	1287	715	286	78	13	1		
14	1	14	91	364	1001	2002	3003	3432	3003	2002	1001	364	91	14	1	
15	1	15	105	455	1365	3003	5005	6435	6435	5005	3003	1365	455	105	15	1
16	1	16	120	560	1820	4368	8008	11440	12870	11440	8008	4368	1820	560	120	16
17	1	17	136	680	2380	6188	12376	19448	24310	24310	19448	12376	6188	2380	680	136
18	1	18	153	816	3060	8568	18564	31824	43758	48620	43758	31824	18564	8568	3060	816
19	1	19	171	969	3876	11628	27132	50388	75582	92378	92378	75582	50388	27132	11628	3876
20	1	20	190	1140	4845	15504	38760	77520	125970	167960	184756	167960	125970	77520	38760	15504
21	1	21	210	1330	5985	20349	54264	116280	203490	293930	352716	352716	293930	203490	116280	54264
22	1	22	231	1540	7315	26334	74613	170544	319770	497420	646646	705432	646646	497420	319770	170544
23	1	23	253	1771	8855	33649	100947	245157	490314	817190	1144066	1352078	1352078	1144066	817190	490314
24	1	24	276	2024	10626	42504	134596	346104	735471	1307504	1961256	2496144	2704156	2496144	1961256	1307504
25	1	25	300	2300	12650	53130	177100	480700	1081575	2042975	3268760	4457400	5200300	5200300	4457400	3268760

BINOMIAL PROBABILITIES

$P(r) = {}^nC_r \, p^r q^{n-r}$ (n = Number of Trials, r = Number of Success, p = Probability of Success)

n	r	Probability (p)								
		0.10	0.20	0.30	0.40	0.50	0.60	0.70	0.80	0.90
1	0	0.90000	0.80000	0.70000	0.60000	0.50000	0.40000	0.30000	0.20000	0.10000
	1	0.10000	0.20000	0.30000	0.40000	0.50000	0.60000	0.70000	0.80000	0.90000
2	0	0.81000	0.64000	0.49000	0.36000	0.25000	0.16000	0.09000	0.04000	0.01000
	1	0.18000	0.32000	0.42000	0.48000	0.50000	0.48000	0.42000	0.32000	0.18000
	2	0.01000	0.04000	0.09000	0.16000	0.25000	0.36000	0.49000	0.64000	0.81000
3	0	0.72900	0.51200	0.34300	0.21600	0.12500	0.06400	0.02700	0.00800	0.00100
	1	0.24300	0.38400	0.44100	0.43200	0.37500	0.28800	0.18900	0.09600	0.02700
	2	0.02700	0.09600	0.18900	0.28800	0.37500	0.43200	0.44100	0.38400	0.24300
	3	0.00100	0.00800	0.02700	0.06400	0.12500	0.21600	0.34300	0.51200	0.72900
4	0	0.65610	0.40960	0.24010	0.12960	0.06250	0.02560	0.00810	0.00160	0.00010
	1	0.29160	0.40960	0.41160	0.34560	0.25000	0.15360	0.07560	0.02560	0.00360
	2	0.04860	0.15360	0.26460	0.34560	0.37500	0.34560	0.26460	0.15360	0.04860
	3	0.00360	0.02560	0.07560	0.15360	0.25000	0.34560	0.41160	0.40960	0.29160
	4	0.00010	0.00160	0.00810	0.02560	0.06250	0.12960	0.24010	0.40960	0.65610
5	0	0.59049	0.32768	0.16807	0.07776	0.03125	0.01024	0.00243	0.00032	0.00001
	1	0.32805	0.40960	0.36015	0.25920	0.15625	0.07680	0.02835	0.00640	0.00045
	2	0.07290	0.20480	0.30870	0.34560	0.31250	0.23040	0.13230	0.05120	0.00810

Contd...

Contd...

Appendix

n	r	Probability (p)								
		0.10	0.20	0.30	0.40	0.50	0.60	0.70	0.80	0.90
	3	0.00810	0.05120	0.13230	0.23040	0.31250	0.34560	0.30870	0.20480	0.07290
	4	0.00045	0.00640	0.02835	0.07680	0.15625	0.25920	0.36015	0.40960	0.32805
	5	0.00001	0.00032	0.00243	0.01024	0.03125	0.07776	0.16807	0.32768	0.59049
6	0	0.53144	0.26214	0.11765	0.04666	0.01563	0.00410	0.00073	0.00006	0.00000
	1	0.35429	0.39322	0.30253	0.18662	0.09375	0.03686	0.01021	0.00154	0.00005
	2	0.09842	0.24576	0.32414	0.31104	0.23438	0.13824	0.05954	0.01536	0.00122
	3	0.01458	0.08192	0.18522	0.27648	0.31250	0.27648	0.18522	0.08192	0.01458
	4	0.00122	0.01536	0.05954	0.13824	0.23438	0.31104	0.32414	0.24576	0.09841
	5	0.00005	0.00154	0.01021	0.03686	0.09375	0.18662	0.30253	0.39322	0.35429
	6	0.00000	0.00006	0.00073	0.00410	0.01563	0.04666	0.11765	0.26214	0.53144
7	0	0.47830	0.20972	0.08235	0.02799	0.00781	0.00164	0.00022	0.00001	0.00000
	1	0.37201	0.36700	0.24706	0.13064	0.05469	0.01720	0.00357	0.00036	0.00001
	2	0.12400	0.27525	0.31765	0.26127	0.16406	0.07741	0.02500	0.00430	0.00017
	3	0.02296	0.11469	0.22689	0.29030	0.27344	0.19354	0.09724	0.02867	0.00255
	4	0.00255	0.02867	0.09724	0.19354	0.27344	0.29030	0.22689	0.11469	0.02296
	5	0.00017	0.00430	0.02500	0.07741	0.16406	0.26127	0.31765	0.27525	0.12400
	6	0.00001	0.00036	0.00357	0.01720	0.05469	0.13064	0.24706	0.36700	0.37201
	7	0.00000	0.00001	0.00022	0.00164	0.00781	0.02799	0.08235	0.20972	0.47830
8	0	0.43047	0.16777	0.05765	0.01680	0.00391	0.00066	0.00007	0.00000	0.00000

Contd...

Contd...

n	r	Probability (p)								
		0.10	0.20	0.30	0.40	0.50	0.60	0.70	0.80	0.90
	1	0.38264	0.33554	0.19765	0.08958	0.03125	0.00786	0.00122	0.00008	0.00000
	2	0.14880	0.29360	0.29648	0.20902	0.10938	0.04129	0.01000	0.00115	0.00002
	3	0.03307	0.14680	0.25412	0.27869	0.21875	0.12386	0.04668	0.00918	0.00041
	4	0.00459	0.04588	0.13614	0.23224	0.27344	0.23224	0.13614	0.04588	0.00459
	5	0.00041	0.00918	0.04668	0.12386	0.21875	0.27869	0.25412	0.14680	0.03307
	6	0.00002	0.00115	0.01000	0.04129	0.10938	0.20902	0.29648	0.29360	0.14880
	7	0.00000	0.00008	0.00122	0.00786	0.03125	0.08958	0.19765	0.33554	0.38264
	8	0.00000	0.00000	0.00007	0.00066	0.00391	0.01680	0.05765	0.16777	0.43047
9	0	0.38742	0.13422	0.04035	0.01008	0.00195	0.00026	0.00002	0.00000	0.00000
	1	0.38742	0.30199	0.15565	0.06047	0.01758	0.00354	0.00041	0.00002	0.00000
	2	0.17219	0.30199	0.26683	0.16124	0.07031	0.02123	0.00386	0.00029	0.00000
	3	0.04464	0.17616	0.26683	0.25082	0.16406	0.07432	0.02100	0.00275	0.00006
	4	0.00744	0.06606	0.17153	0.25082	0.24609	0.16722	0.07351	0.01652	0.00083
	5	0.00083	0.01652	0.07351	0.16722	0.24609	0.25082	0.17153	0.06606	0.00744
	6	0.00006	0.00275	0.02100	0.07432	0.16406	0.25082	0.26683	0.17616	0.04464
	7	0.00000	0.00029	0.00386	0.02123	0.07031	0.16124	0.26683	0.30199	0.17219

Contd...

Contd...

n	r	Probability (p)								
		0.10	0.20	0.30	0.40	0.50	0.60	0.70	0.80	0.90
	8	0.00000	0.00002	0.00041	0.00354	0.01758	0.06047	0.15565	0.30199	0.38742
	9	0.00000	0.00000	0.00002	0.00026	0.00195	0.01008	0.04035	0.13422	0.38742
10	0	0.34868	0.10737	0.02825	0.00605	0.00098	0.00010	0.00001	0.00000	0.00000
	1	0.38742	0.26844	0.12106	0.04031	0.00977	0.00157	0.00014	0.00000	0.00000
	2	0.19371	0.30199	0.23347	0.12093	0.04395	0.01062	0.00145	0.00007	0.00000
	3	0.05740	0.20133	0.26683	0.21499	0.11719	0.04247	0.00900	0.00079	0.00001
	4	0.01116	0.08808	0.20012	0.25082	0.20508	0.11148	0.03676	0.00551	0.00014
	5	0.00149	0.02642	0.10292	0.20066	0.24609	0.20066	0.10292	0.02642	0.00149
	6	0.00014	0.00551	0.03676	0.11148	0.20508	0.25082	0.20012	0.08808	0.01116
	7	0.00001	0.00079	0.00900	0.04247	0.11719	0.21499	0.26683	0.20133	0.05740
	8	0.00000	0.00007	0.00145	0.01062	0.04395	0.12093	0.23347	0.30199	0.19371
	9	0.00000	0.00000	0.00014	0.00157	0.00977	0.04031	0.12106	0.26844	0.38742
	10	0.00000	0.00000	0.00001	0.00010	0.00098	0.00605	0.02825	0.10737	0.34868

$e^{-\lambda}$ VALUES FOR CALCULATING POISSON PROBABILITIES for $(0 < \lambda < 1)$

λ	0	1	2	3	4	5	6	7	8	9
0.0	1	0.9900	0.9802	0.9704	0.9608	0.9512	0.9418	0.9324	0.9231	0.9139
0.1	0.9048	0.8958	0.8869	0.8781	0.8694	0.8607	0.8521	0.8437	0.8353	0.8270
0.2	0.8187	0.8106	0.8025	0.7945	0.7866	0.7788	0.7711	0.7634	0.7558	0.7483
0.3	0.7408	0.7334	0.7261	0.7189	0.7118	0.7047	0.6977	0.6907	0.6839	0.6771
0.4	0.6703	0.6637	0.6570	0.6505	0.6440	0.6376	0.6313	0.6250	0.6188	0.6126
0.5	0.6065	0.6005	0.5945	0.5886	0.5827	0.5769	0.5712	0.5655	0.5599	0.5543
0.6	0.5488	0.5434	0.5379	0.5326	0.5273	0.5220	0.5169	0.5117	0.5066	0.5016
0.7	0.4966	0.4916	0.4868	0.4819	0.4771	0.4724	0.4677	0.4630	0.4584	0.4538
0.8	0.4493	0.4449	0.4404	0.4360	0.4317	0.4274	0.4232	0.4190	0.4148	0.4107
0.9	0.4066	0.4025	0.3985	0.3946	0.3906	0.3867	0.3829	0.3791	0.3753	0.3716

AREA UNDER THE STANDARD NORMAL CURVE

Values represent the area under the curve for a Standard Normal Curve for Area between 0 and z

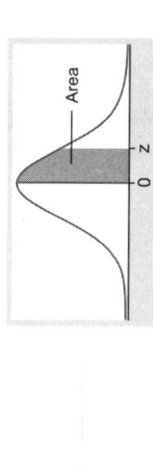

z	0.00	0.01	0.02	0.03	0.04	0.05	0.06	0.07	0.08	0.09
0.0	0.00000	0.00399	0.00798	0.01197	0.01595	0.01994	0.03292	0.02790	0.03188	0.03586
0.1	0.03983	0.04380	0.04776	0.05172	0.05576	0.05962	0.06356	0.06750	0.07142	0.07535
0.2	0.07926	0.08317	0.08706	0.09095	0.09484	0.09871	0.10257	0.10642	0.11026	0.11409
0.3	0.11791	0.12172	0.12552	0.12930	0.13307	0.13683	0.14058	0.14431	0.14803	0.15173
0.4	0.15542	0.15910	0.16276	0.16640	0.17003	0.17365	0.17724	0.18082	0.18439	0.18793
0.5	0.19146	0.19497	0.19847	0.20194	0.20540	0.20884	0.21226	0.21566	0.21904	0.22241
0.6	0.22575	0.22907	0.23237	0.23565	0.23891	0.24215	0.24537	0.24857	0.25175	0.25490
0.7	0.25804	0.26115	0.26424	0.26731	0.27035	0.27337	0.27637	0.27935	0.28231	0.28524
0.8	0.28815	0.29103	0.29389	0.29673	0.29955	0.30234	0.30511	0.30785	0.31057	0.31327
0.9	0.31594	0.31859	0.32121	0.32382	0.32639	0.32894	0.33147	0.33398	0.33646	0.33891
1.0	0.34135	0.34375	0.34614	0.34850	0.35083	0.35314	0.35543	0.35769	0.35993	0.36214
1.1	0.36433	0.36650	0.36864	0.37076	0.37286	0.37493	0.37698	0.37900	0.38100	0.38298
1.2	0.38493	0.38686	0.38877	0.39065	0.39251	0.39435	0.39617	0.39796	0.39973	0.40148
1.3	0.40320	0.40490	0.40658	0.40824	0.40988	0.41149	0.41309	0.41466	0.41621	0.41774
1.4	0.41924	0.42073	0.42220	0.42364	0.42507	0.42647	0.42786	0.42922	0.43056	0.43189
1.5	0.43319	0.43448	0.43574	0.43699	0.43822	0.43943	0.44062	0.44179	0.44295	0.44408
1.6	0.44520	0.44630	0.44738	0.44845	0.44950	0.45053	0.45154	0.45254	0.45352	0.45449
1.7	0.45544	0.45637	0.45728	0.45819	0.45907	0.45994	0.46080	0.46164	0.46246	0.46327
1.8	0.46407	0.46485	0.46562	0.46638	0.46712	0.46784	0.46856	0.46926	0.46995	0.47062

Contd...

Contd...

z	0.00	0.01	0.02	0.03	0.04	0.05	0.06	0.07	0.08	0.09
1.9	0.47128	0.47193	0.47257	0.47320	0.47381	0.47441	0.47500	0.47558	0.47615	0.47671
2.0	0.47725	0.47778	0.47831	0.47882	0.47933	0.47982	0.48030	0.48077	0.48124	0.48169
2.1	0.48214	0.48257	0.48300	0.48341	0.48382	0.48422	0.48461	0.48500	0.48537	0.48574
2.2	0.48610	0.48645	0.48679	0.48713	0.48746	0.48778	0.48809	0.48840	0.48870	0.48899
2.3	0.48928	0.48956	0.48983	0.49010	0.49036	0.49061	0.49086	0.49111	0.49134	0.49158
2.4	0.49180	0.49202	0.49224	0.49245	0.49266	0.49286	0.49305	0.49324	0.49343	0.49361
2.5	0.49379	0.49396	0.49413	0.49430	0.49446	0.49461	0.49477	0.49492	0.49506	0.49520
2.6	0.49534	0.49547	0.49560	0.49573	0.49586	0.49598	0.49609	0.49621	0.49632	0.49643
2.7	0.49653	0.49664	0.49674	0.49683	0.49693	0.49702	0.49711	0.49720	0.49728	0.49737
2.8	0.49745	0.49752	0.49760	0.49767	0.49774	0.49781	0.49788	0.49795	0.49801	0.49807
2.9	0.49813	0.49819	0.49825	0.49831	0.49836	0.49841	0.49846	0.49851	0.49856	0.49861
3.0	0.49865	0.49869	0.49874	0.49878	0.49882	0.49886	0.49889	0.49893	0.49897	0.49900
3.1	0.49903	0.49907	0.49910	0.49913	0.49916	0.49918	0.49921	0.49924	0.49926	0.49929
3.2	0.49931	0.49934	0.49936	0.49938	0.49940	0.49942	0.49944	0.49946	0.49948	0.49950
3.3	0.49952	0.49953	0.49955	0.49957	0.49958	0.49960	0.49961	0.49962	0.49964	0.49965
3.4	0.49966	0.49968	0.49969	0.49970	0.49971	0.49972	0.49973	0.49974	0.49975	0.49976
3.5	0.49977	0.49978	0.49978	0.49979	0.49980	0.49981	0.49982	0.49982	0.49983	0.49984
3.6	0.49984	0.49985	0.49985	0.49986	0.49986	0.49987	0.49987	0.49988	0.49988	0.49989
3.7	0.49989	0.49990	0.49990	0.49990	0.49991	0.49991	0.49992	0.49992	0.49992	0.49993
3.8	0.49993	0.49993	0.49993	0.49994	0.49994	0.49994	0.49994	0.49995	0.49995	0.49995
3.9	0.49995	0.49995	0.49996	0.49996	0.49996	0.49996	0.49996	0.49996	0.49997	0.49997

AREA UNDER THE STANDARD NORMAL CURVE

Values represent the area under the curve for a Standard Normal Curve for Area between -∞ and z, where z ≤ 0.

z	0.00	0.01	0.02	0.03	0.04	0.05	0.06	0.07	0.08	0.09
0.0	0.50000	0.49601	0.49202	0.48803	0.48405	0.48006	0.47608	0.47210	0.46812	0.46414
-0.1	0.46017	0.45620	0.45224	0.44828	0.44433	0.44038	0.43644	0.43251	0.42858	0.42465
-0.2	0.42074	0.41683	0.41294	0.40905	0.40517	0.40129	0.39743	0.39358	0.38974	0.38591
-0.3	0.38209	0.37828	0.37448	0.37070	0.36693	0.36317	0.35942	0.35569	0.35197	0.34827
-0.4	0.34458	0.34090	0.33724	0.33360	0.32997	0.32636	0.32276	0.31918	0.31561	0.31207
-0.5	0.30854	0.30503	0.30153	0.29806	0.29460	0.29116	0.28774	0.28434	0.28096	0.27760
-0.6	0.27425	0.27093	0.26763	0.26435	0.26109	0.25785	0.25463	0.25143	0.24825	0.24510
-0.7	0.24196	0.23885	0.23576	0.23270	0.22965	0.22663	0.22363	0.22065	0.21770	0.21476
-0.8	0.21186	0.20897	0.20611	0.20327	0.20045	0.19766	0.19489	0.19215	0.18943	0.18673
-0.9	0.18406	0.18141	0.17879	0.17619	0.17361	0.17106	0.16853	0.16602	0.16354	0.16109
-1.0	0.15866	0.15625	0.15386	0.15151	0.14917	0.14686	0.14457	0.14231	0.14007	0.13786
-1.1	0.13567	0.13350	0.13136	0.12924	0.12714	0.12507	0.12302	0.12100	0.11900	0.11702
-1.2	0.11507	0.11314	0.11123	0.10935	0.10749	0.10565	0.10383	0.10204	0.10027	0.09853
-1.3	0.09680	0.09510	0.09342	0.09176	0.09012	0.08851	0.08691	0.08534	0.08379	0.08226
-1.4	0.08076	0.07927	0.07780	0.07636	0.07493	0.07353	0.07215	0.07078	0.06944	0.06811
-1.5	0.06681	0.06552	0.06426	0.06301	0.06178	0.06057	0.05938	0.05821	0.05705	0.05592

Contd...

Contd...

z	0.00	0.01	0.02	0.03	0.04	0.05	0.06	0.07	0.08	0.09
-1.6	0.05480	0.05370	0.05262	0.05155	0.05050	0.04947	0.04846	0.04746	0.04648	0.04551
-1.7	0.04457	0.04363	0.04272	0.04182	0.04093	0.04006	0.03920	0.03836	0.03754	0.03673
-1.8	0.03593	0.03515	0.03438	0.03362	0.03288	0.03216	0.03144	0.03074	0.03005	0.02938
-1.9	0.02872	0.02807	0.02743	0.02680	0.02619	0.02559	0.02500	0.02442	0.02385	0.02330
-2.0	0.02275	0.02222	0.02169	0.02118	0.02068	0.02018	0.01970	0.01923	0.01876	0.01831
-2.1	0.01786	0.01743	0.01700	0.01659	0.01618	0.01578	0.01539	0.01500	0.01463	0.01426
-2.2	0.01390	0.01355	0.01321	0.01287	0.01255	0.01222	0.01191	0.01160	0.01130	0.01101
-2.3	0.01072	0.01044	0.01017	0.00990	0.00964	0.00939	0.00914	0.00889	0.00866	0.00842
-2.4	0.00820	0.00798	0.00776	0.00755	0.00734	0.00714	0.00695	0.00676	0.00657	0.00639
-2.5	0.00621	0.00604	0.00587	0.00570	0.00554	0.00539	0.00523	0.00508	0.00494	0.00480
-2.6	0.00466	0.00453	0.00440	0.00427	0.00415	0.00402	0.00391	0.00379	0.00368	0.00357
-2.7	0.00347	0.00336	0.00326	0.00317	0.00307	0.00298	0.00289	0.00280	0.00272	0.00264
-2.8	0.00256	0.00248	0.00240	0.00233	0.00226	0.00219	0.00212	0.00205	0.00199	0.00193
-2.9	0.00187	0.00181	0.00175	0.00169	0.00164	0.00159	0.00154	0.00149	0.00144	0.00139
-3.0	0.00135	0.00131	0.00126	0.00122	0.00118	0.00114	0.00111	0.00107	0.00104	0.00100
-3.1	0.00097	0.00094	0.00090	0.00087	0.00084	0.00082	0.00079	0.00076	0.00074	0.00071
-3.2	0.00069	0.00066	0.00064	0.00062	0.00060	0.00058	0.00056	0.00054	0.00052	0.00050
-3.3	0.00048	0.00047	0.00045	0.00043	0.00042	0.00040	0.00039	0.00038	0.00036	0.00035
-3.4	0.00034	0.00032	0.00031	0.00030	0.00029	0.00028	0.00027	0.00026	0.00025	0.00024

AREA UNDER THE STANDARD NORMAL CURVE

Values represent the area under the curve for a Standard Normal Curve for Area between $-\infty$ and z, where z > 0

z	0.00	0.01	0.02	0.03	0.04	0.05	0.06	0.07	0.08	0.09
0.0	0.50000	0.50399	0.50798	0.51197	0.51595	0.51994	0.52392	0.52790	0.53188	0.53586
0.1	0.53983	0.54380	0.54776	0.55172	0.55567	0.55962	0.56356	0.56749	0.57142	0.57535
0.2	0.57926	0.58317	0.58706	0.59095	0.59483	0.59871	0.60257	0.60642	0.61026	0.61409
0.3	0.61791	0.62172	0.62552	0.62930	0.63307	0.63683	0.64058	0.64431	0.64803	0.65173
0.4	0.65542	0.65910	0.66276	0.66640	0.67003	0.67364	0.67724	0.68082	0.68439	0.68793
0.5	0.69146	0.69497	0.69847	0.70194	0.70540	0.70884	0.71226	0.71566	0.71904	0.72240
0.6	0.72575	0.72907	0.73237	0.73565	0.73891	0.74215	0.74537	0.74857	0.75175	0.75490
0.7	0.75804	0.76115	0.76424	0.76730	0.77035	0.77337	0.77637	0.77935	0.78230	0.78524
0.8	0.78814	0.79103	0.79389	0.79673	0.79955	0.80234	0.80511	0.80785	0.81057	0.81327
0.9	0.81594	0.81859	0.82121	0.82381	0.82639	0.82894	0.83147	0.83398	0.83646	0.83891
1.0	0.84134	0.84375	0.84614	0.84849	0.85083	0.85314	0.85543	0.85769	0.85993	0.86214
1.1	0.86433	0.86650	0.86864	0.87076	0.87286	0.87493	0.87698	0.87900	0.88100	0.88298
1.2	0.88493	0.88686	0.88877	0.89065	0.89251	0.89435	0.89617	0.89796	0.89973	0.90147
1.3	0.90320	0.90490	0.90658	0.90824	0.90988	0.91149	0.91309	0.91466	0.91621	0.91774
1.4	0.91924	0.92073	0.92220	0.92364	0.92507	0.92647	0.92785	0.92922	0.93056	0.93189
1.5	0.93319	0.93448	0.93574	0.93699	0.93822	0.93943	0.94062	0.94179	0.94295	0.94408

Contd...

Contd...

z	0.00	0.01	0.02	0.03	0.04	0.05	0.06	0.07	0.08	0.09
1.6	0.94520	0.94630	0.94738	0.94845	0.94950	0.95053	0.95154	0.95254	0.95352	0.95449
1.7	0.95543	0.95637	0.95728	0.95818	0.95907	0.95994	0.96080	0.96164	0.96246	0.96327
1.8	0.96407	0.96485	0.96562	0.96638	0.96712	0.96784	0.96856	0.96926	0.96995	0.97062
1.9	0.97128	0.97193	0.97257	0.97320	0.97381	0.97441	0.97500	0.97558	0.97615	0.97670
2.0	0.97725	0.97778	0.97831	0.97882	0.97932	0.97982	0.98030	0.98077	0.98124	0.98169
2.1	0.98214	0.98257	0.98300	0.98341	0.98382	0.98422	0.98461	0.98500	0.98537	0.98574
2.2	0.98610	0.98645	0.98679	0.98713	0.98745	0.98778	0.98809	0.98840	0.98870	0.98899
2.3	0.98928	0.98956	0.98983	0.99010	0.99036	0.99061	0.99086	0.99111	0.99134	0.99158
2.4	0.99180	0.99202	0.99224	0.99245	0.99266	0.99286	0.99305	0.99324	0.99343	0.99361
2.5	0.99379	0.99396	0.99413	0.99430	0.99446	0.99461	0.99477	0.99492	0.99506	0.99520
2.6	0.99534	0.99547	0.99560	0.99573	0.99585	0.99598	0.99609	0.99621	0.99632	0.99643
2.7	0.99653	0.99664	0.99674	0.99683	0.99693	0.99702	0.99711	0.99720	0.99728	0.99736
2.8	0.99744	0.99752	0.99760	0.99767	0.99774	0.99781	0.99788	0.99795	0.99801	0.99807
2.9	0.99813	0.99819	0.99825	0.99831	0.99836	0.99841	0.99846	0.99851	0.99856	0.99861
3.0	0.99865	0.99869	0.99874	0.99878	0.99882	0.99886	0.99889	0.99893	0.99896	0.99900
3.1	0.99903	0.99906	0.99910	0.99913	0.99916	0.99918	0.99921	0.99924	0.99926	0.99929
3.2	0.99931	0.99934	0.99936	0.99938	0.99940	0.99942	0.99944	0.99946	0.99948	0.99950
3.3	0.99952	0.99953	0.99955	0.99957	0.99958	0.99960	0.99961	0.99962	0.99964	0.99965
3.4	0.99966	0.99968	0.99969	0.99970	0.99971	0.99972	0.99973	0.99974	0.99975	0.99976

CRITICAL VALUES OF 't'

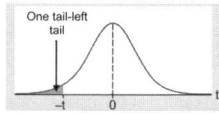
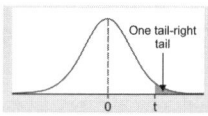
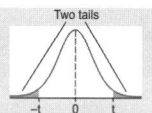

Degrees of Freedom (υ)	Area in One Tail				
	0.10	0.05	0.025	0.010	0.005
	Area in Two Tails				
	0.20	0.10	0.05	0.02	0.01
1	3.0777	6.3138	12.7062	31.8205	63.6567
2	1.8856	2.9200	4.3027	6.9646	9.9248
3	1.6377	2.3534	3.1824	4.5407	5.8409
4	1.5332	2.1318	2.7764	3.7469	4.6041
5	1.4759	2.0150	2.5706	3.3649	4.0321
6	1.4398	1.9432	2.4469	3.1427	3.7074
7	1.4149	1.8946	2.3646	2.9980	3.4995
8	1.3968	1.8595	2.3060	2.8965	3.3554
9	1.3830	1.8331	2.2622	2.8214	3.2498
10	1.3722	1.8125	2.2281	2.7638	3.1693
11	1.3634	1.7959	2.2010	2.7181	3.1058
12	1.3562	1.7823	2.1788	2.6810	3.0545
13	1.3502	1.7709	2.1604	2.6503	3.0123
14	1.3450	1.7613	2.1448	2.6245	2.9768
15	1.3406	1.7531	2.1314	2.6025	2.9467
16	1.3368	1.7459	2.1199	2.5835	2.9208
17	1.3334	1.7396	2.1098	2.5669	2.8982
18	1.3304	1.7341	2.1009	2.5524	2.8784
19	1.3277	1.7291	2.0930	2.5395	2.8609
20	1.3253	1.7247	2.0860	2.5280	2.8453
21	1.3232	1.7207	2.0796	2.5176	2.8314
22	1.3212	1.7171	2.0739	2.5083	2.8188
23	1.3195	1.7139	2.0687	2.4999	2.8073
24	1.3178	1.7109	2.0639	2.4922	2.7969
25	1.3163	1.7081	2.0595	2.4851	2.7874
26	1.3150	1.7056	2.0555	2.4786	2.7787
27	1.3137	1.7033	2.0518	2.4727	2.7707
28	1.3125	1.7011	2.0484	2.4671	2.7633
29	1.3114	1.6991	2.0452	2.4620	2.7564
30	1.3104	1.6973	2.0423	2.4573	2.7500
60	1.2958	1.6706	2.0003	2.3901	2.6603
120	1.2886	1.6577	1.9799	2.3578	2.6174
500	1.2832	1.6479	1.9647	2.3338	2.5857
1000	1.2824	1.6464	1.9623	2.3301	2.5808

CRITICAL VALUES OF CHI SQUARE (χ^2)

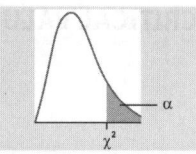

Degree of Freedom	Level of Significance (α)					
	0.100	0.050	0.025	0.010	0.005	0.001
1	2.71	3.84	5.02	6.63	7.88	10.83
2	4.61	5.99	7.38	9.21	10.60	13.82
3	6.25	7.81	9.35	11.34	12.84	16.27
4	7.78	9.49	11.14	13.28	14.86	18.47
5	9.24	11.07	12.83	15.09	16.75	20.52
6	10.64	12.59	14.45	16.81	18.55	22.46
7	12.02	14.07	16.01	18.48	20.28	24.32
8	13.36	15.51	17.53	20.09	21.95	26.12
9	14.68	16.92	19.02	21.67	23.59	27.88
10	15.99	18.31	20.48	23.21	25.19	29.59
11	17.28	19.68	21.92	24.72	26.76	31.26
12	18.55	21.03	23.34	26.22	28.30	32.91
13	19.81	22.36	24.74	27.69	29.82	34.53
14	21.06	23.68	26.12	29.14	31.32	36.12
15	22.31	25.00	27.49	30.58	32.80	37.70
16	23.54	26.30	28.85	32.00	34.27	39.25
17	24.77	27.59	30.19	33.41	35.72	40.79
18	25.99	28.87	31.53	34.81	37.16	42.31
19	27.20	30.14	32.85	36.19	38.58	43.82
20	28.41	31.41	34.17	37.57	40.00	45.31
21	29.62	32.67	35.48	38.93	41.40	46.80
22	30.81	33.92	36.78	40.29	42.80	48.27
23	32.01	35.17	38.08	41.64	44.18	49.73
24	33.20	36.42	39.36	42.98	45.56	51.18
25	34.38	37.65	40.65	44.31	46.93	52.62
26	35.56	38.89	41.92	45.64	48.29	54.05
27	36.74	40.11	43.19	46.96	49.64	55.48
28	37.92	41.34	44.46	48.28	50.99	56.89
29	39.09	42.56	45.72	49.59	52.34	58.30

Contd...

Contd...

Degree of Freedom	Level of Significance (α)					
	0.100	0.050	0.025	0.010	0.005	0.001
30	40.26	43.77	46.98	50.89	53.67	59.70
40	51.81	55.76	59.34	63.69	66.77	73.40
50	63.17	67.50	71.42	76.15	79.49	86.66
60	74.40	79.08	83.30	88.38	91.95	99.61
70	85.53	90.53	95.02	100.43	104.21	112.32
80	96.58	101.88	106.63	112.33	116.32	124.84
90	107.57	113.15	118.14	124.12	128.30	137.21
100	118.50	124.34	129.56	135.81	140.17	149.45

CRITICAL VALUES OF F (0.05 OR 5%)

Degree of Freedom (Denominator)	Degree of Freedom (Numerator)							
	1	2	3	4	5	6	7	8
1	161.447	199.500	215.707	224.583	230.161	233.986	236.768	238.882
2	18.512	19.0000	19.1643	19.2468	19.2964	19.3295	19.3532	19.3710
3	10.128	9.5521	9.2766	9.1172	9.0135	8.9406	8.8867	8.8452
4	7.7086	6.9443	6.5914	6.3882	6.2561	6.1631	6.0942	6.0410
5	6.6079	5.7861	5.4095	5.1922	5.0503	4.9503	4.8759	4.8183
6	5.9874	5.1433	4.7571	4.5337	4.3874	4.2839	4.2067	4.1468
7	5.5914	4.7374	4.3468	4.1203	3.9715	3.8660	3.7870	3.7257
8	5.3177	4.4590	4.0662	3.8379	3.6875	3.5806	3.5005	3.4381
9	5.1174	4.2565	3.8625	3.6331	3.4817	3.3738	3.2927	3.2296
10	4.9646	4.1028	3.7083	3.4780	3.3258	3.2172	3.1355	3.0717
11	4.8443	3.9823	3.5874	3.3567	3.2039	3.0946	3.0123	2.9480
12	4.7472	3.8853	3.4903	3.2592	3.1059	2.9961	2.9134	2.8486
13	4.6672	3.8056	3.4105	3.1791	3.0254	2.9153	2.8321	2.7669
14	4.6001	3.7389	3.3439	3.1122	2.9582	2.8477	2.7642	2.6987
15	4.5431	3.6823	3.2874	3.0556	2.9013	2.7905	2.7066	2.6408
16	4.4940	3.6337	3.2389	3.0069	2.8524	2.7413	2.6572	2.5911
17	4.4513	3.5915	3.1968	2.9647	2.8100	2.6987	2.6143	2.5480
18	4.4139	3.5546	3.1599	2.9277	2.7729	2.6613	2.5767	2.5102
19	4.3807	3.5219	3.1274	2.8951	2.7401	2.6283	2.5435	2.4768
20	4.3512	3.4928	3.0984	2.8661	2.7109	2.5990	2.5140	2.4471
21	4.3248	3.4668	3.0725	2.8401	2.6848	2.5727	2.4876	2.4205
22	4.3009	3.4434	3.0491	2.8167	2.6613	2.5491	2.4638	2.3965
23	4.2793	3.4221	3.0280	2.7955	2.6400	2.5277	2.4422	2.3748
24	4.2597	3.4028	3.0088	2.7763	2.6207	2.5082	2.4226	2.3551
25	4.2417	3.3852	2.9912	2.7587	2.6030	2.4904	2.4047	2.3371
26	4.2252	3.3690	2.9752	2.7426	2.5868	2.4741	2.3883	2.3205
27	4.2100	3.3541	2.9604	2.7278	2.5719	2.4591	2.3732	2.3053
28	4.1960	3.3404	2.9467	2.7141	2.5581	2.4453	2.3593	2.2913
29	4.1830	3.3277	2.9340	2.7014	2.5454	2.4324	2.3463	2.2783
30	4.1709	3.3158	2.9223	2.6896	2.5336	2.4205	2.3343	2.2662
40	4.0847	3.2317	2.8387	2.6060	2.4495	2.3359	2.2490	2.1802
50	4.0343	3.1826	2.7900	2.5572	2.4004	2.2864	2.1992	2.1299
60	4.0012	3.1504	2.7581	2.5252	2.3683	2.2541	2.1665	2.0970
70	3.9778	3.1277	2.7355	2.5027	2.3456	2.2312	2.1435	2.0737
80	3.9604	3.1108	2.7188	2.4859	2.3287	2.2142	2.1263	2.0564
90	3.9469	3.0977	2.7058	2.4729	2.3157	2.2011	2.1131	2.0430
100	3.9361	3.0873	2.6955	2.4626	2.3053	2.1906	2.1025	2.0323

Degree of Freedom (Denominator)	Degree of Freedom (Numerator)							
	9	10	11	12	13	14	15	20
1	240.543	241.881	242.983	243.906	244.689	245.364	245.949	248.013
2	19.3848	19.3959	19.4050	19.4125	19.4189	19.4244	19.4291	19.4458
3	8.8123	8.7855	8.7633	8.7446	8.7287	8.7149	8.7029	8.6602
4	5.9988	5.9644	5.9358	5.9117	5.8911	5.8733	5.8578	5.8025
5	4.7725	4.7351	4.7040	4.6777	4.6552	4.6358	4.6188	4.5581
6	4.0990	4.0600	4.0274	3.9999	3.9764	3.9559	3.9381	3.8742
7	3.6767	3.6365	3.6030	3.5747	3.5503	3.5292	3.5107	3.4445
8	3.3881	3.3472	3.3130	3.2839	3.2590	3.2374	3.2184	3.1503
9	3.1789	3.1373	3.1025	3.0729	3.0475	3.0255	3.0061	2.9365
10	3.0204	2.9782	2.9430	2.9130	2.8872	2.8647	2.8450	2.7740
11	2.8962	2.8536	2.8179	2.7876	2.7614	2.7386	2.7186	2.6464
12	2.7964	2.7534	2.7173	2.6866	2.6602	2.6371	2.6169	2.5436
13	2.7144	2.6710	2.6347	2.6037	2.5769	2.5536	2.5331	2.4589
14	2.6458	2.6022	2.5655	2.5342	2.5073	2.4837	2.4630	2.3879
15	2.5876	2.5437	2.5068	2.4753	2.4481	2.4244	2.4034	2.3275
16	2.5377	2.4935	2.4564	2.4247	2.3973	2.3733	2.3522	2.2756
17	2.4943	2.4499	2.4126	2.3807	2.3531	2.3290	2.3077	2.2304
18	2.4563	2.4117	2.3742	2.3421	2.3143	2.2900	2.2686	2.1906
19	2.4227	2.3779	2.3402	2.3080	2.2800	2.2556	2.2341	2.1555
20	2.3928	2.3479	2.3100	2.2776	2.2495	2.2250	2.2033	2.1242
21	2.3660	2.3210	2.2829	2.2504	2.2222	2.1975	2.1757	2.0960
22	2.3419	2.2967	2.2585	2.2258	2.1975	2.1727	2.1508	2.0707
23	2.3201	2.2747	2.2364	2.2036	2.1752	2.1502	2.1282	2.0476
24	2.3002	2.2547	2.2163	2.1834	2.1548	2.1298	2.1077	2.0267
25	2.2821	2.2365	2.1979	2.1649	2.1362	2.1111	2.0889	2.0075
26	2.2655	2.2197	2.1811	2.1479	2.1192	2.0939	2.0716	1.9898
27	2.2501	2.2043	2.1655	2.1323	2.1035	2.0781	2.0558	1.9736
28	2.2360	2.1900	2.1512	2.1179	2.0889	2.0635	2.0411	1.9586
29	2.2229	2.1768	2.1379	2.1045	2.0755	2.0500	2.0275	1.9446
30	2.2107	2.1646	2.1256	2.0921	2.0630	2.0374	2.0148	1.9317
40	2.1240	2.0772	2.0376	2.0035	1.9738	1.9476	1.9245	1.8389
50	2.0734	2.0261	1.9861	1.9515	1.9214	1.8949	1.8714	1.7841
60	2.0401	1.9926	1.9522	1.9174	1.8870	1.8602	1.8364	1.7480
70	2.0166	1.9689	1.9283	1.8932	1.8627	1.8357	1.8117	1.7223
80	1.9991	1.9512	1.9105	1.8753	1.8445	1.8174	1.7932	1.7032
90	1.9856	1.9376	1.8967	1.8613	1.8305	1.8032	1.7789	1.6883
100	1.9748	1.9267	1.8857	1.8503	1.8193	1.7919	1.7675	1.6764

Degree of Freedom (Denominator)	Degree of Freedom (Numerator)							
	30	40	50	60	70	80	90	100
1	250.095	251.143	251.774	252.195	252.497	252.723	252.900	253.041
2	19.462	19.4707	19.4757	19.4791	19.4814	19.4832	19.4846	19.4857
3	8.6166	8.5944	8.5810	8.5720	8.5656	8.5607	8.5569	8.5539
4	5.7459	5.7170	5.6995	5.6877	5.6793	5.6730	5.6680	5.6641
5	4.4957	4.4638	4.4444	4.4314	4.4220	4.4150	4.4095	4.4051
6	3.8082	3.7743	3.7537	3.7398	3.7298	3.7223	3.7164	3.7117
7	3.3758	3.3404	3.3189	3.3043	3.2939	3.2860	3.2798	3.2749
8	3.0794	3.0428	3.0204	3.0053	2.9944	2.9862	2.9798	2.9747
9	2.8637	2.8259	2.8028	2.7872	2.7760	2.7675	2.7609	2.7556
10	2.6996	2.6609	2.6371	2.6211	2.6095	2.6008	2.5939	2.5884
11	2.5705	2.5309	2.5066	2.4901	2.4782	2.4692	2.4622	2.4566
12	2.4663	2.4259	2.4010	2.3842	2.3720	2.3628	2.3556	2.3498
13	2.3803	2.3392	2.3138	2.2966	2.2841	2.2747	2.2673	2.2614
14	2.3082	2.2664	2.2405	2.2229	2.2102	2.2006	2.1931	2.1870
15	2.2468	2.2043	2.1780	2.1601	2.1472	2.1373	2.1296	2.1234
16	2.1938	2.1507	2.1240	2.1058	2.0926	2.0826	2.0748	2.0685
17	2.1477	2.1040	2.0769	2.0584	2.0450	2.0348	2.0268	2.0204
18	2.1071	2.0629	2.0354	2.0166	2.0030	1.9927	1.9846	1.9780
19	2.0712	2.0264	1.9986	1.9795	1.9657	1.9552	1.9470	1.9403
20	2.0391	1.9938	1.9656	1.9464	1.9323	1.9217	1.9133	1.9066
21	2.0102	1.9645	1.9360	1.9165	1.9023	1.8915	1.8830	1.8761
22	1.9842	1.9380	1.9092	1.8894	1.8751	1.8641	1.8555	1.8486
23	1.9605	1.9139	1.8848	1.8648	1.8503	1.8392	1.8305	1.8234
24	1.9390	1.8920	1.8625	1.8424	1.8276	1.8164	1.8076	1.8005
25	1.9192	1.8718	1.8421	1.8217	1.8069	1.7955	1.7866	1.7794
26	1.9010	1.8533	1.8233	1.8027	1.7877	1.7762	1.7672	1.7599
27	1.8842	1.8361	1.8059	1.7851	1.7700	1.7584	1.7493	1.7419
28	1.8687	1.8203	1.7898	1.7689	1.7535	1.7418	1.7326	1.7251
29	1.8543	1.8055	1.7748	1.7537	1.7382	1.7264	1.7171	1.7096
30	1.8409	1.7918	1.7609	1.7396	1.7240	1.7121	1.7027	1.6950
40	1.7444	1.6928	1.6600	1.6373	1.6205	1.6077	1.5975	1.5892
50	1.6872	1.6337	1.5995	1.5757	1.5580	1.5445	1.5337	1.5249
60	1.6491	1.5943	1.5590	1.5343	1.5160	1.5019	1.4906	1.4814
70	1.6220	1.5661	1.5300	1.5046	1.4857	1.4711	1.4594	1.4498
80	1.6017	1.5449	1.5081	1.4821	1.4628	1.4477	1.4357	1.4259
90	1.5859	1.5284	1.4910	1.4645	1.4448	1.4294	1.4171	1.4070
100	1.5733	1.5151	1.4772	1.4504	1.4303	1.4146	1.4020	1.3917

CRITICAL VALUES OF F (0.01 OR 1%)

Degree of Freedom (Denominator)	Degree of Freedom (Numerator)							
	1	2	3	4	5	6	7	8
1	4052.18	4999.50	5403.35	5624.58	5763.65	5858.99	5928.36	5981.07
2	98.5025	99.0000	99.1662	99.2494	99.2993	99.3326	99.3564	99.3742
3	34.1162	30.8165	29.4567	28.7099	28.2371	27.9107	27.6717	27.4892
4	21.1977	18.0000	16.6944	15.9770	15.5219	15.2069	14.9758	14.7989
5	16.2582	13.2739	12.0600	11.3919	10.9670	10.6723	10.4555	10.2893
6	13.7450	10.9248	9.7795	9.1483	8.7459	8.4661	8.2600	8.1017
7	12.2464	9.5466	8.4513	7.8466	7.4604	7.1914	6.9928	6.8400
8	11.2586	8.6491	7.5910	7.0061	6.6318	6.3707	6.1776	6.0289
9	10.5614	8.0215	6.9919	6.4221	6.0569	5.8018	5.6129	5.4671
10	10.0443	7.5594	6.5523	5.9943	5.6363	5.3858	5.2001	5.0567
11	9.6460	7.2057	6.2167	5.6683	5.3160	5.0692	4.8861	4.7445
12	9.3302	6.9266	5.9525	5.4120	5.0643	4.8206	4.6395	4.4994
13	9.0738	6.7010	5.7394	5.2053	4.8616	4.6204	4.4410	4.3021
14	8.8616	6.5149	5.5639	5.0354	4.6950	4.4558	4.2779	4.1399
15	8.6831	6.3589	5.4170	4.8932	4.5556	4.3183	4.1415	4.0045
16	8.5310	6.2262	5.2922	4.7726	4.4374	4.2016	4.0259	3.8896
17	8.3997	6.1121	5.1850	4.6690	4.3359	4.1015	3.9267	3.7910
18	8.2854	6.0129	5.0919	4.5790	4.2479	4.0146	3.8406	3.7054
19	8.1849	5.9259	5.0103	4.5003	4.1708	3.9386	3.7653	3.6305
20	8.0960	5.8489	4.9382	4.4307	4.1027	3.8714	3.6987	3.5644
21	8.0166	5.7804	4.8740	4.3688	4.0421	3.8117	3.6396	3.5056
22	7.9454	5.7190	4.8166	4.3134	3.9880	3.7583	3.5867	3.4530
23	7.8811	5.6637	4.7649	4.2636	3.9392	3.7102	3.5390	3.4057
24	7.8229	5.6136	4.7181	4.2184	3.8951	3.6667	3.4959	3.3629
25	7.7698	5.5680	4.6755	4.1774	3.8550	3.6272	3.4568	3.3239
26	7.7213	5.5263	4.6366	4.1400	3.8183	3.5911	3.4210	3.2884
27	7.6767	5.4881	4.6009	4.1056	3.7848	3.5580	3.3882	3.2558
28	7.6356	5.4529	4.5681	4.0740	3.7539	3.5276	3.3581	3.2259
29	7.5977	5.4204	4.5378	4.0449	3.7254	3.4995	3.3303	3.1982
30	7.5625	5.3903	4.5097	4.0179	3.6990	3.4735	3.3045	3.1726
40	7.3141	5.1785	4.3126	3.8283	3.5138	3.2910	3.1238	2.9930
50	7.1706	5.0566	4.1993	3.7195	3.4077	3.1864	3.0202	2.8900
60	7.0771	4.9774	4.1259	3.6490	3.3389	3.1187	2.9530	2.8233
70	7.0114	4.9219	4.0744	3.5996	3.2907	3.0712	2.9060	2.7765
80	6.9627	4.8807	4.0363	3.5631	3.2550	3.0361	2.8713	2.7420
90	6.9251	4.8491	4.0070	3.5350	3.2276	3.0091	2.8445	2.7154
100	6.8953	4.8239	3.9837	3.5127	3.2059	2.9877	2.8233	2.6943

Degree of Freedom (Denominator)	Degree of Freedom (Numerator)							
	9	10	11	12	13	14	15	20
1	6022.47	6055.85	6083.32	6106.32	6125.86	6142.67	6157.28	6208.73
2	99.3881	99.3992	99.4083	99.4159	99.4223	99.4278	99.4325	99.4492
3	27.3452	27.2287	27.1326	27.0518	26.9831	26.9238	26.8722	26.6898
4	14.6591	14.5459	14.4523	14.3736	14.3065	14.2486	14.1982	14.0196
5	10.1578	10.0510	9.9626	9.8883	9.8248	9.7700	9.7222	9.5526
6	7.9761	7.8741	7.7896	7.7183	7.6575	7.6049	7.5590	7.3958
7	6.7188	6.6201	6.5382	6.4691	6.4100	6.3590	6.3143	6.1554
8	5.9106	5.8143	5.7343	5.6667	5.6089	5.5589	5.5151	5.3591
9	5.3511	5.2565	5.1779	5.1114	5.0545	5.0052	4.9621	4.8080
10	4.9424	4.8491	4.7715	4.7059	4.6496	4.6008	4.5581	4.4054
11	4.6315	4.5393	4.4624	4.3974	4.3416	4.2932	4.2509	4.0990
12	4.3875	4.2961	4.2198	4.1553	4.0999	4.0518	4.0096	3.8584
13	4.1911	4.1003	4.0245	3.9603	3.9052	3.8573	3.8154	3.6646
14	4.0297	3.9394	3.8640	3.8001	3.7452	3.6975	3.6557	3.5052
15	3.8948	3.8049	3.7299	3.6662	3.6115	3.5639	3.5222	3.3719
16	3.7804	3.6909	3.6162	3.5527	3.4981	3.4506	3.4089	3.2587
17	3.6822	3.5931	3.5185	3.4552	3.4007	3.3533	3.3117	3.1615
18	3.5971	3.5082	3.4338	3.3706	3.3162	3.2689	3.2273	3.0771
19	3.5225	3.4338	3.3596	3.2965	3.2422	3.1949	3.1533	3.0031
20	3.4567	3.3682	3.2941	3.2311	3.1769	3.1296	3.0880	2.9377
21	3.3981	3.3098	3.2359	3.1730	3.1187	3.0715	3.0300	2.8796
22	3.3458	3.2576	3.1837	3.1209	3.0667	3.0195	2.9779	2.8274
23	3.2986	3.2106	3.1368	3.0740	3.0199	2.9727	2.9311	2.7805
24	3.2560	3.1681	3.0944	3.0316	2.9775	2.9303	2.8887	2.7380
25	3.2172	3.1294	3.0558	2.9931	2.9389	2.8917	2.8502	2.6993
26	3.1818	3.0941	3.0205	2.9578	2.9038	2.8566	2.8150	2.6640
27	3.1494	3.0618	2.9882	2.9256	2.8715	2.8243	2.7827	2.6316
28	3.1195	3.0320	2.9585	2.8959	2.8418	2.7946	2.7530	2.6017
29	3.0920	3.0045	2.9311	2.8685	2.8144	2.7672	2.7256	2.5742
30	3.0665	2.9791	2.9057	2.8431	2.7890	2.7418	2.7002	2.5487
40	2.8876	2.8005	2.7274	2.6648	2.6107	2.5634	2.5216	2.3689
50	2.7850	2.6981	2.6250	2.5625	2.5083	2.4609	2.4190	2.2652
60	2.7185	2.6318	2.5587	2.4961	2.4419	2.3943	2.3523	2.1978
70	2.6719	2.5852	2.5122	2.4496	2.3953	2.3477	2.3055	2.1504
80	2.6374	2.5508	2.4777	2.4151	2.3608	2.3131	2.2709	2.1153
90	2.6109	2.5243	2.4513	2.3886	2.3342	2.2865	2.2442	2.0882
100	2.5898	2.5033	2.4302	2.3676	2.3132	2.2654	2.2230	2.0666

Degree of Freedom (Denominator)	Degree of Freedom (Numerator)							
	30	40	50	60	70	80	90	100
1	6260.65	6286.78	6302.52	6313.03	6320.55	6326.20	6330.59	6334.11
2	99.4658	99.4742	99.4792	99.4825	99.4849	99.4867	99.4881	99.4892
3	26.5045	26.4108	26.3542	26.3164	26.2892	26.2688	26.2530	26.2402
4	13.8377	13.7454	13.6896	13.6522	13.6254	13.6053	13.5896	13.5770
5	9.3793	9.2912	9.2378	9.2020	9.1763	9.1570	9.1420	9.1299
6	7.2285	7.1432	7.0915	7.0567	7.0318	7.0130	6.9984	6.9867
7	5.9920	5.9084	5.8577	5.8236	5.7991	5.7806	5.7662	5.7547
8	5.1981	5.1156	5.0654	5.0316	5.0073	4.9890	4.9748	4.9633
9	4.6486	4.5666	4.5167	4.4831	4.4589	4.4407	4.4264	4.4150
10	4.2469	4.1653	4.1155	4.0819	4.0577	4.0394	4.0252	4.0137
11	3.9411	3.8596	3.8097	3.7761	3.7518	3.7335	3.7192	3.7077
12	3.7008	3.6192	3.5692	3.5355	3.5111	3.4928	3.4784	3.4668
13	3.5070	3.4253	3.3752	3.3413	3.3168	3.2984	3.2839	3.2723
14	3.3476	3.2656	3.2153	3.1813	3.1567	3.1381	3.1235	3.1118
15	3.2141	3.1319	3.0814	3.0471	3.0224	3.0037	2.9890	2.9772
16	3.1007	3.0182	2.9675	2.9330	2.9082	2.8893	2.8745	2.8627
17	3.0032	2.9205	2.8694	2.8348	2.8097	2.7908	2.7759	2.7639
18	2.9185	2.8354	2.7841	2.7493	2.7241	2.7050	2.6900	2.6779
19	2.8442	2.7608	2.7093	2.6742	2.6488	2.6296	2.6145	2.6023
20	2.7785	2.6947	2.6430	2.6077	2.5822	2.5628	2.5476	2.5353
21	2.7200	2.6359	2.5838	2.5484	2.5227	2.5032	2.4878	2.4755
22	2.6675	2.5831	2.5308	2.4951	2.4693	2.4496	2.4342	2.4217
23	2.6202	2.5355	2.4829	2.4471	2.4210	2.4013	2.3857	2.3732
24	2.5773	2.4923	2.4395	2.4035	2.3773	2.3573	2.3417	2.3291
25	2.5383	2.4530	2.3999	2.3637	2.3373	2.3173	2.3015	2.2888
26	2.5026	2.4170	2.3637	2.3273	2.3008	2.2806	2.2647	2.2519
27	2.4699	2.3840	2.3304	2.2938	2.2672	2.2469	2.2309	2.2180
28	2.4397	2.3535	2.2997	2.2629	2.2361	2.2157	2.1997	2.1867
29	2.4118	2.3253	2.2714	2.2344	2.2074	2.1869	2.1707	2.1577
30	2.3860	2.2992	2.2450	2.2079	2.1808	2.1601	2.1439	2.1307
40	2.2034	2.1142	2.0581	2.0194	1.9911	1.9694	1.9522	1.9383
50	2.0976	2.0066	1.9490	1.9090	1.8797	1.8571	1.8393	1.8248
60	2.0285	1.9360	1.8772	1.8363	1.8061	1.7828	1.7644	1.7493
70	1.9797	1.8861	1.8263	1.7846	1.7537	1.7298	1.7109	1.6954
80	1.9435	1.8489	1.7883	1.7459	1.7144	1.6901	1.6707	1.6548
90	1.9155	1.8201	1.7588	1.7158	1.6838	1.6591	1.6393	1.6231
100	1.8933	1.7972	1.7353	1.6918	1.6594	1.6342	1.6141	1.5977

FACTORS FOR CONSTRUCTION OF CONTROL CHARTS (STATISTICAL QUALITY CONTROL)

Sample Size	Mean Chart				Standard Deviation Chart						Range Chart				
	Factors for Control Limit			Factors for Central Line	Factors for Control Limits					Factors for Central Line	Factors for Control Limit			
n	A	A_1	A_2	c_2	B_1	B_2	B_3	B_4			d_2	D_1	D_2	D_3	D_4
2	2.121	3.760	1.880	0.564	0	1.843	0	3.267			1.128	0	3.686	0	3.267
3	1.132	2.194	1.021	0.724	0	1.858	0	2.568			1.693	0	4.358	0	2.575
4	1.500	1.880	0.729	0.798	0	1.808	0	2.266			2.059	0	4.698	0	2.282
5	1.342	1.596	0.577	0.841	0	1.756	0	2.089			2.326	0	4.918	0	2.115
6	1.225	1.410	0.483	0.869	0.026	1.711	0.030	1.970			2.534	0	5.978	0	2.004
7	1.134	1.277	0.419	0.888	0.105	1.672	0.118	1.882			2.704	0.205	5.203	0.076	1.924
8	1.061	1.175	0.373	0.903	0.167	1.638	0.185	1.811			2.847	0.387	5.307	0.136	1.861
9	1.000	1.094	0.337	0.914	0.219	1.669	0.239	1.761			2.970	0.546	5.394	0.184	1.816
10	0.949	1.028	0.308	0.923	0.262	1.584	0.284	1.716			3.078	0.687	5.469	0.223	1.777
11	0.905	0.973	0.285	0.930	0.299	1.561	0.321	1.679			3.173	0.812	5.534	0.256	1.744
12	0.866	0.925	0.266	0.936	0.331	1.541	0.354	1.646			3.258	0.924	5.592	0.284	1.710
13	0.832	0.884	0.249	0.941	0.359	1.523	0.382	1.618			3.336	1.026	5.646	0.308	1.692
14	0.802	0.848	0.235	0.945	0.381	1.507	0.406	1.594			3.407	1.121	5.693	0.329	1.671
15	0.775	0.816	0.223	0.949	0.406	1.492	0.428	1.572			3.472	1.207	5.737	0.348	1.632
16	0.750	0.788	0.212	0.952	0.427	1.478	0.448	1.552			3.532	1.285	5.779	0.364	1.636
17	0.728	0.762	0.203	0.955	0.445	1.465	0.466	1.534			3.588	1.359	5.817	0.379	1.621

Contd...

Contd...

Sample Size	Mean Chart			Standard Deviation Chart						Range Chart		
	Factors for Control Limit		Factors for Central Line	Factors for Control Limits					Factors for Central Line	Factors for Control Limit		
18	0.707	0.738	0.194	0.958	0.461	1.454	0.482	1.518	3.649	1.426	5.845	0.392	1.608
19	0.688	0.717	0.187	0.960	0.477	1.443	0.497	1.503	3.689	1.490	5.888	0.404	1.596
20	0.671	0.697	0.180	0.962	0.491	1.433	0.510	1.490	3.735	1.548	5.922	0.414	1.586
21	0.655	0.679	0.173	0.964	0.504	1.424	0.523	1.477	3.778	1.606	5.950	0.425	1.575
22	0.649	0.662	0.167	0.966	0.516	1.415	0.534	1.466	3.819	1.659	5.979	0.434	1.566
23	0.626	0.647	0.162	0.967	0.527	1.407	0.545	1.455	3.838	1.710	6.006	0.443	1.557
24	0.612	0.632	0.157	0.968	0.538	1.399	0.555	1.445	3.895	1.759	6.031	0.452	1.548
25	0.600	0.619	0.153	0.969	0.548	1.392	0.565	1.435	3.931	1.804	6.058	0.459	1.541

STUDENTIZED RANGE: CRITICAL VALUES
Level of Significance 0.05

Degrees of Freedom	Number of Treatments (k)									
	2	3	4	5	6	7	8	9	10	11
1	17.97	26.98	32.82	37.08	40.41	43.12	45.40	47.36	49.07	50.59
2	6.085	8.331	9.798	10.881	11.734	12.44	13.03	13.54	13.99	14.39
3	4.501	5.910	6.825	7.502	8.037	8.478	8.852	9.177	9.462	9.717
4	3.926	5.040	5.757	6.287	6.706	7.053	7.347	7.602	7.826	8.027
5	3.635	4.602	5.218	5.673	6.033	6.330	6.582	6.801	6.995	7.167
6	3.460	4.339	4.896	5.305	5.628	5.895	6.122	6.319	6.493	6.649
7	3.344	4.165	4.681	5.060	5.359	5.606	5.815	5.997	6.158	6.302
8	3.261	4.041	4.529	4.886	5.167	5.399	5.596	5.767	5.918	6.053
9	3.199	3.948	4.415	4.755	5.024	5.244	5.432	5.595	5.738	5.867
10	3.151	3.877	4.327	4.654	4.912	5.124	5.304	5.460	5.598	5.722
11	3.113	3.820	4.256	4.574	4.823	5.028	5.202	5.353	5.486	5.605
12	3.081	3.773	4.199	4.508	4.750	4.950	5.119	5.265	5.395	5.510
13	3.055	3.734	4.151	4.453	4.690	4.884	5.049	5.192	5.318	5.431
14	3.033	3.701	4.111	4.407	4.639	4.829	4.990	5.130	5.253	5.364
15	3.014	3.673	4.076	4.367	4.595	4.782	4.940	5.077	5.198	5.306
16	2.998	3.649	4.046	4.333	4.557	4.741	4.896	5.031	5.150	5.256
17	2.984	3.628	4.020	4.303	4.524	4.705	4.858	4.991	5.108	5.212
18	2.971	3.609	3.997	4.276	4.494	4.673	4.824	4.955	5.071	5.173
19	2.960	3.593	3.977	4.253	4.468	4.645	4.794	4.924	5.037	5.139
20	2.950	3.578	3.958	4.232	4.445	4.620	4.768	4.895	5.008	5.108
21	2.941	3.565	3.942	4.213	4.424	4.597	4.743	4.870	4.981	5.081
22	2.933	3.553	3.927	4.196	4.405	4.577	4.722	4.847	4.957	5.056
23	2.926	3.542	3.914	4.180	4.388	4.558	4.702	4.826	4.935	5.033
24	2.919	3.532	3.901	4.166	4.373	4.541	4.684	4.807	4.915	5.012
25	2.913	3.523	3.890	4.153	4.358	4.526	4.667	4.789	4.897	4.993
26	2.907	3.514	3.880	4.141	4.345	4.511	4.652	4.773	4.880	4.975
27	2.902	3.506	3.870	4.130	4.333	4.498	4.638	4.758	4.864	4.959
28	2.897	3.499	3.861	4.120	4.322	4.486	4.625	4.745	4.850	4.944
29	2.892	3.493	3.853	4.111	4.311	4.475	4.613	4.732	4.837	4.930
30	2.888	3.486	3.845	4.102	4.301	4.464	4.601	4.720	4.824	4.917
35	2.871	3.461	3.814	4.066	4.261	4.421	4.555	4.671	4.773	4.863
40	2.858	3.442	3.791	4.039	4.232	4.388	4.521	4.634	4.735	4.824
60	2.829	3.399	3.737	3.977	4.163	4.314	4.441	4.550	4.646	4.732
80	2.814	3.377	3.711	3.947	4.129	4.277	4.402	4.509	4.603	4.686
120	2.800	3.356	3.685	3.917	4.096	4.241	4.363	4.468	4.560	4.641
∞	2.772	3.314	3.633	3.858	4.030	4.170	4.286	4.387	4.474	4.552

Degrees of Freedom	Number of Treatments (k)								
	12	13	14	15	16	17	18	19	20
1	51.957	53.194	54.323	55.361	56.32	57.212	58.044	58.824	59.558
2	14.749	15.076	15.375	15.650	15.905	16.143	16.365	16.573	16.769
3	9.946	10.155	10.346	10.522	10.686	10.838	10.98	11.114	11.240
4	8.208	8.373	8.524	8.664	8.793	8.914	9.027	9.133	9.233
5	7.323	7.466	7.596	7.716	7.828	7.932	8.030	8.122	8.208
6	6.789	6.917	7.034	7.143	7.244	7.338	7.426	7.508	7.586
7	6.431	6.550	6.658	6.759	6.852	6.939	7.020	7.097	7.169
8	6.175	6.287	6.389	6.483	6.571	6.653	6.729	6.801	6.869
9	5.983	6.089	6.186	6.276	6.359	6.437	6.510	6.579	6.643
10	5.833	5.935	6.028	6.114	6.194	6.269	6.339	6.405	6.467
11	5.713	5.811	5.901	5.984	6.062	6.134	6.202	6.265	6.325
12	5.615	5.710	5.797	5.878	5.953	6.023	6.089	6.151	6.209
13	5.533	5.625	5.711	5.789	5.862	5.931	5.995	6.055	6.112
14	5.463	5.554	5.637	5.714	5.785	5.852	5.915	5.973	6.029
15	5.403	5.492	5.574	5.649	5.719	5.785	5.846	5.904	5.958
16	5.352	5.439	5.519	5.593	5.662	5.726	5.786	5.843	5.896
17	5.306	5.392	5.471	5.544	5.612	5.675	5.734	5.790	5.842
18	5.266	5.351	5.429	5.501	5.567	5.629	5.688	5.743	5.794
19	5.231	5.314	5.391	5.462	5.528	5.589	5.647	5.701	5.752
20	5.199	5.282	5.357	5.427	5.492	5.553	5.610	5.663	5.714
21	5.170	5.252	5.327	5.396	5.460	5.520	5.576	5.629	5.679
22	5.144	5.225	5.299	5.368	5.431	5.491	5.546	5.599	5.648
23	5.121	5.201	5.274	5.342	5.405	5.464	5.519	5.571	5.620
24	5.099	5.179	5.251	5.319	5.381	5.439	5.494	5.545	5.594
25	5.079	5.158	5.230	5.297	5.359	5.417	5.471	5.522	5.570
26	5.061	5.139	5.211	5.277	5.339	5.396	5.450	5.500	5.548
27	5.044	5.122	5.193	5.259	5.320	5.377	5.430	5.480	5.528
28	5.029	5.106	5.177	5.242	5.302	5.359	5.412	5.462	5.509
29	5.014	5.091	5.161	5.226	5.286	5.342	5.395	5.445	5.491
30	5.001	5.077	5.147	5.211	5.271	5.327	5.379	5.429	5.475
35	4.945	5.020	5.088	5.151	5.209	5.264	5.315	5.362	5.408
40	4.904	4.977	5.044	5.106	5.163	5.216	5.266	5.313	5.358
60	4.808	4.878	4.942	5.001	5.056	5.107	5.154	5.199	5.241
80	4.761	4.829	4.892	4.949	5.003	5.052	5.099	5.142	5.183
120	4.714	4.781	4.842	4.898	4.950	4.998	5.043	5.086	5.126
∞	4.622	4.685	4.743	4.796	4.845	4.891	4.934	4.974	5.012

STUDENTIZED RANGE: CRITICAL VALUES
Level of Significance 0.01

Degrees of Freedom	Number of Treatments (k)								
	k = 2	3	4	5	6	7	8	9	10
1	90.024	135.041	164.258	185.575	202.21	215.769	227.166	236.966	245.542
2	14.036	19.019	22.294	24.717	26.629	28.201	29.53	30.679	31.689
3	8.260	10.619	12.170	13.324	14.241	14.998	15.641	16.199	16.691
4	6.511	8.120	9.173	9.958	10.583	11.101	11.542	11.925	12.264
5	5.702	6.976	7.804	8.421	8.913	9.321	9.669	9.971	10.239
6	5.243	6.331	7.033	7.556	7.972	8.318	8.612	8.869	9.097
7	4.949	5.919	6.542	7.005	7.373	7.678	7.939	8.166	8.367
8	4.745	5.635	6.204	6.625	6.959	7.237	7.474	7.680	7.863
9	4.596	5.428	5.957	6.347	6.657	6.915	7.134	7.325	7.494
10	4.482	5.270	5.769	6.136	6.428	6.669	6.875	7.054	7.213
11	4.392	5.146	5.621	5.970	6.247	6.476	6.671	6.841	6.992
12	4.320	5.046	5.502	5.836	6.101	6.320	6.507	6.670	6.814
13	4.260	4.964	5.404	5.726	5.981	6.192	6.372	6.528	6.666
14	4.210	4.895	5.322	5.634	5.881	6.085	6.258	6.409	6.543
15	4.167	4.836	5.252	5.556	5.796	5.994	6.162	6.309	6.438
16	4.131	4.786	5.192	5.489	5.722	5.915	6.079	6.222	6.348
17	4.099	4.742	5.140	5.430	5.659	5.847	6.007	6.147	6.270
18	4.071	4.703	5.094	5.379	5.603	5.787	5.944	6.081	6.201
19	4.046	4.669	5.054	5.334	5.553	5.735	5.889	6.022	6.141
20	4.024	4.639	5.018	5.293	5.510	5.688	5.839	5.970	6.086
21	4.004	4.612	4.986	5.257	5.470	5.646	5.794	5.924	6.038
22	3.986	4.588	4.957	5.225	5.435	5.608	5.754	5.882	5.994
23	3.970	4.566	4.931	5.195	5.403	5.573	5.718	5.844	5.955
24	3.955	4.546	4.907	5.168	5.373	5.542	5.685	5.809	5.919
25	3.942	4.527	4.885	5.144	5.347	5.513	5.655	5.778	5.886
26	3.930	4.510	4.865	5.121	5.322	5.487	5.627	5.749	5.856
27	3.918	4.495	4.847	5.101	5.300	5.463	5.602	5.722	5.828
28	3.908	4.481	4.830	5.082	5.279	5.441	5.578	5.697	5.802
29	3.898	4.467	4.814	5.064	5.260	5.420	5.556	5.674	5.778
30	3.889	4.455	4.799	5.048	5.242	5.401	5.536	5.653	5.756
35	3.852	4.404	4.739	4.980	5.169	5.323	5.453	5.566	5.666
40	3.825	4.367	4.695	4.931	5.114	5.265	5.392	5.502	5.599
60	3.762	4.282	4.594	4.818	4.991	5.133	5.253	5.356	5.447
80	3.732	4.241	4.545	4.763	4.931	5.069	5.185	5.284	5.372
120	3.702	4.200	4.497	4.709	4.872	5.005	5.118	5.214	5.299
∞	3.643	4.120	4.403	4.603	4.757	4.882	4.987	5.078	5.157

Degrees of Freedom	Number of Treatments (k)									
	11	12	13	14	15	16	17	18	19	20
1	253.15	259.98	266.17	271.81	277.00	281.80	286.26	290.43	294.33	298.00
2	32.589	33.398	34.134	34.806	35.426	36.000	36.534	37.034	37.502	37.943
3	17.130	17.526	17.887	18.217	18.522	18.805	19.068	19.315	19.546	19.765
4	12.567	12.84	13.09	13.318	13.53	13.726	13.909	14.081	14.242	14.394
5	10.479	10.696	10.894	11.076	11.244	11.400	11.545	11.682	11.811	11.932
6	9.300	9.485	9.653	9.808	9.951	10.084	10.208	10.325	10.434	10.538
7	8.548	8.711	8.860	8.997	9.124	9.242	9.353	9.456	9.553	9.645
8	8.027	8.176	8.311	8.436	8.552	8.659	8.760	8.854	8.943	9.027
9	7.646	7.784	7.910	8.025	8.132	8.232	8.325	8.412	8.495	8.573
10	7.356	7.485	7.603	7.712	7.812	7.906	7.993	8.075	8.153	8.226
11	7.127	7.250	7.362	7.464	7.560	7.648	7.731	7.809	7.883	7.952
12	6.943	7.060	7.166	7.265	7.356	7.441	7.520	7.594	7.664	7.730
13	6.791	6.903	7.006	7.100	7.188	7.269	7.345	7.417	7.484	7.548
14	6.663	6.772	6.871	6.962	7.047	7.125	7.199	7.268	7.333	7.394
15	6.555	6.660	6.756	6.845	6.927	7.003	7.074	7.141	7.204	7.264
16	6.461	6.564	6.658	6.744	6.823	6.897	6.967	7.032	7.093	7.151
17	6.380	6.480	6.572	6.656	6.733	6.806	6.873	6.937	6.997	7.053
18	6.309	6.407	6.496	6.579	6.655	6.725	6.791	6.854	6.912	6.967
19	6.246	6.342	6.430	6.510	6.585	6.654	6.719	6.780	6.837	6.891
20	6.190	6.285	6.370	6.449	6.523	6.591	6.654	6.714	6.770	6.823
21	6.140	6.233	6.317	6.395	6.467	6.534	6.596	6.655	6.710	6.762
22	6.095	6.186	6.269	6.346	6.417	6.482	6.544	6.602	6.656	6.707
23	6.054	6.144	6.226	6.301	6.371	6.436	6.497	6.553	6.607	6.658
24	6.017	6.105	6.186	6.261	6.330	6.394	6.453	6.510	6.562	6.612
25	5.983	6.070	6.150	6.224	6.292	6.355	6.414	6.469	6.522	6.571
26	5.951	6.038	6.117	6.190	6.257	6.319	6.378	6.432	6.484	6.533
27	5.923	6.008	6.087	6.158	6.225	6.287	6.344	6.399	6.450	6.498
28	5.896	5.981	6.058	6.129	6.195	6.256	6.314	6.367	6.418	6.465
29	5.871	5.955	6.032	6.103	6.168	6.228	6.285	6.338	6.388	6.435
30	5.848	5.932	6.008	6.078	6.142	6.202	6.258	6.311	6.361	6.407
35	5.755	5.835	5.908	5.976	6.038	6.096	6.150	6.200	6.248	6.293
40	5.685	5.764	5.835	5.900	5.961	6.017	6.069	6.118	6.165	6.208
60	5.528	5.601	5.667	5.728	5.784	5.837	5.886	5.931	5.974	6.015
80	5.451	5.521	5.585	5.644	5.698	5.749	5.796	5.840	5.881	5.920
120	5.375	5.443	5.505	5.561	5.614	5.662	5.708	5.750	5.790	5.827
∞	5.227	5.290	5.348	5.400	5.448	5.493	5.535	5.574	5.611	5.645

PEARSON'S CORRELATION COEFFICIENT (r): CRITICAL VALUES

Degrees of Freedom	Level of Significance: Two-tailed Test					
	0.20	0.10	0.05	0.02	0.01	0.001
	Level of Significance: One-tailed Test					
	0.10	0.05	0.025	0.01	0.005	0.0005
1	0.9510	0.9880	0.9970	0.9995	0.9999	0.99999
2	0.800	0.900	0.950	0.980	0.990	0.999
3	0.687	0.805	0.878	0.934	0.959	0.991
4	0.608	0.729	0.811	0.882	0.917	0.974
5	0.551	0.669	0.755	0.833	0.875	0.951
6	0.507	0.621	0.707	0.789	0.834	0.925
7	0.472	0.582	0.666	0.750	0.798	0.898
8	0.443	0.549	0.632	0.715	0.765	0.872
9	0.419	0.521	0.602	0.685	0.735	0.847
10	0.398	0.497	0.576	0.658	0.708	0.823
11	0.380	0.476	0.553	0.634	0.684	0.801
12	0.365	0.457	0.532	0.612	0.661	0.780
13	0.351	0.441	0.514	0.592	0.641	0.760
14	0.338	0.426	0.497	0.574	0.623	0.742
15	0.327	0.412	0.482	0.558	0.606	0.725
16	0.317	0.400	0.468	0.542	0.590	0.708
17	0.308	0.389	0.456	0.529	0.575	0.693
18	0.299	0.378	0.444	0.515	0.561	0.679
19	0.291	0.369	0.433	0.503	0.549	0.665
20	0.284	0.360	0.423	0.492	0.537	0.652
21	0.277	0.352	0.413	0.482	0.526	0.640
22	0.271	0.344	0.404	0.472	0.515	0.629
23	0.265	0.337	0.396	0.462	0.505	0.618
24	0.260	0.330	0.388	0.453	0.496	0.607
25	0.255	0.323	0.381	0.445	0.487	0.597
26	0.250	0.317	0.374	0.437	0.479	0.588
27	0.245	0.311	0.367	0.430	0.471	0.579
28	0.241	0.306	0.361	0.423	0.463	0.570
29	0.237	0.301	0.355	0.416	0.456	0.562
30	0.233	0.296	0.349	0.409	0.449	0.554
40	0.202	0.257	0.304	0.358	0.393	0.490
60	0.165	0.211	0.250	0.295	0.325	0.408
120	0.117	0.150	0.178	0.210	0.232	0.294
500	0.057	0.073	0.087	0.103	0.114	0.146

SPEARMAN RANK CORRELATION COEFFICIENT (ρ): CRITICAL VALUES

Number of Pairs of Observations (N)	Level of Significance: Two-tailed			
	0.01	0.05	0.10	0.20
	Level of Significance: One-tailed			
	0.005	0.025	0.05	0.10
4	–	–	0.800	0.800
5	–	0.900	0.800	0.700
6	0.943	0.829	0.771	0.600
7	0.893	0.745	0.679	0.536
8	0.857	0.691	0.595	0.476
9	0.817	0.683	0.583	0.467
10	0.782	0.636	0.552	0.442
11	0.755	0.609	0.527	0.418
12	0.727	0.580	0.497	0.399
13	0.698	0.555	0.478	0.379
14	0.675	0.534	0.459	0.363
15	0.654	0.518	0.443	0.350
16	0.632	0.500	0.427	0.338
17	0.615	0.485	0.412	0.326
18	0.598	0.472	0.399	0.315
19	0.583	0.458	0.390	0.307
20	0.568	0.445	0.379	0.298
21	0.555	0.435	0.369	0.291
22	0.543	0.424	0.360	0.283
23	0.531	0.415	0.352	0.277
24	0.520	0.406	0.344	0.270
25	0.510	0.398	0.336	0.265
26	0.500	0.389	0.330	0.259
27	0.492	0.382	0.324	0.254
28	0.483	0.375	0.318	0.249
29	0.474	0.369	0.311	0.244
30	0.467	0.362	0.306	0.240
35	0.433	0.335	0.283	0.222
40	0.405	0.313	0.264	0.207
45	0.382	0.294	0.248	0.194
50	0.363	0.279	0.235	0.184

RUN TEST: ACCEPTANCE REGION FOR VALUES OF 'R'

a = Lower boundary (Left Side Critical Value); b = Upper boundary (Right Side Critical Value)

Two-sided	0.1		0.05		0.02		0.01	
One-sided	0.05		0.025		0.01		0.005	
$n_1 = n_2$	a	b	a	b	a	b	a	b
5	3	9			2	10		
6	3	11			2	12		
7	4	12			3	13		
8	5	13			4	14		
9	6	14			4	16		
10	6	16			5	17		
11	7	17	7	16	6	18	5	18
12	8	18	7	18	7	19	6	19
13	9	19	8	19	7	21	7	20
14	10	20	9	20	8	22	7	22
15	11	21	10	21	9	23	8	23
16	11	23	11	22	10	24	9	24
17	12	24	11	24	10	26	10	25
18	13	25	12	25	11	27	10	27
19	14	26	13	26	12	28	11	28
20	15	27	14	27	13	29	12	29
21	16	28			14	30		
22	17	29			14	32		
23	17	31			15	33		
24	18	32			16	34		
25	19	33	18	33	17	35	16	35
26	20	34			18	36		
27	21	35			19	37		
28	22	36			19	39		
29	23	37			20	40		
30	24	38	22	39	21	41	20	41
35	28	43	27	44	25	46	24	47
40	33	48	31	50	30	51	29	52
45	37	54	36	55	34	57	33	58
50	42	59	40	61	38	63	37	64
55	46	65	45	66	43	68	42	69
60	51	70	49	72	47	74	46	75
65	56	75	54	77	52	79	50	81
70	60	81	58	83	56	85	55	86
75	65	86	63	88	61	90	59	92
80	70	91	68	93	65	96	64	97
90	79	102	77	104	74	107	73	108
100	88	117	80	115	84	113	82	119

SIGN TEST: CRITICAL VALUES OF 'T'

	Level of Significance			
Two-sided	0.1	0.05	0.02	0.01
One-sided	0.05	0.025	0.01	0.005
n				
1	–	–	–	–
2	–	–	–	–
3	–	–	–	–
4	–	–	–	–
5	5	–	–	–
6	6	6	–	–
7	7	7	7	–
8	6	8	8	8
9	7	7	9	9
10	8	8	10	10
11	7	9	9	11
12	8	8	10	10
13	7	9	11	11
14	8	10	10	12
15	9	9	11	11
16	8	10	12	12
17	9	9	11	13
18	8	10	12	12
19	9	11	11	13
20	10	10	12	14
21	9	11	13	13
22	10	12	12	14
23	9	11	13	15
24	10	12	14	14
25	11	11	13	15
26	10	12	14	14
27	11	13	13	15
28	10	12	14	16
29	11	13	15	15
30	10	12	14	16

Contd...

Contd...

	Level of Significance			
Two-sided	0.1	0.05	0.02	0.01
One-sided	0.05	0.025	0.01	0.005
n				
31	11	13	15	17
32	12	14	16	16
33	11	13	15	17
34	12	14	16	16
35	11	13	15	17
36	12	14	16	18
37	11	13	17	17
38	12	14	16	18
39	13	15	17	17
40	12	14	16	18
45	13	15	17	19
50	14	16	18	20
55	15	17	19	21
60	14	18	20	22
65	15	17	21	23
70	16	18	22	24
75	17	19	23	25
80	16	20	22	24
90	18	20	24	26
100	18	22	26	28

KOLMOGOROV-SMIRNOV TEST: CRITICAL VALUES OF 'D'

	Level of Significance				
n	0.20	0.15	0.10	0.05	0.01
1	0.900	0.925	0.950	0.975	0.995
2	0.684	0.726	0.776	0.842	0.929
3	0.565	0.597	0.642	0.708	0.823
4	0.494	0.525	0.564	0.624	0.733
5	0.446	0.474	0.510	0.565	0.669
6	0.410	0.436	0.470	0.521	0.618
7	0.381	0.405	0.438	0.486	0.577
8	0.358	0.381	0.411	0.457	0.543
9	0.339	0.360	0.388	0.432	0.514
10	0.322	0.342	0.368	0.410	0.490
11	0.307	0.326	0.352	0.391	0.468
12	0.295	0.313	0.338	0.375	0.450
13	0.284	0.302	0.325	0.361	0.433
14	0.274	0.292	0.314	0.349	0.418
15	0.266	0.283	0.304	0.338	0.404
16	0.258	0.274	0.295	0.328	0.392
17	0.250	0.266	0.286	0.318	0.381
18	0.244	0.259	0.278	0.309	0.371
19	0.237	0.252	0.272	0.301	0.363
20	0.231	0.246	0.264	0.294	0.356
25	0.210	0.220	0.240	0.270	0.320
30	0.190	0.200	0.220	0.240	0.290
35	0.180	0.190	0.210	0.230	0.270
Over 35	$\dfrac{1.07}{\sqrt{n}}$	$\dfrac{1.14}{\sqrt{n}}$	$\dfrac{1.22}{\sqrt{n}}$	$\dfrac{1.36}{\sqrt{n}}$	$\dfrac{1.63}{\sqrt{n}}$

WILCOXON-WILCOX TWO-SIDED TEST: CRITICAL VALUES

n	k →	3	4	5	6	7	8	9	10
\multicolumn{10}{c}{Level of Significance (α = 0.01)}
1		4.1	5.7	7.3	8.9	10.5	12.2	13.9	15.6
2		5.8	8.0	10.3	12.6	14.9	17.3	19.7	22.1
3		7.1	9.8	12.6	15.4	18.3	21.2	24.1	27.0
4		8.2	11.4	14.6	17.8	21.1	24.4	27.8	31.2
5		9.2	12.7	16.3	19.9	23.6	27.3	31.1	34.9
6		10.1	13.9	17.8	21.8	25.8	29.9	34.1	38.2
7		10.9	15.0	19.3	23.5	27.9	32.3	36.8	41.3
8		11.7	16.1	20.6	25.2	29.8	34.6	39.3	44.2
9		12.4	17.1	21.8	26.7	31.6	36.6	41.7	46.8
10		13.0	18.0	23.0	28.1	33.4	38.6	44.0	49.4
11		13.7	18.9	24.1	29.5	35.0	40.5	46.1	51.8
12		14.3	19.7	25.2	30.8	36.5	42.3	48.2	54.1
13		14.9	20.5	26.2	32.1	38.0	44.0	50.1	56.3
14		15.4	21.3	27.2	33.3	39.5	45.7	52.0	58.4
15		16.0	22.0	28.2	34.5	40.8	47.3	53.9	60.5
16		16.5	22.7	29.1	35.6	42.2	48.9	55.6	62.5
17		17.0	23.4	30.0	36.7	43.5	50.4	57.3	64.4
18		17.5	24.1	30.9	37.8	44.7	51.8	59.0	66.2
19		18.0	24.8	31.7	38.8	46.0	53.2	60.6	68.1
20		18.4	25.4	32.5	39.8	47.2	54.6	62.2	69.8
21		18.9	26.0	33.4	40.9	48.3	56.0	63.7	71.6
22		19.3	26.7	34.1	41.7	49.5	57.3	65.2	73.2
23		19.8	27.3	34.9	42.7	50.6	58.6	66.7	74.9
24		20.2	27.8	35.7	43.6	51.7	59.8	68.1	76.5
25		20.6	28.4	36.4	44.5	52.7	61.1	69.5	78.1
\multicolumn{10}{c}{Level of Significance (α = 0.05)}
n	k →	3	4	5	6	7	8	9	10
1		3.3	4.7	6.1	7.5	9	10.5	12	13.5
2		4.7	6.6	8.6	10.7	12.7	14.8	17	19.2
3		5.7	8.1	10.6	13.1	15.6	18.2	20.8	23.5
4		6.6	9.4	12.2	15.1	18	21	24	27.1
5		7.4	10.5	13.6	16.9	20.1	23.5	26.9	30.3
6		8.1	11.5	14.9	18.5	22.1	25.7	29.4	33.2
7		8.8	12.4	16.1	19.9	23.9	27.8	31.8	35.8

Contd...

Contd...

		Level of Significance ($\alpha = 0.05$)							
n	k →	3	4	5	6	7	8	9	10
	8	9.4	13.3	17.3	21.3	25.5	29.7	34	38.3
	9	9.9	14.1	18.3	22.6	27	31.5	36	40.6
	10	10.5	14.8	19.3	23.8	28.5	33.2	38	42.8
	11	11.0	15.6	20.2	25.0	29.9	34.8	39.8	44.9
	12	11.5	16.2	21.1	26.1	31.2	36.4	41.6	46.9
	13	11.9	16.9	22.0	27.2	32.5	37.9	43.3	48.8
	14	12.4	17.5	22.8	28.2	33.7	39.3	45.0	50.7
	15	12.8	18.2	23.6	29.2	34.9	40.7	46.5	52.5
	16	13.3	18.8	24.4	30.2	36.0	42.0	48.1	54.2
	17	13.7	19.3	25.2	31.1	37.1	43.3	49.5	55.9
	18	14.1	19.9	25.9	32.0	38.2	44.5	51.0	57.5
	19	14.4	20.4	26.6	32.9	39.3	45.8	52.4	59.0
	20	14.8	21.0	27.3	33.7	40.3	47.0	53.7	60.6
	21	15.2	21.5	28.0	34.6	41.3	48.1	55.1	62.1
	22	15.5	22.0	28.6	35.4	42.3	49.2	56.4	63.5
	23	15.9	22.5	29.3	36.2	43.2	50.3	57.6	65.0
	24	16.2	23.0	29.9	36.9	44.1	51.4	58.9	66.4
	25	16.6	23.5	30.5	37.7	45.0	52.5	60.1	67.7
		Level of Significance ($\alpha = 0.10$)							
n	k →	3	4	5	6	7	8	9	10
	1	2.9	4.2	5.5	6.8	8.2	9.6	11.1	12.5
	2	4.1	5.9	7.8	9.7	11.6	13.6	15.6	17.7
	3	5.0	7.2	9.5	11.9	14.2	16.7	19.1	21.7
	4	5.8	8.4	11.0	13.7	16.5	19.3	22.1	25.0
	5	6.5	9.4	12.3	15.3	18.4	21.5	24.7	28.0
	6	7.1	10.2	13.5	16.8	20.2	23.6	27.1	30.6
	7	7.7	11.1	14.5	18.1	21.8	25.5	29.3	33.1
	8	8.2	11.8	15.6	19.4	23.3	27.2	31.3	35.4
	9	8.7	12.5	16.5	20.5	24.7	28.9	33.2	37.5
	10	9.2	13.2	17.4	21.7	26.0	30.4	35.0	39.5
	11	9.6	13.9	18.2	22.7	27.3	31.9	36.7	41.5
	12	10.1	14.5	19.0	23.7	28.5	33.4	38.3	43.3
	13	10.5	15.1	19.8	24.7	29.7	34.7	39.9	45.1
	14	10.9	15.7	20.6	25.6	30.8	36.0	41.4	46.8

Contd...

Contd...

		Level of Significance ($\alpha = 0.10$)							
n	k→	3	4	5	6	7	8	9	10
15		11.2	16.2	21.3	26.5	31.9	37.3	42.8	48.4
16		11.6	16.7	22.0	27.4	32.9	38.5	44.2	50.0
17		12.0	17.2	22.7	28.2	33.9	39.7	45.6	51.5
18		12.3	17.7	23.3	29.1	34.9	40.9	46.9	53.0
19		12.6	18.2	24.0	29.9	35.9	42.0	48.2	54.5
20		13.0	18.7	24.6	30.6	36.8	43.1	49.4	55.9
21		13.3	19.2	25.2	31.4	37.7	44.1	50.7	57.3
22		13.6	19.6	25.8	32.1	38.6	45.2	51.9	58.6
23		13.9	20.1	26.4	32.8	39.5	46.2	53.0	60.0
24		14.2	20.5	26.9	33.6	40.3	47.2	54.2	61.2
25		14.5	20.9	27.5	34.2	41.1	48.1	55.3	62.5

MANN-WHITNEY TEST (U TEST): CRITICAL VALUES OF 'U'

		\multicolumn{12}{c}{Level of Significance—One Tail : 0.025 and Two Tail : 0.05}

n_1	$n_2 \rightarrow$	9	10	11	12	13	14	15	16	17	18	19	20
2		0	0	0	1	1	1	1	1	2	2	2	2
3		2	3	3	4	4	5	5	6	6	7	7	8
4		4	5	6	7	8	9	10	11	11	12	13	13
5		7	8	9	11	12	13	14	15	17	18	19	20
6		10	11	13	14	16	17	19	21	22	24	25	27
7		12	14	16	18	20	22	24	26	28	30	32	34
8		15	17	19	22	24	26	29	31	34	36	38	41
9		17	20	23	26	28	31	34	37	39	42	45	48
10		20	23	26	29	33	36	39	42	45	48	52	55
11		23	26	30	33	37	40	44	47	51	55	58	62
12		26	29	33	37	41	45	49	53	57	61	66	69
13		28	33	37	41	45	50	54	59	63	67	72	76
14		31	36	40	45	50	55	59	64	67	74	78	83
15		34	39	44	49	54	59	64	70	75	80	85	90
16		37	42	47	53	59	64	70	75	81	86	92	98
17		39	45	51	57	63	67	75	81	87	93	99	105
18		42	48	55	61	67	74	80	86	93	99	106	112
19		45	52	58	65	72	78	85	92	99	106	113	119
20		48	55	62	69	76	83	90	98	105	112	119	127

		\multicolumn{12}{c}{Level of Significance—One Tail : 0.05 and Two Tail : 0.10}

n_1	$n_2 \rightarrow$	9	10	11	12	13	14	15	16	17	18	19	20
2		1	1	1	2	2	2	3	3	3	4	4	4
3		3	4	5	5	6	7	7	8	9	9	10	11
4		6	7	8	9	10	11	12	14	15	16	17	18
5		9	11	12	13	15	16	18	19	20	22	23	25
6		12	14	16	17	19	21	23	25	26	28	30	32
7		15	17	19	21	24	26	28	30	33	35	37	39
8		18	20	23	26	28	31	33	36	39	41	44	47
9		21	24	27	30	33	36	39	42	45	48	51	54
10		24	27	31	34	37	41	44	48	51	55	58	62
11		27	31	34	38	42	46	50	54	57	61	65	69
12		30	34	38	42	47	51	55	60	64	68	72	77
13		33	37	42	47	51	56	61	65	70	75	80	84
14		36	41	46	51	56	61	66	71	77	82	87	92
15		39	44	50	55	61	66	72	77	83	88	94	100
16		42	48	54	60	65	71	77	83	89	95	101	107
17		45	51	57	64	70	77	83	89	96	102	109	115
18		48	55	61	68	75	82	88	95	102	109	116	123
19		51	58	65	72	80	87	94	101	109	116	123	130
20		54	62	69	77	84	92	100	107	115	123	130	138

CRITICAL VALUES OF τ: KENDALL RANK CORRELATION SIGNIFICANCE TEST

	Level of Significance (α)			
Two-sided	0.10	0.05	0.02	0.05
One-sided	0.05	0.025	0.01	0.005
n				
4	6	—	—	—
5	8	10	10	—
6	11	13	13	15
7	13	15	17	19
8	16	18	20	22
9	18	20	24	26
10	21	23	27	29
11	23	27	31	33
12	26	30	36	38
13	28	34	40	44
14	33	37	43	47
15	35	41	49	53
16	38	46	52	58
17	42	50	58	64
18	45	53	63	69
19	49	57	67	75
20	52	62	72	80
21	56	66	78	86
22	61	71	83	91
23	65	75	89	99
24	68	80	94	104
25	72	86	100	110

WILCOXON SIGNED RANK TEST 'W': LOWER AND UPPER CRITICAL VALUES

One Tail	$\alpha = 0.05$	$\alpha = 0.25$	$\alpha = 0.01$	$\alpha = 0.005$
Two Tail	$\alpha = 0.10$	$\alpha = 0.05$	$\alpha = 0.02$	$\alpha = 0.01$
n	\multicolumn{4}{c}{Lower, Upper}			
5	0, 15	—, —	—, —	—, —
6	2, 19	0, 21	—, —	—, —
7	3, 25	2, 26	0, 28	—, —
8	5, 31	3, 33	1, 35	0, 36
9	8, 37	5, 40	3, 42	1, 44
10	10, 45	8, 47	5, 50	3, 52
11	13, 53	10, 56	7, 59	5, 61
12	17, 61	13, 65	10, 68	7, 71
13	21, 70	17, 74	12, 79	10, 81
14	25, 80	21, 84	16, 89	13, 92
15	30, 90	25, 95	19, 101	16, 104
16	35, 101	29, 107	23, 113	19, 117
17	41, 112	34, 119	27, 126	23, 130
18	47, 124	40, 131	32, 139	27, 144
19	53, 137	46, 144	37, 153	32, 158
20	60, 150	52, 158	43, 167	37, 173

COMMON LOGARITHMS ($\log_{10} x$)

x	0	1	2	3	4	5	6	7	8	9	Mean Difference									
											1	2	3	4	5	6	7	8	9
10	0	43	86	128	170	212	253	294	334	374	4	8	12	17	21	25	29	33	37
11	414	453	492	531	569	607	645	682	719	755	4	6	11	15	19	23	26	30	34
12	792	828	864	899	934	969	1004	1038	1072	1106	3	7	10	14	17	21	24	28	31
13	1139	1173	1206	1239	1271	1303	1335	1367	1399	1430	3	6	10	13	16	19	23	26	29
14	1461	1492	1523	1553	1584	1614	1644	1673	1703	1732	3	6	9	12	15	18	21	24	27
15	1761	1790	1818	1847	1875	1903	1931	1959	1987	2014	3	6	8	11	14	17	20	22	25
16	2041	2068	2095	2122	2148	2175	2201	2227	2253	2279	3	5	8	11	13	16	18	21	24
17	2304	2330	2355	2380	2405	2430	2455	2480	2504	2529	3	5	8	10	12	15	17	20	22
18	2553	2577	2601	2625	2648	2672	2695	2718	2742	2765	2	5	7	9	12	14	16	19	21
19	2788	2810	2833	2856	2878	2900	2923	2945	2967	2989	2	4	7	9	11	13	16	18	20
20	3010	3032	3054	3075	3096	3118	3139	3160	3181	3201	2	4	7	8	11	13	15	17	19
21	3222	3243	3263	3284	3304	3324	3345	3365	3385	3404	2	4	6	8	10	12	14	16	18
22	3424	3444	3464	3483	3502	3522	3541	3560	3579	3598	2	4	6	8	10	12	14	15	17
23	3617	3636	3655	3674	3692	3711	3729	3747	3766	3784	2	4	6	7	9	11	13	15	17
24	3802	3820	3838	3856	3874	3892	3909	3927	3945	3962	2	4	5	7	9	11	12	14	16
25	3979	3997	4014	4031	4048	4065	4082	4099	4116	4133	2	3	5	7	9	10	12	14	15
26	4150	4166	4183	4200	4216	4232	4249	4265	4281	4298	2	3	5	7	8	10	11	13	15

Contd...

Appendix

Contd...

x	0	1	2	3	4	5	6	7	8	9	Mean Difference								
											1	2	3	4	5	6	7	8	9
27	4314	4330	4346	4362	4378	4393	4409	4425	4440	4456	2	3	5	6	8	9	11	13	14
28	4472	4487	4502	4518	4533	4548	4564	4579	4594	4609	2	3	5	6	8	9	11	12	14
29	4624	4639	4654	4669	4683	4698	4713	4728	4742	4757	1	3	4	6	7	9	10	12	13
30	4771	4786	4800	4814	4829	4843	4857	4871	4886	4900	1	3	4	6	7	9	10	12	13
31	4914	4928	4942	4955	4969	4983	4997	5011	5024	5038	1	3	4	6	7	8	10	11	12
32	5051	5065	5079	5092	5105	5119	5132	5145	5159	5172	1	3	4	5	7	8	9	11	12
33	5185	5198	5211	5224	5237	5250	5263	5276	5289	5302	1	3	4	5	6	8	9	10	12
34	5315	5328	5340	5353	5366	5378	5391	5403	5416	5428	1	3	4	5	6	8	9	10	11
35	5441	5453	5465	5478	5490	5502	5514	5527	5539	5551	1	2	4	5	6	7	9	10	11
36	5563	5575	5587	5599	5611	5623	5635	5647	5658	5670	1	2	4	5	6	7	9	10	11
37	5682	5694	5705	5717	5729	5740	5752	5763	5775	5786	1	2	3	5	6	7	8	9	10
38	5798	5809	5821	5832	5843	5855	5866	5877	5888	5899	1	2	3	5	6	7	8	9	10
39	5911	5922	5933	5944	5955	5966	5977	5988	5999	6010	1	2	3	4	5	7	8	9	10
40	6021	6031	6042	6053	6064	6075	6085	6096	6107	6117	1	2	3	4	5	6	8	9	10
41	6128	6138	6149	6160	6170	6180	6191	6201	6212	6222	1	2	3	4	5	6	7	8	9
42	6232	6243	6253	6263	6274	6284	6294	6304	6314	6325	1	2	3	4	5	6	7	8	9
43	6335	6345	6355	6365	6375	6385	6395	6405	6415	6425	1	2	3	4	5	6	7	8	9

Contd...

Contd...

x	0	1	2	3	4	5	6	7	8	9	Mean Difference								
											1	2	3	4	5	6	7	8	9
44	6435	6444	6454	6464	6474	6484	6493	6503	6513	6522	1	2	3	4	5	6	7	8	9
45	6532	6542	6551	6561	6571	6580	6590	6599	6609	6618	1	2	3	4	5	6	7	8	9
46	6628	6637	6646	6656	6665	6675	6684	6693	6702	6712	1	2	3	4	5	6	7	8	9
47	6721	6730	6739	6749	6758	6767	6776	6785	6794	6803	1	2	3	4	5	6	7	7	8
48	6812	6821	6830	6839	6848	6857	6866	6875	6884	6893	1	2	3	4	4	5	6	7	8
49	6902	6911	6920	6928	6937	6946	6955	6964	6972	6981	1	2	3	4	4	5	6	7	8
50	6990	6998	7007	7016	7024	7033	7042	7050	7059	7067	1	2	3	3	4	5	6	7	8
51	7076	7084	7093	7101	7110	7118	7126	7135	7143	7152	1	2	3	3	4	5	6	7	8
52	7160	7168	7177	7185	7193	7202	7210	7218	7226	7235	1	2	2	3	4	5	6	7	7
53	7243	7251	7259	7267	7275	7284	7292	7300	7308	7316	1	2	2	3	4	5	6	6	7
54	7324	7332	7340	7348	7356	7364	7372	7380	7388	7396	1	2	2	3	4	5	6	6	7
55	7404	7412	7419	7427	7435	7443	7451	7459	7466	7474	1	2	2	3	4	5	5	6	7
56	7482	7490	7497	7505	7513	7520	7528	7536	7543	7551	1	2	2	3	4	5	5	6	7
57	7559	7566	7574	7582	7589	7597	7604	7612	7619	7627	1	2	2	3	4	5	5	6	7
58	7634	7642	7649	7657	7664	7672	7679	7686	7694	7701	1	1	2	3	4	4	5	6	7
59	7709	7716	7723	7731	7738	7745	7752	7760	7767	7774	1	1	2	3	4	4	5	6	7
60	7782	7789	7796	7803	7810	7818	7825	7832	7839	7846	1	1	2	3	4	4	5	6	6

Contd...

Contd...

x	0	1	2	3	4	5	6	7	8	9	Mean Difference								
											1	2	3	4	5	6	7	8	9
61	7853	7860	7868	7875	7882	7889	7896	7903	7910	7917	1	1	2	3	4	4	5	6	6
62	7924	7931	7938	7945	7952	7959	7966	7973	7980	7987	1	1	2	3	3	4	5	6	6
63	7993	8000	8007	8014	8021	8028	8035	8041	8048	8055	1	1	2	3	3	4	5	5	6
64	8062	8069	8075	8082	8089	8096	8102	8109	8116	8122	1	1	2	3	3	4	5	5	6
65	8129	8136	8142	8149	8156	8162	8169	8176	8182	8189	1	1	2	3	3	4	5	5	6
66	8195	8202	8209	8215	8222	8228	8235	8241	8248	8254	1	1	2	3	3	4	5	5	6
67	8261	8267	8274	8280	8287	8293	8299	8306	8312	8319	1	1	2	3	3	4	5	5	6
68	8325	8331	8338	8344	8351	8357	8363	8370	8376	8382	1	1	2	3	3	4	4	5	6
69	8388	8395	8401	8407	8414	8420	8426	8432	8439	8445	1	1	2	2	3	4	4	5	6
70	8451	8457	8463	8470	8476	8482	8488	8494	8500	8506	1	1	2	2	3	4	4	5	6
71	8513	8519	8525	8531	8537	8543	8549	8555	8561	8567	1	1	2	2	3	4	4	5	5
72	8573	8579	8585	8591	8597	8603	8609	8615	8621	8627	1	1	2	2	3	4	4	5	5
73	8633	8639	8645	8651	8657	8663	8669	8675	8681	8686	1	1	2	2	3	4	4	5	5
74	8692	8698	8704	8710	8716	8722	8727	8733	8739	8745	1	1	2	2	3	4	4	5	5
75	8751	8756	8762	8768	8774	8779	8785	8791	8797	8802	1	1	2	2	3	3	4	5	5
76	8808	8814	8820	8825	8831	8837	8842	8848	8854	8859	1	1	2	2	3	3	4	5	5
77	8865	8871	8876	8882	8887	8893	8899	8904	8910	8915	1	1	2	2	3	3	4	4	5

Contd...

Contd...

x	0	1	2	3	4	5	6	7	8	9	Mean Difference								
											1	2	3	4	5	6	7	8	9
78	8921	8927	8932	8938	8943	8949	8954	8960	8965	8971	1	1	2	2	3	3	4	4	5
79	8976	8982	8987	8993	8998	9004	9009	9015	9020	9025	1	1	2	2	3	3	4	4	5
80	9031	9036	9042	9047	9053	9058	9063	9069	9074	9079	1	1	2	2	3	3	4	4	5
81	9085	9090	9096	9101	9106	9112	9117	9122	9128	9133	1	1	2	2	3	3	4	4	5
82	9138	9143	9149	9154	9159	9165	9170	9175	9180	9186	1	1	2	2	3	3	4	4	5
83	9191	9196	9201	9206	9212	9217	9222	9227	9232	9238	1	1	2	2	3	3	4	4	5
84	9243	9248	9253	9258	9263	9269	9274	9279	9284	9289	1	1	2	2	3	3	4	4	5
85	9294	9299	9304	9309	9315	9320	9325	9330	9335	9340	1	1	2	2	3	3	4	4	5
86	9345	9350	9355	9360	9365	9370	9375	9380	9385	9390	1	1	2	2	3	3	4	4	5
87	9395	9400	9405	9410	9415	9420	9425	9430	9435	9440	0	1	1	2	2	3	3	4	4
88	9445	9450	9455	9460	9465	9469	9474	9479	9484	9489	0	1	1	2	2	3	3	4	4
89	9494	9499	9504	9509	9513	9518	9523	9528	9533	9538	0	1	1	2	2	3	3	4	4
90	9542	9547	9552	9557	9562	9566	9571	9576	9581	9586	0	1	1	2	2	3	3	4	4
91	9590	9595	9600	9605	9609	9614	9619	9624	9628	9633	0	1	1	2	2	3	3	4	4
92	9638	9643	9647	9652	9657	9661	9666	9671	9675	9680	0	1	1	2	2	3	3	4	4
93	9685	9689	9694	9699	9703	9708	9713	9717	9722	9727	0	1	1	2	2	3	3	4	4

Contd...

Contd...

x	0	1	2	3	4	5	6	7	8	9	\multicolumn{9}{c}{Mean Difference}

x	0	1	2	3	4	5	6	7	8	9	1	2	3	4	5	6	7	8	9
94	9731	9736	9741	9745	9750	9754	9759	9763	9768	9773	0	1	1	2	2	3	3	4	4
95	9777	9782	9786	9791	9795	9800	9805	9809	9814	9818	0	1	1	2	2	3	3	4	4
96	9823	9827	9832	9836	9841	9845	9850	9854	9859	9863	0	1	1	2	2	3	3	4	4
97	9868	9872	9877	9881	9886	9890	9894	9899	9903	9908	0	1	1	2	2	3	3	4	4
98	9912	9917	9921	9926	9930	9934	9939	9943	9948	9952	0	1	1	2	2	3	3	4	4
99	9956	9961	9965	9969	9974	9978	9983	9987	9991	9996	0	1	1	2	2	3	3	3	4

ANTILOGARITHMS

x	0	0.001	0.002	0.003	0.004	0.005	0.006	0.007	0.008	0.009	\multicolumn{9}{c}{Mean Difference}								
											1	2	3	4	5	6	7	8	9
0.00	1000	1002	1005	1007	1009	1012	1014	1016	1019	1021	0	0	1	1	1	1	2	2	2
0.01	1023	1026	1028	1030	1033	1035	1038	1040	1042	1045	0	0	1	1	1	1	2	2	2
0.02	1047	1050	1052	1054	1057	1059	1062	1064	1067	1069	0	0	1	1	1	1	2	2	2
0.03	1072	1074	1076	1079	1081	1084	1086	1089	1091	1094	0	0	1	1	1	2	2	2	2
0.04	1096	1099	1102	1104	1107	1109	1112	1114	1117	1119	0	1	1	1	1	2	2	2	2
0.05	1122	1125	1127	1130	1132	1135	1138	1140	1143	1146	0	1	1	1	1	2	2	2	2
0.06	1148	1151	1153	1156	1159	1161	1164	1167	1169	1172	0	1	1	1	1	2	2	2	2
0.07	1175	1178	1180	1183	1186	1189	1191	1194	1197	1199	0	1	1	1	1	2	2	2	3
0.08	1202	1205	1208	1211	1213	1216	1219	1222	1225	1227	0	1	1	1	1	2	2	2	3
0.09	1230	1233	1236	1239	1242	1245	1247	1250	1253	1256	0	1	1	1	1	2	2	2	3
0.10	1259	1262	1265	1268	1271	1274	1276	1279	1282	1285	0	1	1	1	1	2	2	2	3
0.11	1288	1291	1294	1297	1300	1303	1306	1309	1312	1315	0	1	1	1	2	2	2	2	3
0.12	1318	1321	1324	1327	1330	1334	1337	1340	1343	1346	0	1	1	1	2	2	2	2	3
0.13	1349	1352	1355	1358	1361	1365	1368	1371	1374	1377	0	1	1	1	2	2	2	2	3
0.14	1380	1384	1387	1390	1393	1396	1400	1403	1406	1409	0	1	1	1	2	2	2	3	3
0.15	1413	1416	1419	1422	1426	1429	1432	1435	1439	1442	0	1	1	1	2	2	2	3	3
0.16	1445	1449	1452	1455	1459	1462	1466	1469	1472	1476	0	1	1	1	2	2	2	3	3

Contd...

Appendix

Contd...

| x | 0 | 0.001 | 0.002 | 0.003 | 0.004 | 0.005 | 0.006 | 0.007 | 0.008 | 0.009 | Mean Difference |||||||||
											1	2	3	4	5	6	7	8	9
0.17	1479	1483	1486	1489	1493	1496	1500	1503	1507	1510	0	1	1	1	2	2	2	3	3
0.18	1514	1517	1521	1524	1528	1531	1535	1538	1542	1545	0	1	1	1	2	2	2	3	3
0.19	1549	1552	1556	1560	1563	1567	1570	1574	1578	1581	0	1	1	1	2	2	2	3	3
0.20	1585	1589	1592	1596	1600	1603	1607	1611	1614	1618	0	1	1	1	2	2	3	3	3
0.21	1622	1626	1629	1633	1637	1641	1644	1648	1652	1656	0	1	1	2	2	2	3	3	3
0.22	1660	1663	1667	1671	1675	1679	1683	1687	1690	1694	0	1	1	2	2	2	3	3	3
0.23	1698	1702	1706	1710	1714	1718	1722	1726	1730	1734	0	1	1	2	2	2	3	3	4
0.24	1738	1742	1746	1750	1754	1758	1762	1766	1770	1774	0	1	1	2	2	2	3	3	4
0.25	1778	1782	1786	1791	1795	1799	1803	1807	1811	1816	0	1	1	2	2	2	3	3	4
0.26	1820	1824	1828	1832	1837	1841	1845	1849	1854	1858	0	1	1	2	2	3	3	3	4
0.27	1862	1866	1871	1875	1879	1884	1888	1892	1897	1901	0	1	1	2	2	3	3	3	4
0.28	1905	1910	1914	1919	1923	1928	1932	1936	1941	1945	0	1	1	2	2	3	3	4	4
0.29	1950	1954	1959	1963	1968	1972	1977	1982	1986	1991	0	1	1	2	2	3	3	4	4
0.30	1995	2000	2004	2009	2014	2018	2023	2028	2032	2037	0	1	1	2	2	3	3	4	4
0.31	2042	2046	2051	2056	2061	2065	2070	2075	2080	2084	0	1	1	2	2	3	3	4	4
0.32	2089	2094	2099	2104	2109	2113	2118	2123	2128	2133	0	1	1	2	2	3	3	4	4
0.33	2138	2143	2148	2153	2158	2163	2168	2173	2178	2183	0	1	1	2	2	3	3	4	4

Contd...

Contd...

x	0	0.001	0.002	0.003	0.004	0.005	0.006	0.007	0.008	0.009	Mean Difference								
											1	2	3	4	5	6	7	8	9
0.34	2188	2193	2198	2203	2208	2213	2218	2223	2228	2234	1	1	2	2	3	3	4	4	5
0.35	2239	2244	2249	2254	2259	2265	2270	2275	2280	2286	1	1	2	2	3	3	4	4	5
0.36	2291	2296	2301	2307	2312	2317	2323	2328	2333	2339	1	1	2	2	3	3	4	4	5
0.37	2344	2350	2355	2360	2366	2371	2377	2382	2388	2393	1	1	2	2	3	3	4	4	5
0.38	2399	2404	2410	2415	2421	2427	2432	2438	2443	2449	1	1	2	2	3	3	4	4	5
0.39	2455	2460	2466	2472	2477	2483	2489	2495	2500	2506	1	1	2	2	3	3	4	5	5
0.40	2512	2518	2523	2529	2535	2541	2547	2553	2559	2564	1	1	2	2	3	4	4	5	5
0.41	2570	2576	2582	2588	2594	2600	2606	2612	2618	2624	1	1	2	2	3	4	4	5	5
0.42	2630	2636	2642	2649	2655	2661	2667	2673	2679	2685	1	1	2	2	3	4	4	5	6
0.43	2692	2698	2704	2710	2716	2723	2729	2735	2742	2748	1	1	2	3	3	4	4	5	6
0.44	2754	2761	2767	2773	2780	2786	2793	2799	2805	2812	1	1	2	3	3	4	4	5	6
0.45	2818	2825	2831	2838	2844	2851	2858	2864	2871	2877	1	1	2	3	3	4	5	5	6
0.46	2884	2891	2897	2904	2911	2917	2924	2931	2938	2944	1	1	2	3	3	4	5	5	6
0.47	2951	2958	2965	2972	2979	2985	2992	2999	3006	3013	1	1	2	3	3	4	5	5	6
0.48	3020	3027	3034	3041	3048	3055	3062	3069	3076	3083	1	1	2	3	4	4	5	6	6
0.49	3090	3097	3105	3112	3119	3126	3133	3141	3148	3155	1	1	2	3	4	4	5	6	6
0.50	3162	3170	3177	3184	3192	3199	3206	3214	3221	3228	1	1	2	3	4	4	5	6	7

Contd...

Contd...

x	0	0.001	0.002	0.003	0.004	0.005	0.006	0.007	0.008	0.009	Mean Difference								
											1	2	3	4	5	6	7	8	9
0.51	3236	3243	3251	3258	3266	3273	3281	3289	3296	3304	1	2	2	3	4	5	5	6	7
0.52	3311	3319	3327	3334	3342	3350	3357	3365	3373	3381	1	2	2	3	4	5	5	6	7
0.53	3388	3396	3404	3412	3420	3428	3436	3443	3451	3459	1	2	2	3	4	5	6	6	7
0.54	3467	3475	3483	3491	3499	3508	3516	3524	3532	3540	1	2	2	3	4	5	6	6	7
0.55	3548	3556	3565	3573	3581	3589	3597	3606	3614	3622	1	2	2	3	4	5	6	7	7
0.56	3631	3639	3648	3656	3664	3673	3681	3690	3698	3707	1	2	3	3	4	5	6	7	8
0.57	3715	3724	3733	3741	3750	3758	3767	3776	3784	3793	1	2	3	3	4	5	6	7	8
0.58	3802	3811	3819	3828	3837	3846	3855	3864	3873	3882	1	2	3	4	4	5	6	7	8
0.59	3890	3899	3908	3917	3926	3936	3945	3954	3963	3972	1	2	3	4	5	5	6	7	8
0.60	3981	3990	3999	4009	4018	4027	4036	4046	4055	4064	1	2	3	4	5	6	6	7	8
0.61	4074	4083	4093	4102	4111	4121	4130	4140	4150	4159	1	2	3	4	5	6	7	8	9
0.62	4169	4178	4188	4198	4207	4217	4227	4236	4246	4256	1	2	3	4	5	6	7	8	9
0.63	4266	4276	4285	4295	4305	4315	4325	4335	4345	4355	1	2	3	4	5	6	7	8	9
0.64	4365	4375	4385	4395	4406	4416	4426	4436	4446	4457	1	2	3	4	5	6	7	8	9
0.65	4467	4477	4487	4498	4508	4519	4529	4539	4550	4560	1	2	3	4	5	6	7	8	9
0.66	4571	4581	4592	4603	4613	4624	4634	4645	4656	4667	1	2	3	4	5	6	7	9	10
0.67	4677	4688	4699	4710	4721	4732	4742	4753	4764	4775	1	2	3	4	5	7	8	9	10

Contd...

Contd...

x	0	0.001	0.002	0.003	0.004	0.005	0.006	0.007	0.008	0.009	Mean Difference								
											1	2	3	4	5	6	7	8	9
0.68	4786	4797	4808	4819	4831	4842	4853	4864	4875	4887	1	2	3	4	6	7	8	8	9
0.69	4898	4909	4920	4932	4943	4955	4966	4977	4989	5000	1	2	3	5	6	7	8	9	10
0.70	5012	5023	5035	5047	5058	5070	5082	5093	5105	5117	1	2	4	5	6	7	8	9	10
0.71	5129	5140	5152	5164	5176	5188	5200	5212	5224	5236	1	2	4	5	6	7	8	9	11
0.72	5248	5260	5272	5284	5297	5309	5321	5333	5346	5358	1	2	4	5	6	7	8	10	11
0.73	5370	5383	5395	5408	5420	5433	5445	5458	5470	5483	1	3	4	5	6	8	9	10	11
0.74	5495	5508	5521	5534	5546	5559	5572	5585	5598	5610	1	3	4	5	6	8	9	10	11
0.75	5623	5636	5649	5662	5675	5689	5702	5715	5728	5741	1	3	4	5	7	8	9	10	12
0.76	5754	5768	5781	5794	5808	5821	5834	5848	5861	5875	1	3	4	5	7	8	9	11	12
0.77	5888	5902	5916	5929	5943	5957	5970	5984	5998	6012	1	3	4	6	7	8	10	11	12
0.78	6026	6039	6053	6067	6081	6095	6109	6124	6138	6152	1	3	4	6	7	8	10	11	12
0.79	6166	6180	6194	6209	6223	6237	6252	6266	6281	6295	1	3	4	6	7	9	10	11	13
0.80	6310	6324	6339	6353	6368	6383	6397	6412	6427	6442	1	3	4	6	7	9	10	12	13
0.81	6457	6471	6486	6501	6516	6531	6546	6561	6577	6592	2	3	5	6	8	9	11	12	14
0.82	6607	6622	6637	6653	6668	6683	6699	6714	6730	6745	2	3	5	6	8	9	11	12	14
0.83	6761	6776	6792	6808	6823	6839	6855	6871	6887	6902	2	3	5	6	8	9	11	13	14
0.84	6918	6934	6950	6966	6982	6998	7015	7031	7047	7063	2	3	5	6	8	10	11	13	15

Contd...

Contd...

x	0	0.001	0.002	0.003	0.004	0.005	0.006	0.007	0.008	0.009	Mean Difference								
											1	2	3	4	5	6	7	8	9
0.85	7079	7096	7112	7129	7145	7161	7178	7194	7211	7228	2	3	5	7	8	10	12	13	15
0.86	7244	7261	7278	7295	7311	7328	7345	7362	7379	7396	2	3	5	7	8	10	12	13	15
0.87	7413	7430	7447	7464	7482	7499	7516	7534	7551	7568	2	3	5	7	9	10	12	14	16
0.88	7586	7603	7621	7638	7656	7674	7691	7709	7727	7745	2	4	5	7	9	11	12	14	16
0.89	7762	7780	7798	7816	7834	7852	7870	7889	7907	7925	2	4	5	7	9	11	13	14	16
0.90	7943	7962	7980	7998	8017	8035	8054	8072	8091	8110	2	4	6	7	9	11	13	15	17
0.91	8128	8147	8166	8185	8204	8222	8241	8260	8279	8299	2	4	6	8	9	11	13	15	17
0.92	8318	8337	8356	8375	8395	8414	8433	8453	8472	8492	2	4	6	8	10	12	14	15	17
0.93	8511	8531	8551	8570	8590	8610	8630	8650	8670	8690	2	4	6	8	10	12	14	16	18
0.94	8710	8730	8750	8770	8790	8810	8831	8851	8872	8892	2	4	6	8	10	12	14	16	18
0.95	8913	8933	8954	8974	8995	9016	9036	9057	9078	9099	2	4	6	8	10	12	15	17	19
0.96	9120	9141	9162	9183	9204	9226	9247	9268	9290	9311	2	4	6	8	11	13	15	17	19
0.97	9333	9354	9376	9397	9419	9441	9462	9484	9506	9528	2	4	7	9	11	13	15	17	20
0.98	9550	9572	9594	9616	9638	9661	9683	9705	9727	9750	2	4	7	9	11	13	16	18	20
0.99	9772	9795	9817	9840	9863	9886	9908	9931	9954	9977	2	4	7	9	11	14	16	18	20

Appendix

Index

Page numbers followed by *f* refer to figure, and *t* refer to table

A

Abstract 455, 474
Age specific
    death rate 362
    fertility rate 360
Analytical research studies 71
Anecdotal information 463
Animal ethics 34
Antilogarithms 80
Arithmetic progression 79
Assessment, grading of 47
Attribute
    association of 228
    theory 225
        terminology of 225

B

Baranowicze head injury experiment 15
Basic ethical principles 31
Bayes principle 445
Bayes theorem 139
Bell curve 164
Bibliography 474
Binomial distribution 147, 148, 148*f*, 150*f*
    constants of 151
    properties of 149
Bioethics 13, 14
    main branches of 14
Biographical sketches 474
Biomedical research, broad fields of 62
Biostatistics 52
Bivariate frequency table 94*t*
Bradford Hill causality criteria 332
Budget 475

C

Calculating index numbers, method of 414
C-control chart 435, 435*f*
Central tendency, measure of 95
Central value 321
    concept 95
Chi-square test 296
Civil registration system 358
Clostridium perfringens 16
Clostridium tetani 16
Cluster sampling 181
Cochran Q test 318
Cohort group 367
Cohort life table 368
Colligation, coefficient of 229
Complete life table 371
Conclusion 459
    report and reporting of 69
Contingency, coefficient of 230
Continuous frequency
    distribution 90, 90*t*, 91*t*
Continuous random variable, expectation
    of 145
Control charts 430, 431
Convenience sampling 182
Correlation 191, 232
    measures 192
Cover letter 476
Crude birth rate 360, 361
Curve
    fitting method 385
    types of 125*f*
Cyclical component 383, 396

D

Dachau and Auschwitz concentration camp
    hypothermia experiments 15
Dachau potable sea water experiment 16
Data
    classification of 87
    collection 67
    consistency of 227
    sources of 358
    types of 321
Decision
    criteria 442
    making, principles of 444
    methods 443
    theory 440
    tree analysis 449, 450*f*
    under certainty 443
    under uncertainty 444
Declaration, promotion of 30
Defects, proportion of 436
Determination, multiple coefficient of 219
Development, grading of 47
Discrete frequency distribution 90*t*
Discrete random variable, expectation of 144

Dispersion
   coefficient of 120
   measures of 114
Dynamic life table 368

E

Errors, propagation of 353
Estimates, theory of 238
Estimation 236, 238
   methods of 239
Ethics, major divisions of 13
Etymology 14, 39, 49
Evidence-based practice, components of 39
Expectation method 228
Experimental approach 59
Exploratory research studies 71
Exponential distribution 176
   curves 177f
Exponential trend 388
Exposure disease association 333

F

Face validity 343
False negative rate 348
F-distribution, constants and properties of 269
Fertility rates 360
Fisher's z-transformation 265
Fitting Poisson distribution 160
Fixed size sampling 185
Frequency distribution 87, 89, 105
Friedman test 314, 315
F-test 269-271, 273

G

General fertility rate 360
Generation life table 368
Geometric distribution 147, 156
   constants of 156
Geometric mean 107
   demerits of 109
   merits of 109
Geometric progression 79
Graphic method 400
Graphs 462
Greek alphabets 74
Gross reproduction rate 361

H

Helsinki, declaration of 22
Homoscedasticity 321
H-test 313
Human subjects and institutional review boards, protection of 473
Hypergeometric distribution 147, 157

Hypothesis 68, 241, 243
   testing 71, 235, 236, 240

I

Index number 411, 414
   construction 415
Infant mortality rate 362
Interpolation, methods of 400

J

Josef Mengele's experiments on twins 15
Judgment sampling 182
Justice 32

K

Kaplan-Meier curves 374, 378f
Karl Pearson's coefficient of correlation 193
Kelly's coefficient of distribution 124
Kendall's rank correlation 202
Kolmogorov-Smirnov test 307, 308
Kruskal-Wallis rank sum test 313
Kurtosis 125, 125f
   measure of 126

L

Lagrange's method 404
Large sample tests 249
Least squares method 386, 386f, 388f
Lexis diagram 367
Life expectancy 365
Life table 357, 367, 370
   calculations 370
   construction 368
   types of 368
   uses of 374
Linear regression analysis 207
Literacy rates 364
Literature, types of 65
Logarithms 80
Logistic regression 212
Lognormal distribution 176
Log-rank test 378
Lorenz curve 119, 119f
   graph of 119

M

Malaria experiments 16
Mann-Whitney U test 310
Marginal distribution 94
Maternal mortality rate 363
Maximax principle 445
Maximin principle 444
Mcnemar's test 317
Mean and range control chart 432, 434f

Mean deviation 116
Medical ethics 14
Minimax principle 445
Mode, demerits of 107
Modulus 76
Mortality rates 361, 367
Moving averages method 384
Multinomial distribution 147, 152
Multiple correlation 217
Multiple determination
    adjusted coefficient of 220
    coefficient of 219
Mustard gas experiments 16

N

National Sample Survey Organization 358
Natural increase rate 363
Negative binomial distribution 147, 153
    constants of 154
Negative likelihood ratio 349
Negative predictive value 347
Neonatal death rate 362
Net migration rate 363
Net reproduction rate 361
Newton's backward difference
    interpolation 402
Newton's divided difference method 404
Newton's forward interpolation
    method 401
Nonclinical biomedical research 25
Nonparametric tests 301, 325
Nonprobability sampling methods 182
Nontherapeutic biomedical research 25
Normal curve 123, 164, 166, 166f, 168f
    fitting of 172, 174f
    properties of 165
Normal distribution 163
    fitting of 164
    properties of 164
Null hypothesis 240
Nuremberg code, ten points of 20

O

Odds ratio 334, 350
One and two-tailed tests 245

P

Paired T-test 261
Parabolic curve fitting method 401
Parabolic trends 387
Parametric tests 323
Partial and multiple correlation and
    regression 216, 218
P-control chart 436
Pearson's beta and gamma coefficients 126

Percentile deviation 116
Physicians, duties of 25, 26
Poisson distribution 147, 158
    curves 159f
Population
    attributable risk 335
    vital index of 364
Positive likelihood ratio 348
Positive predictive value 347
Post-neonatal mortality rate 362
Preliminary studies 471
Price index number 414
    construction 414
    types 414
Probability based random sampling
    methods 180
Probability distribution 142
Probability method 228
Probability theory 131
    terminology in 132
Proofing 462
Proportion method 229
P-value 267

Q

Qualitative approach 59
Quartile/decile/percentile, computation of
    103, 104f
Quota sampling 182

R

Random component 383, 396
Ravensbrück sulfonamide
    experiments 15, 16
Raw data 88
R-chart 433
Recommendations, grading of 47
Rectangular distribution 174, 175f
References 459
Regression
    analysis 205
        components of 205
    coefficient 209
    types of 206
Reproductive rates 361
Research
    design 69, 471
        preparation of 66
        types of 66, 70
    ethics 31
    methodology 57, 60
    objectives of 57
    plan 72
        uses of 72
    process 61, 65

question, formulation of 65
report writing 453
report, organization of 453
types of 58
Robert Allen (Laud) Humphreys' tearoom sex study 17
Root mean square deviation 117
Run test 302

S

Sample control chart 431f
Sample design 67
Sample registration system 358
Sample size 236
    calculations 331
Sampling 437
    advantages and disadvantages of 182
    distribution 236
    errors 185
    methods, types of 180
    theory 178, 186
Scatter diagram 193, 193f, 194f, 196f
Scientific method, basic postulates of 61
Sensitivity 346
Sex ratio 364
Simple aggregative method 414
Simple arithmetic mean 97
Simple averages method 391
Simple random sampling 180
Skewed curve for distribution 123f
Skewness 122
    coefficient of 124, 125
    measures of 123
Small sample tests 256
Spearman's and Kendall rank correlation, significance of 317
Spearman's rank correlation 199
    significance test of 201
Specific fertility rate 360
Standard deviation 117, 166, 255
    control chart 435
Standard error 186, 210, 237
Standard normal curve 166, 166f, 168f
Static life table 368
Statistical analysis, types of 52
Statistical quality control 429, 430
Statistical significance 212
Statistical tests, summary of 322
Sterilization experiments 16
Student's T distribution 257
    curve 257f
Student's T test 248, 256, 258
Summary 459
Survival analysis 374
    notations 376
Systematic sampling 180

T

Tchebysheff's theorem 118
Testing hypothesis 246
Theoretical analysis 456
Theoretical continuous probability distributions 163
Theoretical discrete
    distributions, types of 147
    probability distributions 142
Theoretical distribution 146
Theoretical survival function curve 376f
Time series
    analysis 381
    components 382
Title 454, 474
    page 454
Total fertility rate 361, 366
Tree diagram 449
True negative test 346
Tuskegee syphilis study 18

U

Union, axiom of 136
Unit 731 17
Universal declaration 26

V

Validity, assessment of 344
Venn diagram 76, 78, 78f, 139f
Vibrio cholerae 328
Vital statistics 357
    basics of 357
    measures of 359

W

Wilcoxon signed rank test 304
Wilcoxon-Wilcox test 311
Working hypothesis, development of 66
Writing materials and methods section 456

Y

Yates' correction 299
Yules' coefficient of association 229

Z

Z test 248, 254, 255